Study Guide

Chemistry

EIGHTH EDITION

Steven S. Zumdahl
University of Illinois at Urbana-Champaign

Susan Arena Zumdahl
University of Illinois at Urbana-Champaign

Written by

Paul Kelter
Northern Illinois University

BROOKS/COLE
CENGAGE Learning™

Australia • Brazil • Japan • Korea • Mexico • Singapore • Spain • United Kingdom • United States

For product information and technology assistance, contact us at **Cengage Learning Customer & Sales Support, 1-800-354-9706**

For permission to use material from this text or product, submit all requests online at **www.cengage.com/permissions** Further permissions questions can be emailed to **permissionrequest@cengage.com**

ISBN-13: 978-0-547-16872-2
ISBN-10: 0-547-16872-1

Brooks/Cole
10 Davis Drive
Belmont, CA 94002-3098
USA

Cengage Learning is a leading provider of customized learning solutions with office locations around the globe, including Singapore, the United Kingdom, Australia, Mexico, Brazil, and Japan. Locate your local office at: **www.cengage.com/international**

Cengage Learning products are represented in Canada by Nelson Education, Ltd.

To learn more about Brooks/Cole, visit **www.cengage.com/brookscole**

Purchase any of our products at your local college store or at our preferred online store **www.ichapters.com**

Printed in the United States of America
1 2 3 4 5 6 7 12 11 10 09

Table of Contents

PREFACE

My thirty years of experience in general chemistry have shown me that there are several characteristics that successful students have in common:

- They know why they are in college;
- They keep up with the work on a daily basis;
- They study until they understand concepts deeply;
- They look for different ways to learn;
- They raise questions and seek answers to gain clarity.

This *Study Guide* can't help address the first item, but it can support you as you work through the other four parts of the list. The guide is written for the eighth edition of Steve Zumdahl's *Chemistry* textbook; my section descriptions and table and figure references match his. However there are a few instances in the guide when my approach will differ from that used by Dr. Zumdahl. Both ways have proven successful and we want you to use the method that works best for you.

There are over 1500 problems, many of them worked out, in the guide. Ideally, you will read the textbook, do the problems, then read this *Study Guide* and do its problems. The guide is not a substitute for the text. The textbook has a richness of information that the guide cannot approach. However, the guide can enrich and deepen your understanding. Please use it with that in mind.

Acknowledgments

To Mary LeQuesne;
To Richard Stratton, simply the best;
To Rebecca Berardy Schwartz, who has it all together;
To Charlie Hartford, with or without Clemens, Bagwell and The Babe.
To Barb, how pleasant to share our lives

Chapter 1

Chemical Foundations

This chapter will help guide your study of **what it means to think like a scientist in terms of strategy and measurement.**

1.1   Chemistry: An Overview

This section considers the nature of chemistry and elements. There are several key points here:

1. Chemistry has a historical and current impact on us.
2. All of what we are and what we do is determined by the interactions of only about 100 different types of atoms.
3. There is a "chemist's shorthand"—a way of expressing chemical processes symbolically. We will be learning many parts of this shorthand in the study of chemistry.
4. Science is systematic, rather than being haphazard. It is also a thoughtful undertaking, rather than one in which we follow along blindly. Creativity in experimentation, thinking, and asking questions are the hallmarks of our discipline.

1.2   The Scientific Method

When you finish this section you will be able to:

*   List and define each of the major steps in the **scientific method**.
*   Identify the limitations of the **scientific method**.

The **scientific method** is a general procedure by which scientists gain understanding of our universe. Structured thinking is as applicable to biology, physics, or geology as it is to chemistry. Thoughtful work is important to any scientific experiment.

As explained in your textbook, the **scientific method** is characterized by the following steps:

1. Making **observations** (gathering data).
2. Formulating a **hypothesis**. Your hypothesis gives a possible explanation of **how** a process occurs. It is not the same as a **fact**, which is some observation or process that is so correct that it is unreasonable to dispute. If your hypothesis survives scrutiny by many experiments, it becomes a **theory**.
3. Performing experiments. This is done to determine if your hypothesis has **predictive value**.

Example 1.2   Scientific Method

One of the most important laws in chemistry, called Boyle's Law, states that **as the pressure (P) is increased on a gas, the volume (V) that the gas occupies will shrink proportionately, so that at a constant temperature, $PV = $ a constant**. A chemistry student, unaware that this pressure-volume relationship exists, sought to "discover" it. His procedure was:

a. Hypothesize that, at constant temperature, pressure increases as volume increases.
b. Run one experiment where he observes the pressure on a gas as he changes the volume.
c. His result was inconclusive, so he proposed a theory that pressure and volume of a gas are unrelated.
d. He put away his equipment and went on to his next project.

List and explain those things that were wrong with this student's "scientific method."

Solution

1. He made a hypothesis based on no previous observations. A hypothesis can only be reasonable if it is based on some real-life experience.

2. As we will see in **Section 1.4**, one experiment gives very little information. The student should have performed many experiments so that sources of random error could have been eliminated.

3. He called his conclusion a theory. Hypotheses become theories only after long periods of testing to see if the hypothesis is correct. In addition, by putting away his equipment and going on to the next project, he did not test his own conclusions. It is always better if you can find your own mistakes before someone else does.

Your textbook points out that scientists, just as everyone else, have biases and prejudices. Science is self-correcting. If our hypotheses are invalid, we will eventually run enough tests to find out. The best that we chemists and chemistry students can do is to keep asking questions about theories. Questioning represents the best "scientific method."

1.3   *Units of Measurement*

When you finish this section, you will be able to:

- List the fundamental SI units of measurement.
- List the important SI prefixes.
- Differentiate between fundamental and derived units.

The United States is the last country in the industrial world that uses the English system. To be consistent with every other country, we are slowly adopting the **International System** (le Systéme International in French), or **SI System**.

The SI system is based on the metric system. We see examples of the SI system when we go to the grocery store to buy soda. The volume is listed as 12 fluid ounces (English system) and 354 milliliters (SI system).

You should know the 7 fundamental SI units. These are given in Table 1.1 of your textbook. It is important to develop "a feel" for units in the SI system. For example, in Earth's gravitational field, if you weigh **220 pounds**, your mass is **100 kilograms (kg)**. If you are **5.00 feet tall**, your height is **1.52 meters (m)**.

Measurement in meters is fine if you are measuring how tall you are. On the other hand, if you are measuring the distance from the Earth to Jupiter the value of 1,000,000,000,000 m gets rather cumbersome. For large distances, we often use **kilo**meters, or **km**. Kilo is the prefix meaning 1000, or 1×10^3. Jupiter is about 1 billion km from Earth.

We use prefixes with SI units to denote small values as well as large ones. The mass of a styrofoam peanut that is used as a packing material is about 0.040 g. It is easier to describe it as 40 milligrams (mg).

Table 1.2 in your textbook lists SI prefixes. You should know the ones that are listed in color.

Example 1.3 A   *Practice with Prefixes*

Put the following prefixes in order, from smallest to largest:

centi   mega   milli   deka   deci   nano   kilo   micro

_____   _____   _____   _____   _____   _____   _____   _____

smallest   largest

Solution

These prefixes and their values need to be committed to memory. You will get used to working with them soon enough.

nano	micro	milli	centi	deci	deka	kilo	mega
smallest							largest

Distance (in meters) is a **fundamental SI unit**. Speed can be measured as distance per unit time, or **m/s**. Because speed is derived from fundamental units, it is called a **derived unit**. Volume, in m³, is another example of a derived unit, because it is derived from a fundamental unit, length.

Example 1.3 B   *Practice With Units*

Based on your reading from Section 1.3, what is the fundamental or derived unit in which each of the following measurements would be expressed?

    a.   graduated cylinder
    b.   tape measure
    c.   laboratory balance
    d.   ammeter

Solution

Table 1.1 in your textbook help us to see the following answers.

    a.   milliliters
    b.   meters
    c.   grams
    d.   amperes

1.4   Uncertainty in Measurement

When you finish this section, you will be able to:

- Define accuracy and precision.
- Determine if a set of measurements is accurate and/or precise.
- State the difference between implied and estimated error.

Among the most important skills that chemists have is to measure quantities. We can measure some things better than others. For example, taking into account the table of contents, the index, appendices and so forth, you can "measure" the number of pages in your textbook. If you can count separate units (such as pages of a book or eggs in a dozen) you can get an **exact number**.

Now pick up your text. How much does it weigh? On a bathroom scale the book weighs 5 pounds, or a little over 2 kg. That is **not an exact number** because the bathroom scale is hard to read and is not very consistent in its values. Using a bathroom scale to measure weight gives rise to **uncertainty in your measurement**. The better your scale, the better your measurement. If you have an excellent scale, or have access to a high-quality balance, your answer will be very accurate.

- **Accuracy: How close you are to the true value.** If you take repeated measurements, your average value is likely to become accurate.
- **Precision: How close your values are to one another (internal consistency).**

<u>Figure 1.8 in your textbook</u> illustrates the possible combinations of accuracy and precision. The accompanying text discusses **random and systematic errors**. Know how these terms relate to accuracy and precision.

Example 1.4   Accuracy and Precision

Each of 4 general chemistry students measured the mass of a chemistry textbook. They each weighed the book 4 times. Knowing that the true mass is 2.31 kg, which student weighed the book:

- a.   accurately and precisely
- b.   inaccurately but precisely
- c.   accurately but imprecisely
- d.   inaccurately and imprecisely

weighing	student #1	student #2	student #3	student #4
1	2.38 kg	2.06 kg	2.32 kg	2.71 kg
2	2.23	1.94	2.30	2.63
3	2.07	2.09	2.31	2.66
4	2.55	2.40	2.32	2.68
Average:	2.31±0.16 kg	2.12±0.14 kg	2.31±0.01kg	2.67±0.03 kg

Strategy

You have to ask yourself two questions about each data set:

1. Is the average close to the accepted (true) value? If it is, then the result is accurate.
2. Is the average deviation small relative to the actual value? If it is, then the result is precise.

Keep in mind that accuracy and precision are relative terms. What is considered accurate and precise in one procedure may not be good enough under other experimental conditions.

Solution

Data set #1 - **accurate but imprecise**
Data set #2 - **inaccurate and imprecise**
Data set #3 - **accurate and precise**
Data set #4 - **inaccurate but precise**

In the previous example, each of the calculations had a calculated average deviation with it. When we don't have enough information for such a calculation, as may be the case with a single calculation, we estimate that the error is probably ±1 in the last digit. For example:

value	estimated error
5 grams	±1 gram
3.006 mL	±0.001 mL
3.8×10^2 K	$±0.1 \times 10^2$ K

1.5   *Significant Figures and Calculations*

When you finish this section you will be able to:

- Determine the number of significant figures in a single value.
- Calculate results of mathematical expressions to the proper number of significant figures.

When you report a measured value, it is assumed that all the figures are correct except for the last one, where there is an uncertainty of ±1. If your value is expressed in proper **exponential notation**, all of the figures in the pre-exponential value are significant, with the **last digit** being the **least significant figure (LSF).**

"7.143×10^{-3} grams" contains 4 significant figures (SF)
$\uparrow$
(LSF)

If that value is expressed as 0.007143, it still has 4 significant figures. Zeros, **in this case,** are place holders. **If you are ever in doubt about the number of significant figures in a value, write it in exponential notation.**

The rules for counting significant figures are given in your textbook. Let's use those rules in the following exercise.

Example 1.5 A   *Counting Significant Figures*

Give the number of significant figures in the following values:

a. 38.4703 mL
b. 0.00052 g
c. 0.05700 s
d. 6.19×10^8 years

Helpful Hint

Convert to exponential form if you are not certain as to the proper number of significant figures.

Solution

a. 38.4703 mL = **6 SF**.
b. 0.00052 g = 5.2×10^{-4} g = **2 SF**. The leading zeros are not significant.
c. 0.05700 s = 5.700×10^{-2} s = **4 SF**. Again, the leading zeros are not significant. The trailing zeros are.
d. 6.19×10^8 years = **3 SF**.

There is uncertainty in all measured values (we assume that exact numbers are not "measured," but rather "counted"). Performing mathematical operations with measured values NEVER decreases uncertainty. Such operations only exacerbate error (make it more severe).

At the start of **Section 1.5** your textbook states the rules for retaining the proper number of significant figures in your calculations.

A very important idea is that you **DO NOT ROUND OFF YOUR ANSWER UNTIL THE VERY END OF THE PROBLEM**. Rounding off in the middle just leads to greater error.

Example 1.5 B   Significant Figures in Calculations

Perform the indicated calculations on the following measured values, giving the final answer with the correct number of significant figures.

   a. $12.734 - 3.0$
   b. 61×0.00745

   c. $\dfrac{5 \times 10^{16}}{(4.78 - 2.314)}$

   d. $(6.022 \times 10^{23} + 4.14 \times 10^{17}) \times (8.31 \times 10^{-11} - 9.2 \times 10^{-9})$

Helpful Hint

Recall that you never round off intermediate answers. Only round off at the very end of the problem.

Solution

   a.   12.734 has **3** figures past the decimal point. 3.0 has only **1** figure past the decimal point. Therefore, your final result, **where only addition or subtraction is involved**, should round off to **one figure past the decimal point**.

$$12.734$$
$$\underline{-3.0}$$
$$9.734 \xrightarrow{\text{round off to}} \mathbf{9.7}$$

   b.   In multiplication and division, your answer should have the **same number of significant figures as the least precise measurement**. (The one with the fewest significant figures will be the least precise.)

$$61 \times 0.00745 = 0.45445 = \mathbf{0.45}$$

(2 SF)   (3 SF)   (rounded
least precise   to 2 SF)

   c.   You will often have to solve problems where there is a combination of mathematical operations. To get reasonable answers, you need to recall your **order of operations** given the following table:

Order of Operations

   1.   parentheses
   2.   exponents and logs
   3.   multiplication and division
   4.   addition and subtraction

Therefore, in our problem, we might think that we do the subtraction last. However, the numbers are **inside parentheses**. Therefore, the subtraction becomes our **highest** priority. Our problem really is:

$$5 \times 10^{16} / 2.46_6 = 2.028 \times 10^{16}$$

not really a SF

In this division, 5×10^{16}, having only 1 significant figure, is the least precise (compared to 3 significant figures from our subtraction). Therefore, our answer must round to $\mathbf{2 \times 10^{16}}$.

d. Here again, the parentheses dictate our order of operations, working left to right,

$$(6.022 \times 10^{23} + 4.14 \times 10^{17}) = 6.022 \times 10^{23}$$

Your calculator did not make a mistake! The value of 4.14×10^{17} is so small relative to 6.022×10^{23} that it is insignificant and can be ignored. Therefore, in this special case, you are entitled to retain 4 significant figures from this operation.

In the next operation,

$$(8.31 \times 10^{-11} - 9.2 \times 10^{-9}) = -9.1169 \times 10^{-9}$$

Although this rounds off to $\mathbf{-9.1 \times 10^{-9}}$ (2 SF) we will retain the entire value **until the very end**.

Finally,

$$6.022 \times 10^{23} \times -9.1169 \times 10^{-9} = -5.4902 \times 10^{15},$$

which rounds to 2 significant figures to equal $\mathbf{-5.5 \times 10^{15}}$.

You must not rush with significant figure calculations. Make sound, careful decisions, and your answers will be reasonable.

1.6   *Dimensional Analysis*

When you finish this section, you will be able to:

- Convert between English and metric units.
- Convert values from one prefix to another.

Dimensional analysis is the single most valuable mathematical technique that you will use in general chemistry. The method involves **using conversion factors to cancel units until you have the proper unit in the proper place**.

When you are setting up problems using dimensional analysis, you are more concerned with **units** than with numbers. Let's illustrate this by finding out the mass of a 125-pound box (1 kg = 2.2046 pounds).

Problem Solving Steps

1. List the relevant **conversion factors**.

2. Set up the problem as follows:

$$
\text{kg} \quad \| \quad \underset{\text{↓}}{\overset{\text{exact number}^*}{\frac{1\,\text{kg}}{2.2046\,\text{pounds}}}} \times \frac{125\,\text{pounds}}{1} = \mathbf{56.7\ kg}
$$

the unit(s) you want on the left hand side	double lines mean "solution to follow"	conversion factor with desired unit in proper place (kg in numerator)	your original value, in pounds, needs to be in the numerator to cancel pounds from the conversion factor

3. Multiply all the values in the numerator, and divide by all those in the denominator.
4. Double check that your units cancel properly. If they do, your numerical answer is probably correct. If they don't, your answer is certainly wrong.

Once you understand what the problem is asking for, then

UNITS ARE THE KEY TO PROBLEM SOLVING!

Let's extend the method to the following example.

Example 1.6 A   *Practice With Dimensional Analysis*

It takes, including other ingredients, exactly 1 egg to make 8 pancakes. A pancake eating contest was held at which the winner ate 74 pancakes in 6.0 minutes. At this rate, **how many eggs** (in the pancakes) would be eaten by the winner **in one hour**?

Strategy

Following our problem solving procedure,

Step 1. Conversion Factors:

a. 1 egg = 8 pancakes. Keep in mind that this is **exactly the same** as 8 pancakes = 1 egg. You can therefore either use

$$
\frac{1\,\text{egg}}{8\,\text{pancakes}} \quad \text{or} \quad \frac{8\,\text{pancakes}}{1\,\text{egg}}
$$

However, it is **NOT CORRECT** to use $\dfrac{8\,\text{eggs}}{1\,\text{pancake}}$ or $\dfrac{1\,\text{pancake}}{8\,\text{eggs}}$. When you flip units, the numbers must flip with them.

b. Although it is not stated in the problem, you need a conversion factor from minutes to hours.

$$
\frac{60\,\text{minutes}}{1\,\text{hour}} \quad \text{or} \quad \frac{1\,\text{hour}}{60\,\text{minutes}} \quad (\text{exactly})
$$

---

*Exact number means perfect integer, that is, it has infinite significance and doesn't affect significant figures in the calculation.

c. 74 pancakes per 6.0 minutes can be expressed as

$$\frac{74 \text{ pancakes}}{6.0 \text{ minutes}} \quad \text{or} \quad \frac{6.0 \text{ minutes}}{74 \text{ pancakes}}$$

Step 2. Setting up the problem:

The units that you want go on the left hand side.

$$\frac{\text{eggs}}{\text{hour}} \quad \| \quad \frac{1 \text{ egg}}{8 \text{ pancakes}} \quad \times \quad \frac{74 \text{ pancakes}}{6.0 \text{ minutes}} \quad \times \quad \frac{60 \text{ minutes}}{1 \text{ hour}}$$

read: "eggs per hour" | has proper unit (eggs) in numerator | cancels pancakes, puts time in denominator | cancels minutes, puts hours in denominator

Solution and Double Check

Step 3. Multiply in the numerator and denominator and then divide by the denominator:

$$\frac{74 \times 60}{8 \times 6} = \frac{4440}{48} = \frac{92.5 \text{ eggs}}{\text{hour}}, \text{ rounds to } \frac{92 \text{ eggs}}{\text{hour}}$$

Step 4. Double check your units:

$$\frac{\text{eggs}}{\text{pancake}} \times \frac{\text{pancakes}}{\text{minutes}} \times \frac{\text{minutes}}{\text{hour}} = \frac{\text{eggs}}{\text{hour}}$$

Dimensional analysis will often involve interconversion among prefixes of the same unit. You must be very careful to **think** about the validity of your final value.

Example 1.6 B   *Practice With Prefixes*

How many μm are there in 1 km?

Critical Concept

Keep in mind that the prefix "**micro**" means 10^{-6}. It is a very small number. There are many μm in a meter. You can write the conversion between μm and m as follows:

$$\text{(A) } 1 \text{ μm} = 10^{-6} \text{ m}$$
$$\text{or (B) } 10^{6} \text{ μm} = 1 \text{ m.}$$

You want μm.

$$\text{km} \longrightarrow \text{m} \longrightarrow \text{μm.}$$
The bridge between prefixes

$$\text{μm} = 1 \text{ km} \times \frac{1 \times 10^{3} \text{ m}}{1 \text{ km}} \times \frac{1 \text{ μm}}{1 \times 10^{-6} \text{ m}} = \mathbf{1 \times 10^{9} \text{ μm in 1 km}}$$

Does the Answer Make Sense?

A μm is very small. A km is very large. Therefore, we would expect that there would be many μm in a km. Carelessly inverting prefix conversions (such as incorrectly stating that $1\ m = 10^3\ km$) is among the major sources of incorrect answers in general chemistry.

Example 1.6 C   Summing it All Up

In the vacuum of space, light travels at a speed of 186,000 miles per second. How many centimeters can light travel in a year?

Strategy

We have $\dfrac{\text{miles}}{\text{second}}$. We want $\dfrac{\text{cm}}{\text{year}}$.

Step 1.   Conversion factors: (You know some of these. You may have to check the inside back cover of your textbook for others.)

186,000 miles per 1 second
60 seconds per minute
60 minutes per hour
24 hours per 1 day
365 days per 1 year
1 mile per 1.6093 kilometers
1×10^3 m per kilometer
1×10^{-2} m per 1 cm

Step 2.   Setting up the problem:

$$\frac{\text{mi}}{\text{s}} \longrightarrow \frac{\text{km}}{\text{s}} \longrightarrow \frac{\text{m}}{\text{s}} \longrightarrow \frac{\text{cm}}{\text{s}} \longrightarrow \frac{\text{cm}}{\text{min}} \longrightarrow \frac{\text{cm}}{\text{hr}} \longrightarrow \frac{\text{cm}}{\text{day}} \longrightarrow \frac{\text{cm}}{\text{year}}$$

start — top is converted to cm — finish

$$\frac{\text{cm}}{\text{yr}} = \frac{186{,}000\ \text{mi}}{1\ \text{s}} \times \frac{1.6093\ \text{km}}{1\ \text{mi}} \times \frac{1\times10^3\ \text{m}}{1\ \text{km}} \times \overset{(*)}{\frac{1\ \text{cm}}{1\times10^{-2}\ \text{m}}} \times \frac{60\ \text{s}}{1\ \text{min}} \times \frac{60\ \text{min}}{1\ \text{hr}} \times \frac{24\ \text{hr}}{1\ \text{day}} \times \frac{365\ \text{days}}{1\ \text{year}}$$

(*) = Note that identical terms don't have to be <u>next</u> to each other to cancel. As long as a term appears in both the numerator and denominator, it cancels.

Step 3.   Multiply numerator and divide by denominator:

$$\frac{(186{,}000)(1.6093)(1\times10^3)(1)(60)(60)(24)(365)}{(1)(1)(1)(1\times10^{-2})(1)(1)(1)(1)} = \frac{9.44\times10^{15}}{1\times10^{-2}} = \mathbf{9.44 \times 10^{17}\ cm/year}$$

Step 4.   Double check your units:

$$\frac{\text{mi}}{\text{s}} \times \frac{\text{km}}{\text{mi}} \times \frac{\text{m}}{\text{km}} \times \frac{\text{cm}}{\text{m}} \times \frac{\text{s}}{\text{min}} \times \frac{\text{min}}{\text{hr}} \times \frac{\text{hr}}{\text{day}} \times \frac{\text{days}}{\text{yr}} = \frac{\text{cm}}{\text{year}}$$

1.7   Temperature

When you finish this section you will be able to interconvert among Fahrenheit, Celsius, and Kelvin temperature scales.

Your textbook gives information regarding the Fahrenheit, Celsius, and Kelvin temperature scales. Among the things you should be able to state regarding these three temperature scales are:

- how to convert among them,
- the freezing and boiling points of water on each scale, and
- the units of each one: °F, °C and K (no <u>degree</u> sign in Kelvin).

A helpful hint - you can determine the conversion factor between °C and °F (and vice versa) just by remembering that the boiling point of water can be expressed as 100°C or 212°F. The best way to get from 100°C to 212°F, and include a factor of 9/5 is:

$$100°C \times \frac{9°F}{5°C} = 180 + 32 = 212°F$$

$$\text{or} \quad °C \times \frac{9}{5} + 32 = °F$$

Let's try some practice problems.

Example 1.7 A   Conversions Among Temperature Scales

The boiling point of water on top of Long's Peak in Colorado (14,255 feet above sea level) is about 86°C. What is the boiling point in:

a.   Kelvin,
b.   degrees Fahrenheit?

Solution

a. $T_K = T_C + 273.15$

substituting,

$$T_K = 86 + 273.15 = 359.15 \text{ K}$$

which rounds to **359 K**.

b. $T_F = T_C \times \frac{9°F}{5°C} + 32°F$

substituting,

$$T_F = 86 \times 9/5 + 32 = 154.8 + 32 = 186.8°F$$

which rounds to **187°F**.

In the next problem, we convert °C to °F using an alternate method.

Example 1.7 B   Conversions Among Temperature Scales

New materials can act as superconductors at temperatures above 250 K. Convert 250 K to degrees Fahrenheit.

Solutions

There are two ways of solving this problem. Both involve converting

$$K \longrightarrow {}^\circ C \longrightarrow {}^\circ F$$

Method 1:

Convert Kelvin to degrees Celsius:

$$T_C = T_K - 273.15 = 250 - 273.15 = \mathbf{-23.15^\circ C}$$

We don't round off because this is an **intermediate** calculation. Now convert degrees Celsius to degrees Fahrenheit as in the last problem.

$$T_F = T_C \times \frac{9^\circ F}{5^\circ C} + 32^\circ F = -23.15 \times \frac{9}{5} + 32 = -9.67^\circ F$$

This rounds to **−10.°F**. Note that if we had rounded to −23°C, our answer would have been −9°F. That's not far off, but it's not correct either.

Method 2:

Convert Kelvin to degrees Celsius as above. Convert degrees Celsius to degrees Fahrenheit as given in your textbook.

$$\frac{T_F + 40}{T_C + 40} = \frac{9^\circ F}{5^\circ C}$$

Substituting in degrees Celsius from above, solve for T_F.

$$\frac{T_F + 40}{-23.15 + 40} = \frac{T_F + 40}{16.85} = \frac{9}{5} \quad \Longrightarrow \quad T_F + 40 = 30.33$$

$$T_F = -9.67^\circ F, \quad \text{or} \quad \mathbf{T_F = -10^\circ F}$$

Either method is correct for converting degrees Fahrenheit to degrees Celsius.

1.8   Density

When you finish this section, you will be able to perform density calculations.

As discussed in your textbook,

$$\mathbf{density} = \frac{\text{mass}}{\text{volume}} .$$

It is a derived unit, most often expressed as grams/cm^3 or grams/mL. When solving a density problem, the key is to **keep it simple**. Find out **how many grams** you have, determine the **total volume** your mass occupies, calculate the **ratio** of the two, and report your answer using the proper units.

Example 1.8 A   *Density Calculations*

The density of mercury is 13.6 g/cm^3. How many pounds (1 pound = 453.6 g) would 1 liter of mercury weigh?

Solution

There are 13.6 g per cm^3. There are 1000 cm^3 in 1 liter. Therefore, there are

$$\frac{13.6 \text{ g}}{\text{cm}^3} \times \frac{1000 \text{ cm}^3}{1 \text{ L}} = \frac{13{,}600 \text{ g}}{1 \text{ L}}$$

Now convert g/L to pound/L:

$$\frac{13{,}600 \text{ g}}{1 \text{ L}} \times \frac{1 \text{ pound}}{453.6 \text{ g}} = \frac{29.982 \text{ pounds}}{1 \text{ L}} = \frac{\textbf{30.0 pounds of mercury}}{\textbf{1 L}}$$

An alternative strategy is to set the entire problem up using dimensional analysis:

$$\text{pounds} \;\bigg\|\; \frac{1 \text{ pound}}{453.6 \text{ g}} \times \frac{13.6 \text{ g}}{1 \text{ cm}^3} \times \frac{1000 \text{ cm}^3}{1 \text{ L}} = \frac{\textbf{30.0 pounds of mercury}}{\textbf{1 L}}$$

Example 1.8 B   *Density Calculations*

An unknown metal having a mass of 287.8 g was added to a graduated cylinder that contained 31.47 mL of water. After the addition of the metal, the water level rose to 56.85 mL. Using this information and Table 1.5 in your textbook, identify the unknown metal.

Strategy

As with any density problem, your goal is to determine the ratio of mass/volume. You know the mass of the metal (287.8 g). The volume is equal to the amount that the water has been displaced.

Solution

Mass of the metal = 287.8 g

Displacement of water = 56.85 mL − 31.47 mL = **25.38 mL**

                        water   original
                          +   volume
                          metal   of water

$$\text{Density of the metal} = \frac{287.8 \text{ g}}{25.38 \text{ mL}} = \textbf{11.34 g/mL}$$

According to Table 1.5 in your textbook this metal is **lead**.

1.9   *Classification of Matter*

This study section will help guide you through some of the terms and techniques that are discussed in this section of your textbook.

Terms you should be able to define:

1. Matter
   a. solid
   b. liquid
   c. gas

2. Mixtures
   a. homogeneous (solution)
   b. heterogeneous

3. Physical and Chemical Changes

4. Compounds and Elements

You should be able to describe the following separation techniques:

1. Distillation

2. Filtration

3. Chromatography

Exercises

Section 1.2

1. In a simulated boxing match done on a computer several years ago, Joe Louis (champion in the 1940s) defeated Muhammad Ali (champion in the 1960s) in 15 rounds. The computer simulation took into account the fighters' heights, weights, styles, punching power, and speed. Was the simulation realistic? What other variables would have to be included to make the simulation more valid?

2. If your chemistry teacher said she **knew** that the Earth is flat and challenged you to prove, using scientific tests, that she was wrong, what tests might you perform?

Section 1.3

3. List the fundamental units that you would combine to get the following derived units (you may need to look up the meaning of some of the terms):

   a. velocity   d. specific heat
   b. acceleration   e. density
   c. volume   f. pressure

4. Put the following prefixes in order, from smallest to largest: femto, tera, mega, deci, kilo, atto, exa.

5. Put the following prefixes in order, from smallest to largest: giga, hecto, micro, peta, milli, nano, pico.

Section 1.4

6. Three students weighed the **same** sample of copper shot five times. Their results were as follows:

weighing	student #1	student #2	student #3
1	17.516 g	15.414 g	13.893 g
2	17.888 g	16.413 g	13.726 g
3	19.107 g	14.408 g	13.994 g
4	21.456 g	15.637 g	13.810 g
5	19.983 g	15.210 g	13.476 g

   a. Calculate the mean (average) mass of the sample determined by each student.

   b. The **average deviation (a.d.)** is a measure of **precision**. It is calculated by the formula:

$$\text{a.d.} = \frac{\sum\limits_{i=1}^{n} \left| x_i - \bar{x} \right|}{n}$$

   That means that you take each of the values in a data set (5 total values), subtract the mean, and take the absolute value. Add all 5 absolute values and divide the total by 5. That number is your a.d. Calculate the average deviation for each student's data.

   c. If the true mass of copper shot is **15.384 g**, which student was most accurate? Which was most precise? What could be the possible sources of error in the determinations?

7. A student weighed 15 pennies on a balance and recorded the following masses:

3.078	3.055	3.060	3.066	3.102
2.107	3.121	2.518	3.052	2.476
2.546	3.050	3.073	3.080	3.128

   a. Calculate the mean mass.
   b. What might cause the difference in weights?

8. Define the following terms:

   a. accuracy
   b. precision

   c. random error
   d. systematic error

Section 1.5

9. How many significant figures are there in each of the following values?

   a. 6.07×10^{-15}
   b. 0.003840
   c. 17.00

   d. 8×10^8
   e. 463.8052

10. How many significant figures are there in each of the following values?

   a. 1406.20
   b. 0.0007

   c. 1600.0
   d. 0.0261140

   e. 1.250×10^{-3}

11. Which value has more significant figures, 7.63×10^{-11} or 0.00076?

12. Use exponential notation to express the number 37,100,000 with:

   a. one sig fig
   b. two sig figs

   c. three sig figs
   d. six sig figs

13. Perform the indicated calculations on the following measured values, giving the final answer with the correct number of significant figures.

   a. $16.81 + 3.2257$
   b. 324.6×815.991
   c. $(3.8 \times 10^{-12} + 4 \times 10^{-13})/(4 \times 10^{12} + 6.3 \times 10^{13})$
   d. $3.14159 \times 68 / (5.18 \times 10^{-11} - 6 \times 10^{-4})$

14. Perform the indicated calculations on the following measured values, giving the final answer with the correct number of significant figures.

   a. $2.85 + 3.4621 + 1.3$
   b. $7.442 - 7.429$
   c. 1.65×14

   d. $27/4.148$
   e. $[(3.901 - 3.887)/3.901] \times 100.0$
   f. $6.404 \times 2.91 \times (18.7 - 17.1)$

Section 1.6

15. Using the conversion factors in the inside back cover of your textbook, convert 3.5 quarts to:

   a. liters (L)
   b. milliliters (mL)

   c. microliters (μL)
   d. cubic centimeters (cm^3)

16.   Using the conversion factors in the inside back cover of your textbook, convert 4.2 yards to:

    a.   meters   c.   micrometers
    b.   centimeters   d.   kilometers

17.   If you put 8 gallons of gas in your car and it cost you a total of $19.04, what is the cost of gas per liter?

18.   A runner can run a 5.0 kilometer race in a time of 21 minutes and 22 seconds.

    a.   What is the runner's speed in miles per hour?
    b.   How long, on the average, did it take for the runner to run one mile?

19.   A student made the 27.0 kilometer drive to school in 16 minutes.

    a.   How many miles did the student drive?
    b.   If the speed limit is 55 mph, was the student speeding? How fast was the student driving?

20.   A radio station broadcasts at a frequency of 107.9 megahertz (MHz), (cycles per second).

    a.   How many seconds per cycle are there in 107.9 MHz?
    b.   What is the broadcast frequency in gigahertz (GHz)?

21.   A 5-lb bag of flour costs $0.89. What is the cost of flour per kilogram?

22.   A "joule," like a calorie, is a unit of energy. There are 4.184 joules per calorie. How many joules of energy are available in one ounce of Frosted Flakes®, which has 120,000 calories? (Note: a "food calorie" (called a Calorie, with a capital "C") = 1000 energy calories.)

23.   If you have to eat 20 tablespoonfuls of cereal to eat the entire ounce of dry Frosted Flakes® (discussed in the previous problem) and there are 3 teaspoon measures in one tablespoon, how many food calories of Frosted Flakes® are in each teaspoonful?

24.   During a recent baseball game, a pitcher threw a fastball that had a velocity of 93.7 mph.

    a.   Calculate the velocity in meters/second.
    b.   Calculate how long it took this pitch to travel from the mound to home plate (60 ft 6 in).

25.   Which of the following is greater:

    a.   35 kg or 3500 g?
    b. 6×10^4 mL or 6×10^3 L?

26.   If a student weighs 185 lb, what is his mass in µg?

Section 1.7

27.   Which is the higher temperature, 42°C or 92°F?

28.   Perform the following temperature conversions:

    a.   300. K to °F   c.   −40.°F to °C   e.   1555 K to °C
    b.   300.°F to K   d.   −100.°C to K   f.   0.0 K to °F

29.   Perform the following temperature conversions:

    a.   16°C to °F   c.   0.0°F to °C   e.   −45°C to K
    b.   305 K to °F   d.   150°F to K   f.   920 K to °C

Section 1.8

30. A sample of motor oil with a mass of 440 g occupies 500 mL. What is the density of the motor oil?

31. A worker at the United States Mint wants to know if a batch of 100 pennies was minted before 1982 (100% copper) or after 1982 (3% copper and 97% zinc). Assuming that both groups of pennies had the same dimensions, would she get her answer by weighing the coins? Explain.

32. The density of an object is 1.63 g/mL. Its volume is 0.27 L. What is the mass of the object?

33. An object weighing 4.0 lbs occupies 1700 mL. What is the density of the object in g/mL?

34. The density of the Earth is about 3.5 g/cm^3. If the Earth has a radius of 7000 miles, what is its mass? (volume = $[4\pi r^3/3]$)

35. Which of the following is less:

    a. 8.7 g/mL or 6.1 μg/μL?
    b. 4×10^{-2} kg/cm^3 or 4×10^{-1} mg/cm^3?

Section 1.9

36. Define the following:

    a. mixture
    b. pure substance
    c. homogeneous
    d. heterogeneous

37. List five physical methods of separating mixtures.

Multiple Choice Questions

38. Which of the following is the correct order of steps to establish a theory?

    A. Conducting experimental work, collecting observations, making a hypothesis, establishing a theory.
    B. Establishing a theory, conducting experimental work, collecting observations, making a hypothesis.
    C. Making a hypothesis, collecting observations, establishing a theory, conducting experimental work.
    D. Collecting observations, making a hypothesis, conducting experimental work, establishing a theory.

39. Which of the following statements is not an observation?

    A. There are two hydrogen atoms and an oxygen atom in a water molecule.
    B. The Earth is the third planet of our solar system.
    C. The Universe might have begun with a Big Bang explosion that took place approximately 15 billion years ago.
    D. In a total absence of water, plants cannot survive.

40. Which one of the following is not one of the seven basic SI units?

    A. Mass   B. Volume   C. Time   D. Mole

41. The basic unit for amount of substance is:

    A. Mole   B. Liter   C. Kilogram   D. Gram

42. A picometer is:

    A. Greater than a micrometer
    B. Equal to a micrometer
    C. Less than a micrometer
    D. Twice as large as a micrometer

43. Which of the following units would best describe the distance between two asteroids that are 8 deca hecto kilo miles apart?

    A. Gigamile   B. Teramile   C. Macromile   D. Megamile

44. The number of zeroes following and preceding the decimal point for the following two prefixes, femto and exa, are

    A. 18 and 18   B. 14 and 18   C. 15 and 18   D. 15 and 19

45. Which of the following cannot be an exact number?

    A. Three eggs
    B. 12 mL of water measured in a 20-mL cylinder
    C. 12 roses
    D. 2400 air flights

46. Which of the following is the estimated error for 12.3008 g of sugar?

    A. ± 0.001   B. ± 0.0001   C. ± 0.0008   D. ± 0.00001

47. Which number has the greatest uncertainty?

    A. 1.02 ± 0.01 B. 4.60 ± 0.01 C. 100.0 ± 10.00 D. 1.00 ± 0

48. Round 20.589958 to 4, 3, and 2 significant figures, respectively.

    A. 20.59, 20.5, 20   B. 20.58, 20.5, 20   C. 20.60, 20.6, 21   D. 20.59, 20.6, 21

49. The solution to (22.41 + 0.464) × 999/18.465 is:

    A. 1237.32   B. 1.24×10^3 C. 1.2×10^4 D. 1237

50. The solution to 9.99/22.41 × (18.465 + 0.464) is

    A. 8.44   B. 8   C. 8.4   D. 8.438

51. Assume mass and weight to be equivalent, i.e., 28.4 g = 1.00 oz. Calculate the weight of the Earth in lbs if its mass is 3.7×10^{24} kg.

    A. 8.1×10^{24} lbs   B. 8.3×10^{38} lbs   C. 4.0×10^{20} lbs   D. 1.7×10^{69} lbs

52. What are the dimensions, in metric units, for a linebacker who is 6' 4.0" and weighs 245 lbs?

    A. 193 cm and 6,900 g
    B. 1.93 m and 111 kg
    C. 0.193 m and 100 kg
    D. 1.11 m and 193 kg

53. A rectangular tile, 15 by 18 inches, can be converted into square meters by which one of the following conversion setups?

    A. (15 in × 18 in)(2.54 cm/1in)(1 m/100 cm)
    B. (15 in × 18 in)(2.54 cm/1in)^2(1 m/100 cm)
    C. (15 in × 18 in)(2.54 cm/1 in)^2(1 m/100 cm)^2
    D. (15 in × 18 in)(2.54 cm/1 in)(1 m/100 cm)^2

54. A parsec is an astronomical unit of distance. 1 parsec = 3.26 light years (or the distance traveled by light in one year). Light speed = 186,000 miles per second. An object travels 9.6 parsecs. Calculate this distance in cm.

    A. 2.0×10^{13} cm   B. 3.0×10^{19} cm   C. 9.6×10^8 cm   D. 3.7×10^{15} cm

55.   How many cubic feet are there in a cube whose edge is 6.0×10^{21} miles in length?

    A. 2.2×10^{63} ft^3 B. 3.2×10^{76} ft^3 C. 1.0×10^{27} ft^3 D. 3.6×10^{42} ft^3

56.   Ultra-deka light is ten times the speed of light (speed of light = 186,000 miles per second). How many feet will a space vessel travel at ultra-deka light in 10 seconds?

    A. 9.8×10^{10} ft   B. 6.7×10^{7} ft   C. 1.9×10^{9} ft   D. 3.9×10^{9} ft

57.   8 quarts = 1 peck, and 4 pecks = 1 bushel. How many quarts are there in half a megabushel?

    A. 1.6×10^{7} qt   B. 1.3×10^{4} qt   C. 4.0×10^{6} qt   D. 6.5×10^{7} qt

58.   At what temperature does °C = 0.5 (°F)?

    A. °C = 60 and °F = 120   C. °C = 45 and °F = 90
    B. °C = 160 and °F = 320   D. °C = 0 and °F = 0

59.   The average daytime temperatures on Earth and Jupiter are 72°F and 313 K, respectively. Calculate the difference in temperature, in °C, between these two planets.

    A. 18°C   B. 32°C   C. 20°C   D. 193°C

60.   A column of liquid is found to expand linearly on heating 5.25 cm for a 10°F rise in temperature. If the initial temperature of the liquid is 98.6°F, what will the final temperature be in °C if the liquid has expanded by 18.5 cm?

    A. 37.0°C   B. 72.2°C   C. 19.6°C   D. 56.6°C

61.   Calculate the density, in kg/L, of a block of wood 2.5 feet by 18 inches by 1 yard that weighs 646 lbs.

    A. 0.92 kg/L   B. 9.2 kg/L   C. 1.1 kg/L   D. 4.9 kg/L

62.   The specific gravity of benzene is 0.865 and the density of water is 0.996 g/cm^3 at 25°C. Specific gravity is defined as the ratio of the density of some material to the density of some standard material such as water. Calculate the density of benzene at 25°C.

    A. 0.865 g/cm^3 B. 0.862 g/cm^3 C. 1.000 g/cm^3 D. 0.996 g/cm^3

63.   If the volume occupied by an electron equals that occupied by a proton, what is the ratio of densities of proton to electron? You can consult your textbook to find the mass of the electron and that of the proton.

    A. 1840:1   B. 1:1000   C. 1000:1   D. 1.00:1.00

64.   Which of the following is a pure substance?

    A. An egg   B. Sea water   C. Bronze   D. Copper

65.   Which of the following is a homogeneous mixture?

    A. An egg   C. Oil and vinegar
    B. Copper   D. An unused piece of photocopy paper

66.   Which of the following techniques is the least desirable for separating sand and water?

    A. Freezing   B. Decanting   C. Evaporation   D. Extraction

Answers to Exercises

1.   The simulation was not especially realistic. The simulation could not take into account such variables as the mental condition of the fighters, whether they overtrained, whether they had to struggle to make the weight, or the differences in styles between the 1940s and 1960s fighters and equipment.

2.   You might travel around the Earth and end up at the same point (if you had the money), or use a pendulum, among other possibilities.

3.   a.   m/s   c. m^3 e.   g/mL (or g/cm^3)
     b. m/s^2 d.   J/g K (or J/g°C)   f. $kg/m\ s^2$

4.   atto, femto, deci, kilo, mega, tera, exa

5.   pico, nano, micro, milli, hecto, giga, peta

6.   a.   means: student #1 = 19.190 g
                 student #2 = 15.416 g
                 student #3 = 13.780 g
     b.   average deviations: student #1 = 1.224 g
                              student #2 = 0.487 g
                              student #3 = 0.143 g
     c.   Student #2 is the most accurate. Student #3 is the most precise. Among the sources of error could be some spillage of the sample, an inaccurate balance, or a wet beaker.

7.   a.   2.901
     b.   Beginning in 1982, "copper" pennies were made from zinc with a copper coating outside (97% Zn, 3% Cu).

8.   a.   the agreement of a particular value with the true value
     b.   the degree of agreement among several measurements of the same quantity
     c.   an error that has an equal probability of being high or low
     d.   an error that always occurs in the same direction and is likely to have a procedural cause

9.   a.   3   c.   4   e.   7
     b.   4   d.   1

10. a.   6   c.   5   e.   4
     b.   1   d.   6

11. 7.63×10^{-11} has 3 significant figures. The other value, 0.00076, has 2 significant figures.

12. a. 4×10^7 b. 3.7×10^7 c. 3.71×10^7 d. 3.71000×10^7

13. a.   20.04   b. 2.649×10^5 c. 6.3×10^{-26} d. -4×10^5

14. a.   7.6   c.   23   e.   0.36
     b.   0.013   d.   6.5   f.   30.

15. a.   3.3 L   b. 3.3×10^3 mL   c. 3.3×10^6 μL   d. $3.3 \times 10^3\ cm^3$

16. a.   3.8 m   b. 3.8×10^2 cm   c. 3.8×10^6 μm   d.   0.0038 km

17. $0.63/liter

18. a. 8.7 mi/hr   b. 6 minutes and 53 seconds

19. a. 16.8 miles   b. yes, roughly 63 mph

20. a. 9.268×10^{-9} s/cycle b. 0.1079 GHz

21. $0.39/kg

22. There are 5.0×10^5 joules (2 significant figures) of available energy in a bowl of Frosted Flakes.

23. There are 2 food calories (2000 energy calories) in each teaspoonful of Frosted Flakes.

24. a. 41.9 m/s   b. 0.440 seconds

25. a. 35 kg   b. 6×10^3 L

26. 8.39×10^{10} µg

27. 42°C equals 108°F, so it is the higher temperature.

28. a. 80.3°F   c. −40°C   e. 1282°C
     b. 422 K   d. 173 K   f. −460°F

29. a. 61°F   c. −18°C   e. 228 K
     b. 89.3°F   d. 339 K   f. 647°C

30. 0.88 g/mL

31. Yes, weighing the pennies would work. If the volume of both groups is the same (equal dimensions), any difference in mass must result from the density difference between the groups of pennies.

32. 440 g

33. 1.1 g/mL

34. 2.1×10^{28} g

35. a. 6.1 µg/µL   b. 4×10^{-1} mg/cm^3

36. a. substance that has a variable composition
     b. substance that has a constant composition
     c. mixtures that are the same throughout
     d. mixtures containing regions with different properties

37. distillation, filtration, liquid-column chromatography, gas chromatography, paper chromatography

38. D	39. C	40. B	41. A	42. C	43. D
44. B	45. B	46. B	47. C	48. D	49. B
50. A	51. A	52. B	53. C	54. B	55. B
56. A	57. A	58. B	59. A	60. D	61. A
62. B	63. A	64. D	65. D	66. A	

Chapter 2

Atoms, Molecules, and Ions

This chapter will help guide your study of **the nature of atoms, molecules, and ions—how they combine and how we name them**.

2.1   The Early History of Chemistry

The first part of this section considers the work of the ancient Greeks. The book points out that the Greeks did not use experiments as a way to establish the veracity of their ideas. It can be added that according to experts on the culture, there was, in effect, a "pecking order" in which the aristocracy of society, including Plato, and later, Socrates and Aristotle (as in "*aristoc*racy"), felt that experimentation was beneath them. Rather, such activities were the purview of the "non-intellectuals" in the society.

This section also discusses *alchemy*. In addition to comments on the positive aspects of this "pseudo-science," we can note that the opening of trade routes throughout Europe and Asia in the reign of Alexander the Great led to an important intercultural sharing of alchemical ideas among the Greeks, Egyptians, Indians, and Chinese. The word "alchemy" itself is probably derived from a combination of the Arabic definite article **al** with the Chinese **chin-I**, meaning "gold-making juice." Then again, the Greek word for casting or pouring metals is **cheo**. However, the Egyptian word for black (the color of metals in preparation for alchemical treatment) is **khem**. Chemistry truly had international origins!

The section ends with the advent of experimental chemistry at the hands of Robert Boyle. The quantitative aspects of chemistry were especially refined with the work of Antoine Lavoisier, the focus of the first part of Section 2.2.

2.2   Fundamental Chemical Laws

When you finish this section you will be able to use:

- The Law of Conservation of Mass
- The Law of Definite Proportion
- The Law of Multiple Proportions

Your textbook defines and discusses **the law of conservation of mass** and **the law of definite proportion**. Read that material, then try the following exercise.

Example 2.2 A   The Law of Conservation of Mass

In a combustion reaction, 46.0 g of ethanol reacts with 96.0 g of oxygen to produce water and carbon dioxide. If 54.0 g of water is produced, how much carbon dioxide is produced?

Solution

If the law of conservation of mass holds, the total of the masses of reactants must equal the product masses. Therefore,

$$\textbf{Total mass reacted } = \textbf{ Total mass produced}$$
$$46.0 \text{ g} + 96.0 \text{ g } = \text{ } 54.0 \text{ g} + \text{g carbon dioxide}$$
$$142.0 \text{ g} - 54.0 \text{ g } = \textbf{ 88.0 g carbon dioxide produced}$$

The law of definite proportion is described in your textbook. After reading that paragraph, try the following problem.

Example 2.2 B   The Law of Definite Proportion

A sample of chloroform is found to contain 12.0 g of carbon, 106.4 g of chlorine, and 1.01 g of hydrogen. If a second sample of chloroform is found to contain 30.0 g of carbon, how many grams of chlorine and grams of hydrogen does it contain?

Strategy

Assuming that the law of definite proportion is true, the **ratios of the masses of the elements in chloroform** are constant. Therefore, if the amount of carbon is increased by a factor of 2.5 (30.0 g/12.0 g, as above) then the same must hold true for chlorine and hydrogen.

Solution

Carbon is increased by a factor of 2.5 (30.0 g/12.0 g). Therefore,

$$\text{g chlorine} = 106.4 \text{ g} \times 2.5 = \textbf{266. g chlorine}$$
$$\text{g hydrogen} = 1.01 \text{ g} \times 2.5 = \textbf{2.53 g hydrogen}$$

The law of multiple proportions is discussed in your textbook. The key idea with this law is that, given a compound such as H_2O, if H_2O_2 is formed, the **ratio of the masses of oxygens** that combine with one gram of hydrogen will be a small whole number. The following exercise illustrates this.

Example 2.2 C   The Law of Multiple Proportions

Water, H_2O, contains 2.02 g of hydrogen and 16.0 g of oxygen. Hydrogen peroxide, H_2O_2, contains 2.02 g of hydrogen and 32.0 g of oxygen. Show how these data illustrate the law of multiple proportions.

Solution

In H_2O, 7.92 g (16.0/2.02) of oxygen combines with each 1.0 g of hydrogen. In H_2O_2, 15.84 g (32.0/2.02) of oxygen combines with each 1.0 g of hydrogen. The **ratio of the masses** of oxygen in the two compounds is:

$$\frac{15.84 \text{ g}}{7.92 \text{ g}} = 2.00$$

This is a "small whole number" that illustrates the law of multiple proportions.

2.3   Dalton's Atomic Theory

When you finish this section you will be able to select those ideas in Dalton's atomic theory that apply to specific chemical problems.

Read the four statements relating to Dalton's atomic theory that are given in your textbook. Use the following exercise to test your understanding of the theory.

Example 2.3   Dalton's Atomic Theory

Match the chemical statement on the left side with the atomic theory statement on the right side.

Chemistry

a. Although graphite and diamond have different properties (due to the nature of interatomic bonding), they are both composed solely of carbon. The carbon atoms are identical.

b. $2H_2 + O_2 \rightarrow 2H_2O$, not CS_2 or $NaCl$

c. There are 6.02×10^{23} atoms in 55.85 g of iron.

d. $C + O_2 \rightarrow CO_2$. CO_2 is not CO, $NaHCO_3$ or Fe_2O_3.

Atomic Theory

1. Each element is made up of tiny particles called atoms.

2. The atoms of a given element are identical; the atoms of different elements are different in some fundamental way or ways.

3. Chemical compounds are formed when atoms combine with each other. A given compound always has the same relative numbers and types of atoms.

4. Chemical reactions involve reorganization of the atoms - changes in the way they are bound together. The atoms themselves are not changed in a chemical reaction.

Solution

a. 2
b. 4
c. 1
d. 3

The statements in Dalton's atomic theory are **not necessarily mutually exclusive**. As a result, statement "b" can fit #3, and statement "d" can fit #4.

In reading this section in your text, don't neglect the discussion of **Avogadro's hypothesis**! The stated relationship between volumes and numbers of particles of gases sets the stage for realizing the *diatomic* nature of gases such as O_2, H_2, and Cl_2.

2.4   *Early Experiments to Characterize the Atom*

The focus here is the structure of the atom itself. The experiments of Thomson that determined the existence of electrons are described. The magnitude of the charge on an electron, as well as its mass, was determined by Millikan. The surprising results of Rutherford, described in <u>The Nuclear Atom subsection in your textbook</u>, explained the nature of the atom as having a small, dense, positively-charged nucleus surrounded by electrons moving around it at a relatively large distance.

2.5   *The Modern View of Atomic Structure: An Introduction*

When you finish this section you will be able to:

- Assign the correct number of protons, neutrons, and electrons to an atom,
- Write the symbol for an atom, and
- Do the above operations for any isotopes of an atom.

The names, sizes, and masses of the parts of the atom are discussed in your textbook. Pay special attention to <u>Table 2.1 in your textbook</u>.

The key point is that although the mass of an atom is concentrated in its nucleus, most of the volume of the atom is taken up by electron movement. Since electrons are involved in bonding, we need a way to symbolize the number of protons, neutrons, and electrons in an atom. Given

$$_Z^A X$$

X = The element symbol, as read from the periodic table.
Z = The number of protons.
A = The number of protons plus neutrons. (The number of neutrons = A − Z.)

The number of **protons** determines the element. **Potassium** has **19 protons**. If an element has **20 protons**, it is **calcium**. The number of neutrons and electrons do not determine what element you have. Only the number of protons is important. For example, phosphorus

$$_{15}^{31} P$$

contains **15 protons, 15 electrons** (in an <u>atom</u>, which is electrically neutral, the number of protons equals the number of electrons), and **16 neutrons** (31−15). The atomic mass is 31. If instead, we had

$$_{15}^{30} P$$

everything would be the same as in the previous example except for the <u>number of neutrons</u>, which would be **15 neutrons** (30−15). This phosphorus atom is therefore an **isotope** of the previous phosphorus atom.

Example 2.5   *Protons, Neutrons, Electrons, and Symbols*

Fill in the missing information in the following table:

Symbol	Element	Protons	Neutrons	Electrons
$^{14}_{6}C$	_____	_____	_____	_____
$^{235}_{92}$ __	_____	_____	_____	_____
__ __	U	_____	146	_____
^{55} __	_____	_____	_____	25

Helpful Hints

1. # of protons = # of electrons **in an atom**.
2. The element is determined by the number of protons.
3. In the periodic table, the elements are placed in order of increasing number of protons (although the groupings are really much more involved and meaningful).

Solution

Symbol	Element	Protons	Neutrons	Electrons
$^{14}_{6}C$	C	6	8	6
$^{235}_{92}U$	U	92	143	92
$^{238}_{92}U$	U	92	146	92
$^{55}_{25}Mn$	Mn	25	30	25

Determining the number of each kind of particle becomes more challenging when **ions** are introduced. Both your textbook and this study guide deal with these ideas in our next sections.

2.6   *Molecules and Ions*

When you finish this section you will be able to:

- Determine if a species is an atom, molecule, or ion.
- Determine the charge and the number of protons, neutrons, and electrons in monatomic ions.

Some very fundamental ideas and terms are discussed in your textbook. Read this section and be able to define:

1. covalent bonds
2. molecule
3. chemical formula
4. structural formula
5. ion
6. cation
7. anion
8. ionic bond
9. polyatomic ion

The key point of this section is that **molecules contain atoms** rather than ions. The molecule SF_6 contains 7 atoms (6 fluorine and 1 sulfur).

Certain atoms tend to ionize. For example, CaF_2 contains 3 ions: a Ca^{2+} and 2 F^-. We call CaF_2 an **ionic solid** or a **salt**. The positive and negative charges in salts exactly balance. In CaF_2, there are 2+ and 2− charges, yielding a neutral salt.

In future chapters you will learn how to predict *whether* molecules or salts will form. At this point, it is only expected that you know one when you see one.

Example 2.6 A   Molecule, Atom, or Ion?

Which of the following is an **atom**, an **ion**, or a **molecule**?

$C_6H_{12}O_6$ _____

N_2 _____

$CO_3{}^{2-}$ _____

Ag _____

Fe^{3+} _____

$NH_4{}^+$ _____

Solution

$C_6H_{12}O_6$	molecule
N_2	molecule (contains more than one atom)
$CO_3{}^{2-}$	ion (polyatomic anion)
Ag	atom
Fe^{3+}	ion (cation)
$NH_4{}^+$	ion (polyatomic cation)

For the next exercise, please keep in mind that the **total charge** of an ion is determined by whether the ion has **more protons** (net positive charge) **or electrons** (net negative charge).

Fe^{3+} means that iron has **3 more protons than electrons** (26 protons, 23 electrons)
O^{2-} means that oxygen has **2 more electrons than protons** (8 protons, 10 electrons)

Example 2.6 B   Protons, Neutrons, and Electrons in Ions

Fill in the missing numbers in the following table:

Symbol	Protons	Neutrons	Electrons	Charge
$^{32}_{16}S^{-2}$	____	____	____	____
$^{\_\_}_{\_\_}\_\_^{-}$	56	81	54	____
$_{\_\_}Cl^-$	____	20	____	−1
$^{52}_{24}\_\_^{7+}$	____	____	____	____

Solution

Symbol	Protons	Neutrons	Electrons	Charge
$^{32}_{16}S^{-2}$	16	16	18	−2
$^{137}_{56}Ba^{2+}$	56	81	54	+2
$^{37}_{17}Cl^-$	17	20	18	−1
$^{52}_{24}Cr^{7+}$	24	28	17	+7

2.7   An Introduction to the Periodic Table

When you finish this section you will be able to state whether an element is a metal or nonmetal and list what group it belongs to.

This section points out that the periodic table is organized based on the **properties elements have in common** with one another. You can combine the section reading with your own experience to identify the essential properties of metals vs. nonmetals.

Look at the periodic table in this section of your textbook. Please note:

1. The bold line from boron to polonium that separates metals from nonmetals.
2. Group names. You will be using these designations all year.
3. The various properties of each group, as described in the text.
4. **Groups** go **down**, **periods** go **across**.

Note also that the symbols of several elements are based on original Latin or Greek names. These elements, and their original names, are given in Table 2.2 in your textbook.

Example 2.7   Identify the Element

Given the following information, identify each element:

   a.   This is the only metal in Group 6A.
   b.   This alkali metal is in the same period as iodine.
   c.   Two atoms of this element, which is in the same period as magnesium, combine with a calcium ion to form a salt.

Solution

   a.   The only metal in Group 6A is **polonium.** It is to the left of the bold line on the periodic table. Elements that are next to that line are sometimes called metalloids because they have some properties of both metals and nonmetals.

   b.   The only alkali metal that is in the same period as iodine is **rubidium.**

   c.   Calcium forms Ca^{2+}. Therefore, if two atoms combine with it to make an electrically neutral salt, they each must have a −1 charge. **Halogens** readily form ions with a −1 charge. The halogen that is in the same period as magnesium is **chlorine.**

2.8 Naming Simple Compounds

When you finish this section you will be able to name, or give formulas for, the following classes of compounds:

- binary salts
- salts with polyatomic ions
- binary covalent compounds
- acids

There are really two keys to naming, or assigning formulas to compounds. The first key is to **know the names of the elements** (typically, knowing #1 - #54 is sufficient). The second key is to **correctly assign charges to ions**.

IUPAC (International Union of Pure and Applied Chemistry) nomenclature standards are used in your textbook and in this study guide. That way, any chemist (or chemistry student) should know exactly what you are talking about when you name, or give the formula of, a compound.

A. Naming Compounds

1. **Binary Salts: A binary salt contains only one kind of cation and one kind of anion.**

   a. **Cations:** Ions with positive charges fall into two general classes. **The first class** contains elements that **ionize to only one oxidation state**, as given in the following list:

Elements that Ionize to a Single Oxidation State (Cation)	Oxidation State	Examples
All Group 1A (Alkali metals)	+1	Na^+, Rb^+
All Group 2A (Alkaline earth metals)	+2	Mg^{2+}, Ba^{2+}
All Group 3A	+3	Al^{3+}, Ga^{3+}

   Such cations form what your book calls **"Type I binary ionic compounds."** When a cation forms from an element in Groups 1A, 2A, or 3A, we **give the ion the name of the element**, and add "ion." For example, K^+ is called **potassium ion**, and Be^{+2} is called **beryllium ion**. The reverse process is also true. **Calcium ion** is labeled Ca^{2+}.

   The **second class** of cations contains elements that can ionize to one of several oxidation states. Your textbook calls these **"Type II binary ionic compounds."** For example, iron, Fe, can form stable oxidation states of Fe^{+3}, Fe^{+2}, and Fe^{+6}. (At this point, you are not expected to predict which one will form. You should just be able to know one when you see one, and tell what you see.) **Most transition elements form several positive oxidation states. Some halogens and noble gases also form positive oxidation states.**

   We use **Roman numerals in parentheses after the element** to name cations of elements that can form several positive oxidation states. Some examples are given in the following table:

Element	Oxidation State	Symbol	Name
iron	+3	Fe^{3+}	iron(III) ion
iron	+2	Fe^{2+}	iron(II) ion
chromium	+7	Cr^{7+}	chromium(VII) ion
chromium	+3	Cr^{3+}	chromium(III) ion
nitrogen	+5	N^{5+}	nitrogen(V) ion

(Your textbook points out that although silver is a transition element, virtually all silver compounds contain the cation as Ag^+. Therefore, we normally omit the Roman numeral, and use the "Type I" designation, i.e.; AgBr = silver bromide, **NOT** silver(I) bromide. Zinc is another example of a metal that has only one nonzero oxidation state, Zn^{2+}. So, $ZnCl_2$ = zinc chloride, NOT zinc(II) chloride.) Try the following example regarding cation nomenclature

Example 2.8 A   *Naming Cations*

Fill in the following table:

Element	Oxidation State	Symbol	Name
phosphorus	+5	P^{5+}	_____
		Mn^{7+}	_____
_____	_____	_____	rubidium ion
_____	_____	_____	bromine(V) ion
vanadium	+2	_____	_____
strontium	+2	_____	_____

Strategy

Keep in mind that certain elements form cations of **only one oxidation state**. These **do not** require Roman numerals. All others do.

Solution

Element	Oxidation State	Symbol	Name
phosphorus	+5	P^{5+}	phosphorus(V) ion
manganese	+7	Mn^{7+}	manganese(VII) ion
rubidium	+1	Rb^+	rubidium ion
bromine	+5	Br^{5+}	bromine(V) ion
vanadium	+2	V^{2+}	vanadium(II) ion
strontium	+2	Sr^{2+}	strontium ion

b. **Anions:** Negative ions (anions) that are made up of a single element are all named in the same fashion. You simply **strip off the ending and add "ide."** For example,

Symbol	Element	Anion Name
Cl^-	chlorine	chlor**ide**
H^-	hydrogen	hydr**ide**
S^{2-}	sulfur	sul**fide**
N^{3-}	nitrogen	nit**ride**

Some anions can have several oxidation states. For purposes of naming the anion part of compounds, the oxidation state is unimportant. The "ide" rule always applies. Similarly, when you see an "ide" in a name, it indicates an anion.

In order to determine the oxidation state of anions, you can use the following general rule (which will be justified in Chapter 7 of your textbook):

THE OXIDATION STATE OF AN ANION IS EQUAL TO
THE GROUP NUMBER MINUS EIGHT

For example, if nitrogen (Group 5A) acts as an anion (it can also be a cation in many cases),

$$\text{oxidation state} = 5 - 8 = -3, \text{ or } N^{3-}$$

group #

Looking at bromine (if it acts as an anion),

$$\text{oxidation state} = 7 - 8 = -1 \text{ or } Br^-.$$

The following exercise will help you test your ability to go from anion to name and name to anion.

Example 2.8 B   *Naming Anions*

Fill in the following table:

Element	Oxidation State	Symbol	Anion Name
oxygen	-2		
		Se^{2-}	
iodine			
			phosphide

Solution

Element	Oxidation State	Symbol	Anion Name
oxygen	-2	O^{2-}	oxide
selenium	-2	Se^{2-}	selenide
iodine	-1	I^-	iodide
phosphorus	-3	P^{3-}	phosphide

c. **Naming binary salts:** The process of naming binary salts involves combining the names of the cation and anion parts. The only complication comes in determining whether or not the **cation** requires a Roman numeral. As you name your compounds, keep in mind:

  1. The cation **always comes first** in a name (it is written on the left hand side).
  2. The total **charge** for the compound **must equal zero**.

Many students find the second point to be troublesome, but you don't have to. Just remember that the total positive count must equal the total negative count.

Let's name **CaF₂** together, working with the anion side first. The fact that there are two fluorines rather than one or three is important so that we can establish a total negative charge due to fluorine (−1 for each fluorine × 2 fluorines = −2 total). However, the **name** of the anion is still **fluoride**.

Let's consider the cation (left) side. Because the fluorine ions contribute a total charge of −2, the calcium ion must be +2 to make a neutral compound. The cation is Ca^{+2}. (Actually, we knew this already, because calcium is an alkaline earth metal, and can only oxidize to a +2 state!) The name for the cation is therefore **calcium ion**, not calcium(II) ion, because calcium ion can have only one oxidation state. Combining the anion and cation gives us

calcium + fluoride = **calcium fluoride**

cation   anion

For a different problem, let's name **PbS₂.** Working on the anion side as before, we have two **sulfides**. Because sulfur is in Group **6A**, its oxidation state is **−2** (6 − 8). Therefore, the two sulfides contribute **−4** to the compound.

On the cation side, the **lead** can have more than one oxidation state. The oxidation state on the lead must be **+4**, because the positive and negative charges must balance, and there is only one lead atom in the compound. Therefore, the cation name is **lead(IV)** ion.

Combining the cation and anion parts, the compound is named **lead(IV) sulfide**.

Example 2.8 C   Naming Binary Salts

Name the following binary compounds:

a. KCl   b. $SnCl_4$ c. IF_5 d. Fe_2O_3

Strategy

1. Name the anion.
2. Establish the total charge due to the anion.
3. Name the cation.
4. Determine the charge on each cation. (Remember that the compound must be neutral.)
5. If the cation is not in Groups 1A, 2A, or 3A, its oxidation should be specified with Roman numerals.
6. Combine the cation and anion to get the name of the compound.

Solution

a. **potassium chloride** (potassium is in Group 1A)
b. **tin(IV) chloride**
c. **iodine(V) fluoride** (Iodine can act as a cation! We will see later that we can also call this iodine pentafluoride.)
d. **iron(III) oxide**

2. **Salts With Polyatomic Ions**

a. **Conventional polyatomic ions:** <u>Table 2.5 in your textbook</u> contains formulas for, and names of, the polyatomic ions you should know. The same rules are followed when naming compounds involving polyatomic ions as when naming binary salts. Keep in mind that you do not change the name of the polyatomic ions when you form the compound name. Let's work together to name the compound whose formula is $NaC_2H_3O_2$. The **Na^+** is **sodium** ion. The **$C_2H_3O_2^-$** is the **acetate** ion. The compound name is therefore **sodium acetate**.

b. **Oxyanions:** Many polyatomic anions, including several in Table 2.5, contain **different numbers of oxygen atoms** combined with an atom of a different element. These are called **oxyanions**. An oxyanion series normally contains either 2 or 4 members. In a two-member series, the anion with **more oxygens** gets the suffix **"ate,"** and the one with **fewer oxygens** gets the suffix **"ite."**

$$NO_2^- = \text{nitrite} \qquad SO_3^{2-} = \text{sulfite}$$
$$NO_3^- = \text{nitrate} \qquad SO_4^{2-} = \text{sulfate}$$

When there are four oxygen atoms in the series (as in many halogen-containing oxyanions), the sequence is:

$$IO^- = \textbf{hypoiodite}$$
$$IO_2^- = \textbf{iodite}$$
$$IO_3^- = \textbf{iodate}$$
$$IO_4^- = \textbf{periodate}$$

Just remember to approach the naming in a systematic fashion (as in Example 2.8 C) and relax - you are not being timed.

Example 2.8 D   *Naming Compounds With Polyatomic Ions*

Name the following compounds:

a. $KMnO_4$ d. NaH_2PO_4 g. KIO_3

b. $Ba(OH)_2$ e. $(NH_4)_2Cr_2O_7$ h. $Ca(OCl)_2$

c. $Fe(OH)_3$ f. $NaBrO_4$ i. $Cr(NO_3)_3$

Solution

a. potassium permanganate

b. barium hydroxide (Barium is in Group 2A - it is not a transition element and doesn't require a Roman numeral.)

c. iron(III) hydroxide (Iron is a transition metal. A Roman numeral is required.)

d. sodium dihydrogen phosphate

e. ammonium dichromate

f. sodium perbromate (BrO_4^- is part of an oxyanion series.)

g. potassium iodate (IO_3^- is part of an oxyanion series.)

h. calcium hypochlorite (OCl^- is also part of an oxyanion series.)

i. chromium(III) nitrate (Chromium is a transition element.)

3. **Binary Covalent Compounds**

Such compounds are formed between **two nonmetals**. Your book calls these "**Type III: Covalent Compounds**." We have already considered IUPAC naming of such compounds (see IF_5 in Example 2.8 C in this study guide). However, your book proposes a more common, and equally acceptable, nomenclature system. In Table 2.6 in your textbook, prefixes for you to use in naming both the cation and anion parts of covalent compounds are presented. Note that "**mono**" is *not* put in front of a **single cation** or **anion** - the prefix is assumed.

The binary covalent compound IF_5, which was previously called iodine(V) fluoride, can also be named **iodine pentafluoride**. Your book uses common systematic names for binary covalent compounds, and we will do that in this study guide as well.

(REMEMBER, COMMON SYSTEMATIC NAMES ARE TO BE USED ONLY WITH BINARY COVALENT COMPOUNDS!!)

Example 2.8 E   *Naming Binary Covalent Compounds*

Name the following compounds:

a. CO_2 c. N_2O_3 e. SO_3

b. P_2O_5 d. Cl_2O_7 f. BrF_3

Solution

a. carbon dioxide

b. diphosphorus pentoxide (drop the "a" in penta - the o or a in the prefix is normally dropped when the element begins with a vowel.)

c. dinitrogen trioxide

d. dichlorine heptoxide

e. sulfur trioxide

f. bromine trifluoride

4. Acids

You will learn about several different types of acids in Chapters 14 and 15. At this early point, **for purposes of nomenclature only**, you may consider an **acid** to be any compound in which hydrogen is the only cation. This is a very liberal definition. We will modify and tighten this definition in later chapters. HCl is an acid. HNO_3 is an acid. NH_4OH is not, because NH_4^+ is the cation, not just H^+. The names and formulas of some important acids are given in <u>Tables 2.7 and 2.8 in your textbook</u>.

To name an acid, determine if the anion part is polyatomic or not.

IF THE ANION PART IS MONATOMIC (such as Cl^- or S^{2-}):

    a.   change the "**ide**" ending to "**ic acid**," and
    b.   add the prefix "**hydro**" to the beginning of the name.

For example, given H_2S

    a. S^{2-} is sulfide. This part gets changed to **sulfuric acid**.
    b.   Add "**hydro**" to the front.

H_2S is **hydrosulfuric acid**.

The one exception to the rule is HCN. Even though the anion, CN^-, is polyatomic, the acid is known as hydrocyanic acid.

IF THE ANION PART IS POLYATOMIC (such as NO_3^- or $C_2H_3O_2^-$):

    a.   If the anion suffix is "**ite**," change it to "**ous acid**."
    b.   If the anion suffix is "**ate**," change it to "**ic acid**."

For example, HIO_4 contains the anion IO_4^-, periodate. We change the "**ate**" to "**ic acid**," giving **periodic acid**.

Example 2.8 F   *Naming Acids*

Name the following acids:

    a. HF   c. $HBrO_3$ e. HI
    b. $HC_2H_3O_2$ d. HBrO   f. HNO_2

Solution

    a. hydrofluoric acid (monatomic anion)
    b. acetic acid (polyatomic anion)
    c. bromic acid (polyatomic anion)
    d. hypobromous acid (the anion is hypobromate)
    e. hydroiodic acid (monatomic anion)
    f. nitrous acid (the anion is nitrite)

Example 2.8 G   *Tying it All Together*

Name the following compounds:

    a. PCl_5 d. Sb_2S_3 g. $NaC_2H_3O_2$
    b. $HClO_2$ e. XeF_4 h. $NaHCO_3$
    c. $Ni(NO_3)_2$ f. NH_4OH i. LiH

Solution

a. phosphorus pentachloride
b. chlorous acid (The anion is chlorite.)
c. nickel(II) nitrate (The nickel is a transition metal.)
d. antimony(III) sulfide (3 sulfurs × −2 each = −6; each antimony must have +6/2 = +3 oxidation state.)
e. xenon(IV) fluoride or xenon tetrafluoride
f. ammonium hydroxide (two polyatomic ions)
g. sodium acetate
h. sodium bicarbonate
i. lithium hydride (Hydrogen is an anion here.)

B. Getting Formulas From Names

The key to getting **formulas** from **names** is to remember that chemical names contain both a **cation** and an **anion** part. In virtually every case, one of the two parts will unequivocally tell you what its oxidation state is. Remembering that the compound is neutral, the oxidation state on one part of the compound dictates the oxidation state on the other part.

Let's try **magnesium nitride** together. Magnesium is in Group 2A, and as such, can only form the Mg^{2+} ion.

Nitrogen is found in many oxidation states, depending on the compound. However, when it exists as an **anion** in an ionic binary compound, we need to use our rule that the oxidation state equals the group number minus eight. For the nitride,

$$\text{oxidation state} = 5 - 8 = -3.$$

Nitrogen is present as the N^{3-} ion.

The simplest way for a +2 ion to combine with a −3 ion to maintain charge neutrality is to have **3 Mg^{2+}** and **2 N^{3-}**, or **Mg_3N_2**.

The same thinking applies to deriving the formulas of acids, such as **nitric acid**. An **acid** has hydrogen for a cation. Recall the rules for **anions of acids**. The ending **"ic acid"** came from an anion with an **"ate"** ending. The anion was nitr**ate**, or NO_3^-. Therefore, the formula is **HNO_3**.

As a final exercise in this section, let's work out the formulas of compounds from their names.

Example 2.8 H   Formulas from Names

Write the formulas for each compound:

a. sodium chloride
b. calcium fluoride
c. iron(III) nitrate
d. copper(I) chloride

e. hypoiodous acid
f. tin(IV) oxide
g. dinitrogen tetroxide
h. ammonium acetate

Solution

a. NaCl
b. CaF_2
c. $Fe(NO_3)_3$
d. CuCl

e. HIO (IO^- is the hypoiodite ion)
f. SnO_2
g. N_2O_4
h. $NH_4C_2H_3O_2$

Exercises

Section 2.2

1. In an exothermic (heat producing) reaction, chlorine reacts with 2.0200 g of hydrogen to form 72.926 g of hydrogen chloride gas. How many grams of chlorine reacted with hydrogen?

2. Sulfur and oxygen can react to form both sulfur dioxide and sulfur trioxide. In sulfur dioxide, there are 32.06 g of sulfur and 32.00 g of oxygen. In sulfur trioxide, 32.06 g of sulfur are combined with 48.00 g of oxygen.

    a. What is the ratio of the weights of oxygen that combine with 32.06 g of sulfur?
    b. How do these data illustrate the law of multiple proportions?

3. By experiment it has been found that 2.18 g of zinc metal combines with oxygen to yield 2.71 g of zinc oxide. How many grams of oxygen reacted with zinc metal?

4. A sample of H_2SO_4 contains 2.02 g of hydrogen, 32.07 g of sulfur, and 64 g of oxygen. How many grams of sulfur and grams of oxygen are present in a second sample of H_2SO_4 containing 7.27 g of hydrogen?

Section 2.3

5. Describe what part of Dalton's atomic theory each chemical statement relates to.

    a. $H_2 + Cl_2 \rightarrow 2HCl$
    b. There are 3.01×10^{23} atoms in 20.04 g of calcium.
    c. Lead does not change to chromium when it forms lead hydroxide.

Section 2.5

6. Identify each of the following elements:

    a. $^{91}_{40}X$ d. $^{85}_{36}X$
    b. $^{108}_{47}X$ e. $^{51}_{23}X$
    c. $^{33}_{16}X$ f. $^{133}_{55}X$

7. Identify each of the following elements:

    a. $^{98}_{43}X$ d. $^{14}_{6}X$
    b. $^{186}_{75}X$ e. $^{40}_{19}X$
    c. $^{75}_{33}X$ f. $^{131}_{54}X$

8. How many protons and neutrons are in each of the following elements?

    a. ^{89}Y d. ^{238}U
    b. ^{73}Ge e. $^{35}Cl^-$
    c. $^{24}Mg^{2+}$ f. ^{65}Zn

9. How many protons and neutrons are in each of the following elements?

    a. ^{227}Ac d. ^{251}Cf
    b. ^{70}Ga e. ^{239}Pu
    c. ^{11}B f. ^{64}Cu

Section 2.6

10. How many protons, neutrons, and electrons are in each of the following ions?

   a. $^{56}Fe^{3+}$

   b. $^{40}Ca^{2+}$

   c. $^{19}F^-$

   d. $^{31}P^{3-}$

   e. $^{127}I^-$

   f. $^{127}I^{7+}$

11. How many protons, neutrons, and electrons are in each of the following?

   a. $^{195}Pt^+$

   b. ^{93}Nb

   c. $^{40}Ar^-$

   d. $^{16}O^{2-}$

   e. $^{122}Sb^{2+}$

   f. $^{56}Fe^{2+}$

   g. ^{184}W

   h. $^{133}Cs^+$

   i. $^{28}Si^{3-}$

12. Fill in the missing information in the following table:

Symbol	Protons	Neutrons	Electrons	Charge
$^{80}_{35}Br^-$	___	___	___	___
___$^{5+}$	35	45	___	+5
^{137}___	56	___	54	___
$^{108}_{47}Ag^+$	___	___	___	___
$^{51}_{23}$___$^{5+}$	___	___	___	___
___Co___	___	32	___	+2

13. Fill in the missing information in the following table:

Symbol	Protons	Neutrons	Electrons	Charge
^{27}___ ___	13	___	10	___
$^{88}_{38}$___	___	___	___	+1
___$^{2+}$	30	35	___	___
^{35}___	___	18	18	___
___Te^{2-}	___	76	___	___
^{85}Rb___	___	___	___	+1

Section 2.7

14. Name the family to which each of the following elements belongs:

    a. Fe
    b. Cl

    c. Ar
    d. Sr

    e. Rb
    f. Nd

15. Are the following elements metals or nonmetals?

    a. Mg
    b. P
    c. Hg

    d. Br
    e. O
    f. Bi

    g. Co
    h. Mo
    i. Xe

16. Name the family to which each of the following elements belongs:

    a. Es
    b. I
    c. Au

    d. Yb
    e. Kr

    f. Fr
    g. Ca

17. Given the position in the periodic table, what is the most likely oxidation state that each element will have when forming an ion?

    a. Cs
    b. N

    c. Br
    d. K

    e. Al
    f. S

Section 2.8

18. Would you expect the following atoms to gain or lose electrons when forming an ion? If so, how many would be gained or lost?

    a. Be
    b. Cl
    c. Al

    d. O
    e. F

    f. Li
    g. P

19. An element combines with 2 atoms of chlorine to form an ionic compound. The element has 20 neutrons in its most abundant form. Write the formula of the compound.

20. Predict the formula and state the name of a compound likely to be formed from the following pairs of elements:

    a. sodium and fluorine

    b. aluminum and oxygen

21. Predict the formula and state the name of the compound likely to be formed from the following substances:

    a. calcium and phosphate ion

    b. potassium and nitrate ion

22. Name each of the following compounds:

    a. PbI_2
    b. NH_4Cl
    c. Fe_2O_3
    d. LiH

    e. $CsCl$
    f. OsO_4
    g. $Cr(OH)_3$
    h. $NaC_2H_3O_2$

    i. $K_2Cr_2O_7$
    j. Na_2SO_4
    k. KH_2PO_4

23. Name each of the following compounds:

    a. $MgSO_4$
    b. N_2O_3
    c. Ce_2O_3

    d. $KMnO_4$
    e. NiO
    f. $BaSO_4$

    g. $Fe(IO_4)_3$
    h. SO_3
    i. $KClO_4$

24.   Name each of the following compounds:

    a. NI_3 c. CO   e. N_2O_4
    b. PCl_5 d. P_4O_{10} f. NH_3

25.   Name each of the following compounds:

    a. P_4O_6 d. $AgNO_3$ f. AgCl
    b. KOH   e. BF_3 g. $KHCO_3$
    c. N_2

26.   Name each of the following compounds:

    a. HIO_3 d. HCN   f. K_2SO_3
    b. HBr   e. $NaNO_2$ g. $NaHSO_3$
    c. HNO_2

27.   Name each of the following compounds:

    a. UF_6 d. SF_6 f. $SnCl_2$
    b. $Cu(NO_3)_2$ e. $Mg(OH)_2$ g. Na_2CO_3
    c. H_3PO_4

28.   Write formulas for each of the following compounds:

    a. sodium cyanide   d. lead(II) nitrate   f. calcium phosphate
    b. tin(II) fluoride   e. iron(III) oxide   g. sodium bromate
    c. sodium hydrogen sulfate

29.   Write formulas for each of the following compounds:

    a. sodium sulfate   d. potassium hypochlorite   g. magnesium oxide
    b. manganese(IV) oxide   e. lithium aluminum hydride   h. copper(I) oxide
    c. potassium chlorate   f. barium chloride

30.   Write formulas for each of the following compounds:

    a. potassium carbonate   g. rubidium nitrate   l. sulfurous acid
    b. magnesium hydroxide   h. potassium chlorate   m. potassium hydrogen phosphate
    c. dinitrogen tetroxide   i. carbon tetrachloride   n. ammonium acetate
    d. hypoiodous acid   j. sodium iodate   o. ammonium dichromate
    e. iron(III) chloride   k. potassium permanganate   p. hydrobromic acid
    f. tin(IV) oxide

31.   Give the names of the following acids:

    a. H_2SO_3 c. HBr   e. H_3PO_4
    b. HI   d. HNO_2 f. HCl

32.   Give formulas for the following acids:

    a. nitric acid   c. sulfuric acid   e. hydrosulfuric acid
    b. hydrofluoric acid   d. hydrocyanic acid   f. acetic acid

33.   Give the alternate or common name for each of the following compounds or cations:

    a. sodium hydrogen carbonate ($NaHCO_3$)   d. iron(II) (Fe^{2+})
    b. dinitrogen monoxide (N_2O)   e. tin(IV) (Sn^{4+})
    c. nitrogen monoxide (NO)   f. lead(II) (Pb^{2+})

Multiple Choice Questions

34. The masses of an apple, orange, grape, and banana are 800, 750, 72, and 650 g, respectively. Determine the combined mass of 10 apples, 6 oranges, 20 grapes, and 5 bananas.

A. 17190 g

B. 8595 g

C. 2272 g

D. 95200 g

35. A pound cake consists of 1.0 lb of butter, 1.25 lb of flour, 1.0 lb of sugar, 6 eggs (1.25 lb in mass) and 0.5 lb of milk. After the cake has baked and cooled, it weighs 5.25 lbs. Which of the following statements is true?

A. The law of conservation of mass has been violated by a gain in 0.25 lbs.
B. The law of conservation of mass is conserved since 0.25 lbs of gas were produced during baking.
C. The law of conservation of mass has been violated by a gain in 12.0 oz.
D. The law of conservation of mass has been violated by a gain in 8.0 oz.

36. The oxides of CO and CO_2 must have the following carbon-to-oxygen mass ratio:

A. 12:16, 12:32

B. 12:12, 12:16

C. 12:8, 12:4

D. 12:12, 12:24

37. When silicone and oxygen combine to form silicon dioxide, silicon and oxygen

A. Fuse together to yield a new atom
B. Retain their identities
C. Duplicate their mass
D. Have some atoms that retain their individual identities, and some that do not

38. Every atom contains

A. As many neutrons as electrons
B. As many protons as neutrons

C. As many nuclei as neutrons
D. As many electrons as protons

39. The atomic number represents

A. The number of nuclei in that atom
B. The number of protons in that atom

C. The number of neutrons in that atom
D. The number of electrons in that atom

40. Which of the following elements has $Z = 68$ and $A = 167$?

A. Erbium

B. Californium

C. Calcium

D. Dysprosium

41. The atomic number and atomic mass, respectively, for vanadium, are:

A. 23, 51

B. 51, 23

C. 46, 102

D. 46, 51

42. Atom A has 30 protons, 32 neutrons, and 30 electrons. Atom B has 30 protons, 28 neutrons, and 30 electrons. Atoms A and B are

A. Isotopes

B. Isobars

C. Isomers

D. Isoneutrons

43. How many electrons and protons, respectively, are there in Ra^{2+}?

A. 88, 88

B. 86, 88

C. 224, 226

D. 228, 224

44. How many total protons are found in two molecules of $C_{20}H_{30}O$?

A. 102

B. 316

C. 302

D. 600

45. What is the charge of an ion with 29 protons and 28 neutrons?

A. 0

B. +1

C. +2

D. Unknown

46. What is the charge of an ion with 38 electrons, 38 neutrons, and 35 protons?

    A. 0   B. +3   C. −3   D. −5

47. How many electrons does an ion with mass number 210, with 125 neutrons, and a charge of −2 have?

    A. 85   B. 83   C. 87   D. 89

48. An ion has a charge of +3 and 55 electrons. Which of the following elements can form such an ion?

    A. Th   B. Ce   C. Mn   D. Co

49. Atom A loses 1 electron and atom B gains 2 electrons. What formula results if these two ions combine to produce a neutral compound?

    A. AB   B. A_2B C. AB_2 D. A_2B_3

50. Which one of the following chemical symbols does not represent an element?

    A. SO   B. Gd   C. Am   D. Au

51. Which of the following symbols represents an element?

    A. IF   B. HI   C. AU   D. C

52. Which group of elements belongs to the <u>Transition Metal</u> family?

    A. Ru, C, Hg, Ir   B. Pd, Ir, Ac, Re   C. Bi, Sc, Pu, Rn   D. Ti, Sc, Au, Fr

53. The $(SO_4)^{2-}$ ion is called

    A. Sulfite ion   B. Sulfate ion   C. Sulfur tetroxide ion   D. Sulfur oxide

54. The formula for iron(III) carbonate is

    A. $FeCO_3$ B. $Fe_2(CO)_3$ C. $Fe_2(CO_3)_3$ D. $Fe_3(CO)_3$

55. The compound $Co_2(CO_3)_3$ is named

    A. Cobalt(III) carbonate   C. Cobalt(II) carbonate
    B. Cobalt carbonate   D. Cobalt carbon trioxide

56. The ion SCN^- is named

    A. Sulfocyano ion   B. Thiocyano ion   C. Cyano ion   D. Thiocyanate ion

57. Which of the following is the formula for barium chloride?

    A. BaCl   B. Ba_2Cl C. Ba_2Cl_2 D. $BaCl_2$

58. A compound in which the nitrite-to-metal ion ratio is 2:1 has the following formula:

    A. $M(NO_2)_2$ B. $M(NO_2)_4$ C. M_2NO_2 D. MNO_2

59. Which one of the following is not a polyatomic anion?

    A. Sulfate   B. Hydride   C. Nitrite   D. Carbonate

60. The formula for hydrosulfuric acid is

    A. H_2S B. H_2SO_3 C. H_2SO_4 D. HSO_4

61. The name for BrO_3^- is

    A. Bromite   B. Perbromate   C. Bromate   D. Hypobromite

62. The oxoacid HIO is named

    A. Hypoiodous   B. Iodic   C. Iodous   D. Periodic

63. What is the formula of ammonium perchlorate?

    A. NH_3ClO_3 B. NH_4ClO_4 C. NH_3ClO D. NH_3ClO_2

64. When hydrogen ions in sulfuric acid are replaced by an iron (3+) ion, the resulting compound has the following formula:

    A. $Fe_2(SO_4)_3$ B. Fe_3SO_4 C. FeSO   D. $Fe_3(SO_4)_2$

65. When a calcium ion is combined with a sulfate ion, the following neutral compound is produced:

    A. $CaSO_4$ B. CaS   C. $Ca_2(SO_4)_3$ D. $Ca(SO_4)_2$

Answers to Exercises

1. 70.906 g

2. a. The ratio is 1.5 to 1, or 3 to 2.
    b. The ratios are whole numbers.

3. 0.53 grams

4. 115.4 g of sulfur and 230.34 g of oxygen

5. a. Chemical reactions involve reorganization of the atoms.
    b. Each element is made up of tiny particles called atoms.
    c. Chemical reactions involve reorganization of the atoms.

6. a. Zr   c. S   e. V
    b. Ag   d. Kr   f. Cs

7. a. Tc   c. As   e. K
    b. Re   d. C   f. Xe

8. a. Y = 39 p, 50 n   c. Mg^{2+} = 12 p, 12 n   e. Cl^- = 17 p, 18 n
    b. Ge = 32 p, 41 n   d. U = 92 p, 146 n   f. Zn = 30 p, 35 n

9. a. Ac = 89 p, 138 n   c. B = 5 p, 6 n   e. Pu = 94 p, 145 n
    b. Ga = 31 p, 39 n   d. Cf = 98 p, 153 n   f. Cu = 29 p, 35 n

10.

	protons	neutrons	electrons
a.	26	30	23
b.	20	20	18
c.	9	10	10
d.	15	16	18
e.	53	74	54
f.	53	74	46

11.

	protons	neutrons	electrons
a.	78	117	77
b.	41	52	41
c.	18	22	19
d.	8	8	10
e.	51	71	49
f.	26	30	24
g.	74	110	74
h.	55	78	54
i.	14	14	17

12.

Symbol	Protons	Neutrons	Electrons	Charge
$^{80}_{35}Br^-$	35	45	36	−1
$^{80}_{35}Br^{5+}$	35	45	30	+5
$^{137}_{56}Ba^{2+}$	56	81	54	+2
$^{108}_{47}Ag^+$	47	61	46	+1
$^{51}_{23}V^{5+}$	23	28	18	+5
$^{59}_{27}Co^{2+}$	27	32	25	+2

13.

Symbol	Protons	Neutrons	Electrons	Charge
$^{27}_{13}Al^{3+}$	13	14	10	+3
$^{88}_{38}Sr^+$	38	50	37	+1
$^{65}_{30}Zn^{2+}$	30	35	28	+2
$^{35}_{17}Cl^-$	17	18	18	−1
$^{128}_{52}Te^{2-}$	52	76	54	−2
$^{85}_{37}Rb^+$	37	48	36	+1

14. a. transition metal   c. noble gas   e. alkali metal
     b. halogen   d. alkaline earth metal   f. lanthanide

15. a. metal   d. nonmetal   g. metal
     b. nonmetal   e. nonmetal   h. metal
     c. metal   f. metal   i. nonmetal

16. a. actinides   d. lanthanide   f. alkali metals
     b. halogens   e. noble gases   g. alkaline earth metal
     c. transition metals

17. a. 1+   c. 1−   e. 3+
     b. 3−   d. 1+   f. 2−

18. a. lose 2   d. gain 2   f. lose 1
     b. gain 1   e. gain 1   g. gain 3
     c. lose 3

19. $CaCl_2$

20. a. NaF, sodium fluoride   b. Al_2O_3, aluminum oxide

21. a. $Ca_3(PO_4)_2$, calcium phosphide   b. KNO_3, potassium nitrate

22. a. lead(II) iodide   g. chromium(III) hydroxide
     b. ammonium chloride   h. sodium acetate
     c. iron(III) oxide   i. potassium dichromate
     d. lithium hydride   j. sodium sulfate
     e. cesium chloride   k. potassium dihydrogen phosphate
     f. osmium(VIII) oxide or osmium tetroxide

23. a. magnesium sulfate
    b. dinitrogen trioxide
    c. cerium(III) oxide
    d. potassium permanganate
    e. nickel(II) oxide
    f. barium sulfate
    g. iron(III) periodate
    h. sulfur trioxide
    i. potassium perchlorate

24. a. nitrogen triiodide
    b. phosphorus pentachloride
    c. carbon monoxide
    d. tetraphosphorus decoxide
    e. dinitrogen tetroxide
    f. ammonia

25. a. tetraphosphorus hexoxide
    b. potassium hydroxide
    c. molecular nitrogen
    d. silver nitrate
    e. boron trifluoride
    f. silver chloride
    g. potassium hydrogen carbonate

26. a. iodic acid
    b. hydrobromic acid
    c. nitrous acid
    d. hydrocyanic acid
    e. sodium nitrite
    f. potassium sulfite
    g. sodium bisulfite

27. a. uranium(VI) fluoride
    b. copper(II) nitrate
    c. phosphoric acid
    d. sulfur hexafluoride
    e. magnesium hydroxide
    f. tin(II) chloride
    g. sodium carbonate

28. a. $NaCN$
    b. SnF_2
    c. $NaHSO_4$
    d. $Pb(NO_3)_2$
    e. Fe_2O_3
    f. $Ca_3(PO_4)_2$
    g. $NaBrO_3$

29. a. Na_2SO_4
    b. MnO_2
    c. $KClO_3$
    d. $KClO$
    e. $LiAlH_4$
    f. $BaCl_2$
    g. MgO
    h. Cu_2O

30. a. K_2CO_3
    b. $Mg(OH)_2$
    c. N_2O_4
    d. HIO
    e. $FeCl_3$
    f. SnO_2
    g. $RbNO_3$
    h. $KClO_3$
    i. CCl_4
    j. $NaIO_3$
    k. $KMnO_4$
    l. H_2SO_3
    m. K_2HPO_4
    n. $NH_4C_2H_3O_2$
    o. $(NH_4)_2Cr_2O_7$
    p. HBr

31. a. sulfurous acid
    b. hydroiodic acid
    c. hydrobromic acid
    d. nitrous acid
    e. phosphoric acid
    f. hydrochloric acid

32. a. HNO_3
    b. HF
    c. H_2SO_4
    d. HCN
    e. H_2S
    f. $HC_2H_3O_2$

33. a. sodium bicarbonate
    b. nitrous oxide
    c. nitric oxide
    d. ferrous ion
    e. stannic ion
    f. plumbous ion

34. A	35. B	36. A	37. B	38. D	39. B
40. A	41. A	42. A	43. B	44. B	45. D
46. C	47. C	48. B	49. B	50. A	51. D
52. B	53. B	54. C	55. A	56. D	57. D
58. A	59. B	60. A	61. C	62. A	63. B
64. A	65. A				

Chapter 3

Stoichiometry

This chapter will help guide your study of the **mathematics of chemical reactions**.

3.1   Counting by Weighing

When you finish this section you will be able to solve problems related to determining the number of atoms you have based on total mass and average mass.

The section reminds us that atoms are so small that it is not possible to count them individually. However, as with a sample of jelly beans whose average mass we know, we can count atoms by knowing their average mass and the number of atoms we want.

Here is another illustration. Airline companies make calculations about the amount of fuel they need based on several measures, including the number of passengers on board a plane and the average weight of those passengers. From 1938 until 2003, the average weight of each male passenger on U.S. airlines was assumed to be 170 pounds. If there were 150 male passengers on an aircraft, the total weight can be calculated,

$$150 \text{ passengers} \times \frac{170 \text{ pounds}}{\text{passenger}} = 25,500 \text{ pounds}$$

We can also say that if, for a given aircraft, the total weight of the male passengers is 51,000 pounds, we can calculate the number of male passengers, knowing that their average weight is 170 pounds.

$$51,000 \text{ pounds} \times \frac{1 \text{ passenger}}{170 \text{ pounds}} = 300 \text{ male passengers}$$

Example 3.1   *Counting by Weighing*

The average male U.S. male airline passenger is heavier than in the past, with a current assumed weight by the airlines of 184 pounds per person. If the total male passenger weight on an aircraft is 55,200 pounds, how many male passengers are on the aircraft?

Solution

$$55,200 \text{ pounds} \times \frac{1 \text{ male passenger}}{184 \text{ pounds}} = 300 \text{ passengers}$$

Does the Answer Make Sense?

Passengers weigh more these days, so the total weight for the same number of passengers as on flights in the past should be higher. Our answer makes sense.

3.2   Atomic Masses

When you finish this section you will be able to:

- Calculate average atomic masses from isotopic data.
- Calculate relative isotope abundances from pertinent mass spectral data.

The **Mass Spectrometer** is the best instrument for measuring the masses of atoms and ions (charged particles). The essential operation of the mass spectrometer is given in your textbook.

Example 3.2 A   *Calculation of Average Atomic Masses from Isotopic Data*

A sample of metal "M" is vaporized and injected into a mass spectrometer. The mass spectrum tells us that 60.10% of the metal is present as ^{69}M and 39.90% is present as ^{71}M. The mass values for ^{69}M and ^{71}M are 68.93 amu and 70.92 amu, respectively.

- What is the **average atomic** mass of the element?
- What is the element?

Strategy

The basic question here is "how much does each isotope contribute to the overall atomic mass of the element?"

^{69}M contributes 60.10%
^{71}M contributes 39.90%

For every 100 atoms of M, on average

60.10 are ^{69}M
39.90 are ^{71}M

The total mass of 100 atoms of M is, therefore, a weighted average of the atomic mass of ^{69}M and ^{71}M.

^{69}M has a mass of 68.93 amu's
^{71}M has a mass of 70.92 amu's

The mass of **one atom** is the total mass of 100 atoms divided by 100.

Solution

Mass of 100 atoms of M $\quad = $ Mass of ^{69}M + Mass of ^{71}M

$$= 60.10 \text{ atoms} \times 68.93 \ \frac{\text{amu}}{\text{atom}} + 39.90 \text{ atoms} \times 70.92 \ \frac{\text{amu}}{\text{atom}}$$

$$= 6972 \text{ amu}$$

Average atomic mass of M $\quad = $ Mass of 1 atom of M

$$= \frac{\text{Mass of 100 atoms of M}}{100 \text{ atoms}}$$

$$= \frac{6972 \text{ amu}}{100 \text{ atoms}}$$

$$= \textbf{69.72 amu/atom}$$

Look at your periodic table. The element with an average atomic mass of 69.72 amu/atom is **Ga**.

Does the Answer Make Sense?

Many more of the isotopes are present as ^{69}M than ^{71}M. We would, therefore, expect the average atomic mass to be closer to 69 than to 71.

Example 3.2 B   Calculation of Relative Isotope Abundances

The element indium exists naturally as two isotopes. ^{113}In has a mass of 112.9043 amu, and ^{115}In has a mass of 114.9041 amu. The average atomic mass of indium is 114.82 amu. Calculate the **percent relative abundance** of the two isotopes of indium.

Solution

We have 2 unknowns:

Unknown 1 = relative abundance of ^{113}In (expressed as a fraction between 0 and 1)
Unknown 2 = relative abundance of ^{115}In (also expressed as a fraction between 0 and 1)

Because we have two unknowns, we need two equations to solve the problem. The first equation expresses the "weighted average" of the two isotopes, as in Example 3.2 A.

Equation #1:

atomic mass of ^{113}In × relative abundance of ^{113}In + atomic mass of ^{115}In × relative abundance of ^{115}In

= **average atomic mass of In**

The second equation indicates that the sum of the relative abundances of the isotopes of indium equals 1. (The total of the two isotopic abundances must be 100% because all the isotopes present must be accounted for. This total, expressed as a whole number rather than as a percentage, is equal to 1.)

Equation #2:

relative abundance of ^{113}In + relative abundance of ^{115}In = 1

The hard part of this problem is coming up with the proper equations.

Solution

Let the relative abundance of ^{113}In $= X$
Let the relative abundance of ^{115}In $= Y$

The two simultaneous equations for two unknowns:

Equation #1: $112.9043(X) + 114.9041(Y) = 114.82$
Equation #2: $\qquad X + \qquad Y = 1$

Make the coefficients in front of Y in the two equations equal by multiplying equation #2 by 114.9041, so we can solve for X. Our two simultaneous equations become:

Equation #1: $112.9043(X) + 114.9041(Y) = 114.82$
Equation #2: $114.9041(X) + 114.9041(Y) = 114.9041$

Subtract equation #2 from equation #1, and solve for X.

$$-1.9998(X) + 0(Y) = -0.0841$$

$$X = \frac{0.0841}{1.9998}$$

$$= \mathbf{0.042}$$

To solve for Y, plug the answer for X back into the original equation #2.

$$0.042 + Y = 1$$
$$Y = 1 - 0.042$$
$$= \mathbf{0.958}$$

To calculate the percent relative abundance of each isotope, multiply X and Y by 100%. Thus the answer to the problem is:

% Relative abundance of ^{113}In = 4.2%
% Relative abundance of ^{115}In = 95.8%

Does the Answer Make Sense?

Since the average atomic mass of indium is very close to the mass of ^{115}In, we would expect most of the indium to be in that form. Be careful not to accidentally put the wrong percentage with the wrong isotope.

3.3   *The Mole*

When you finish this section, you will be able to interconvert between **moles**, **mass**, and the **number of particles** of a given element.

The **mole** is the key to many chemical calculations. A **mole** is defined as **the number equal to the number of carbon atoms in exactly 12 grams of pure ^{12}C**.

1 mole of anything = 6.022×10^{23} units of that thing
1 mole of airline passengers = 6.022×10^{23} airline passengers
1 mole of compact discs = 6.022×10^{23} compact discs

One mole of **ANY** substance contains **Avogadro's number** of particles.

$$\textbf{Avogadro's number} = \textbf{6.022} \times \textbf{10}^{23} \textbf{ particles}$$

One mole of airline passengers is heavier than 1 mole of compact discs, **BUT** 1 mole of airline passengers contain the **same** (Avogadro's) number of units as 1 mole of compact discs. The masses of 1 mole of various elements are given in Table 3.1 in your textbook.

Key Problem Solving Relationship: The average mass of **one atom** of a substance expressed in **amu** is the same number as the mass of **one mole** of a substance expressed in **grams**.

$$\textbf{1 atom of } {}^{20}\text{Ne} = \textbf{20.18 amu}$$
$$\textbf{1 mole of } {}^{20}\text{Ne} = \textbf{20.18 grams}$$
$$\textbf{1 mole of } {}^{20}\text{Ne} = \textbf{6.022} \times \textbf{10}^{23} \textbf{ atoms of Ne}$$
$$\textbf{1 gram (exactly)} = \textbf{6.022} \times \textbf{10}^{23} \textbf{ amu}$$

Example 3.3 A   *Conversion Among Grams, Atoms, and AMU's*

How many grams does a sample containing 34 atoms of neon weigh?

Strategy

A good way to approach this problem is with a "**flowchart**":

$$34 \text{ Ne atoms} \xrightarrow{\text{convert to}} \text{amu's in 34 Ne atoms} \xrightarrow{\text{convert to}} \text{grams in 34 Ne atoms}$$
Start   Finish

As we get to more difficult problems, the skillful use of **conversion factors** will become more and more important.

Solution

$$34 \text{ Ne atoms} \times \frac{20.18 \text{ amu}}{1 \text{ Ne atom}} \times \frac{1 \text{ g}}{6.022 \times 10^{23} \text{ amu}} = \textbf{1.139} \times \textbf{10}^{-21} \textbf{ g}$$

Does the Answer Make Sense?

Having so few atoms of a substance should lead to a **very** low mass. Be careful not to accidentally **invert** conversion factors so that your answer comes out to be unfortunately large!

Example 3.3 B   *Conversion Between Atoms, Moles, and Mass*

A sample of elemental silver (Ag) has a mass of 21.46 g.

- How many **moles** of silver are in the sample?
- How many **atoms** of silver are in the sample?

Solution

We can find the number of moles of silver by multiplying mass by atomic weight. If you are not certain of this relationship, the **units** in the problem tell you that this must be so.

Once we calculate the number of moles of Ag, we can use **Avogadro's number** to find the number of atoms. (Make sure your units cancel properly!)

$$21.46 \text{ g Ag} \times \frac{1 \text{ mol Ag}}{107.9 \text{ g Ag}} = \textbf{0.1989 moles Ag}$$

$$0.1989 \text{ moles Ag} \times \frac{6.022 \times 10^{23} \text{ atoms Ag}}{1 \text{ mole Ag}} = \textbf{1.198} \times \textbf{10}^{\textbf{23}} \textbf{ atoms Ag}$$

Example 3.3 C   Practice With Conversions

What is the weight of 7.81×10^{22} atoms of calcium?

Strategy

You want **grams**. You are given **atoms**. The central relationship between the mass and the number of particles is the **mole**. We will go through moles to get to grams.

$$\text{atoms Ca} \xrightarrow{\text{Avogadro's number}} \text{moles Ca} \xrightarrow{\text{atomic mass}} \text{grams Ca}$$

CAUTION: MAKE SURE YOUR UNITS CANCEL PROPERLY!

Solution

$$\text{g Ca} \parallel 7.81 \times 10^{22} \text{ atoms Ca} \times \frac{1 \text{ mole Ca}}{6.022 \times 10^{23} \text{ atoms Ca}} \times \frac{40.08 \text{ g Ca}}{1 \text{ mole Ca}} = \textbf{5.20 g Ca}$$

Does the Answer Make Sense?

You have 7.81×10^{22} atoms of Ca, which is a little over 10% of a mole. Therefore, we would expect a mass of a little over 10% of the atomic weight.

3.4   Molar Mass

When you finish this section, you will be able to:

- Calculate molar masses.
- Interconvert among molar mass, moles, mass, and number of particles in a given sample.

Recall that a **molecule** is a **covalently bonded collection of atoms**. Carbon dioxide (CO_2), benzene (C_6H_6), and ethanol (C_2H_6O) are all examples of molecules. **Ionic compounds** contain ions, such as K^+ electrically bound with Cl^-. A single unit of the resulting ionic compound, KCl, is not a molecule, but rather a **formula unit** that is part of a much larger group of K^+ and Cl^- ions that interact forming, in this case, a KCl crystal structure. Examples of individual formula units of ionic compounds include $K_2Cr_2O_7$, NaCl, $NaNO_3$, and $LiCO_3$.

MOLAR MASS OF A MOLECULE = The sum of the masses of the individual atoms in the molecule

In most chemical calculations, we express molar masses in **grams**, using prefixes as necessary (mg, μg, kg, etc.).

Example 3.4 A   Calculation of the Molar Mass of a Compound

Calculate the molar mass of potassium dichromate, $K_2Cr_2O_7$.

Strategy

To calculate the molar mass of a compound, you **total** the atomic weights of **EVERY ATOM** in the compound. If an atom appears more than once (as all do in our example), you must multiply its atomic weight by the number of times it appears in the compound (represented by its **subscript**; if there is no subscript, it is present only once).

Solution

atom	subscript		atomic mass		total mass
K	2	×	39.10	=	78.20
Cr	2	×	52.00	=	104.00
O	7	×	16.00	=	112.00
		Molar Mass of $K_2Cr_2O_7$		=	**294.20 g/mole**

Once you are comfortable calculating molar masses, you can use the conversion factor of **grams/mole** (in our previous example 294.20 g $K_2Cr_2O_7$/1 mole $K_2Cr_2O_7$) to solve a host of fundamental chemistry problems.

Example 3.4 B   Converting Among Molar Mass, Moles, Mass, and Number of Particles

How many µg are there in 3.82×10^{-7} moles of pyrogallol, $C_6H_6O_3$?

Strategy

This is a two-step conversion:

$$\text{moles pyrogallol} \xrightarrow[\text{mass}]{\text{molar}} \text{g pyrogallol} \xrightarrow[\text{conversion}]{\text{unit}} \text{µg pyrogallol}$$

(Be careful to use the proper unit conversion from g to µg!)

Solution

The molar mass of pyrogallol (pyrgl) is:

C: 6 × 12.01 =   72.06 g
H: 6 × 1.008 =   6.048g
O: 3 × 16.00 =   48.00 g
molar mass   = 126.11 g/mole pyrogallol

$$\text{µg pyrgl} \parallel 3.82 \times 10^{-7} \text{ moles pyrgl} \times \frac{126.11 \text{ g pyrgl}}{1 \text{ mole pyrgl}} \times \frac{\overset{\text{exactly}}{\downarrow} 1 \times 10^6 \text{ µg pyrgl}}{1 \text{ g pyrgl}} = \textbf{48.2 µg pyrogallol}$$

Example 3.4 C   Practice With Conversions

Freon-12, which has the formula CCl_2F_2, is used as a refrigerant in air conditioners and as a propellant in aerosol cans. Given a 5.56 mg sample of Freon-12:

   a.   Calculate the number of molecules of Freon-12 in that sample.
   b.   How many mg of chlorine are in the sample?

Strategy

Use **moles** as your **bridge** between milligrams and molecules. Also, always keep in mind that a conversion factor is valid in two ways, for example:

$$\frac{6.022 \times 10^{23} \text{ molecules}}{1 \text{ mole}} \quad \text{and} \quad \frac{1 \text{ mole}}{6.022 \times 10^{23} \text{ molecules}} \quad \text{are both correct.}$$

You must be careful to use the conversion factor in the correct fashion so the units cancel.

$$\text{mg Freon-12} \xrightarrow[\text{molar mass}]{\text{mg to grams, and}} \text{moles Freon-12} \xrightarrow{\text{Avogadro's number}} \text{molecules Freon-12}$$

The mass of 2 chlorine atoms **divided by** the molar mass of Freon-12 gives the **fraction** of the total mass that is chlorine. Multiplying the **chlorine fraction** by the **total sample size** gives the mg chlorine in the sample.

Solution

The molar mass of Freon-12 (Fr-12) is:

C:	1×12.01	$=$	12.01 g
Cl:	2×35.45	$=$	70.90 g
F:	2×19.00	$=$	38.00 g
molar mass		$=$	**120.91 g/mole Freon-12**

$$5.56 \text{ mg Fr-12} \times \frac{1 \text{ g Fr-12}}{1000 \text{ mg Fr-12}} \times \frac{1 \text{ mole Fr-12}}{120.91 \text{ g Fr-12}} \times \frac{6.022 \times 10^{23} \text{ molecules Fr-12}}{1 \text{ mole Fr-12}}$$

$$= \textbf{2.77} \times \textbf{10}^{\textbf{19}} \textbf{ molecules Fr-12}$$

From the molar mass calculation above, Cl = 70.90 g of the molar mass of Freon-12.

$$\text{fraction of Freon-12 that is Cl} = \frac{70.90 \text{ g Cl}}{120.91 \text{ g Fr-12}} = \textbf{0.586}$$

$$\text{mg Cl in 5.56 mg of Freon-12} = 0.586 \times 5.56 \text{ mg} = \textbf{3.26 mg Cl}$$

3.5   *Learning to Solve Problems*

The key messages in this section are:
- Solving problems in a flexible, creative way is called **conceptual problem solving**.
- The textbook authors will work with you as you become an independent problem solver.
- With a new problem, you need to:
  a.   Decide on a final goal.
  b.   Work backward from the final goal.
  c.   Ask yourself "Does the answer make sense?", as we do throughout this study guide.

3.6   *Percent Composition of Compounds*

When you finish this section, you will be able to calculate the mass percent of each element in a compound.

There are 3 steps to calculating the **mass percent** of each element in a compound.

   a. Compute the **molecular mass** of the compound.

   b. Calculate how much of the molecular mass comes from each element.

   c. Divide each element's mass contribution by the total molecular mass, and multiply by 100 to convert to percent.

Example 3.6 A   Determination of the Mass Percent in a Compound

Calculate the mass percent of each element in glucose, $C_6H_{12}O_6$.

Solution

Using the **general strategy** outlined at the start of this section,

   a. Molar mass:

$$
\begin{aligned}
C: &\ 6 \times 12.01 &=&\ \ 72.06 \text{ g} \\
H: &\ 12 \times 1.008 &=&\ \ 12.096 \text{ g} \\
O: &\ 6 \times 16.00 &=&\ \ \underline{96.00 \text{ g}} \\
&\textbf{molar mass} &=&\ \ \textbf{180.156 g glucose}
\end{aligned}
$$

(Note: We retain all figures in the molar mass because this is an intermediate calculation.)

   b. Contribution of each element:

As shown in the molar mass calculation,

      C contributes **72.06 g**
      H contributes **12.096 g**
      O contributes **96.00 g**

   c. Mass percent of each element:

$$
\text{Mass percent of C} = \frac{72.06 \text{ g C}}{180.156 \text{ g glucose}} \times 100\% = \textbf{40.00\% C}
$$

$$
\text{Mass percent of H} = \frac{12.096 \text{ g H}}{180.156 \text{ g glucose}} \times 100\% = \textbf{6.71\% H}
$$

$$
\text{Mass percent of O} = \frac{96.00 \text{ g O}}{180.156 \text{ g glucose}} \times 100\% = \textbf{53.29\% O}
$$

Does the Answer Make Sense?

The best way to check your answer is to add up the percentages. If they total 100.00% (as they do here), your answers are fine (assuming your molar mass is correct).

COMMENT: Although it would have been correct to calculate the mass percent of O by subtracting (%C + %H) from 100% in the previous problem, this eliminates your ability to double check your answer, because you have, in effect, already "guaranteed" that the mass percents must equal 100%. We therefore recommend against this procedure.

Example 3.6 B   Practice with Mass Percents

Calculate the mass percent of each element in potassium ferricyanide, $K_3Fe(CN)_6$.
Solution

Using the same strategy as in the previous example,

a.   K: 3×39.10 =   117.30 g
     Fe: 1×55.85 =   55.85 g
     C: 6×12.01 =   72.06 g
     N: 6×14.01 = $\underline{\quad 84.06\ g}$
     molar mass   =   **329.27 g/mole $K_3Fe(CN)_6$**

b.   Contribution of each element:

K = 117.30 g;   Fe = 55.85 g;   C = 72.06 g;   N = 84.06 g

c.   Mass percent of each element:

$$\text{Mass percent of K} \ = \ \frac{117.30\ \text{g K}}{329.27\ \text{g K}_3\text{Fe(CN)}_6} \times 100\% = \textbf{35.62\% K}$$

$$\text{Mass percent of Fe} = \ \frac{55.85\ \text{g Fe}}{329.27\ \text{g K}_3\text{Fe(CN)}_6} \times 100\% = \textbf{16.96\% Fe}$$

$$\text{Mass percent of C} \ = \ \frac{72.06\ \text{g C}}{329.27\ \text{g K}_3\text{Fe(CN)}_6} \times 100\% = \textbf{21.88\% C}$$

$$\text{Mass percent of N} \ = \ \frac{84.06\ \text{g N}}{329.27\ \text{g K}_3\text{Fe(CN)}_6} \times 100\% = \textbf{25.53\% N}$$

Checking the Figures

Adding up the percentages gives **99.99%**, not 100.00%. This is an example of the kind of problem you will see from time to time, where rounding off to 2 significant figures after the decimal point leads to a **loss** of 0.01%. In such cases, we say the answers are correct "within round-off error."

3.7   *Determining the Formula of a Compound*

When you finish this section you will be able to:
  • Determine the empirical formula of a compound.
  • Calculate the molecular formula of a compound.

The **empirical formula** is represented by the **simplest whole number ratio** of atoms in a compound. Examples of empirical formulas are:

$$CH,\ CH_4,\ CH_2O,\ K_2Cr_2O_7,\ K_3Fe(CN)_6.$$

The **molecular formula** is the **actual ratio** of atoms in a compound. Examples of molecular formulas are:

$$C_6H_6,\ CH_4,\ C_6H_{12}O_6,\ K_2Cr_2O_7,\ K_3Fe(CN)_6.$$

The empirical and molecular formulas **can** be the same. Review the methods for experimentally determining the empirical and molecular formulas in your text, and note especially the summary at the end of this section in your textbook. Then try the following problems.

Example 3.7 A   Determination of the Empirical Formula of a Compound

The analysis of a rocket fuel showed that it contained 87.4% nitrogen and 12.6% hydrogen by weight. Mass spectral analysis showed the fuel to have a molar mass of 32.05 grams. What are the empirical and molecular formulas of the fuel?

Strategy

The problems in this section involve just the opposite strategy as those in Section 3.4. In that section, you found what percentage of each element made up a known compound. Here, you **know** the percentages, and you **need** the compound.

To solve empirical formula problems that do not involve combustion (added oxygen):

1. Assume the compound has a mass of exactly 100 grams. You can therefore convert percentage to grams.
2. Calculate moles of each kind of atom present.
3. Determine simplest whole number ratios by dividing the moles of each compound by the smallest calculated mole value.

To determine the molecular formula, divide the molar mass by the empirical formula mass. This will give the number of empirical formula units in the actual molecule. For example, if the **empirical formula is CH**, and you determine that the **molar mass is 6 times** the empirical formula mass, the **molecular formula is C_6H_6.**

Solution

Assuming the compound has a mass of 100 grams (we will drop this assumption after calculating the empirical formula),

N = 87.4 g (87.4% of 100)
H = 12.6 g (12.6% of 100)

$$\text{moles of N} \quad \Big\| \quad 87.4 \text{ g N} \times \frac{1 \text{ mole N}}{14.01 \text{ g N}} = \textbf{6.24 moles N}$$

$$\text{moles of H} \quad \Big\| \quad 12.6 \text{ g H} \times \frac{1 \text{ mole H}}{1.008 \text{ g H}} = \textbf{12.5 moles H}$$

$$\text{Simplest whole number ratio} = \text{N} = \frac{6.24}{6.24} = \textbf{1.0}$$

$$\text{H} = \frac{12.5}{6.24} = \textbf{2.0}$$

The **empirical formula is NH_2**

To find the molecular formula,

mass of empirical formula = (NH_2) = 16.026 g
mass of compound = 32.05 g

$$\text{units of empirical formula in compound} = \frac{32.05 \text{ g}}{16.026 \text{ g}} = \textbf{2.00 units}$$

Therefore, **the molecular formula is N_2H_4** (hydrazine).
(**A note on precision**: Empirical formula determinations are not among the most precise experiments in chemistry. You may expect variations of up to about 5% in your values. Be flexible!)

Example 3.7 B   Determination of the Empirical Formula of a Compound

Determine the empirical and molecular formulas for a deadly nerve gas that gives the following mass percent analysis:

C = 39.10%   H = 7.67%   O = 26.11%   P = 16.82%   F = 10.30%

The molar mass is known to be 184.1 grams.

Solution

We will proceed as we did with the previous example that is, by assuming a molar mass of exactly 100 grams **(for purposes of determining the empirical formula only)**.

C = 39.10 g   H = 7.67 g   O = 26.11 g   P = 16.82 g   F = 10.30 g

moles of C $\|$ $39.10 \text{ g C} \times \dfrac{1 \text{ mole C}}{12.01 \text{ g C}}$ = **3.26 moles C**

moles of H $\|$ $7.67 \text{ g H} \times \dfrac{1 \text{ mole H}}{1.008 \text{ g H}}$ = **7.61 moles H**

moles of O $\|$ $26.11 \text{ g O} \times \dfrac{1 \text{ mole O}}{16.00 \text{ g O}}$ = **1.63 moles O**

moles of P $\|$ $16.82 \text{ g P} \times \dfrac{1 \text{ mole P}}{30.97 \text{ g P}}$ = **0.543 moles P**

moles of F $\|$ $10.30 \text{ g F} \times \dfrac{1 \text{ mole F}}{19.00 \text{ g F}}$ = **0.542 moles F**

Dividing through by smallest number:

$$C_{\frac{3.26}{0.542}} \quad H_{\frac{7.61}{0.542}} \quad O_{\frac{1.63}{0.542}} \quad P_{\frac{0.543}{0.542}} \quad F_{\frac{0.542}{0.542}}$$

Gives an empirical formula of

$$C_{6.01} \quad H_{14.04} \quad O_{3.01} \quad P_{1.00} \quad F_1 = \mathbf{C_6H_{14}O_3PF}$$

The empirical formula mass = **184.1 grams. Therefore, the molecular formula and the empirical formula are the same.**

Example 3.7 C   Determining the Empirical Formula from Combustion Data

A combustion device was used to determine the empirical formula of a compound containing **ONLY carbon, hydrogen, and oxygen**. A 0.6349 g sample of the unknown produced 1.603 g of CO_2 and 0.2810 g H_2O. Determine the empirical formula of the compound.

Solution

The primary assumption is that the **combustion was complete**. Therefore, **all** the carbon was converted to CO_2. All the hydrogen was converted to H_2O. Then, if you calculate the **grams of carbon in 1.603 g CO_2** and the **grams of hydrogen in 0.2810 g H_2O**, you know how much C and H was in your original compound. You can calculate the grams of oxygen in the original compound as:

grams O = total grams − (grams C + grams H).

a. Molar mass of CO_2 = 44.01 grams.

Mass fraction of C in CO_2 = $\dfrac{12.01}{44.01}$ = 0.2729 (0.2729 g C in every g of CO_2)

g C in 1.603 g CO_2 = 0.2729 × 1.603 g = **0.4374 g C** in CO_2 (and therefore in our original compound).

b. Molar mass of H_2O = 18.016 grams.

Mass fraction of H in H_2O = $\dfrac{2.016}{18.016}$ = 0.1119 (0.1119 g H in every g of H_2O)

g H in 0.2810 g H_2O = 0.1119 × 0.2810 g = **0.03144 g H** in H_2O (and therefore in our original compound).

c. grams O in original compound = 0.6349 g − (0.4374 g C + 0.03144 g H) = **0.1661 g O** in original compound.

d. Converting each to moles,

moles C ‖ 0.4374 g C × $\dfrac{1 \text{ mole C}}{12.01 \text{ g C}}$ = **0.03642 mole C**

moles H ‖ 0.03144 g H × $\dfrac{1 \text{ mole H}}{1.008 \text{ g H}}$ = **0.03119 mole H**

moles O ‖ 0.1661 g O × $\dfrac{1 \text{ mole O}}{16.00 \text{ g O}}$ = **0.01038 mole O**

e. dividing by smallest number

$C_{\frac{0.03642}{0.01038}}$ $H_{\frac{0.03119}{0.01038}}$ $O_{\frac{0.01038}{0.01038}}$ = $C_{3.5}H_3O_1$

The empirical formula (simplest WHOLE number ratio) = $C_7H_6O_2$. Two step-by-step methods for obtaining the molecular formula of a compound are summarized in a <u>Problem Solving Strategy Box in your textbook</u>.

3.8   *Chemical Equations*

When you finish this section you will be able to:

- State the meaning of each part of a chemical reaction.
- State a variety of different relationships that are inferred from a chemical equation.

Chemical equations are all of the form

Reactants $\longrightarrow$ Products

Chemical equations describe chemical changes that occur in a reaction. The physical states that can be present (with their symbols) in a reaction are described in your textbook. The general information that can be gotten from chemical reactions is given in Table 3.2 in your textbook.

Example 3.8 A   Interpreting a Chemical Equation

State, completely, what is happening in the reaction given below. Include in your answer the physical states of each component of the reaction.

$$Na_2CO_3(aq) + 2HCl(aq) \rightarrow CO_2(g) + H_2O(l) + 2NaCl(aq)$$

Solution

In this reaction, one mole of aqueous sodium carbonate is reacting with two moles of aqueous hydrochloric acid to produce one mole of gaseous carbon dioxide, one mole of water, and two moles of aqueous sodium chloride.

In this problem you will note that the equation is **balanced**. That is, there are the same number of atoms of each element on each side of the equation.

Example 3.8 B   Relating Reactions and Products of a Chemical Equation

Show, by means of molar masses, that matter is neither created nor destroyed in the equation given below.

$$C_2H_6O(l) + 3O_2(g) \rightarrow 2CO_2(g) + 3H_2O(g)$$

Strategy

The law of conservation of matter says that matter is neither created nor destroyed in a chemical reaction. Therefore,

number of grams of reactant = number of grams of product

Remember to take the **number of moles** of each compound into account when doing your calculations!

Solution

Compound	Molar Mass	Number of moles in equation	Total grams in equation
C_2H_6O	46.068 g	1	46.068 g
O_2	32.00 g	3	96.00 g
CO_2	44.01 g	2	88.02 g
H_2O	18.016 g	3	54.048 g

So, the **total grams of reactant** = 46.068 + 96.00 = **142.07 g**
the **total grams of product** = 88.02 + 54.048 = **142.07 g**
grams of reactant = grams of product

This relationship must hold in every chemical reaction.

3.9   Balancing Chemical Equations

When you finish this section you will be able to balance many chemical equations.

A **balanced chemical equation** means that the **number of atoms of each element on the reactants' side equals the number of atoms of each element on the products' side.**

Example 3.9 A   Balancing Chemical Equations

Balance the following chemical equation

$$NaOH(aq) + H_3PO_4(aq) \rightarrow Na_3PO_4(aq) + H_2O(l)$$

Strategy

Chemists have many different strategies for balancing chemical equations. There is no one best way. There is an excellent strategy given just before Example 3.13 in your textbook.

An alternate strategy is:

1. Never change a molecular structure. Only use coefficients.
2. Find the atoms that are in only one compound on the reactants' side. Balance those first.
3. In general, leave oxygen and hydrogen until the very end. They usually appear many times, and balancing other atoms will often force O and H to become balanced.
4. Always double check **AFTER** balancing to make sure that the atoms of each element are equal on the reactant and product sides.

Solution

1. Multiply **NaOH by 3** to balance sodium atoms, as there are 3 sodium atoms in Na_3PO_4.

$$3NaOH + H_3PO_4 \rightarrow Na_3PO_4 + H_2O$$

Totals:		
	3 Na atoms	3 Na atoms
	6 H atoms	2 H atoms
	1 P atom	1 P atom
	7 O atoms	5 O atoms

2. Phosphorus is already balanced. There are 6 hydrogen atoms on the left side and 2 on the right. Multiply **H₂O by 3** to balance hydrogen.

$$3NaOH + H_3PO_4 \rightarrow Na_3PO_4 + 3H_2O$$

Totals:		
	3 Na atoms	3 Na atoms
	6 H atoms	6 H atoms
	1 P atom	1 P atom
	7 O atoms	7 O atoms

The equation is now balanced.

Does the Answer Make Sense?

Yes, because the law of conservation of mass holds here. The number of atoms of each element on the left side of the reaction equals the number of atoms of each element on the right side.

Example 3.9 B   Practice With Balancing

Balance the following combustion reaction:

$$C_4H_{10}(g) + O_2(g) \rightarrow CO_2(g) + H_2O(g)$$

Strategy

Combustion represents a very common class of reactions. We will proceed as before, balancing carbon first and saving hydrogen and oxygen for later. See if you can keep a running tally of atoms as we proceed.

Solution

a.   Multiply **CO_2 by 4** to balance carbons.
b.   Multiply **H_2O by 5** to balance hydrogens.

   This leaves us with 4 carbons and 10 hydrogens on both sides. We have **2 oxygens on the left side** and **13 oxygens on the right**.

c.   Multiply **O_2 by 13/2** to balance oxygens.

   Our balanced equation becomes:

$$C_4H_{10}(g) + 13/2 O_2(g) \rightarrow 4CO_2(g) + 5H_2O(g)$$

   We have:

$$4C, 10H, 13O \rightarrow 4C, 10H, 13O$$

   Fractional coefficients **are** used from time to time in chemistry. For clarity, however, we prefer to use simple **whole number ratios**. Therefore:

d.   Multiply both sides of the reaction by 2. This converts the 13/2 to 13, a whole number. The number of atoms of each element on both sides of the equation are **still equal** because we have multiplied both sides by 2.

   Final answer:

$$\mathbf{2C_4H_{10}(g) + 13O_2(g) \rightarrow 8CO_2(g) + 10H_2O(g)}$$

3.10   Stoichiometric Calculations: Amounts of Reactants and Products

When you finish this section you will be able to:

- Calculate the amount of products obtained from a given amount of reactant.
- Calculate the amount of reactants required to generate a desired amount of product.

The most important reason for learning to balance chemical equations is (as in Section 3.9) to establish reactant-product mole (and therefore mass) relationships. For instance, in Example 3.9 B, we found that:

$$\textbf{2 moles of } C_4H_{10} + \textbf{13 moles of } O_2 \rightarrow \textbf{8 moles of } CO_2 + \textbf{10 moles of } H_2O$$

In terms of the grams of O_2 to grams of CO_2 relationship,

$$13 \text{ moles } O_2 \times \frac{32.00 \text{ g } O_2}{1 \text{ mole } O_2} = \textbf{416 g } O_2 \text{ reacts with two moles of } C_4H_{10} \text{ to yield 10 moles of } H_2O,$$

and $8 \text{ moles } CO_2 \times \dfrac{44.01 \text{ g } CO_2}{1 \text{ mole } CO_2} = \textbf{352 g } CO_2.$

The **key relationship** in problems such as these is the **moles of reactant to moles of product** conversion factor. Note that

$$\frac{13 \text{ moles } O_2}{8 \text{ moles } CO_2} \text{ and } \frac{8 \text{ moles } CO_2}{13 \text{ moles } O_2} \text{ are both correct.}$$

Example 3.10 A   Amount of Product from a Given Amount of Reactant

Given the following reaction:

$$Na_2S(aq) + AgNO_3(aq) \rightarrow Ag_2S(s) + NaNO_3(aq)$$

How many grams of Ag_2S can be generated from the reaction of 3.94 g of $AgNO_3$ with excess Na_2S?

Strategy

1. **Always balance the chemical equation!**
2. The **mole ratio conversion factor** acts as the **bridge** between $AgNO_3$ and Ag_2S. This leads to the following strategy:

$$\text{g } AgNO_3 \xrightarrow[\text{mass}]{\text{molar}} \text{moles } AgNO_3 \xrightarrow[\text{ratio}]{\text{mole}} \text{moles } Ag_2S \xrightarrow[\text{mass}]{\text{molar}} \text{g } Ag_2S$$

It is recommended that you solve such problems with one set of equations, instead of splitting the calculation up into separate steps. Many calculations may lead to confusion and errors.

Solution

The balanced chemical equation is:

$$Na_2S(aq) + 2AgNO_3(aq) \rightarrow Ag_2S(s) + 2NaNO_3(aq).$$

Then use one equation:

$$\text{g } Ag_2S \ \bigg\| \ 3.94 \text{ g } AgNO_3 \times \frac{1 \text{ mole } AgNO_3}{169.88 \text{ g } AgNO_3} \times \frac{1 \text{ mole } Ag_2S}{2 \text{ moles } AgNO_3} \times \frac{247.8 \text{ g } Ag_2S}{1 \text{ mole } Ag_2S}$$

$$= \textbf{2.87 g } Ag_2S$$

Does the Answer Make Sense?

Although the answer looks reasonable, the best way to check your answer is by double-checking your use of units. If they cancel properly, then your answer is likely to be correct.

Example 3.10 B   Amount of Reactant Needed to Produce a Product

Aspirin (acetylsalicylic acid) is prepared by the reaction of salicylic acid ($C_7H_6O_3$) and acetic anhydride ($C_4H_6O_3$). How many grams of salicylic acid (sal) are needed to make 500 aspirin tablets weighing 1.00 g each (assuming 100% yield)?

$$C_7H_6O_3 + C_4H_6O_3 \rightarrow C_9H_8O_4 + HC_2H_3O_2$$

salicylic   acetic   aspirin   acetic
acid   anhydride   acid

Strategy

We will work this problem in basically the same fashion as we did the previous example. The only difference is that we are going from **product** to **reactant**.

number of aspirin tablets $\xrightarrow{\text{mass per tablet}}$ grams of aspirin $\xrightarrow{\text{molar mass}}$ moles of aspirin
(start)

grams of salicylic acid $\xleftarrow{\text{molar mass}}$ moles of salicylic acid $\xleftarrow{\text{mole ratio}}$

Solution

$$\text{g salicylic acid} \; \Big\| \; 500 \text{ aspirin tablets} \times \frac{1.00 \text{ g aspirin}}{1 \text{ aspirin tablet}} \times \frac{1 \text{ mole aspirin}}{180.15 \text{ g aspirin}} \times \frac{1 \text{ mole sal}}{1 \text{ mole aspirin}}$$

$$\times \frac{138.12 \text{ g sal}}{1 \text{ mole sal}} = \textbf{383 g salicylic acid}$$

Remember that the most important relationships in all of chemistry are between **moles of reactants** and **moles of products**.

3.11   The Concept of Limiting Reagent

When you finish this section, you will be able to:

- Identify the limiting reactant.
- Solve problems involving a limiting reactant.
- Calculate percent yield of a product.

In the previous section, we were given the amount of a reactant and asked to find the amount of product formed. The **assumption** then was that **we had as much of the other reactants as we required to make the reactant go completely**. This will not always be the case. For example, given the following reaction:

$$2H_2(g) + O_2(g) \rightarrow 2H_2O(g),$$

We need **2 moles of H_2** to completely react with **1 mole of O_2**, forming **2 moles of H_2O**. If we have only **1.5 moles of H_2**, only **0.75 moles of O_2** will react, forming **1.5 moles of H_2O**. That means that 1 mole O_2 − 0.75 mole O_2 or **0.25 moles of O_2** will be left over, or **in excess**.

H_2 is said to be the **limiting reactant** in this reaction because it **limits the amount of product that can form**.

Example 3.11 A   Calculations Involving a Limiting Reactant

Sodium hydroxide reacts with phosphoric acid to give sodium phosphate and water. If 17.80 g of NaOH is mixed with 15.40 g of H_3PO_4,

  a. How many grams of Na_3PO_4 can be formed?
  b. How many grams of the excess reactant remain unreacted?
  c. If the <u>actual yield</u> of Na_3PO_4 was 15.00 g, what is the percent yield of Na_3PO_4?

Strategy

We must first write a balanced chemical equation for this reaction:

$$3NaOH(aq) + H_3PO_4(aq) \rightarrow Na_3PO_4(aq) + 3H_2O(l)$$

a. The basic question to be addressed in limiting reactant problems is **"which compound limits the amount of product formed?"** It is therefore recommended that you calculate the amount of product formed by **each** reactant. **The reactant that forms less product is limiting**, and that total amount of product will be formed:

g of reactant $\xrightarrow{\text{molar mass}}$ moles of reactant $\xrightarrow{\text{mole ratio}}$ moles of product $\rule{2cm}{0.4pt}$

                     g of product $\xleftarrow{\text{molar mass}}$

b. To determine grams of excess reactant, you must calculate **how many moles of excess reactant were actually used**.

      # moles excess = # moles original − # moles used

Then convert moles to grams.

c. The percent yield $= \dfrac{\text{actual yield}}{\text{theoretical yield}} \times 100\%$.

Solution

a. Determination of limiting reactant

$$\begin{array}{c} \text{g Na}_3\text{PO}_4 \\ \text{from NaOH} \end{array} \left\| \quad 17.80 \text{ g NaOH} \times \frac{1 \text{ mole NaOH}}{40.00 \text{ g NaOH}} \times \frac{1 \text{ mole Na}_3\text{PO}_4}{3 \text{ moles NaOH}} \times \frac{163.94 \text{ g Na}_3\text{PO}_4}{1 \text{ mole Na}_3\text{PO}_4} \right.$$

     = **24.32 g Na_3PO_4 from NaOH**

$$\begin{array}{c} \text{g Na}_3\text{PO}_4 \\ \text{from H}_3\text{PO}_4 \end{array} \left\| \quad 15.40 \text{ g H}_3\text{PO}_4 \times \frac{1 \text{ mole H}_3\text{PO}_4}{98.00 \text{ g H}_3\text{PO}_4} \times \frac{1 \text{ mole Na}_3\text{PO}_4}{1 \text{ mole H}_3\text{PO}_4} \times \frac{163.94 \text{ g Na}_3\text{PO}_4}{1 \text{ mole Na}_3\text{PO}_4} \right.$$

     = **25.76 g Na_3PO_4 from H_3PO_4**

Therefore, **NaOH is the limiting reactant, and 24.32 g Na_3PO_4 are formed**.

b. **Determination of grams of excess reactant**

H_3PO_4 is in excess. If 24.32 g Na_3PO_4 is formed, the **moles of H_3PO_4 used** is given by:

$$\text{moles } H_3PO_4 \text{ used} \ \bigg\| \ 24.32 \text{ g } Na_3PO_4 \times \frac{1 \text{ mole } Na_3PO_4}{163.94 \text{ g } Na_3PO_4} \times \frac{1 \text{ mole } H_3PO_4}{1 \text{ mole } Na_3PO_4}$$

$$= \textbf{0.1483 moles } \mathbf{H_3PO_4} \textbf{ used}$$

The number of **moles of H_3PO_4 originally present** is:

$$15.40 \text{ g } H_3PO_4 \times \frac{1 \text{ mole } H_3PO_4}{98.00 \text{ g } H_3PO_4} = \textbf{0.1571 moles } \mathbf{H_3PO_4} \textbf{ originally present}$$

$$\begin{aligned} \text{moles } H_3PO_4 \text{ excess} &= \text{moles } H_3PO_4 \text{ originally present} - \text{moles } H_3PO_4 \text{ used} \\ &= 0.1571 \text{ moles} - 0.1483 \text{ moles} \\ &= \textbf{0.0088 moles } \mathbf{H_3PO_4} \textbf{ excess} \end{aligned}$$

$$\text{grams } H_3PO_4 \text{ excess} \ \bigg\| \ 0.0088 \text{ moles } H_3PO_4 \times \frac{98.00 \text{ g } H_3PO_4}{1 \text{ mole } H_3PO_4} = \textbf{0.86 g } \mathbf{H_3PO_4} \textbf{ excess}$$

c. **Determination of percent yield**

$$\text{percent yield} = \frac{\text{actual yield}}{\text{theoretical yield}} \times 100\% = \frac{15.00 \text{ g}}{24.32 \text{ g}} \times 100\% = \textbf{61.68\% yield}$$

Example 3.11 B *Tying it All Together*

The Space Shuttle environmental control system handles excess CO_2 (which the astronauts breathe out - it is 4% by mass of exhaled air) by reacting it with lithium hydroxide, LiOH, pellets to form lithium carbonate, Li_2CO_3, and water. If there are 7 astronauts on board the shuttle, and each exhales 20 liters of air per minute, how long could clean air be generated if there were 25,000 g of LiOH pellets available for each shuttle mission? Assume the density of air is 0.0010 g/mL.

Solution

$$\frac{\text{g } CO_2 \text{ generated}}{\text{minute}} \ \bigg\| \ \frac{0.0010 \text{ g air}}{\text{mL}} \times \frac{20,000 \text{ mL air}}{\text{min - astronaut}} \times 7 \text{ astronauts} \times \frac{4 \text{ g } CO_2}{100 \text{ g air}}$$

$$= \textbf{5.6 g } \mathbf{CO_2} \textbf{ / minute}$$

The reaction of CO_2 with LiOH is

$$CO_2 + 2LiOH \rightarrow Li_2O_3 + H_2O$$

Grams of CO_2 that can react with 25,000 g of LiOH:

$$\text{g } CO_2 \ \bigg\| \ 2.5 \times 10^4 \text{ g LiOH} \times \frac{1 \text{ mol LiOH}}{23.949 \text{ g LiOH}} \times \frac{1 \text{ mol } CO_2}{2 \text{ mol LiOH}} \times \frac{44.01 \text{ g } CO_2}{1 \text{ mol } CO_2}$$

$$= \textbf{22,971 g } \mathbf{CO_2}$$

Number of days clean air can be generated:

$$22,971 \text{ g } CO_2 \times \frac{1 \text{ minute}}{5.6 \text{ g } CO_2} = 4102 \text{ minutes} = \textbf{2.85 days of clean air}$$

Exercises

Section 3.2

1. An element "E" is present as ^{10}E with a mass value of 10.01 amu, and as ^{11}E with a mass value of 11.01 amu. The natural abundances of ^{10}E and ^{11}E are 19.78% and 80.22% respectively. What is the average atomic mass of the element? What is the element?

2. Naturally occurring sulfur consists of four isotopes, ^{32}S (95.0%), ^{33}S (0.76%), ^{34}S (4.22%), and ^{36}S (0.014%). Using these data, calculate the atomic weight of naturally occurring sulfur. The masses of the isotopes are given in the table below.

Isotope	Atomic mass (amu)
^{32}S	31.97
^{33}S	32.97
^{34}S	33.97
^{36}S	35.97

3. An unknown sample of mystery element "T" is injected into the mass spectrometer. According to the mass spectrum, 7.42% of the element is present as ^6T, and 92.58% is present as ^7T. The mass values are 6.02 amu for ^6T and 7.02 amu for ^7T. Calculate the average atomic mass, and identify the mystery element.

4. A noble gas consists of three isotopes of masses 19.99 amu, 20.99 amu, and 21.99 amu. The relative abundance of these isotopes is 90.92%, 0.257%, and 8.82% respectively. What is the average atomic mass of this noble gas? What noble gas is this?

5. Chlorine has two stable isotopes. The mass of one isotope is 34.97 amu. Its relative abundance is 75.53%. What is the mass of the other stable isotope?

6. Complete the following table of isotopic information for the element neon (Ne).

Isotope	Mass (amu)	Abundance
^{20}Ne	19.99	_____
^{21}Ne	20.99	0.257%
^{22}Ne	21.99	_____

7. Silicon has three stable isotopes in nature as shown in the table below. Fill in the missing information.

Isotope	Mass (amu)	Abundance
^{28}Si	27.98	_____
^{29}Si	_____	4.70%
^{32}Si	29.97	3.09%

8. Gallium has two stable isotopes of masses 68.93 amu (^{69}Ga) and 70.92 amu (^{71}Ga). What are the relative abundances of the two isotopes?

9. Magnesium exists as three isotopes in nature. One isotope (^{25}Mg) has a mass of 24.99 amu and a relative abundance of 10.13%. The other two isotopes have masses of 23.99 amu (^{24}Mg) and 25.98 amu (^{26}Mg). What are their relative abundances? (atomic mass Mg = 24.305 amu)

10. An element "X" has 5 major isotopes, listed below along with their abundances. What is the element? Does the atomic mass that you calculate based on these data agree with that listed in your periodic table?

Isotope	% Natural Abundance	Atomic Mass
^{46}X	8.0%	45.95269
^{47}X	7.3%	46.951764
^{48}X	73.8%	47.947947
^{49}X	5.5%	48.947841
^{50}X	5.4%	49.944792

Section 3.3

11. How many moles are in a sample of 300 atoms of nitrogen? How many grams?

12. How many atoms of gold does it take to make 1 gram of gold?

13. If you buy 38.9 moles of M&M's®, how many M&M's® do you have? (1 mole of M&M's® = 6.022×10^{23} M&M's®)

14. A sample of sulfur has a mass of 5.37 g. How many moles are in the sample? How many atoms?

15. Give the number of moles of each element present in 1.0 mole of each of the following substances:

    a. Hg_2I_2 c. $PbCO_3$ e. $RbOH \cdot 2H_2O$
    b. LiH d. $Ba_3(AsO_4)_2$ f. H_2SiF_6

16. How many grams of zinc are in 1.16×10^{22} atoms of zinc?

17. How many amu are in 3.68 moles of iron?

Section 3.4

18. Calculate the molar masses of each of the following:

    a. Cu_2SO_4 c. $C_{10}H_{16}O$ e. $Ca_2Fe(CN)_6 \cdot 12H_2O$
    b. NH_4OH d. $Zr(SeO_3)_2$ f. $Cr_4(P_2O_7)_3$

19. Calculate the molar mass of

    a. $Zn(CN)_4$
    b. $Cu(NH_3)_4 \cdot 8H_2O$

20. What is the mass of 4.28×10^{22} molecules of water?

21. What is the mass of 4.89×10^{23} atoms of the element "X" described in problem # 10?

22. How many milligrams of Br_2 are in 4.8×10^{20} molecules of Br_2?

23. How many sodium ions are present in each of the following:

    a. 2 moles of sodium phosphate
    b. 5.8 grams of sodium chloride
    c. a mixture containing 14.2 grams of sodium sulfate and 2.9 grams of sodium chloride?

24.   How many potassium ions are present in each of the following:

    a.   3 moles of potassium chloride
    b.   6.2 grams of potassium nitrate
    c.   a mixture containing 12.6 grams of potassium phosphate and 5.4 grams of potassium chloride?

25.   What is the weight in grams of

    a.   0.4 moles of CH_4
    b.   11 moles of SO_4^{2-}
    c.   5 moles of $Mg(OH)_2$?

26.   Determine the molar mass of $KAl(SO_4)_2 \cdot 12H_2O$.

27.   How many moles of cadmium bromide, $CdBr_2$, are in a 39.25 g sample?

28.   A sample of calcium chloride, $CaCl_2$, has a mass of 23.8 g. How many moles of calcium chloride is this?

29.   If 0.172 moles of baking soda, $NaHCO_3$, were used to bake a chocolate cherry cake, how many grams of baking soda would the recipe call for?

30.   How many moles are there in a sample of barium sulfate, $BaSO_4$, weighing 9.90×10^{-7} g?

31.   How many grams are there in 0.36 moles of cobalt (III) acetate, $Co(C_2H_3O_2)_3$? How many grams of cobalt are in this sample? How many atoms of cobalt?

32.   How many milligrams of chlorine are there in a sample of 3.9×10^{19} molecules of chlorine gas, Cl_2? How many atoms of chlorine?

33.   Bauxite, the principle ore used in the production of aluminum cans, has a molecular formula of $Al_2O_3 \cdot 2H_2O$.

    a.   Determine the molar mass of bauxite.
    b.   How many grams of Al are in 0.58 moles of bauxite?
    c.   How many atoms of Al are in 0.58 moles of bauxite?
    d.   What is the mass in grams of 2.1×10^{24} formula units of bauxite?

Section 3.6

34.   Calculate the mass percent of Cl in each of the following compounds:

    a.   ClF
    b. $HClO_2$
    c. $CuCl_2$
    d.   PuOCl

35.   Calculate the mass percent of each element in $C_5H_{10}O$.

36.   Calculate the mass percent of each element in potassium ferricyanide, $K_3Fe(CN)_6$.

37.   Calculate the mass percent of each element in barium sulfite, $BaSO_3$.

38.   Calculate the mass percent of each element in natural lucite, $KAlSi_2O_6$.

39.   Calculate the mass percent of silver in each of the following compounds:

    a.   AgCl
    b.   AgCN
    c. $AgNO_3$

40. Chlorophyll **a** is essential for photosynthesis. It contains 2.72% magnesium by mass. What is the molar mass of chlorophyll **a** assuming there is one atom of magnesium in every molecule of chlorophyll **a**?

41. Calculate the mass percent of each of the elements in Nicotine, $C_{10}H_{14}N_2$.

Section 3.7

42. Which of the following formulas can be empirical?

    a. CH_4
    b. CH_2
    c. $KMnO_4$

    d. N_2O_5
    e. B_2H_6
    f. NH_4Cl

    g. Sb_2S_3
    h. N_2O_4
    i. CH_2O

43. Determine the empirical and molecular formulas of a compound that has a mass of 31.04 g/mole and contains the following percentages of elements by mass:

    $$C = 38.66\%, \quad H = 16.24\%, \quad N = 45.10\%$$

44. The analysis of a rocket fuel showed that it contained 87.4% nitrogen and 12.6% hydrogen by weight. Mass spectral analysis showed the fuel to have a molar mass of 32.05 grams. What are the empirical and molecular formulas of the fuel?

45. A compound is found, by mass spectral analysis, to contain the following percentages of elements by mass:

    $$C = 49.67\%, \quad Cl = 48.92\%, \quad H = 1.39\%$$

    The molar mass of the compound is 289.9 g/mole. Determine the empirical and molecular formula of the compound.

46. Vanillin, the pleasant smelling ingredient used to bake chocolate chip cookies, is often used in the production of vanilla extract. Vanillin has a mass of 152.08 g/mole and contains the following percentages of elements by mass:

    $$C = 63.18\%, \quad H = 5.26\%, \quad O = 31.56\%$$

    Determine the empirical and molecular formula of vanillin.

47. Determine the empirical formula of a compound that contains the following percentages of elements by mass:

    $$Mo = 43.95\%, \quad O = 7.33\%, \quad Cl = 48.72\%$$

48. A molecule with a molecular weight of approximately 110 g/mole is analyzed. The results show that it contains 10.05% of carbon, 0.84% of hydrogen, and 89.10% of chloride. Calculate the molecular formula of this compound.

49. Using the data provided, calculate the empirical formulas for the compounds indicated:

    a. an oxide of nitrogen, a sample of which contains 6.35 g of nitrogen and 3.65 g of oxygen
    b. an oxide of copper, one gram of which contains 0.7989 g of copper
    c. an oxide of carbon that contains 42.84% carbon
    d. a compound of potassium, chloride, and oxygen containing K = 31.97%, O = 39.34%
    e. a compound of hydrogen, carbon, and nitrogen containing H = 3.70%, C = 44.44%, and N = 51.85%.

Section 3.8

50. How many grams of product are formed in each of the following reactions?

    a.   Two moles of H_2 react with one mole of O_2.

    b.   One mole of silver nitrate reacts with one mole of sodium chloride.

    c.   Three moles of sodium hydroxide react with one mole of phosphoric acid.

51. The following reaction was performed:

$$Fe_2O_3(s) + 2X(s) \rightarrow 2Fe(s) + X_2O_3(s)$$

It was found that 79.847 g of Fe_2O_3 reacted with "X" to form 55.847 g of Fe and 50.982 g of X_2O_3. Identify element X.

52. Do these equations follow the conservation of matter?

    1. $Na_2SiO_3 + 6HF \rightarrow SiF_4 + 2NaF + 3H_2O$

    2. $3N_2O_4 + 2H_2O \rightarrow 4HNO_3 + 2NO_2$

Section 3.9

53. Fill in the blanks to balance the following chemical equations:

    a.   ___AgI + ___Na_2S $\rightarrow$ ___Ag_2S + ___NaI

    b.   ___$(NH_4)_2Cr_2O_7$ $\rightarrow$ ___Cr_2O_3 + ___N_2 + ___H_2O

    c.   ___Na_3PO_4 + ___HCl $\rightarrow$ ___NaCl + ___H_3PO_4

    d.   ___$TiCl_4$ + ___H_2O $\rightarrow$ ___TiO_2 + ___HCl

    e.   ___Ba_3N_2 + ___H_2O $\rightarrow$ ___$Ba(OH)_2$ + ___NH_3

    f.   ___HNO_2 $\rightarrow$ ___HNO_3 + ___NO + ___H_2O

54. Balance the following equation:

___$NH_4OH(l)$ + ___$KAl(SO_4)_2 \cdot 12H_2O$ $\rightarrow$ ___$Al(OH)_3(s)$ + ___$(NH_4)_2SO_4(aq)$ + ___$KOH(aq)$ + ___$H_2O(l)$

55. Balance the following equation:

___$Fe(s)$ + ___$HC_2H_3O_2(aq)$ $\rightarrow$ ___$Fe(C_2H_3O_2)_3(aq)$ + ___$H_2(g)$

56. Complete the following reactions (making sure they are balanced):

    a. HNO_3 + _____ $\rightarrow$ H_2O + KNO_3

    b.   _____ + Na_3PO_4 $\rightarrow$ $Ca_3(PO_4)_2$ + NaCl

    c. $Mg(OH)_2$ + HCl $\rightarrow$ $MgCl_2$ + _____

    d.   _____ + Cl_2 $\rightarrow$ NaCl   + Br_2

57. Balance the following equations:

    a.   _____ Ca + _____ C + _____ O_2 $\rightarrow$ _____ $CaCO_3$

    b.   _____ FeS + _____ O_2 $\rightarrow$ _____ Fe_2O_3 + _____ SO_2

    c.   _____ HNO_2 $\rightarrow$ _____ NO_2 + _____ H_2O + _____ NO

    d.   _____ PCl_5 + _____ H_2O $\rightarrow$ _____ H_3PO_4 + _____ HCl

Section 3.10

58. How many grams of water vapor can be generated from the combustion of 18.74 g of ethanol?

$$C_2H_6O(g) + O_2(g) \rightarrow CO_2(g) + H_2O(g) \textbf{ (unbalanced)}$$

59. How many grams of sodium hydroxide are required to form 51.63 g of lead hydroxide?

$$Pb(NO_3)_2(aq) + NaOH(aq) \rightarrow Pb(OH)_2(s) + NaNO_3(aq) \textbf{ (unbalanced)}$$

60. How many grams of potassium iodide are necessary to completely react with 20.61 g of mercury (II) chloride?

$$HgCl_2(aq) + KI(aq) \rightarrow HgI_2 + KCl(aq) \textbf{ (unbalanced)}$$

61. How many grams of oxygen are necessary to completely react with 22.8 grams of methane, CH_4? (Please write the entire reaction.)

62. If, in the previous problem, only 25.9 grams of water vapor were formed, how much methane actually reacted with oxygen?

63. What mass of calcium carbonate ($CaCO_3$) would be formed if 248.6 g of carbon dioxide (CO_2) were exhaled into limewater, $Ca(OH)_2$? How many grams of calcium would be needed to form that amount of calcium carbonate? Assume 100% yield in each reaction.

64. The following reaction is used to form lead iodide crystals. What mass of crystal (PbI_2) could be formed from 1.0×10^3 g of lead (II) acetate [$Pb(C_2H_3O_2)_2$]?

$$Pb(C_2H_3O_2)_2(aq) + 2KI(aq) \rightarrow PbI_2(s) + 2KC_2H_3O_2(aq)$$

65. How many grams of precipitate (Hg_2Cl_2) would be formed from a solution containing 102.9 g of mercury ions that are reacted with chloride ions as follows?

$$2Hg^+(aq) + 2Cl^-(aq) \rightarrow Hg_2Cl_2(s)$$

66. You were hired by a laboratory to recycle 6 moles of silver ions. You were given 150 g of copper. How many grams of silver can you recover? Is this enough copper to recycle 6 moles of silver ions?

$$2Ag^+ + Cu \rightarrow 2Ag + Cu^{2+}$$

67. Fermentation converts sugar into ethanol and carbon dioxide. If you were to ferment a bushel of apples containing 235 g of sugar, what is the maximum amount of ethanol in grams that would be produced?

$$C_6H_{12}O_6 \rightarrow 2C_2H_6O + 2CO_2$$

68. The reaction between potassium chlorate and red phosphorus is highly exothermic and takes place when you strike a match on a matchbox. If you were to react 52.9 g of potassium chlorate ($KClO_3$) with red phosphorus, how many grams of tetraphosphorus decaoxide (P_4O_{10}) would be produced?

$$KClO_3(s) + P_4(s) \rightarrow P_4O_{10}(s) + KCl(s) \textbf{ (unbalanced)}$$

Section 3.11

69. A reaction combines 133.484 g of lead(II) nitrate with 45.010 g of sodium hydroxide (see problem 59).

   a. How much lead(II) hydroxide is formed?
   b. Which reactant is limiting? Which is in excess?
   c. How much of the excess reactant is left over?
   d. If the actual yield of lead(II) hydroxide were 80.02 g, what was the percent yield?

70. A reaction combines 64.81 grams of silver nitrate with 92.67 grams of potassium bromide.

$$AgNO_3(aq) + KBr(aq) \rightarrow AgBr(s) + KNO_3(aq)$$

   a. How much silver bromide is formed?
   b. Which reactant is limiting? Which is in excess?
   c. How much of the excess reactant is left over?
   d. If the actual yield of silver bromide were 14.77 g, what was the percent yield?

71. A reaction proceeds between 94.6 g of $KClO_3$ and 65.3 g of P_4 (see problem 68).

   a. How much potassium chloride is formed?
   b. Which reactant is limiting? Which is in excess?
   c. How much of the excess reactant is left over?
   d. If the actual yield of potassium chloride were 21.0 g, what was the percent yield?

72. DDT, an insecticide harmful to fish, birds, and humans, is produced by the following reaction:

$$2C_6H_5Cl + C_2HOCl_3 \rightarrow C_{14}H_9Cl_5 + H_2$$
$$\text{chlorobenzene} \quad \text{chloral} \qquad \text{DDT}$$

   In a government lab 1142 g of chlorobenzene were reacted with 485 g of chloral.

   a. How much DDT is formed?
   b. Which reactant is limiting? Which is in excess?
   c. How much of the excess reactant is left over?
   d. If the actual yield of DDT were 200.0 g, what was the percent yield?

Multiple Choice Questions

73. Which one of the following elements has been selected as the current atomic weight standard?

   A. O   B. C   C. H   D. Na

74. Bromine is composed of two isotopes. One of the isotopes, Br-xx.x, makes up 49.7% of the total, while the other, Br-78.9, makes up 50.3% of the total. Calculate the atomic mass of Br-xx.x.

   A. 80.0   B. 80.9   C. 89.7   D. 78.9

75. How many years would 1.0 mole of seconds make up? Do not consider leap years.

   A. 1.9×10^{16} B. 1.1×10^9 C. 3.0×10^4 D. 3.5×10^{17}

76. How many neutrons are present in 50 atoms of chlorine that have a 50-50 mixture of Cl-35 and Cl-37?

   A. 38   B. 1.7×10^3 C. 8×10^2 D. 9.5×10^2

77. Four beakers containing potassium nitrate dissolved in water are allowed to evaporate to dryness. Beakers 1 through 4 contain 2.3, 1.91, 5.985, and 0.52 g of dry potassium nitrate respectively. How many moles of potassium nitrate were recovered after the water evaporated?

   A. 0.106   B. 0.212   C. 0.500   D. 2.35

78. How many atoms of uranium (U) are present in 1 nanogram of uranium?

   A. 2.5×10^{20} B. 5.0×10^{10} C. 2.5×10^{12} D. 5.0×10^{30}

79. Water has a density of 1.00 g/cm³. How many moles of water are present in ½ cup of water? 4 cups = 1 qt.

    A. 2.0   B. 3.05   C. 0.65   D. 6.65

80. Calculate how many moles of electrons are found in 401.4 g of P^{3-}.

    A. 194   B. 402   C. 154   D. 233.2

81. 18.0 cm³ water (assume density 1.00 g/cm³) contains 1.0 moles of water molecules. If a beaker containing 0.9 cm³ of water is allowed to evaporate over a period of 24 hours, how many molecules of water evaporate per second? Assume a constant rate of evaporation.

    A. 7.0×10^{14} B. 1.5×10^{23} C. 3.0×10^{23} D. 3.5×10^{17}

82. Calculate the percent composition of $Na_2S_2O_3$.

    A. 29% Na, 41% S, 30% O   C. 49% Na, 34% S, 17% O
    B. 37% Na, 45% S, 23% O   D. 17% Na, 47% S, 36% O

83. Calculate the percent composition of hydrogen in sucrose ($C_{12}H_{22}O_{11}$).

    A. 6%   B. 10%   C. 33%   D. 20%

84. Two of the three forms of vitamin B6 are pyridoxine ($C_8H_{11}NO_2$) and pyridoxamine ($C_8H_{12}N_2O$). Calculate the respective percentage of nitrogen in each compound.

    A. 8.8%, 17%   B. 9.3%, 18%   C. 9.2%, 17%   D. 18%, 9.2%

85. Calculate the molecular formula of a compound that has 48.8% Cd, 20.8% C, 2.62 % H, 27.8 % O, and a molecular weight of 460.8 g/mol.

    A. $Cd_2C_8H_{12}O_8$ B. $CdC_4H_6O_4$ C. $CdCHO$ D. $Cd_2C_6H_6O_6$

86. Determine the molecular formula of a compound that contains 26.7% P, 12.1% N, 61.2% Cl, and a molecular weight of 580 g/mol.

    A. $(PNCl)_3$ B. $(PNCl_2)_5$ C. $(P_2NCl_2)_5$ D. $PNCl_2$

87. Calculate the empirical formula for a compound that contains 32.2% Ca and 67.8% N by mass.

    A. $Ca(N_3)_2$ B. CaN_2 C. CaN D. CaN_4

88. A 20 mg sample of C_xH_y is burned in oxygen to produce 60 mg of carbon dioxide and 32 mg of water. Calculate x and y by using these data.

    A. 2, 4   B. 3, 6   C. 1, 4   D. 3, 8

89. Find the empirical formula of the compound that contains 15.8% Al, 28.1% S, and 56.1% O.

    A. $Al_2(SO_4)_3$ B. $AlSO_2$ C. $AlSO$ D. Al_2SO_3

90. Find the identity of the element X in the following equation

    $$C_3H_6X_3 + 3X_2 \rightarrow 3CX_2 + 3H_2O$$

    A. O   B. H   C. Cl   D. Br

91. 505 grams of KOH are required to completely react with 4.50 moles of sulfuric acid. How many moles of products are produced?

    A. 9 moles   B. 4.50 moles   C. 13.5 moles   D. 5.50 moles

92. The proper set of coefficients for the following equation are

$$\_\_C_3H_6O_3 + \_\_O_2 \rightarrow \_\_CO_2 + \_\_H_2O$$

A. 1, 3, 3, 3   B. 2, 4, 3, 3   C. 1, 2, 3, 3   D. 1, 6, 6, 6

93. The proper set of coefficients for the following equation are

$$\_\_AgCl + \_\_HNO_3 \rightarrow \_\_AgNO_3 + \_\_HCl$$

A. 1, 1, 1, 1   B. 1, 2, 2, 1   C. 2, 3, 1, 1   D. 2, 2, 2, 2

94. The proper set of coefficients for the following equation are

$$\_\_C_3H_8 + \_\_F_2 \rightarrow \_\_C_3F_8 + \_\_HF$$

A. 1, 1, 1, 1   B. 1, 3, 1, 3   C. 2, 16, 2, 16   D. 1, 8, 1, 8

95. For every liter of sea water that evaporates, 3.7 g of magnesium hydroxide are produced. How many liters of sea water must evaporate to produce 5.00 moles of magnesium hydroxide?

A. 78.4   B. 50   C. 143   D. 18.5

96. A solution of copper sulfate is treated with zinc metal. How many grams of copper are produced if 2.9 g of zinc are consumed?

$$CuSO_4 + Zn \rightarrow ZnSO_4 + Cu$$

A. 2.9 g   B. 2.8 g   C. 5.7 g   D. 3.7 g

97. How many grams of carbon dioxide are produced from the burning of 1368 g of sucrose according to the following equation?

$$C_{12}H_{22}O_{11} + 12O_2 \rightarrow 12CO_2 + 11H_2O$$

A. 342 g   B. 176 g   C. 1056 g   D. 2111 g

98. How many grams of sulfur dioxide are produced when 90.0 g of thionyl chloride reacts with excess water according to the following equation?

$$SOCl_2 + H_2O \rightarrow 2HCl + SO_2$$

A. 96.8   B. 90.0   C. 24.2   D. 48.5

99. Calcium oxide is a basic oxide that is not very soluble in water solutions. Calcium oxide can react with carbon dioxide to form calcium carbonate (according to the equation below). Calcium carbonate is an insoluble salt that forms stalactites and stalagmites. How many moles of carbon dioxide are removed from water if a 400.0 lb stalagmite is formed?

$$CaO + CO_2 \rightarrow CaCO_3$$

A. 1813   B. 908   C. 4000   D. 2258

100. Calculate the number of grams of $TiOCl_2$ required to react with 134 g of carbon.

$$2TiOCl_2 + 2C \rightarrow 2Ti + CO_2 + CCl_4$$

A. 134   B. 1.51×10^3 C. 536   D. 67

101. Calculate the number of grams of methane (CH_4) required to react with 25.0 g of chlorine according to the following equation:

$$3CH_4 + 4Cl_2 \rightarrow 2CH_3Cl + CH_2Cl_2 + 4HCl$$

A. 33.3   B. 18.8   C. 2.11   D. 4.23

102. Identify the limiting reagent and calculate the number of grams left over in excess for the following equation:

$$Cr_2O_3 + 3CCl_4 \rightarrow 2CrCl_3 + 3COCl_2$$

The reaction began with 5.00 g of Cr_2O_3 and 12.0 g of CCl_4.

A. Limiting: CCl_4, 1.05 g Cr_2O_3 left over
B. Limiting: Cr_2O_3, 10.0 g CCl_4 left over

C. Limiting: CCl_4, 3.70 g Cr_2O_3 left over
D. Limiting: Cr_2O_3, 6.75 g CCl_4 left over

103. Identify the limiting reagent and the number of grams left over in excess if 1.9 grams each of phosgene and sodium hydroxide are combined in the following reaction:

$$COCl_2 + 2NaOH \rightarrow 2NaCl + H_2O + CO_2$$

A. Limiting: $COCl_2$, 0.588 g NaOH left over
B. Limiting: $COCl_2$, 0.4 g NaOH left over

C. Limiting: NaOH, 0.520 g $COCl_2$ left over
D. Limiting: NaOH, 0.355 g $COCl_2$ left over

Answers to Exercises

1. The average atomic mass is 10.81 amu. The element is boron.

2. The molar mass of sulfur is 32.06 amu.

3. The average atomic mass is 6.95 amu. The element is lithium.

4. The average atomic mass is 20.17 amu. The gas is neon.

5. The mass of the other isotope is 36.93 amu.

6. $^{20}Ne = 90.37\%$; $^{22}Ne = 9.37\%$

7. The abundance of silicon-28 is 92.21%. The atomic mass of silicon-29 is 29.01.

8. gallium-69 = 60.3%; gallium-71 = 39.7%

9. magnesium-24 = 79.13%; magnesium-25 = 10.13%; magnesium-26 = 10.74%

10. 47.88 amu. Yes, it agrees with the periodic table value for Ti

11. 4.98×10^{-22} moles of nitrogen; mass = 6.98×10^{-21} g

12. 3.06×10^{21} atoms of gold

13. 2.34×10^{25} M&M's®

14. 0.167 moles of sulfur; 1.01×10^{23} atoms in the sample

15. a. 2Hg/2I   c. 1Pb/1C/3O   e. 1Rb/3O/5H
    b. 1Li/1H   d. 3Ba/2As/8O   f. 2H/1Si/6F

16. 1.26 g Zn

17. 1.23×10^{26} amu of iron

18. a. 223.1 g/mole   c. 152.2 g/mole   e. 508.3 g/mole
    b. 35.0 g/mole   d. 345.1 g/mole   f. 729.8 g/mole

19. a. 169.46 g/mol   b. 275.87 g/mol

20. Mass = 1.28 g

21. 38.9 g

22. 127 mg Br_2

23. a. 3.6×10^{24} ions   b. 6.0×10^{22} ions   c. 1.5×10^{23} ions

24. a. 1.8×10^{24} ions   b. 3.7×10^{22} ions   c. 1.5×10^{23} ions

25. a. 6.4 g   b. 1100 g   c. 300 g

26. 474.4 g/mole

27. 0.1442 moles of cadmium bromide

28. 0.214 moles of calcium chloride

29. 14.4 g of $NaHCO_3$

30. 4.24×10^{-9} moles

31. 85 g of cobalt(III) acetate. There are 21 g of cobalt. There are 2.2×10^{23} atoms.

32. There are 4.6 mg of chlorine. There are 7.8×10^{19} atoms of chlorine.

33. a. bauxite = 137.99 g/mole   c. 6.99×10^{23} atoms of Al
    b. 31.3 g Al   d. 4.8×10^2 g of bauxite

34. a. 65.11% Cl   c. 52.73% Cl
    b. 51.78% Cl   d. 12.0% Cl

35. Mass percent of C = 69.72%, H = 11.70%, O = 18.58%

36. Mass percent of C = 21.89%, Fe = 16.96%, N = 25.53, K = 35.62%

37. Mass percent of Ba = 63.17%, S = 14.75%, O = 22.08%

38. Mass percent of K = 17.91%, Al = 12.36%, Si = 25.74%, O = 43.99%

39. a. Ag = 75.27%   b. Ag = 80.57%   c. Ag = 63.50%

40. Chlorophyll a = 894 g/mole

41. Percent composition of C = 74.03%, H = 8.70%, N = 17.27%

42. Empirical formulas can be: a, b, c, d, f, g, i

43. Empirical formula = molecular formula = CH_5N

44. Empirical formula = molecular formula = NH_2; molecular formula = N_2H_4

45. Empirical formula = C_3HCl; molecular formula = $C_{12}H_4Cl_4$

46. Empirical formula = molecular formula = $C_8H_8O_3$

47. Empirical formula = $MoOCl_3$

48. $CHCl_3$

49. a. N_2O c. CO   e. HCN
    b. CuO   d. $KClO_3$

50. a. 36 g H_2O
    b. 143 g AgCl + 85 g $NaNO_3$ = 228 g product
    c. 164 grams of sodium phosphate + 54 grams of water = 218 g product

51. X = aluminum

52. a. Yes   b. No

53. a. 2, 1, 1, 2   c. 1, 3, 3, 1   e. 1, 6, 3, 2
     b. 1, 1, 1, 4   d. 1, 2, 1, 4   f. 3, 1, 2, 1

54. $4NH_4OH(aq) + KAl(SO_4)_2 \cdot 12H_2O(s) \rightarrow Al(OH)_3(s) + 2(NH_4)_2SO_4(aq) + KOH(aq) + 12H_2O(l)$

55. $2Fe(s) + 6HC_2H_3O_2(aq) \rightarrow 2Fe(C_2H_3O_2)_3(aq) + 3H_2(g)$

56. a. $HNO_3 + KOH \rightarrow H_2O + KNO_3$
     b. $3CaCl_2 + 2Na_3PO_4 \rightarrow Ca_3(PO_4)_2 + 6NaCl$
     c. $Mg(OH)_2 + 2HCl \rightarrow MgCl_2 + 2H_2O$
     d. $2NaBr + Cl_2 \rightarrow 2NaCl + Br_2$

57. a. 2, 2, 3, 2   c. 2, 1, 1, 1,
     b. 4, 7, 2, 4   d. 1, 4, 1, 5

58. 21.99 g of water vapor

59. 17.12 g of NaOH

60. 25.20 g of KI

61. 91.2 g; $2O_2(g) + CH_4(g) \rightarrow CO_2(g) + H_2O(g)$

62. 0.719 moles or 11.5 grams

63. 565.3 g of $CaCO_3$; 226.4 g of Ca^{2+}

64. 1.4×10^3 g of PbI_2

65. 121.1 g of Hg_2Cl_2

66. 509. g of Ag recovered; no, you could recycle only 4.72 moles of Ag^+

67. 120. g of ethanol

68. 36.8 g of P_4O_{10}; (balanced equation: $10KClO_3(s) + 3P_4(s) \rightarrow 3P_4O_{10}(s) + 10KCl(s)$)

69. a. 97.214 g of lead(II) hydroxide is formed.
     b. Lead nitrate is limiting. Sodium hydroxide is in excess.
     c. 12.771 g of sodium hydroxide is left over.
     d. The yield was 82.31%.

70. a. 71.64 g of silver bromide is formed.
     b. Silver nitrate is limiting. Potassium bromide is in excess.
     c. 47.27 g of potassium bromide is left over.
     d. The yield was 20.62%

71. a. 57.5 g of KCl is formed.
     b. Potassium chlorate is limiting. Red phosphorus is in excess.
     c. 36.6 g of P_4 is left over.
     d. The yield was 36.5%.

72. a.   1166. g of DDT is formed.
     b.   Chloral is limiting. Chlorobenzene is in excess.
     c.   401 g of chlorobenzene is left over.
     d.   The yield was 17.1%.

73. B	74. B	75. A	76. D	77. A	78. C
79. D	80. D	81. D	82. A	83. A	84. C
85. A	86. B	87. A	88. D	89. A	90. A
91. C	92. A	93. A	94. D	95. A	96. B
97. D	98. D	99. A	100. B	101. D	102. A
103. B					

Chapter 4

Types of Chemical Reactions
and Solution Stoichiometry

In this review chapter we discuss, from both a quantitative and qualitative standpoint, what happens in the preparation of aqueous solutions.

4.1   Water, the Common Solvent

When you finish this section you will be able to:

- State why water acts as a common solvent.
- Draw the structure of water, including partial charges.
- Write equations for the dissolution of some ionic salts in water.

This section introduces you to the nature of interactions between atoms and electrons in the water molecule. Please pay special attention to the following ideas:

a. **Water is *not* a linear molecule**. It is bent at an angle of about $105°$.
b. Electrons are not evenly distributed around the atoms in water. Notice the position of the partial charges on the molecule shown in Figure 4.1 of your textbook. The molecule is *polar* because the charges are not distributed symmetrically.
c. Like dissolves like. The following classes of molecules, in general, are miscible:
   - polar and ionic
   - polar and polar
   - nonpolar and nonpolar

Ionic salts dissolve in water. Compounds that contain *only* carbon and hydrogen are nonpolar. Given that information, please try the following example.

Example 4.1 A   Will the Substances Mix?

Predict whether each pair of substances will mix. State why or why not.

    a. $NaNO_3$ and H_2O
    b. C_6H_{14} and H_2O
    c. I_2 and C_6H_{14}
    d. I_2 and H_2O

Solution

    a. Miscible. Sodium nitrate is ionic, and water is polar.
    b. Immiscible. Cyclohexane (C_6H_{14}) is nonpolar, while water is polar.
    c. Miscible. The iodine molecule is composed of two identical atoms. Therefore, the electrons are distributed symmetrically. Iodine is therefore almost completely nonpolar. Cyclohexane is also nonpolar.
    d. Immiscible. The iodine molecule is nonpolar, and water is polar.

The dissociation of simple ionic salts in water is often written as shown in the following equations:

$$NaI(s) \xrightarrow{H_2O(l)} \underset{\text{cation}}{Na^+(aq)} + \underset{\text{anion}}{I^-(aq)}$$

$$K_2Cr_2O_7(s) \xrightarrow{H_2O(l)} \underset{\text{cation}}{2K^+(aq)} + \underset{\text{anion}}{Cr_2O_7^{2-}(aq)}$$

$$Ba(OH)_2(s) \xrightarrow{H_2O(l)} \underset{\text{cation}}{Ba^{2+}(aq)} + \underset{\text{anion}}{2OH^-(aq)}$$

Example 4.1 B   Practice with Equations

Complete each of the following dissociation equations:

    a. $CaCl_2(s) \xrightarrow{H_2O(l)}$
    b. $Fe(NO_3)_3(s) \xrightarrow{H_2O(l)}$
    c. $KBr(s) \xrightarrow{H_2O(l)}$
    d. $(NH_4)_2Cr_2O_7(s) \xrightarrow{H_2O(l)}$

Solution

    a. $CaCl_2(s) \xrightarrow{H_2O(l)} Ca^{2+}(aq) + 2Cl^-(aq)$
    b. $Fe(NO_3)_3(s) \xrightarrow{H_2O(l)} Fe^{3+}(aq) + 3NO_3^-(aq)$
    c. $KBr(s) \xrightarrow{H_2O(l)} K^+(aq) + Br^-(aq)$
    d. $(NH_4)_2Cr_2O_7(s) \xrightarrow{H_2O(l)} 2NH_4^+(aq) + Cr_2O_7^{2-}(aq)$

4.2   The Nature of Aqueous Solutions: Strong and Weak Electrolytes

When you finish this section, you will be able to classify many substances as strong, weak, or nonelectrolytes.

In this section your book introduces **solute** and **solvent**. These terms are discussed in more detail in Chapter 11, but for now we will just define solute as the substance being dissolved and solvent as the dissolving medium. An **aqueous solution** means that **water is the solvent**.

Figure 4.4 in your textbook shows the effect of strong, weak, and nonelectrolytes on the ability to pass a current (conductivity) in an aqueous solution.

Electrolyte	Conductivity	Degree of Dissociation	Examples
strong	high	total	strong acids such as HCl; many salts such as NaCl and $Sr(NO_3)_3$; strong bases such as NaOH, $Ba(OH)_2$, and other Group I and II hydroxides.
weak	low to moderate	partial	weak organic acids such as HCO_2H and $HC_2H_3O_2$; weak bases such as NH_3
non	none	close to zero	sugar, AgCl, Fe_2O_3

Example 4.2   Strong, Weak, or Nonelectrolyte

List whether each of the following is a strong, weak, or nonelectrolyte.

a. $HClO_4$
b. C_6H_{12}
c. LiOH

d. NH_3
e. $CaCl_2$
f. $HC_2H_3O_2$

Solution

a. strong ($HClO_4$ is a strong acid)
b. non (cyclohexane contains only carbon and hydrogen, and is, therefore, not soluble in water!)
c. strong
d. weak
e. strong
f. weak (acetic acid)

4.3   The Composition of Solutions

When you finish this section you will be able to:
*   Determine the molarity of a solution.
*   Calculate the molarity of each ion in a solution.
*   Determine the mass and/or volume of reagents necessary to prepare a solution of a given molarity.
*   Solve problems related to dilution.
*   Solve problems involving parts per million.

We deal here with the *preparation of solutions*. **Molarity (M) is defined as moles of solute per liter of solution**.

$$M = \frac{\text{moles of solute}}{\text{liter of solution}} .$$

Keep in mind that $\dfrac{\text{moles}}{\text{liter}} = \dfrac{\text{millimoles}}{\text{milliliter}} = \dfrac{\text{micromoles}}{\text{microliter}}$, but $\dfrac{\text{moles}}{\text{liter}}$ **DOES NOT EQUAL** $\dfrac{\text{millimoles}}{\text{liter}}$ **OR** $\dfrac{\text{moles}}{\text{microliter}}$. Be very careful with your units!

Example 4.3 A   Calculating Molarity

Calculate the molarity of a solution prepared by dissolving 11.85 g of solid $KMnO_4$ in enough water to make 750. mL of solution.

Strategy

Remember that our units are **moles per liter**. Therefore, you must first convert *grams of $KMnO_4$* to *moles of $KMnO_4$*, then divide by the volume. (Remember to convert volume from mL to L!)

Solution

The molar mass of $KMnO_4 = 158.04$ g/mole.

Method A

In this method, we can proceed in two steps:

1. Convert from grams of $KMnO_4$ to moles of $KMnO_4$.

    $$\text{moles } KMnO_4 \ \Big\| \ \frac{1 \text{ mol } KMnO_4}{158.04 \text{ g } KMnO_4} \times 11.85 \text{ g } KMnO_4 = 0.07498 \text{ mol } KMnO_4$$

2. Divide moles by volume (in liters!) to get molarity.

    $$M = \frac{\text{mol}}{\text{L}} = \frac{0.07498 \text{ mol } KMnO_4}{0.750 \text{ L solution}} = 0.09997 \ M \, KMnO_4 = \mathbf{0.100 \ } \textit{\textbf{M}} \textbf{ KMnO}_4$$

Method B

We can use dimensional analysis to solve the problem in one longer step:

$$\frac{\text{mol } KMnO_4}{\text{L solution}} \ \Big\| \ \frac{11.85 \text{ g } KMnO_4}{0.750 \text{ L soln}} \times \frac{1 \text{ mol } KMnO_4}{158.04 \text{ g } KMnO_4} = \mathbf{0.100 \ } \textit{\textbf{M}} \textbf{ KMnO}_4$$

We will often use dimensional analysis in this study guide. However, it is never the ONLY correct method. Let's change the problem around a bit. Instead of using the mass of the solute to calculate the solution molarity, let's begin with **molarity** to calculate **mass**.

Example 4.3 B   Mass from Molarity

Calculate the mass of NaCl needed to prepare 175. mL of a 0.500 M NaCl solution.

Strategy

From a *dimensional analysis* point of view, the question we have to answer is, "how can we go from molarity (mol/L) to mass(g)?" Note that this means that somewhere along the line, we will have to *cancel volume*.

Solution

$$g\ NaCl \parallel \frac{0.500\ mol\ NaCl}{L\ soln} \times \underset{\underset{\text{convert to grams}}{\uparrow}}{\frac{58.44\ g\ NaCl}{1\ mol\ NaCl}} \times \underset{\underset{\text{cancel volume}}{\uparrow}}{0.175\ L\ soln} = \mathbf{5.11\ g\ NaCl}$$

Our values for molarity and volume had 3 significant figures, so we rounded off accordingly.

Whether you want **mass, molarity,** or **volume**, the essential problem-solving strategy is the same—you can use dimensional analysis to get your units in the proper place. Try the next problem.

Example 4.3 C   Volume from Molarity

How many mL of solution are necessary if we are to have a 2.48 *M* NaOH solution that contains 31.52 g of the dissolved solid?

Strategy

The units must cancel to give mL. Starting off by inverting molarity will put volume in the numerator. Remember to convert L to mL!

Solution

$$mL\ solution \parallel \frac{1.00\ L\ soln}{2.48\ mol\ NaOH} \times \underset{\underset{\substack{\text{convert from} \\ \text{L to mL}}}{\uparrow}}{\frac{1000\ mL}{L}} \times \underset{\underset{\substack{\text{cancel moles} \\ \text{NaOH}}}{\uparrow}}{\frac{1\ mol\ NaOH}{40.00\ g\ NaOH}} \times \underset{\underset{\substack{\text{cancel grams} \\ \text{NaOH}}}{\uparrow}}{31.52\ g\ NaOH}$$

$$= \mathbf{318.\ mL\ solution}$$

Does the Answer Make Sense?

The amount of NaOH, 31.52 g, is about 3/4 of a mole. If the solution is 2.48 *M*, you have about 1/3 as much NaOH as you need to make 1 L, or enough to make about 300 mL, so the answer seems to make sense.

In the previous problems we calculated the molarity of the solute. However, we neglected to take into account the fact that each of the solutes, $KMnO_4$, NaCl, and NaOH is a **strong electrolyte and completely dissociates** in aqueous solution (recall Section 4.2). For example,

$$KMnO_4 \xrightarrow{H_2O(l)} K^+(aq) + MnO_4^-(aq)$$

This means that while it is generally acceptable to discuss your solution concentration as "molarity of $KMnO_4$," it is more correct chemically to discuss "molarity of K^+ ions and molarity of MnO_4^- ions."

A solution that is 0.85 M in $KMnO_4$ is really 0.85 M in K^+ ion and 0.85 M in MnO_4^- ion because $KMnO_4$ completely dissociates, and the dissociation is a 1 to 1 to 1 ratio; that is, *one mole of $KMnO_4$ dissociates into one mole of K^+ ion and one mole of MnO_4^- ion.*

Keeping the stoichiometry of the dissociation equation in mind, try the next problem.

Example 4.3 D   Molarity of Ions in Solution

Calculate the molarity of all the ions in each of the following solutions.

      a.   0.25 M $Ca(OCl_2)$ b.   2 M $CrCl_3$

Helpful Hint

Always write out the dissociation equation so your "ion-to-solute" mole ratio will be clear.

Solution

    a. $Ca(OCl)_2(s) \xrightarrow{\text{H}_2\text{O}(l)} Ca^{2+}(aq) + 2OCl^-(aq)$
          molarity of Ca^{2+} = molarity of $Ca(OCl)_2$ = **0.25 M**
          molarity of OCl^- = twice the molarity of $Ca(OCl)_2$ = **0.50 M**

    b. $CrCl_3(s) \xrightarrow{\text{H}_2\text{O}(l)} Cr^{3+}(aq) + 3Cl^-(aq)$
          molarity of Cr^{3+} = molarity of $CrCl_3$ = **2 M**
          molarity of Cl^- = three times the molarity of $CrCl_3$ = **6 M**

Example 4.3 E   Molarity of Ions in Solution

Determine the molarity of Cl^- ion in a solution prepared by dissolving 9.82 g of $CuCl_2$ in enough water to make 600. mL of solution.

Strategy

Let's finish a strategy that we began to discuss in the last problem. One process to find the ***ion concentration*** (molarity of an ion) of a strong electrolyte is to:

   1.   Calculate the solute concentration (molarity of the solute).
   2.   Determine the ion-to-solute mole ratio by writing the dissociation equation.
   3.   Use that mole ratio, along with the solute concentration, to calculate the ion concentration.

Solution

   1.   **Solute concentration:**

$$\frac{\text{mol } CuCl_2}{\text{L soln}} \;\bigg|\bigg|\; \frac{1 \text{ mol } CuCl_2}{134.45 \text{ g } CuCl_2} \times 9.82 \text{ g } CuCl_2 \times \frac{1}{0.600 \text{ L soln}} = \textbf{0.1217 } \boldsymbol{M} \textbf{ CuCl}_2$$

We retain all our figures because this is an *intermediate* calculation.

2. **Ion to solute ratio:**

$$CuCl_2(s) \xrightarrow{H_2O(l)} Cu^{2+}(aq) + 2Cl^-(aq)$$

Therefore, the ratio $\dfrac{\text{ion}}{\text{solute}} = \dfrac{2 \text{ mol } Cl^-}{1 \text{ mol } CuCl_2}$.

3. **The final molarity of Cl^- in solution (also expressed as "$[Cl^-]$"):**

$$[Cl^-] = \frac{0.1217 \text{ mol } CuCl_2}{\text{L soln}} \times \frac{2 \text{ moles } Cl^-}{1 \text{ mole } CuCl_2} = \textbf{0.243 } \textbf{\textit{M}}$$

We rounded off because we were at the end of the problem. Three significant figures is correct because both mass and volume were given to three figures. Note that if we had rounded off earlier we would have calculated that $[Cl^-] = 0.122 \ M \times 2 = 0.244 \ M$, which is not strictly correct.

ALTERNATIVELY, we can use dimensional analysis to solve the entire problem with one equation.

$$\frac{\text{mol } Cl^-}{\text{L soln}} \ \| \ \frac{2 \text{ mol } Cl^-}{1 \text{ mol } CuCl_2} \times \underset{\substack{\uparrow \\ \text{from dissociation} \\ \text{equation}}}{} \frac{1 \text{ mol } CuCl_2}{134.45 \text{ g } CuCl_2} \times \underset{\substack{\uparrow \\ \text{molar mass}}}{} 9.82 \text{ g } CuCl_2 \times \underset{\substack{\uparrow \\ \text{given}}}{} \frac{1}{0.600 \text{ L soln}} \underset{\substack{\uparrow \\ \text{given}}}{} = \textbf{0.243 } \textbf{\textit{M}}$$

Example 4.3 F   Practice With Ion Concentration

Determine the molarity of Fe^{3+} ions and SO_4^{2-} ions in a solution prepared by dissolving 48.05 g of $Fe_2(SO_4)_3$ in enough water to make 800. mL of solution.

Solution

The dissociation equation is:

$$Fe_2(SO_4)_3(s) \xrightarrow{H_2O(l)} 2Fe^{3+}(aq) + 3SO_4^{2-}(aq)$$

So the mole ratio of $\dfrac{\text{ion}}{\text{solute}} = \dfrac{2 \text{ mol } Fe^{3+}}{1 \text{ mol } Fe_2(SO_4)_3}$ and $\dfrac{3 \text{ mol } SO_4^{2-}}{1 \text{ mol } Fe_2(SO_4)_3}$

Let's work with $[Fe^{3+}]$.

$$\frac{\text{mol } Fe^{3+}}{\text{L soln}} \ \| \ \frac{2 \text{ mol } Fe^{3+}}{1 \text{ mol } Fe_2(SO_4)_3} \times \frac{1 \text{ mol } Fe_2(SO_4)_3}{399.9 \text{ g } Fe_2(SO_4)_3} \times 48.05 \text{ g } Fe_2(SO_4)_3 \times \frac{1}{0.800 \text{ L soln}}$$

$$= \textbf{0.300 } \textbf{\textit{M}} = [Fe^{3+}]$$

Looking at the **mole ratios**, can you tell that the ratio of $\dfrac{[SO_4^{2-}]}{[Fe^{3+}]}$ must equal $\dfrac{3 \text{ moles}}{2 \text{ moles}}$?

Therefore, since the volume is constant, $[SO_4^{2-}] = \dfrac{3}{2}[Fe^{3+}] = \dfrac{3}{2}(0.300 \ M)$

$$[SO_4^{2-}] = \textbf{0.450 } \textbf{\textit{M}}$$

An important part of your chemistry experience is to be able to *prepare dilute solutions from more concentrated ("stock") solutions.* Your textbook points out that the most important idea in diluting solutions is that

moles of solute after dilution = moles of solute before dilution

$$\text{If molarity} = \frac{\text{moles of solute}}{\text{liter of solution}}, \text{ then}$$

$$\text{moles of solute} = \frac{\text{moles of solute}}{\text{liter of solution}} \times \text{liters of solution} = M \times V.$$

If the **moles of solute remain identical** before and after dilution (only the amount of water changes), then

$$M_iV_i = M_fV_f$$

where M_i = molarity of concentrated (initial) solution
   V_i = volume of concentrated solution that you add to water to dilute . . . this will often be your "unknown volume"
   M_f = molarity of dilute (final) solution
   V_f = total volume of your dilute solution

Try the following introductory example.

Example 4.3 G   Preparation of a Dilute Solution

What volume of 12 M hydrochloric acid must be used to prepare 600 mL of a 0.30 M HCl solution?

Strategy

Be aware that no matter how much you dilute your acid, **the number of moles of acid in the solution will remain the same**. Your molarity will change, but not the total number of moles. $M_iV_i = M_fV_f$ is your best bet in solving these types of problems.

Solution

Let M_i = 12 M HCl V_i = unknown
     M_f = 0.30 M HCl V_f = 600. mL (or 0.600 L). Because the units of molarity
                                           cancel, you may use any volume unit that you want.

$$12\,M \times V_i = 0.30\,M\,(600.\text{ mL}) = \frac{0.30\text{ M }(600.\text{ mL})}{12\,M} = V_i = \textbf{15 mL of 12 } \textit{\textbf{M}} \textbf{ HCl}$$

Double Check

$$\frac{12\text{ mol}}{L}\,(0.015\text{ L}) = \frac{0.30\text{ mol}}{L}\,(0.600\text{ L})$$

$$0.18\text{ mol} = 0.18\text{ mol}$$

The number of moles, 0.18 mol, is the same on each side.

Example 4.3 H   More Practice Preparing Dilute Solutions

What volume of 9.0 M sodium hydroxide must be used to prepare 1.2 L of a 1.0 M NaOH solution?

Solution

We proceed as before.

Let M_i = 9.0 M NaOH V_i = unknown
M_f = 1.0 M NaOH V_f = 1.2 L

$$9.0\ M \times V_i = 1.0\ M\,(1.2\ L) = \frac{1.0\ M\,(1.2\ L)}{9.0\ M} = V_i = \textbf{0.13 L of 9.0 } M \textbf{ NaOH}$$

Upon checking, using the non-rounded value of 0.1333 L, the number of moles on each side is identical.

Although not discussed in Chapter 4 of your textbook, some of the challenge problems in your text use the term **parts per million**.

$$1 \text{ part per million of "X" (ppm)} = \frac{1 \text{ part X}}{1 \times 10^6 \text{ parts solution}},$$

$$= \frac{1\text{ g X}}{1 \times 10^6 \text{ g soln}} = \frac{1\ \mu\text{g X}}{1 \text{ g soln}}$$

If the solution is water (whose density = 1.00 g/mL),

$$1 \text{ ppm X} = \frac{\textbf{1}\ \mu\textbf{g X}}{\textbf{1 mL soln}} = \frac{\textbf{1 mg X}}{\textbf{1 L soln}}$$

Note that this differs from molarity in that the units are *mass per volume*, not moles per volume. To set the stage for your work with the challenge problems, consider the following problem.

Example 4.3 I   Parts per Million

An aqueous solution with a total volume of 750 mL contains 14.38 mg of Cu^{2+}. What is the concentration of Cu^{2+} in parts per million?

Helpful Hint

The use of ppm DOES NOT necessarily require you to calculate moles on your way to your answer. The dimensional analysis unit of choice is **mg/L**.

Solution

$$\frac{14.38 \text{ mg Cu}^{2+}}{0.750 \text{ L soln}} = \textbf{19.2 ppm Cu}^{2+}$$

The next problem requires you to deal with moles.

Example 4.3 J   Molarity to Parts per Million

A solution is 3×10^{-7} M in manganese(VII) ion. What is the Mn^{7+} concentration in ppm?

Solution

You want $\dfrac{\text{mg Mn}^{7+}}{\text{L soln}}$, and you have $\dfrac{\text{moles Mn}^{7+}}{\text{L soln}}$. Dimensional analysis is a good approach.

$$\frac{\text{mg Mn}^{7+}}{\text{L soln}} \quad \| \quad \frac{3\times10^{-7} \text{ mol Mn}^{7+}}{\text{L soln}} \times \frac{54.94 \text{ g Mn}^{7+}}{1 \text{ mol Mn}^{7+}} \times \frac{1000 \text{ mg Mn}^{7+}}{1 \text{ g Mn}^{7+}}$$

$$= \frac{0.0164 \text{ mg Mn}^{7+}}{\text{L soln}}, \text{ or } \textbf{0.02 ppm Mn}^{7+} \text{ (rounded off)}$$

4.4   Types of Chemical Reactions

This section points out that reactions are divided into precipitation reactions, acid-base reactions, and oxidation-reduction reactions. The remainder of the chapter considers reactions that fall into these classes.

4.5   Precipitation Reactions

When you finish this section you will be able to predict the products of many reactions that occur in solution.

The major ideas in this section are that:

1. Many salts dissociate into ions in aqueous solution.
2. If a solid forms from a combination of selected ions in solution, the solid must contain an anion part and cation part, and the net charge on the solid must be zero.
3. There are some simple solubility rules you can use that can help you predict the products of reactions in aqueous solutions.

Table 4.1 in your textbook lists solubility rules for salts in water. The information is important enough to reprint below.

Simple Rules for the Solubility of Salts in Water:

1. Most nitrate (NO_3^-) salts are soluble.
2. Most salts containing the alkali metal ions (Li^+, Na^+, K^+, Cs^+, Rb^+) and the ammonium ion (NH_4^+) are soluble.
3. Most chloride, bromide, and iodide salts are soluble. Notable exceptions are salts containing the ions Ag^+, Pb^+, and Hg_2^{2+}.
4. Most sulfate salts are soluble. Notable exceptions are $BaSO_4$, $PbSO_4$, $HgSO_4$, and $CaSO_4$.
5. Most hydroxides are only slightly soluble. The important soluble hydroxides are NaOH and KOH. The compounds $Ba(OH)_2$, $Sr(OH)_2$, and $Ca(OH)_2$ are marginally soluble.
6. Most sulfide (S^{2-}), carbonate (CO_3^{2-}), chromate (CrO_4^{2-}), and phosphate (PO_4^{3-}) salts are only slightly soluble, except for those containing the cations in rule 2.

If two soluble substances (call them AX and BZ) are combined, you can assume that the products will be AZ and BX.

$$AX(aq) + BZ(aq) \rightarrow AZ + BX$$

Your goal is to determine, based on your knowledge of solubility rules, whether AZ or BX will form a solid (precipitate). Let's look at the following reaction.

$$AgNO_3(aq) + Na_2S(aq) \rightarrow \text{products}$$

The reactants are electrolytes that will dissociate to form the ions

$$Ag^+(aq) + NO_3^-(aq) + 2Na^+(aq) + S^{2-}(aq)$$

If a solid forms, and if it is to have zero charge, it can be either **Ag₂S** or **NaNO₃**. According to **solubility rule #1**, $NaNO_3$ is soluble. According to **solubility rule #6**, Ag_2S is insoluble and will therefore precipitate. The correct (balanced) reaction, therefore, is

$$2AgNO_3(aq) + Na_2S(aq) \rightarrow Ag_2S(s) + 2NaNO_3(aq)$$

Example 4.5   *Predicting Precipitates*

Complete and balance the following reactions, determining, in each case, if a precipitate is formed.

   a. $KCl(aq) + Pb(NO_3)_2(aq) \rightarrow$
   b. $AgNO_3(aq) + MgBr_2(aq) \rightarrow$
   c. $Ca(OH)_2(aq) + FeCl_3(aq) \rightarrow$
   d. $NaOH(aq) + HCl(aq) \rightarrow$

Solution

   a. $2KCl(aq) + Pb(NO_3)_2(aq) \rightarrow$ **PbCl₂(s)** $+ 2KNO_3(aq)$
   b. $2AgNO_3(aq) + MgBr_2(aq) \rightarrow$ **2AgBr(s)** $+ Mg(NO_3)_2(aq)$
   c. $3Ca(OH)_2(aq) + 2FeCl_3(aq) \rightarrow$ **2Fe(OH)₃(s)** $+ 3CaCl_2(aq)$
   d. $NaOH(aq) + HCl(aq) \rightarrow H_2O(l) + NaCl(aq)$

↑
water is the
solvent, hence "(l)"

4.6   *Describing Reactions in Solution*

When you finish this section you will be able to write formula, ionic, and net ionic equations to describe reactions in solution.

This section discusses the three kinds of equations that are used to describe reactions in aqueous solution. Specific definitions for the **formula**, **complete ionic**, and **net ionic equations** are given in your textbook. Let's look at how the aqueous reaction of *silver nitrate* with *sodium sulfide* can be expressed with each type of equation.

   a.   **Formula**: This gives the overall reaction. While it does give information on stoichiometry it gives no information on whether or not compounds really exist as ions in solution. Formula form:

$$2AgNO_3(aq) + Na_2S(aq) \rightarrow Ag_2S(s) + 2NaNO_3(aq)$$

   b.   **Complete Ionic**: This gives the equation including all ions in solution. Because **all** compounds and ions are present, some information may be redundant. Complete ionic form:

$$2Ag^+(aq) + 2NO_3^-(aq) + 2Na^+(aq) + S^{2-}(aq) \rightarrow Ag_2S(s) + 2Na^+(aq) + 2NO_3^-(aq)$$

   c.   **Net Ionic**: This only gives information on those species that undergo a chemical change. Ions that appear in the same form on both sides of the complete ionic equation are called **spectator ions** and are not included in the net ionic equation. In our sample equation, Na^+ and NO_3^- are unaltered during the reactions. They are, therefore, omitted from the net ionic equation. Net ionic form:

$$2Ag^+(aq) + S^{2-}(aq) \rightarrow Ag_2S(s)$$

The **formula** and **net ionic** forms of equations are the most commonly used. The **complete ionic** form helps us determine the net ionic form.

Example 4.6   Formula, Complete Ionic, and Net Ionic Equations

Write the formula, complete ionic, and net ionic forms for each of the following equations.

    a.   Aqueous nickel(II) chloride reacts with aqueous sodium hydroxide to give a nickel(II) hydroxide precipitate and aqueous sodium chloride.

    b.   Solid potassium metal reacts with water to give aqueous potassium hydroxide and hydrogen gas.

    c.   Aqueous sodium hydroxide reacts with aqueous phosphoric acid to give water and aqueous sodium phosphate.

Keep in Mind

Remember to <u>balance</u> the **formula** equation before you write the **complete and net ionic** forms.

Solution

a.   **formula**: $NiCl_2(aq) + 2NaOH(aq) \rightarrow Ni(OH)_2(s) + 2NaCl(aq)$
    complete ionic: $Ni^{2+}(aq) + 2Cl^-(aq) + 2Na^+(aq) + 2OH^-(aq)$
$$\rightarrow Ni(OH)_2(s) + 2Na^+(aq) + 2Cl^-(aq)$$
    net ionic: $Ni^{2+}(aq) + 2OH^-(aq) \rightarrow Ni(OH)_2(s)$

b.   **formula**: $2K(s) + 2H_2O(l) \rightarrow 2KOH(aq) + H_2(g)$
    complete ionic: $2K(s) + 2H_2O(l) \rightarrow 2K^+(aq) + 2OH^-(aq) + H_2(g)$
    net ionic: $2K(s) + 2H_2O(l) \rightarrow 2K^+(aq) + 2OH^-(aq) + H_2(g)$

Note that in <u>part b</u>, **every** reactant undergoes some chemical change. Therefore, the complete and net ionic equations are the same.

c.   **formula**: $3NaOH(aq) + H_3PO_4(aq) \rightarrow 3H_2O(l) + Na_3PO_4(aq)$
    complete ionic: $3Na^+(aq) + 3OH^-(aq) + 3H^+(aq) + PO_4^{3-}(aq)$
$$\rightarrow 3H_2O(l) + 3Na^+(aq) + PO_4^{3-}(aq)$$
    net ionic: $3OH^-(aq) + 3H^+(aq) \rightarrow 3H_2O(l)$

4.7   Stoichiometry of Precipitation Reactions

When you finish this section you will:

- have your first exposure to "railroad" problems in chemistry, and
- be able to solve a variety of problems involving the formation of precipitates.

The material in this section is very important because it combines most of the previous chemical ideas that you have learned up until now. Solving problems involving precipitates from solution makes use of *molarity, solubility rules, balancing equations, and limiting reactant calculations*.

The problems can be quite "wordy," but wordy problems are not necessarily hard. They are just wordy. Each sentence contains some chemical information, and it is useful to jot down information as it appears. Keep in mind that although no two examples are exactly alike, the approach to solving the problems will be similar.

Your textbook suggests **SIX STEPS** to solving solution problems. Let's use these ideas in the following example.

Example 4.7 A   An Introduction to Problems Based on Precipitation Reactions

Calculate the mass of Ag_2S produced when 125. mL of 0.200 M $AgNO_3$ is added to excess Na_2S solution. ("Excess" Na_2S means that you have more than you need—$AgNO_3$ is the limiting reactant.)

Solution

Before we do any calculations, we first have to determine what is happening in the solution. This is best done by writing a **balanced formula equation**.

The problem says that $AgNO_3(aq)$ is being added to $Na_2S(aq)$. According to our solubility rules (Section 4.5), insoluble silver sulfide will be formed.

$$2AgNO_3(s) + Na_2S(aq) \rightarrow Ag_2S(s) + 2NaNO_3(aq)$$

The critical point here is that the *mole ratio* of Ag_2S to $AgNO_3$ is

$$\frac{1 \text{ mol Ag}_2\text{S}}{2 \text{ mol AgNO}_3}$$

Now that we know what is going on in solution, we need to know **how many moles of reactant ($AgNO_3$)** we have. You learned how to do such calculations in <u>Section 4.3 in your textbook and this study guide</u>.

$$\text{moles AgNO}_3 \ \left\| \ \frac{0.200 \text{ mol AgNO}_3}{\text{L solution}} \ \times \ 0.125 \text{ L solution} = \mathbf{0.0250 \text{ mol AgNO}_3}\right.$$

How many moles of product (Ag_2S) can be formed from this many moles of reactant? Given the mole ratio of 2:1 that we determined earlier,

$$\text{moles Ag}_2\text{S} \ \left\| \ 0.0250 \text{ mol AgNO}_3 \ \times \ \frac{1 \text{ mol Ag}_2\text{S}}{2 \text{ mol AgNO}_3} = \mathbf{0.0125 \text{ mol Ag}_2\text{S formed}}\right.$$

To finish up, we need to convert **moles of Ag_2S to grams of Ag_2S**.

$$\text{g Ag}_2\text{S} \ \left\| \ 0.0125 \text{ mol Ag}_2\text{S} \ \times \ \frac{247.9 \text{ g Ag}_2\text{S}}{1 \text{ mol Ag}_2\text{S}} = \mathbf{3.10 \text{ g Ag}_2\text{S formed}}\right.$$

Note that early on, once we wrote down the correct equation, we could have solved the problem with one long dimensional analysis equation:

$$\text{g Ag}_2\text{S} \ \left\| \ \frac{247.9 \text{ g Ag}_2\text{S}}{1 \text{ mol Ag}_2\text{S}} \ \times \ \frac{1 \text{ mol Ag}_2\text{S}}{2 \text{ mol AgNO}_3} \ \times \ \frac{0.200 \text{ mol AgNO}_3}{\text{L solution}} \ \times \ 0.125 \text{ L solution}\right.$$

$$= \mathbf{3.10 \text{ g Ag}_2\text{S formed}}$$

However you choose to solve such problems, a stepwise approach is helpful.

Example 4.7 B Practice With Precipitation Problems

What mass of $Fe(OH)_3$ is produced when 35. mL of a 0.250 M $Fe(NO_3)_3$ solution is mixed with 55 mL of a 0.180 M KOH solution?

Helpful Hint

Remember your systematic procedure for solving precipitation problems. The complication here is that by combining different amounts of reactant, you have a **limiting reactant** problem. Also, remember to **properly balance your chemical equation**!

Solution

The reaction of interest is:

$$Fe(NO_3)_3(aq) + 3KOH(aq) \rightarrow Fe(OH)_3 + 3KNO_3(aq)$$

Using a stepwise approach,

moles $Fe(NO_3)_3$ before reaction $\|$ $\dfrac{0.250 \text{ mol } Fe(NO_3)_3}{\text{L solution}} \times 0.035 \text{ L soln}$

$= \textbf{0.00875 moles } Fe(NO_3)_3$

moles KOH before reaction $\|$ $\dfrac{0.180 \text{ mol KOH}}{\text{L solution}} \times 0.055 \text{ L soln} = \textbf{0.00990 moles KOH}$

According to the balanced equation, 0.00875 moles of $\textbf{Fe(NO}_3)_3$ can yield **0.00875 moles of Fe(OH)₃** (1:1 mole ratio), and 0.00990 moles of **KOH** can yield **0.00330 moles of Fe(OH)₃** (3:1 mole ratio).

Therefore, **KOH** is the **limiting reactant** (see Section 3.10), and **0.00330 moles of Fe(OH)₃(s) will be produced**. Converting from moles of $Fe(OH)_3$ to grams of $Fe(OH)_3$ (F.W. = 106.9 g/mol) gives **0.35 g Fe(OH₃) produced**.

Although not discussed in Chapter 4 of your textbook, the following gravimetric analysis example will help you solve some of the additional problems at the end of the chapter.

Example 4.7 C   Gravimetric Analysis

An ore sample is to be analyzed for sulfur. As part of the procedure, the ore is dissolved, and the sulfur is converted to sulfate ion, SO_4^{2-}. Barium nitrate is added, which causes the sulfate to precipitate out as $BaSO_4$.

The original sample had a mass of 3.187 g. The dried $BaSO_4$ has a mass of 2.005 g. What is the percent of sulfur in the original ore?

Solution

All the sulfur in the original ore sample is (we assume) still present but now in the $BaSO_4$ instead of the ore. Therefore, the key question here is, "How much sulfur is in 2.005 g of the dried $BaSO_4$?"

g S in $BaSO_4$ $\|$ $2.005 \text{ g } BaSO_4 \times \dfrac{1 \text{ mol } BaSO_4}{233.36 \text{ g } BaSO_4} \times \dfrac{1 \text{ mol S}}{1 \text{ mol } BaSO_4} \times \dfrac{32.07 \text{ g S}}{1 \text{ mol S}}$

$= \textbf{0.276 g S}$

$\% S = \dfrac{0.276 \text{ g S}}{3.187 \text{ g in the ore}} \times 100\% = \textbf{8.65\% S in the ore}$

4.8   Acid-Base Reactions

When you finish this section you will be able to solve a variety of problems related to acid-base neutralizations.

The key to solving acid-base problems is to know that they **require the same strategy** as most of the other types of problems in this chapter. **Writing down a balanced chemical equation is always your first, and most important, step.**

AN ACID IS A PROTON DONOR.
A BASE IS A PROTON ACCEPTOR.

Some STRONG ACIDS   Some STRONG BASES
HCl   NaOH
HNO_3 KOH
H_2SO_4 NH_2^-
$HClO_4$

You may assume that the acid-base reactions used in this section go to completion. The steps for solving acid-base problems are given in a <u>Problem Solving Strategy Box in your textbook</u>. Let's solve the following problem together to show how these steps are used.

Example 4.8 A   Neutralization of a Strong Acid

How many mL of a 0.800 *M* NaOH solution is needed to just neutralize 40.00 mL of a 0.600 *M* HCl solution?

Solution

Neutralization is often used in acid-base chemistry. Neutralization of an acid implies stoichiometric addition, i.e., "just enough" of a strong base so that no acid remains. The analogous definition applies to the neutralization of a base by a strong acid.

1. **List the species** present in solution before reaction.

    H^+, Cl^- Na^+, OH^- and H_2O
    from HCl(*aq*)   from NaOH(*aq*)

2. **Write the balanced net ionic reaction**.

    $$H^+(aq) + OH^- \rightarrow H_2O(l)$$

3. Find the **number of moles of acid** we need **to neutralize**.

    moles H^+ $\left\|$ $\dfrac{0.600 \text{ moles } H^+}{1 \text{ L soln}}$ $\times$ 0.04000 L soln $=$ **0.0240 moles H^+**
     change mL to L

4. Because the stoichiometry of the reaction is 1 mole of base to 1 mole of acid, **0.0240 moles OH^- are required to neutralize the acid**.

5. **Determine the volume of OH^-** (from NaOH) **needed** to give that many moles.

    L NaOH soln $\left\|$ $\dfrac{1 \text{ L NaOH soln}}{0.800 \text{ moles NaOH}}$ $\times$ 0.0240 moles NaOH $=$ 0.0300 L soln

    $=$ **30.0 mL NaOH solution**

You should <u>always check</u> to make sure the **moles of acid** equal the **moles of base**.

Example 4.13 in your textbook is really a limiting reactant problem like the one we did in Section 3.11 of this study guide and your textbook. Review that material, and remember, when doing limiting reactant problems, **convert everything to moles**.

The remainder of this section deals with **VOLUMETRIC ANALYSIS**. This kind of analysis uses **precisely measured amounts of liquid** to carry out an analysis. There are several new terms introduced such as **titration, buret, equivalence point, indicator**, and **endpoint**. When solving volumetric analysis problems, the same chemical rules apply as with most other acid-base problems:

- Write down the reaction.
- Convert to moles, and relate moles of acid to moles of base.
- Don't be frightened by "wordy" railroad problems.
- Ignore superfluous information.

See if you can do the next example on your own.

Example 4.8 B   Acid-Base Titration

You wish to determine the molarity of a solution of sodium hydroxide. To do this, you titrate a 25.00-mL aliquot of your sample, which has had 3 drops of phenolphthalein indicator added so that it is pink, with 0.1067 M HCl. The sample turns clear (indicating that the NaOH(aq) has been precisely neutralized by the HCl solution) after the addition of 42.95 mL of the HCl. Calculate the molarity of your NaOH solution.

Solution

1. The net ionic reaction is:

$$\mathbf{H^+}(aq) \ + \ \mathbf{OH^-}(aq) \ \rightarrow \ \mathbf{H_2O}(l)$$

2. The total number of moles of H^+ added to neutralize the NaOH equals:

$$\text{moles H}^+ \ \bigg\| \ \frac{0.1067 \text{ mol H}^+}{\text{L soln}} \ \times \ 0.04295 \text{ L soln} \ = \ \mathbf{4.583 \times 10^{-3} \text{ mol H}^+}$$

3. The net ionic reaction tells us that **the number of moles of H^+ equals the number of moles of OH^-**. Therefore, our OH^- sample of unknown molarity contains **4.583 $\times$ 10^{-3} mol OH$^-$**.

4. We had a 25.00-mL (0.02500 L) aliquot of our NaOH solution. The molarity of the solution is:

$$\frac{\text{mol OH}^-}{\text{L soln}} \ = \ \frac{4.583 \times 10^{-3} \text{ mol OH}^-}{0.02500 \text{ L soln}} \ = \ \mathbf{0.1833 \ M \ NaOH}$$

Does the Answer Make Sense?

You had 25.00 mL of NaOH solution. It took almost twice the volume of 0.1067 M HCl to neutralize. Because you have the **same number of moles in about half the solution volume**, it makes sense that the NaOH solution would be almost twice the concentration.

Example 4.8 C   Acid-Base Titration II

A student wishes to determine the concentration of a sodium hydroxide solution. To do this, the student weighs out a 0.8196 g sample of potassium hydrogen phthalate ($KHC_8H_4O_4$, KHP) and dissolves it in water with an indicator. The titration of KHP requires 35.48 mL of the NaOH solution. Calculate the molarity of the NaOH solution. The molar mass of KHP = 204.22 g/mol.

Strategy

Our strategy is to recognize the 1-to-1 mole ratio of KHP to NaOH. Therefore, the moles of NaOH needed will equal the number of moles of KHP present. The molarity of the NaOH equals the moles of NaOH divided by the volume of NaOH used in the titration.

Solution

1. The net ionic reaction is:

$$HC_8H_4O_4^-(aq) + OH^-(aq) \rightarrow H_2O(l) + C_8H_4O_4^{2-}(aq)$$

2. The number of moles of KHP present equals:

$$\text{moles KHP} \;\Big\|\; 0.8196 \text{ g KHP} \times \frac{1 \text{ mol KHP}}{204.22 \text{ g KHP}} = \mathbf{4.013 \times 10^{-3} \text{ mol KHP}}$$

3. mol NaOH = mol KHP = 4.013×10^{-3} mol NaOH

4. The titration required 35.48 mL of our NaOH solution. The molarity of the solution is:

$$\text{molarity NaOH} = \frac{4.013 \times 10^{-3} \text{ mol NaOH}}{0.03548 \text{ L NaOH soln}} = \mathbf{0.1131 \; M \text{ NaOH}}$$

Example 4.8 D   Molar Mass of an Acid

You want to determine the molar mass of an acid. The acid contains one acidic hydrogen per molecule. You weigh out a 2.879 g sample of the pure acid and dissolve it, along with 3 drops of phenolphthalein indicator, in distilled water. You titrate the sample with 0.1704 M NaOH. The pink endpoint is reached after addition of 42.55 mL of the base. Calculate the molar mass of the acid.

Solution

As always, we write the equation first. Let "HA" be your acid. It has one acidic hydrogen. Therefore, it will react with base in a 1:1 ratio.

$$HA + NaOH \rightarrow H_2O + NaA,$$

or

$$H^+(aq) + OH^-(aq) \rightarrow H_2O(l)$$

We have the number of grams of our acid, **2.879 g**. In order to calculate the molar mass (g/mol) of HA, we need to know **how many moles** of HA we have.

$$\text{moles NaOH} = \text{moles HA}$$

Our task, then, is to find moles of NaOH (present as OH^- in solution).

$$\text{moles } OH^- \;\Big\|\; \frac{0.1704 \text{ moles } OH^-}{\text{L soln}} \times 0.04255 \text{ L soln} = \mathbf{7.251 \times 10^{-3} \text{ mol } OH^-}$$

The reaction stoichiometry is 1 mole of HA per mole of NaOH. Therefore, there are 7.251×10^{-3} moles HA in 2.879 g HA.

$$\text{molar mass of the acid} = \frac{2.879 \text{ g}}{7.251 \times 10^{-3} \text{ mol}} = \mathbf{397.1 \text{ g/mole}}$$

4.9 Oxidation-Reduction Reactions

When you finish this section you will be able to:

- Assign **oxidation states** to atoms in a compound.
- Define **oxidation** and **reduction**.

Both this section and Section 4.10 seem, at first reading, to be unrelated to the general topic of solution chemistry. This section deals with definitions and assignments relating to **electron exchange**, and the next deals with balancing such equations. **You need these skills** before you do electrochemical solution problems because you need to understand how to deal with electron exchange before you can manipulate values to solve problems.

Your textbook defines **oxidation-reduction reactions** as being those in which **one or more electrons are transferred**.

$$H_2(g) + Cl_2(g) \rightarrow 2HCl(g)$$

In this case, electrons are transferred from the hydrogen to the chlorine.

From the last section, we saw that hydrogen can form H^+, or hydrogen with a "+1" oxidation state. It has one more proton than electron. Chlorine can form Cl^-, or chlorine with a "−1" oxidation state. Many elements are more energetically stable when they gain or lose electrons so that they have a "**noble gas electronic configuration**." Sodium forms Na^+, calcium forms Ca^{2+}, and sulfur can form S^{2-}. All these are common **oxidation states**.

There are a set of five rules for **assigning oxidation states** that are given in Table 4.2 in your textbook. You must, for now, memorize these rules. They will become second nature as you do more and more "redox" problems. Use those rules in the following problem.

Remember that the **sum of the oxidation states in a neutral compound must equal zero** and must be equal to the **overall charge** in an **ionic compound**.

Example 4.9 A   Assigning Oxidation States

Assign oxidation states to each of the atoms in the following compounds:

a.	CaF_2	c.	H_2O	e.	$KMnO_4$
b.	C_2H_6	d.	ICl_5	f.	SO_4^{2-}

Solution

a. Fluorine has a greater attraction for electrons than calcium, and it has a charge of **−1 (rule #3)**. Therefore, calcium must be +2, so that the overall charge balances at zero. Notice that calcium is in Group 2A. Metals in Groups 1A and 2A always have an oxidation state equal to their group number when in compounds.

$$CaF_2$$
$$\uparrow \quad \uparrow$$
$$+2 \;\; -1$$

b. Hydrogen is **assigned** an oxidation state of **+1** in covalent compounds (rule #5). Because there are 6 H's, the total charge due to H is +6. In order to balance, each of the two carbons must have an oxidation state of −3.

$$C_2H_6$$
$$\uparrow \quad \uparrow$$
$$-3 \;\; +1$$

c. **Rules #4 and 5** dictate that the oxygen has a charge of -2, and each H has a charge of $+1$. The water molecule therefore comes out to be electronically neutral.

$$\underset{\underset{+1}{\uparrow}\;\underset{-2}{\uparrow}}{H_2O}$$

d. In this compound, Cl is the more "electronegative" (has greater attraction for electrons), and therefore, gets the negative oxidation state as Cl^-. To electrically balance, iodine, which acts as the **cation**, must be I^{5+}.

$$\underset{\underset{+5}{\uparrow}\;\underset{-1}{\uparrow}}{ICl_5}$$

e. Although this compound is not totally covalent, oxygen is the **most electronegative** element here and is, therefore, assigned an oxidation state of -2. With four oxygens, **the total charge due to oxygen is -8**. Potassium is in Group 1A. According to our discussion in part a of this example, it must have an **oxidation state of $+1$**. Mathematically then, manganese must have an **oxidation state of $+7$** to allow the compound to electrically balance.

$$\underset{\underset{+1}{\uparrow}\;\;\;\underset{+7}{\uparrow}\;\;\;\underset{-2}{\uparrow}}{KMnO_4}$$

f. According to **rule #4**, each oxygen is assigned a -2 oxidation state in this covalent ion, for a total of -8 for four oxygens. The sum of the oxidation states in this ion must add up to -2. Therefore, the sulfur must have an oxidation state of $+6$.

$$\underset{\underset{+6}{\uparrow}\;\underset{-2}{\uparrow}}{SO_4{}^{2-}}$$

Additional Notes

1. In <u>Example 4.16 in your textbook</u>, you will notice that Dr. Zumdahl **checks** the math at the end of each part to make sure that the number of atoms of a given oxidation state, when added to all other atoms multiplied by their respective oxidation states, all add up to the total charge on the compound or ion. You should follow this practice with this example and other related examples.

2. Oxidation states can be arbitrary. Don't be frightened by non-integer or awkward states. These things **can** happen.

Example 4.9 B   Practice With Oxidation States

Assign oxidation states to each atom in the equation (treat each compound separately).

$$Fe_2O_3 + 2Al \rightarrow Al_2O_3 + 2Fe$$

Solution

$$Fe_2O_3(s) + 2Al(s) \rightarrow Al_2O_3(s) + 2Fe(s)$$

+3 per Fe	0 per Al	+3 per Al	0 per Fe
−2 per O	(rule #1)	−2 per O	(rule #1)

Let's look at what happened to each element as a result of the reaction.

- Fe went from Fe^{3+} to Fe^0 (gained 3 electrons).
- Al went from Al^0 to Al^{3+} (lost 3 electrons).
- O remained the same.

THE IRON <u>GAINED ELECTRONS</u>. IT HAS BEEN <u>REDUCED</u>.

THE ALUMINUM <u>LOST ELECTRONS</u>. IT HAS BEEN <u>OXIDIZED</u>.

Remember **OIL RIG**:   <u>O</u>xidation <u>I</u>nvolves <u>L</u>oss (of electrons).
                          <u>R</u>eduction <u>I</u>nvolves <u>G</u>ain (of electrons).

This next idea is tricky: Something that is **reduced** is called an **oxidizing agent** (it causes something else to be oxidized). Something that is **oxidized** is called a **reducing agent** (it causes something else to be reduced).

In the previous example:

- Iron was reduced.
- Aluminum was oxidized.
- Iron(III) oxide was the oxidizing agent.
- Aluminum was the reducing agent.

Example 4.9 C   Which Atoms Undergo Redox?

For each reaction, identify the atoms that undergo reduction or oxidation. Also, list the oxidizing and reducing agents.

a. $2H_2(g) + O_2(g) \rightarrow 2H_2O(g)$
b. $Zn(s) + Cu^{2+}(aq) \rightarrow Zn^{2+}(aq) + Cu(s)$
c. $2AgCl(s) + H_2(g) \rightarrow 2H^+(aq) + 2Ag(s) + 2Cl^-(aq)$
d. $2MnO_4^-(aq) + 16H^+(aq) + 5C_2O_4^{2-}(aq) \rightarrow 2Mn^{2+}(aq) + 10CO_2(g) + 8H_2O(l)$

Solution

For purposes of assigning oxidation states, treat each compound or ion by itself.

a. $2H_2(g) + O_2(g) \rightarrow 2H_2O(g)$
      0   0   +1 per H; −2 per O

     oxidized: hydrogen (0 to +1)   oxidizing agent: molecular oxygen
     reduced: oxygen (0 to −2)   reducing agent: molecular hydrogen

b. $Zn(s) + Cu^{2+}(aq) \rightarrow Zn^{2+}(aq) + Cu(s)$
      0   +2   +2   0

     oxidized: zinc (0 to +2)   oxidizing agent: copper
     reduced: copper (+2 to 0)   reducing agent: zinc

c. $2AgCl(s) + H_2(g) \rightarrow 2H^+(aq) + 2Ag(s) + 2Cl^-(aq)$
      +1 −1   0   +1   0   −1

     oxidized: hydrogen (0 to +1)   oxidizing agent: silver chloride
     reduced: silver (+1 to 0)   reducing agent: molecular hydrogen

d. $2MnO_4^-(aq) + 16H^+(aq) + 5C_2O_4^{2-}(aq) \rightarrow 2Mn^{2+}(aq) + 10CO_2(g) + 8H_2O(l)$
      +7 −2   +1   +3 −2   +2   +4 −2   +1 −2

     oxidized: carbon (+3 to +4)   oxidizing agent: permanganate ion (MnO_4^-)
     reduced: manganese (+7 to +2)   reducing agent: oxalate ion ($C_2O_4^{2-}$)

4.10   Balancing Oxidation-Reduction Equations

When you finish this section you will be able to balance redox equations using the half-reaction method in acidic or basic solutions.

In Section 4.9, you learned how to assign oxidation states to atoms in a compound. You also learned that in redox reactions <u>oxidation states on some atoms change</u>. We will use this information now to learn how to **balance redox equations using the oxidation states method in this section and the half-reaction method in Chapter 17**.

Example 4.10 A   Balancing by the Oxidation States Method

Please balance the reaction for respiration.

$$C_6H_{12}O_6 + O_2 \rightarrow CO_2 + H_2O$$

Solution

Following the procedure on in your textbook, the oxidation states are,

$$C_6H_{12}O_6 + O_2 \rightarrow CO_2 + H_2O$$
$$0 \; +1 \; -2 \qquad 0 \qquad +4 \; -2 \quad +1 \; -2$$

Each carbon is oxidized from 0 to +4, while each oxygen (in O_2) is reduced from 0 to −2. To balance the electron exchange, we will therefore need *twice as many* oxygen atoms (from O_2) as carbon atoms. There are 6 carbons, so we will need 12 oxygens from O_2, or $6O_2$.

$$C_6H_{12}O_6 \;\; + \;\; 6O_2 \; \rightarrow \;\; \text{products}$$
$$\uparrow \qquad\qquad \uparrow$$

24 electrons   24 electrons
lost   gained
(6C × 4)   (12O × 2)

Balancing the rest of the equation by inspection,

$$\mathbf{C_6H_{12}O_6 + 6O_2 \rightarrow 6CO_2 + 6H_2O}$$

Example 4.10 B   Practice with Oxidation States Method Balancing

Please balance the equation for the combustion of hydrazine, N_2H_4, with dinitrogen tetroxide, N_2O_4. This reaction helps keep the Space Shuttle in Earth orbit.

$$N_2H_4 \; + \; N_2O_4 \; \rightarrow \; N_2 \; + \; H_2O$$

Solution

$$N_2H_4 \; + \; N_2O_4 \; \rightarrow \; N_2 \; + \; H_2O$$
$$-2 \; +1 \qquad +4 \; -2 \qquad 0 \qquad +1 \; -2$$

The nitrogen in hydrazine is oxidized from −2 to 0.
The nitrogen in dinitrogen tetroxide is reduced from +4 to 0.

The ratio of N_2H_4 to N_2O_4 must therefore be 2 to 1.

$$2N_2H_4 \; + \; N_2O_4 \; \rightarrow \; \text{products}$$

Balancing the remainder of the equation by inspection,

$$\mathbf{2N_2H_4 \; + \; N_2O_4 \; \rightarrow \; 3N_2 \; + \; 4H_2O}$$

Exercises

Section 4.1

1. Write dissociation equations for the following when they are dissolved in water:

   a. $HF(g)$ c. $MgBr_2(s)$ e. $NaNO_3(s)$

   b. $SrBr_2(s)$ d. $NH_4Cl(s)$ f. $Al_2(SO_4)_3(s)$

2. Write dissociation equations for the following when they are dissolved in water:

   a. $Na_2SO_4(s)$ c. $NaOH(s)$ e. $Mg(OH)_2(s)$

   b. $KCl(s)$ d. $Na_2CrO_4(s)$ f. $HCOOH(l)$

3. Which of the following pairs of substances are miscible, and which are not? Give a reason.

   a. CH_3CH_2OH and H_2O c. C_6H_6 and H_2O

   b. C_6H_6 and C_6H_{12} d. $LiBr$ and H_2O

Section 4.2

4. Classify the following as strong, weak, or nonelectrolyte.

   a. CH_3CH_2OH c. HCl e. C_6H_{12}

   b. $C_{12}H_{22}O_{11}$ (sugar)   d. NH_3

5. Determine whether each of the following is a strong, weak, or nonelectrolyte.

   a. $MgCl_2$ c. $Be(OH)_2$ e. CH_3COONa

   b. C_4H_6 d. HNO_3

Section 4.3

6. Calculate the molarity of the following solutions:

   a. 49.73 g H_2SO_4 in enough water to make 500 mL of solution

   b. 4.739 g $RuCl_3$ in enough water to make 1.00 L of solution

   c. 5.035 g $FeCl_3$ in enough water to make 250 mL of solution

   d. 27.74 g $C_{12}H_{22}O_{11}$ in enough water to make 750 mL solution

   e. 218.7 g HCl in enough water to make 500 mL of solution

7. Calculate the molarity of the following solutions:

   a. 18.92 g of HNO_3 in enough water to make 500 mL of solution

   b. 5.761 g of KOH in enough water to make 350 mL of solution

   c. 21.18 g of $Fe(NO_3)_3$ in enough water to make 1.000 L of solution

   d. 72.06 g of $BaCl_2$ in enough water to make 800 mL of solution

8. Calculate the concentrations of each of the ions in the following solutions:

   a. 0.25 M Na_3PO_4 b. 0.15 M $Al_2(SO_4)_3$ c. 0.87 M Na_2CO_3

9. Calculate the concentrations of each of the ions in the following solutions:

   a. 0.62 M $K_2Cr_2O_7$ c. 0.14 M $Co(NO_3)_2$ e. 0.23 M $(NH_4)_2Cr_2O_7$

   b. 0.35 M $NaOH$ d. 0.07 M Na_3PO_4 f. 0.49 M $Al_2(SO_3)_3$

10. Describe how you would prepare the following solutions:

    a. 100. mL of 1.00 M NaCl
    b. 250. mL of 1.00 M Na$_2$SO$_4$
    c. 1.50 L of 0.500 M K$_2$Cr$_2$O$_7$

11. Describe how you would prepare the following solutions:

    a. 400. mL of 0.100 M HCl
    b. 750. mL of 0.350 M Ba(NO$_3$)$_2$
    c. 1.00 L of 1.50 M KMnO$_4$
    d. 250 mL of 0.20 M AgNO$_3$

12. Describe how you would prepare the following solutions:

    a. 500 mL of 1.0 M H$_2$SO$_4$ from 17.8 M H$_2$SO$_4$
    b. 1.5 L of 0.25 M KMnO$_4$ from 1.0 M stock solution
    c. 1.0 L of 0.15 M KBrO$_3$ from solid KBrO$_3$
    d. 100 mL of 0.01 M AgNO$_3$ from 0.5 M stock solution
    e. 1 L of 0.5 M AgNO$_3$ from solid AgNO$_3$

13. Describe how you would prepare the following solutions:

    a. 250 mL of 0.1 M HCl from 12.5 M HCl
    b. 500 mL of 1.5 M NaCl from 7.3 M NaCl
    c. 800 mL of 0.2 M NiCl$_2$ from 4.6 M NiCl$_2$
    d. 750 mL of 0.05 M FeSO$_4$ from 0.1 M FeSO$_4$

14. A standard solution of KHP (C$_8$H$_5$O$_4$K) was made by dissolving 3.697 g of KHP in enough water to make 100.0 mL of solution. Calculate the KHP concentration.

15. A stock solution of sodium hydroxide is prepared by dissolving 120.0 g of NaOH in 500.0 mL of water. What is the molarity of the stock solution?

16. A stock solution of HCl is prepared by adding 30.00 mL of a 12.00 M HCl solution to water and diluting to a final volume of 250.0 mL. What is the molarity of the stock solution?

17. How many moles of KHP are contained in 30.00 mL of the solution in Problem 14?

18. A solution of ammonium acetate (NH$_4$C$_2$H$_3$O$_2$) was made by dissolving 3.85 g of ammonium acetate in enough water to make 500 mL of solution. Calculate the solution concentration.

19. How many moles of ammonium acetate are contained in 17 mL of the solution in problem 18?

20. How many milliliters of 0.136 M NaOH are required to react with the H$_2$SO$_4$ in 10 mL of a 0.202 M solution? The reaction is:

$$2NaOH + H_2SO_4 \rightarrow Na_2SO_4 + 2H_2O$$

21. How many milliliters of 0.50 M Ca(OH)$_2$ are required to react with the HCl in 30 mL of a 0.12 M solution? The reaction is:

$$2HCl + Ca(OH)_2 \rightarrow CaCl_2 + 2H_2O$$

22. A chemistry student wants to verify the molarity of the acid stock solution prepared in problem 16, so she titrates a 25.00 mL aliquot (measured sample) with the NaOH solution described in problem 15. If she wants to use 40.00 mL of the NaOH solution as the titrant, what molarity would her NaOH solution need to be? How would she prepare 100.0 mL of that solution?

23. What is the concentration of the following in ppm?

    a. 1.0×10^{-2} g Cu^{2+} in 2.0 L of solution.
    b. The concentration of Pb^{2+} in 2.1×10^{-5} M $Pb(NO_3)_2$.

24. What is the concentration of the following in ppm?

    a. 6.2×10^{-3} g Be^{+2} in 750 mL of solution.
    b. 255 mg $NaIO_3$ in 1.5 L of solution.

Section 4.5

25. Balance the following reactions:

    a. $C_3H_8 + O_2 \rightarrow CO_2 + H_2O$
    b. $K_2CO_3 + Al_2Cl_6 \rightarrow KCl + Al_2(CO_3)_3$
    c. $Mg_3N_2 + H_2O \rightarrow MgO + NH_3$

    d. $Ca_3(PO_4)_2 + H_2SO_4 \rightarrow CaSO_4 + H_3PO_4$
    e. $KOH + H_3PO_4 \rightarrow K_3PO_4 + H_2O$
    f. $KClO_3 + C_{12}H_{22}O_{11} \rightarrow KCl + CO_2 + H_2O$

26. Complete and balance the following reactions:

    a. $NaCl(aq) + Hg_2(NO_3)_2(aq) \rightarrow$
    b. $Ca(OH)_2(aq) + Na_2CO_3(aq) \rightarrow$
    c. $Na_2S(aq) + FeCl_3(aq) \rightarrow$

27. A solution contains Ag^+, Pb^{2+}, and Fe^{3+}. If you want to precipitate the Pb^{2+} selectively, what anion would you choose?

Section 4.6

28. Write formula, complete ionic, and net ionic equations for the following reactions:

    a. aqueous sodium sulfide reacts with aqueous copper(II) nitrate
    b. aqueous hydrogen fluoride reacts with aqueous potassium hydroxide to give water and aqueous potassium fluoride

Section 4.7

29. What mass of $Mg(OH)_2$ is produced when 100. mL of 0.42 M $Mg(NO_3)_2$ is added to excess NaOH solution?

30. What mass of $BaSO_4$ is produced when 15.0 mL of 3.00 M H_2SO_4 is added to 20.0 mL of 0.100 M $BaCl_2$?

31. Calculate the mass of $CaSO_4$ produced when 10 mL of 6.0 M H_2SO_4 is added to 100 mL of 0.52 M $Ca(NO_3)_2$.

32. What mass of $CaCO_3$ is produced when 250 mL of 6.0 M Na_2CO_3 is added to 750 mL of 1.0 M CaF_2?

33. A 50.00-mL sample containing chloride ion, Cl^-, is combined with 125.0 mL of a 0.02157 M $AgNO_3$ solution. The resulting precipitate weighs 211.6 mg. What was the concentration of the chloride ion in the original sample? Was it useful to know the amount and concentration of the $AgNO_3$ solution? Why or why not?

34. Calculate the mass of Al_2S_3 produced when 100 mL of 0.50 M $AlCl_3$ is added to 100 mL of 0.50 M Na_2S.

35. You are given the equation: $AgBr + 2S_2O_3^{2-} \rightarrow Ag(S_2O_3)_2^{3-} + Br^-$. What mass of AgBr can be dissolved by 750. mL of 0.300 M $Na_2S_2O_3$?

36. Given the following chemical equation, determine the theoretical yield of B_2H_6 if exactly 100.0 g of $LiAlH_4$ was allowed to react with 225 g of BF_3.

$$3LiAlH_4 + 4BF_3 \rightarrow 3LiF + 3AlF_3 + 2B_2H_6$$

Section 4.8

37. What volume of 0.1379 M HCl is required to neutralize 10.0 mL of 0.2789 M NaOH solution?

38. How many mL of 1.50 M NaOH is required to neutralize 275 mL of 0.5 M H_2SO_4?

39. What is the molarity of a solution of HCl if it requires 29.31 mL of a 0.0923 M NaOH solution to reach a phenolphthalein endpoint for the titration of a 10.0 mL aliquot of the HCl solution?

40. Given the following two equations, determine the number of grams of manganese dioxide required to prepare enough chlorine gas to produce 25.0 g of potassium hypochlorite.

$$MnO_2 + 4HCl \rightarrow MnCl_2 + Cl_2 + 2H_2O$$
$$Cl_2 + 2KOH \rightarrow KCl + KClO + H_2O$$

41. A titration is done using 0.1302 M NaOH to determine the molar mass of an acid. The acid contains one acidic hydrogen per molecule. If 1.863 g of the acid require 70.11 mL of the NaOH solution, what is the molar mass of the acid?

42. A 2.000-g sample of silver alloy was dissolved in nitric acid and then precipitated as AgBr. After drying, the sample of silver bromide weighed 2.000 g. Calculate the percentage of silver in the alloy.

43. Complete and balance each acid-base equation (assume complete neutralization):

a. H_2SO_4 + NaOH $\rightarrow$
b. H_3PO_4 + $Mg(OH)_2$ $\rightarrow$

c. H_2SO_3 + NaOH $\rightarrow$
d. $HC_2H_3O_2$ + $Ba(OH)_2$ $\rightarrow$

44. What volume of 0.2 M NaOH is required to neutralize 50 mL of 0.1 M H_2SO_3?

45. How many mL of 2.3 M HNO_3 is needed to neutralize 0.92 L of 0.5 M KOH?

46. The neutralization of a 25.00-mL sample of an unknown base requires 18.34 mL of 0.100 M HCl. Assuming that the acid-base stoichiometry is 1:1, what is the concentration of the unknown base?

47. Pennies made after 1982 contain about 97% zinc by mass. A student wants to prove this by filing the copper outside of a penny until he sees zinc and then putting the penny in a 1.00 M HCl solution. The zinc will be oxidized, and the H^+ (from HCl) will be reduced:

$$Zn + 2H^+ \rightarrow Zn^{2+} + H_2$$

If the entire penny has a mass of 2.80 grams, how many mL of 1.00 M HCl are required to just react with all the zinc? (You would, in reality, add much more to completely surround the penny.)

48. Assuming that the stoichiometry is 1:1, what is the concentration of an unknown acid if a 20.0-mL sample of it is neutralized with precisely 33.4 mL of 0.250 M base?

49. A 25.0-mL sample of an ammonia solution is analyzed by titration with HCl. The reaction is given below. It took 18.96 mL of 0.150 M HCl to titrate the ammonia. What is the concentration of the original ammonia solution?

$$NH_3 + H^+ \rightarrow NH_4^+$$

Section 4.9

50.   Determine the oxidation number for Mn in each of the following:

    a. $KMnO_4$ c. MnO_2 e. Mn_2O_7

    b. $LiMnO_2$ d. K_2MnCl_4

51.   Determine the oxidation number for each atom in the following compounds or ions:

    a. H_3O^+ c. S_8 e. NH_4ClO_4

    b. P_4O_{10} d. H_2CO

52.   Determine the oxidation number for each atom in the following compounds:

    a. $MgBr_2$ c. $Cr_2O_7^{2-}$ e. $NaClF_4$

    b. Na_2SO_4 d. $CaCO_3$ f. HNO_3

Section 4.10

53.   Balance the following oxidation-reduction reactions. Which species in each is the oxidizing agent, reducing agent?

    a. $P + Cl_2 \rightarrow PCl_5$

    b. $Sn^{2+} + Cu^{2+} \rightarrow Sn^{4+} + Cu^+$

    c. $Cu + H^+ + NO_3^- \rightarrow Cu^{2+} + NO_2 + H_2O$

    d. $Br_2 + SO_2 + H_2O \rightarrow H^+ + Br^- + SO_4^{2-}$

    e. $H_2SO_4 + HBr \rightarrow SO_2 + Br_2 + H_2O$

54.   Balance the following redox reactions. Identify the oxidizing agent and the reducing agent.

    a. $ClO^- + H^+ + Cu \rightarrow Cl^- + H_2O + Cu^{2+}$

    b. $H_2O + Cr^{3+} + XeF_6 \rightarrow Cr_2O_7^{2-} + H^+ + Xe + F^-$

    c. $H_2O + SO_3^{2-} + Fe^{3+} \rightarrow SO_4^{2-} + Fe^{2+} + H^+$

Multiple Choice Questions

55.   The hydrogen atoms of a water molecule would be most attracted to which one of the following?

    A. Cl^- B. Mg^{2+} C. CH_4 D. Cl_2

56.   Which of the following solvents would probably be the best one to dissolve NaBr?

    A. CH_4 C. H_2O

    B. CH_3CH_3 D. $CH_3CH_2CH_2CH_2CH_2CH_2CH_2OH$

57.   Which of the following ions would the oxygen atom of a water molecule have the greatest attraction for?

    A. Cl^- B. S^{2-} C. Mg^{2+} D. Na^+

58.   Which of the following is not an electrolyte?

    A. KCl   B. CH_3COOH C. NH_4Cl D. Br_2

59.   A strong base is one that

    A.   Dissociates completely into water molecules.

    B.   Dissociates completely into hydrogen and hydride ions.

    C.   Dissociates completely into hydrogen ions and anions.

    D.   Dissociates completely into hydroxide ions and cations.

60. Calculate the molarity when 18.5 g of nitric acid are dissolved in enough water to prepare 100.0 mL of solution.

   A. 2.94 M   B. 5.78 M   C. 3.51 M   D. 0.287 M

61. Calculate the molarity of a solution when 0.500 pound of silver nitrate is dissolved in enough water to prepare 16.75 L of solution.

   A. 1.34 M   B. 0.0797 M   C. 2.67 M   D. 0.615 M

62. How many grams of nitric acid are present in 250.0 mL of 6.70 M acid solution?

   A. 16.8 g   B. 106 g   C. 335 g   D. 211 g

63. How many liters of water are required to prepare a 0.1590 M silver nitrate solution if 1.00 pound of silver nitrate is used?

   A. 8.38 L   B. 16.8 L   C. 7.61 L   D. 36.9 L

64. Considering that calcium chloride is a strong electrolyte, what is the molarity of the chloride ions you would find in a solution prepared by mixing 8.99 g of calcium chloride with enough water to prepare 150.0 mL of solution?

   A. 0.540 M   B. 0.270   C. 66.00 M   D. 1.08 M

65. When 25.0 mL of a 10.6 M HCl solution is diluted to 200.0 mL, what is the final molarity of this acidic solution?

   A. 1.33 M   B. 2.65 M   C. 3.98 M   D. 3.53 M

66. 25.0 mL of 3.00 M hydrochloric acid solution is diluted to 100.0 mL. 10.0 mL of this solution is then diluted a second time to 100.0 mL. Calculate the molarity of the acidic solution after the second dilution.

   A. 0.0750 M   B. 0.0375 M   C. 1.50 M   D. 0.150 M

67. Which of the following compounds do you expect to precipitate in an aqueous solution?

   A. $AgNO_3$ B. $PbSO_4$ C. LiBr   D. KI

68. With which of the following solutions would you mix a silver nitrate solution to precipitate a silver salt?

   A. Lead sulfate solution   C. Sodium chloride solution
   B. Potassium nitrate solution   D. None of the above

69. Which of the following solutions would be the best to precipitate sodium chloride?

   A. AgCl solution   B. $AgNO_3$ solution   C. KOH solution   D. None of the above

70. Which of the following solutions would form a precipitate when mixed with $Ba(NO_3)_2$?

   A. KCl   B. $Pb(NO_3)_2$ C. KNO_3 D. none of these

71. How many grams of $K_4Fe(CN)_6$ are required to precipitate all of the cadmium (2+) ions as $Cd_2Fe(CN)_6$ from 4.00 mL of 0.15 M cadmium chloride solution according to the following equation?

$$K_4Fe(CN)_6 + 2CdCl_2 \rightarrow 4KCl + Cd_2Fe(CN)_6$$

   A. 0.11 g   B. 7.4 g   C. 2.4 g   D. 0.78 g

72. How many grams of calcium phosphate are precipitated when 25.0 mL of 0.220 M calcium chloride solution are allowed to react with 15.0 mL of 0.880 M phosphoric acid solution?

   A. 8.32 g   B. 1.13 g   C. 4.08 g   D. 0.568 g

73. How many mL of 1.00 M sulfuric acid solution are required to precipitate out all of the barium ion from 40.0 mL of 0.250 M barium chloride solution?

$$H_2SO_4 + BaCl_2 \rightarrow BaSO_4 + 2HCl$$

A. 10.0 mL   B. 8.25 mL   C. 20.4 mL   D. 5.25 mL

74. How many mL of 0.21 M phosphoric acid solution are required to precipitate all of the calcium ions from 5.00 mL of a 0.16 M calcium chloride solution?

$$3Ca^{2+}(aq) + 2PO_4^{3-}(aq) \rightarrow Ca_3(PO_4)_2(s)$$

A. 0.63 mL   B. 1.9 mL   C. 2.5 mL   D. 3.8 mL

75. How many mL of 1.00 M KOH are required to just neutralize 600.0 mL of a 1.5 M HCl solution?

$$KOH + HCl \rightarrow H_2O + KCl$$

A. 900 mL   B. 400 mL   C. 2.50 L   D. 40.0 mL

76. A 30.0-mL sample of an unknown basic solution is neutralized after the addition of 12.0 mL of a 0.15 M HCl solution. What is the molarity of the monoprotic base?

A. 0.00240 M   B. 0.375 M   C. 0.0600 M   D. 0.00180 M

77. 1.122 g of an unknown monoprotic base dissolved in 50.0 mL of water is titrated to endpoint with 20.0 mL of a 1.00 M HCl solution. Identify the base.

A. NaOH   B. NH_3 C. KOH   D. LiOH

78. Calculate the number of mL of 0.250 M LiOH required to neutralize 70.0 mL of 0.155 M H_3AsO_4.

$$3LiOH + H_3AsO_4 \rightarrow 3H_2O + Li_3AsO_4$$

A. 43.4 mL   B. 130 mL   C. 86.8 mL   D. 113 mL

79. Determine the oxidation state of oxygen in KO_2.

A. $-\frac{1}{2}$ B. $+\frac{1}{2}$ C. -2 D. -1

80. Determine the average oxidation number of carbon in $C_6H_{12}O_6$ (glucose).

A. -2 B. $+4$ C. $+2$ D. 0

81. In the following reaction, select the element that has undergone oxidation:

$$5FeCl_2 + KMnO_4 + 8HCl \rightarrow 5FeCl_3 + KCl + MnCl_2 + 4H_2O$$

A. H   B. Cl   C. Mn   D. Fe

82. In the following reaction, select the element that is the oxidizing agent:

$$2MnO_2 + KClO_3 + 2KOH \rightarrow 2KMnO_4 + KCl + H_2O$$

A. Mn   B. K   C. Cl   D. H

83. Balance the following equation:

$$\_H_2S + \_H^+ + \_MnO_4^- \rightarrow \_Mn^{2+} + \_S + \_H_2O$$

A. 5, 6, 2, 2, 5, 8   B. 5, 8, 2, 3, 6, 5   C. 1, 1, 1, 1, 1, 2   D. 1, 1, 1, 1, 1, 1

Answers to Exercises

1. a. $HF(g) \xrightarrow{H_2O(l)} H^+(aq) + F^-(aq)$
    b. $SrBr_2(s) \xrightarrow{H_2O(l)} Sr^{2+}(aq) + 2Br^-(aq)$
    c. $MgBr_2(s) \xrightarrow{H_2O(l)} Mg^{2+}(aq) + 2Br^-(aq)$
    d. $NH_4Cl(s) \xrightarrow{H_2O(l)} NH_4^+(aq) + Cl^-(aq)$
    e. $NaNO_3(s) \xrightarrow{H_2O(l)} Na^+(aq) + NO_3^-(aq)$
    f. $Al_2(SO_4)_3(s) \xrightarrow{H_2O(l)} 2Al^{3+}(aq) + 3SO_4^{2-}(aq)$

2. a. $Na_2SO_4(s) \xrightarrow{H_2O(l)} 2Na^+(aq) + SO_4^{2-}(aq)$
    b. $KCl(s) \xrightarrow{H_2O(l)} K^+(aq) + Cl^-(aq)$
    c. $NaOH(s) \xrightarrow{H_2O(l)} Na^+(aq) + OH^-(aq)$
    d. $Na_2CrO_4(s) \xrightarrow{H_2O(l)} 2Na^+(aq) + CrO_4^{2-}(aq)$
    e. $Mg(OH)_2(s) \xrightarrow{H_2O(l)} Mg^{2+}(aq) + 2OH^-(aq)$
    f. $HCOOH(l) \xrightarrow{H_2O(l)} H^+(aq) + COOH^-(aq)$

3. a. miscible; polar O–H bonds in both
    b. miscible; both nonpolar
    c. immiscible; C_6H_6 is nonpolar, water is polar
    d. miscible; LiBr is ionic, and H_2O is polar

4. a. nonelectrolyte
    b. nonelectrolyte
    c. strong
    d. weak
    e. nonelectrolyte

5. a. strong
    b. nonelectrolyte
    c. strong
    d. strong
    e. strong

6. a. 1.01 M
    b. 0.0228 M
    c. 0.124 M
    d. 0.108 M
    e. 12.0 M

7. a. 0.601 M
    b. 0.293 M
    c. 0.08756 M
    d. 0.433 M

8. a. 0.75 $M\,Na^+$, 0.25 $M\,PO_4^{3-}$
    b. 0.3 $M\,Al^{3+}$, 0.45 $M\,SO_4^{2-}$
    c. 1.74 $M\,Na^+$, 0.87 $M\,CO_3^{2-}$

9. a. 1.24 $M\,K^+$, 0.62 $M\,Cr_2O_7^{2-}$
    b. 0.35 $M\,Na^+$, 0.35 $M\,OH^-$
    c. 0.14 $M\,Co^{2+}$, 0.28 $M\,NO_3^-$
    d. 0.21 $M\,Na^+$, 0.07 $M\,PO_4^{3-}$
    e. 0.46 $M\,NH_4^+$, 0.23 $M\,Cr_2O_7^{2-}$
    f. 0.98 $M\,Al^{3+}$, 1.47 $M\,SO_3^{2-}$

10. a. 5.84 g NaCl in enough water to make 100 mL.
    b. 35.5 g of Na_2SO_4 in enough water to make 250 mL.
    c. 221 g of $K_2Cr_2O_7$ in enough water to make 1.5 L.

11. a. 1.46 g of HCl in enough water to make 400 mL.
    b. 68.6 g of $Ba(NO_3)_2$ in enough water to make 750 mL.
    c. 237 g of $KMnO_4$ in enough water to make 1 L.
    d. 8.5 g of $AgNO_3$ in enough water to make 250 mL.

12. a. 28 mL of 17.8 M H_2SO_4 diluted to 500 mL.
    b. 375 mL of 1 M $KMnO_4$ diluted to 1.5 L.
    c. 25 g of $KBrO_3$ in enough water to make 1.0 L.
    d. 2 mL of 0.5 M $AgNO_3$ diluted to 100 mL.
    e. 85 g $AgNO_3$ in enough water to make 1 L.

13. a. 2 mL of 12.5 M HCl diluted to 250 mL.
    b. 103 mL of 7.3 M NaCl diluted to 500 mL.
    c. 35 mL of 4.6 M $NiCl_2$ diluted to 800 mL.
    d. 375 mL of 0.1 M $FeSO_4$ diluted to 750 mL.

14. 0.1810 M KHP

15. 6.000 M NaOH

16. 1.440 M HCl

17. 5.430×10^{-3} mole KHP

18. 0.0999 M $NH_4C_2H_3O_2$

19. 1.7×10^{-3} mole $NH_4C_2H_3O_2$

20. 29.7 mL NaOH

21. 3.6 mL $Ca(OH)_2$

22. 0.9000 M NaOH solution prepared by taking 15.00 mL of the 6.000 M NaOH and diluting to 100.0 mL.

23. a. 5 ppm Cu^{2+}
    b. 4.4 ppm Pb^{2+}

24. a. 8.3 ppm Be^{2+}
    b. 170 ppm $NaIO_3$

25. a. $C_3H_8 + 5O_2 \rightarrow 3CO_2 + 4H_2O$
    b. $3K_2CO_3 + Al_2Cl_6 \rightarrow 6KCl + Al_2(CO_3)_3$
    c. $Mg_3N_2 + 3H_2O \rightarrow 3MgO + 2NH_3$
    d. $Ca_3(PO_4)_2 + 3H_2SO_4 \rightarrow 3CaSO_4 + 2H_3PO_4$
    e. $3KOH + H_3PO_4 \rightarrow K_3PO_4 + 3H_2O$
    f. $8KClO_3 + C_{12}H_{22}O_{11} \rightarrow 8KCl + 12CO_2 + 11H_2O$

26. a. $2NaCl(aq) + Hg_2(NO_3)_2(aq) \rightarrow Hg_2Cl_2(s) + 2NaNO_3(aq)$
    b. $Ca(OH)_2(aq) + Na_2CO_3(aq) \rightarrow CaCO_3(s) + 2NaOH(aq)$
    c. $3Na_2S(aq) + 2FeCl_3(aq) \rightarrow Fe_2S_3(s) + 6NaCl(aq)$

27. SO_4^{2-}

28. a. formula: $Na_2S(aq) + Cu(NO_3)_2(aq) \rightarrow CuS(s) + 2NaNO_3(aq)$
       complete ionic: $2Na^+(aq) + S^{2-}(aq) + Cu^{2+}(aq) + 2NO_3^-(aq) \rightarrow CuS(s) + 2Na^+(aq) + 2NO_3^-(aq)$
       net ionic: $S^{2-}(aq) + Cu^{2+}(aq) \rightarrow CuS(s)$
    b. formula: $HF(aq) + KOH(aq) \rightarrow H_2O(l) + KF(aq)$
       complete ionic: $H^+(aq) + F^-(aq) + K^+(aq) + OH^-(aq) \rightarrow H_2O(l) + K^+(aq) + F^-(aq)$
       net ionic: $H^+(aq) + OH^-(aq) \rightarrow H_2O(l)$

29. 2.4 g $Mg(OH)_2$

30. 0.467 g $BaSO_4$

31. 7.1 g $CaSO_4$

32. 75 g $CaCO_3$

33. 0.02952 M Cl^-; yes, needed to know that Cl^- is the limiting reactant.

34. 2.5 g Al_2S_3

35. 21.1 g AgBr

36. 45.9 g B_2H_6

37. 20.2 mL HCl

38. 183 mL NaOH

39. 0.271 M HCl

40. 24.0 g MnO_2

41. 204.1 g/mole

42. 57.45% Ag in the alloy

43. a. $H_2SO_4 + 2NaOH \rightarrow 2H_2O + Na_2SO_4$
     b. $2H_3PO_4 + 3Mg(OH)_2 \rightarrow 6H_2O + Mg_3(PO_4)_2$
     c. $H_2SO_3 + 2NaOH \rightarrow 2H_2O + Na_2SO_3$
     d. $2HC_2H_3O_2 + Ba(OH)_2 \rightarrow 2H_2O + Ba(C_2H_3O_2)_2$

44. 50 mL NaOH

45. 200 mL of 2.3 M HNO_3

46. 7.34×10^{-2} M base

47. 83.1 mL of 1.00 M HCl are required.

48. 0.418 M acid

49. 0.114 M NH_3

50. a. +7   c. +4   e. +7
     b. +3   d. +2

51. a. H=+1; O=−2   c. S=0   e. N=−3; H=+1; Cl=+7; O=−2
     b. P=+5; O=−2   d. H=+1; C=0; O=−2

52. a. Mg=+2; Br=−1   c. Cr=+6; O=−2   e. Na=+1; Cl=+3; F=−1
     b. Na=+1; S=+6; O=−2   d. Ca=+2; C=+4; O=−2   f. H=+1; N=+5; O=−2

53. a. $2P + 5Cl_2 \rightarrow 2PCl_5$; ox. agent = Cl_2; red. agent = P
     b. $Sn^{2+} + 2Cu^{2+} \rightarrow Sn^{4+} + 2Cu^+$; ox. agent = Cu^{2+}; red. agent = Sn^{2+}
     c. $Cu + 4H^+ + 2NO_3^- \rightarrow Cu^{2+} + 2NO_2 + 2H_2O$; ox. agent = NO_3^-; red. agent = Cu
     d. $Br_2 + SO_2 + 2H_2O \rightarrow 4H^+ + 2Br^- + SO_4^{2-}$; ox. agent = Br_2; red. agent = SO_2
     e. $H_2SO_4 + 2HBr \rightarrow SO_2 + Br_2 + 2H_2O$; ox. agent = H_2SO_4; red. agent = HBr

54. a. $ClO^- + 2H^+ + Cu \rightarrow Cl^- + H_2O + Cu^{2+}$; ox. agent = ClO^-; red. agent = Cu
     b. $7H_2O + 2Cr^{3+} + XeF_6 \rightarrow Cr_2O_7^{2-} + 14H^+ + Xe + 6F^-$; ox. agent = XeF_6; red. agent = Cr^{3+}
     c. $H_2O + SO_3^{2-} + 2Fe^{3+} \rightarrow SO_4^{2-} + 2Fe^{2+} + 2H^+$; ox. agent = Fe^{3+}; red. agent = SO_3^{2-}

55.	A	56.	C	57.	C	58.	D	59.	D	60.	A
61.	B	62.	B	63.	B	64.	D	65.	A	66.	A
67.	B	68.	C	69.	D	70.	A	71.	A	72.	D
73.	A	74.	C	75.	A	76.	C	77.	C	78.	B
79.	A	80.	D	81.	D	82.	C	83.	A		

Chapter 5

Gases

In this chapter we explore the laws that govern the behavior of gases.

5.1 Pressure

When you finish this section you will be able to:

- Define "pressure."
- State how pressure is different from force.
- Convert among various units of pressure.

Your textbook introduces you to the **manometer** and the **barometer**. In order to really know how they work, you must understand force, pressure, and the units of both.

$$\textbf{Force} = \textbf{mass} \times \textbf{acceleration}$$

When you weigh yourself on a scale, you are measuring the **force** that your body exerts on the scale. Let's say that your body mass is 68 kilograms. Also, the **acceleration** due to **gravity** that the Earth exerts on you is constant 9.8 meters/second^2. The **total force** exerted by your body on the scale is

$$F = m \times a$$

$$F = 68 \text{ kg} \times \frac{9.8 \text{ m}}{\text{s}^2} = \frac{670 \text{ kg m}}{\text{s}^2}$$

The S.I. unit of force is the **Newton (N)**.

$$1 \text{ N} = \frac{1 \text{ kg m}}{\text{s}^2}$$

Therefore, your force on the scale is **670 N**. This corresponds to your weight of **150 pounds** (4.47 N/pound).

Example 5.1 A   The Units of Force

Let's say that you (with your mass of 68 kg) are on the moon, where the acceleration due to gravity is about 1.6 m/s^2. What force would you exert on a scale ("how much would you weigh") in Newtons and in pounds?

Solution

$$F = m \times a = 68 \text{ kg} \times \frac{1.6 \text{ m}}{\text{s}^2} = \frac{108.8 \text{ kg m}}{\text{s}^2} = \textbf{110 N}$$

$$\text{weight in pounds} = 108.8 \text{ N} \times \frac{1 \text{ pound}}{4.47 \text{ N}} = \textbf{24 pounds}$$

$$\textbf{Pressure} = \frac{\text{Force}}{\text{Area}}$$

If you weigh 670 N (670 kg m/s^2) on Earth, and you are standing on a 0.5 m x 0.5 m square, your **area = (0.5 m)^2 = 0.25 m^2**, and the pressure you exert is

$$P = \frac{F}{A} = \frac{670 \text{ kg m/s}^2}{0.25 \text{ m}^2} = \frac{2700 \text{ kg}}{\text{m s}^2} \text{ or } \frac{2700 \text{ N}}{\text{m}^2}$$

The S.I. unit of pressure is the **Pascal (Pa)**.

$$1 \text{ Pa} = \frac{1 \text{ kg}}{\text{m s}^2} = \frac{1 \text{ N}}{\text{m}^2}$$

Therefore, the pressure you exert is **2700 Pa**.

Example 5.1 B   The Units of Pressure

Let's say that someone is wearing high heels with a total area (for both heels) of 1×10^{-4} **m^2**. The force is **670 N**. Calculate the pressure that you exert.

$$P = \frac{F}{A} = \frac{670 \text{ kg m/s}^2}{1 \times 10^{-4} \text{ m}^2} = \frac{6.7 \times 10^6 \text{ N}}{\text{m}^2} = \textbf{6.7} \times \textbf{10}^6 \textbf{ Pa}$$

This pressure is quite high! That is why it hurts if someone wearing a stiletto heel stands on your foot.

Air exerts pressure as well. **The "standard atmosphere" is equal to 100,000 Pa, though most texts still use the old figure of 101,325 Pa**. This is an awkwardly large number. We commonly express air pressure in units of **atmospheres (atm)**, **torr**, or **mm Hg**.

$$\textbf{1 atm} = \textbf{101,325 Pa}$$

$$\textbf{1 atm} = \textbf{760 torr} = \textbf{760 mm Hg}$$

Example 5.1 C   Interconverting the Units of Pressure

The pressure exerted by a gas is measured to be 0.985 atm. Convert this pressure to **torr** and **pascals**.

Solution

$$0.985 \text{ atm} \times \frac{760 \text{ torr}}{\text{atm}} = \textbf{749 torr (749 mm Hg)}$$

$$0.985 \text{ atm} \times \frac{101{,}325 \text{ Pa}}{1 \text{ atm}} = \textbf{98,300 Pa}$$

5.2   *The Gas Laws of Boyle, Charles, and Avogadro*

When you finish this section you will be able to solve problems relating to the fundamental gas laws.

There are **three gas laws** you need to know and be able to use. In the next section, these laws will be combined to give a more general law governing the behavior of gases.

A. Boyle's Law

This law says that **the pressure exerted by a gas is inversely proportional to the volume the gas occupies**. In other words, as you squeeze the gas, it exerts more pressure. As an example, fill a zip-lock freezer bag with air, and then seal it. Try to squeeze it (**reduce the volume**). It is very difficult to squeeze because the **pressure of the gas is increasing** as you reduce the volume of the freezer bag.

$$\textbf{Pressure} \times \textbf{Volume} = \textbf{Constant}$$

$$PV = k$$

If PV is a constant for a gas at constant temperature, then

$$P_1 V_1 = P_2 V_2 \text{ (at constant temperature)}$$

where P_1 = initial pressure
V_1 = initial volume
P_2 = final pressure
V_2 = final volume

Your textbook notes that you should *make sure your problem answer makes physical sense*. This is especially important when doing gas law problems.

Example 5.2 A   Boyle's Law

A gas that has a pressure of 1.3 atm occupies a volume of 27 L. What volume will the gas occupy if the pressure is increased to 3.9 atm at constant temperature?

Predicting the Answer

If the pressure on the gas is **increased** by a factor of 3 (3.9 atm/1.3 atm) (you are squeezing the gas), we would expect the volume to **decrease** proportionately. Therefore, the volume should be 1/3 of its original value.

Solution

$$P_1 = 1.3 \text{ atm} \qquad\qquad\qquad P_2 = 3.9 \text{ atm}$$
$$V_1 = 27 \qquad\qquad\qquad\qquad\quad V_2 = ?$$

$$P_1V_1 = P_2V_2$$

$$1.3 \text{ atm } (27 \text{ L}) = 3.9 \text{ atm } (V_2)$$

$$V_2 = \frac{1.3 \text{ atm } (27 \text{ L})}{3.9 \text{ atm}} = \textbf{9.0 L}$$

The final volume is 9.0 L (1/3 of 27), the answer we predicted. Always consider, *"Does my answer make sense?"*

Example 5.2 B   The Units of P × V

The *PV* constant in the previous example was $35._1$ L atm (1.3 atm × 27 L). Convert this constant to

    a.   L Pa, and
    b.   fundamental SI units.

Solution

    a. $35._1$ L atm $\times \dfrac{101,325 \text{ Pa}}{1 \text{ atm}} = \textbf{3.5}_\textbf{6} \times \textbf{10}^\textbf{6} \textbf{ L Pa}$

    b.   We know from Section 5.1 that **1 Pascal = 1 kg m^{-1} s^{-2}**. 1 L = 1000 mL, or 1000 cm^3, or (10 cm)^3. This equals (0.1 m)^3, or **1 L = 1 x 10^{-3} m^3**. Therefore,

$$3.5_6 \times 10^6 \text{ L Pa} = 3.56 \times 10^6 \text{ L Pa} \times \frac{1 \times 10^{-3} \text{ m}^3}{1 \text{ L}} \times \frac{1 \text{ kg m}^{-1} \text{ s}^{-2}}{1 \text{ Pa}} = \textbf{3.5}_\textbf{6} \times \textbf{10}^\textbf{3} \textbf{ kg m}^\textbf{2} \textbf{ s}^{\textbf{-2}}$$

The unit kg m^2 s^{-2} is the SI unit of **energy** called the **Joule**. *PV, then, is really a unit of energy!* We will use this information in later chapters.

B. Charles's Law

Charles's Law says that **at constant pressure, the volume of a gas is directly proportional to the temperature (in Kelvins) of the gas**.

$$\frac{\textbf{Volume}}{\textbf{Temperature}} = \textbf{Constant}$$

$$\frac{V}{T} = b$$

Using the same reasoning as we did in Boyle's Law, if we change the temperature (at constant pressure), the volume will change so that the ratio of volume to temperature will remain constant.

$$\frac{V_1}{T_1} = \frac{V_2}{T_2} \quad \text{(at constant pressure)}$$

V_1 = initial pressure
T_1 = initial temperature
V_2 = final volume
T_2 = final temperature

Example 5.2 C   Charles's Law

A gas at 30.0°C and 1.00 atm occupies a volume of 0.842 L. What volume will the gas occupy at 60.0°C and 1.00 atm?

Predicting the Answer

The temperature is increased; therefore, according to Charles's Law, the volume occupied should increase as well. We predict that our **final volume will be larger**. (Remember to convert temperature to Kelvins!)

Solution

$$V_1 = 0.842 \text{ L} \qquad\qquad\qquad\qquad\qquad V_2 = ?$$
$$T_1 = 30.0°C + 273 = 303 \text{ K} \qquad\qquad\qquad T_2 = 60.0°C + 273 = 333 \text{ K}$$

$$\frac{V_1}{T_1} = \frac{V_2}{T_2}$$

$$\frac{0.842 \text{ L}}{303 \text{ K}} = \frac{V_2}{333 \text{ K}}$$

$$V_2 = \frac{0.842 \text{ L } (333 \text{ K})}{303 \text{ K}} = \textbf{0.925 L}$$

The final volume agrees with our prediction, so it "makes sense."

C. Avogadro's Law

Avogadro's Law says that **for a gas at constant temperature and pressure the volume is directly proportional to the number of moles of gas**.

$$\text{Volume} = \text{Constant} \times \text{number of moles} \qquad \text{(constant } T, P)$$

$$V = an \qquad \text{(at constant } T, P)$$

If you triple the number of moles of gas (at constant temperature and pressure), the volume will also triple.

$$\frac{V_1}{n_1} = \frac{V_2}{n_2} \qquad \text{(at constant } T, P)$$

V_1 = initial volume
n_1 = initial number of moles
V_2 = final volume
n_2 = final number of moles

Example 5.2 D   Avogadro's Law

A 5.20 L sample at 18.0°C and 2.00 atm pressure contains 0.436 moles of a gas. If we add an **additional** 1.27 moles of the gas at the same temperature and pressure, what will be the **total volume** occupied by the gas?

Predicting the Answer

According to Avogadro's Law, as we increase the number of moles of a gas, the volume should increase proportionately. Therefore, we would predict that our volume would increase.

Solution

$$V_1 = 5.20 \text{ L} \qquad\qquad\qquad V_2 = ?$$
$$n_1 = 0.436 \text{ moles} \qquad\qquad n_2 = 0.436 + 1.27 = 1.70_6 \text{ moles}$$

$$\frac{V_1}{n_1} = \frac{V_2}{n_2}$$

$$\frac{5.20 \text{ L}}{0.426 \text{ mol}} = \frac{V_2}{1.70_6 \text{ mol}}$$

$$V_2 = \frac{5.20 \text{ L } (1.70_6 \text{ mol})}{0.436 \text{ mol}} = \textbf{20.3 L}$$

The answer agrees with our prediction.

Keep in mind that we have assumed that our gases behave **ideally**. That is not always a valid assumption, as your textbook points out.

5.3   The Ideal Gas Law

When you finish this section you will be able to solve a variety of problems relating to the ideal gas law.

The ideal gas law is a combination of Boyle's, Charles's, and Avogadro's laws. It relates <u>pressure</u>, <u>temperature</u>, <u>volume</u> and the <u>number of moles</u> of a gas. The derivation of the ideal gas law is given in your textbook. The equation of interest is

Pressure × Volume = # of moles of the gas × a constant × temperature

$$PV = nRT$$

P = pressure in **atm**
V = volume in **L**
n = number of **moles**
R = 0.08206 **L atm/K mol**
T = temperature in **Kelvin**

Please keep in mind that

1. This relationship assumes that the gas behaves **ideally**. As we will see in <u>Section 5.10</u>, there are certain conditions under which a gas will not behave ideally, and correction factors must be added to the ideal gas law.
2. You need to keep track of your dimensions. Many ideal gas law problems are best solved using **dimensional analysis**.
3. Always **list** what you are given. You may be able to simplify the problem.

Let's try a few examples.

Example 5.3 A   Ideal Gas Law

A sample containing 0.614 moles of a gas at 12.0°C occupies a volume of 12.9 L. What pressure does the gas exert?

Strategy

List the things that you are given, including constants. Once you see what you have and what you are being asked to find, you can decide how to manipulate your equation.

Solution

$$P = ? \qquad\qquad n = 0.614 \text{ moles} \qquad\qquad T = 285 \text{ K}$$
$$V = 12.9 \text{ L} \qquad\qquad R = 0.08206 \text{ L atm/K mol}$$

$$PV = nRT$$

or

$$P = \frac{nRT}{V}$$

$$P = \frac{(0.614 \text{ mol})(0.08206 \text{ l atm/K mol})(285 \text{ K})}{12.9 \text{ L}} = \textbf{1.11 atm}$$

You can check your answer by using dimensional analysis as shown below.

$$\text{atm} \; \Big\| \; \frac{0.08206 \text{ L atm}}{\text{K mol}} \times 0.614 \text{ mol} \times 285 \text{ K} \times \frac{1}{12.9 \text{ L}} = \textbf{1.11 atm}$$

↑
start with "*R*" with
the units where you
want them, then cancel

Example 5.3 B   Practice with Gas Laws

A sample of methane gas (CH_4) at 0.848 atm and 4.0°C occupies a volume of 7.0 L. What volume will
the gas occupy if the pressure is increased to 1.52 atm and the temperature increased to 11.0°C?

Predicting the Answer

According to Section 5.2 in your textbook, if the **pressure is increased**, the **volume should decrease**
(Boyle's Law). If the **temperature is increased**, the **volume should also increase** (Charles's Law).
However, the pressure almost doubles while the temperature increase is relatively small. Therefore the
pressure effect will be dominant. Overall then, we would expect the **volume to decrease**.

Setting up the Equations

You are given two sets of data.

$$P_1 = 0.848 \text{ atm} \qquad\qquad P_2 = 1.52 \text{ atm}$$
$$V_1 = 7.0 \qquad\qquad\qquad V_2 = ?$$
$$T_1 = 277 \text{ K} \qquad\qquad\quad T_2 = 284 \text{ K}$$

$$n_1 = n_2$$
$$R = R$$

$$P_1 V_1 = n_1 R T_1 \qquad \text{and} \qquad P_2 V_2 = n_2 R T_2$$

$$\frac{P_1 V_1}{T_1} = n_1 R \qquad \text{and} \qquad \frac{P_2 V_2}{T_2} = n_2 R$$

$$n_1 R = n_2 R \; \text{(because } n_1 = n_2\text{), so}$$

$$\frac{P_1 V_1}{T_1} = \frac{P_2 V_2}{T_2}$$

This is often called the **combined gas law**. The keys to setting up this equation are to **list what you are
given** and to **know what you are being asked to solve for**.

Solution

By converting the combined gas law to solve for V_2, we find that

$$V_2 = \frac{P_1 V_1 T_2}{P_2 T_1} = \frac{(0.848 \text{ atm})(7.0 \text{ L})(284 \text{ K})}{(1.52 \text{ atm})(277 \text{ K})} = \textbf{4.0 L}$$

This agrees with our prediction.

Example 5.3 C   More Practice with Gas Laws

How many moles of a gas at 104°C would occupy a volume of 6.8 L at a pressure of 270 mm Hg?

Solution

$$P = 270 \text{ mm Hg} \times \frac{1 \text{ atm}}{760 \text{ mm Hg}} = 0.355 \text{ atm}$$

$$V = 6.8 \text{ L} \qquad\qquad T = 377 \text{ K} \qquad\qquad n =?$$

$$n = \frac{PV}{RT} = \frac{(0.355 \text{ atm})(6.8 \text{ L})}{(0.08206 \text{ L atm/K mol})(377 \text{ K})} = \textbf{0.078 moles}$$

Note that the units are in moles. You can use dimensional analysis as an additional check.

$$\text{moles} \; \Big\| \; \frac{1 \text{ K mol}}{0.08206 \text{ L atm}} \times 6.8 \text{ L} \times \frac{1}{377 \text{ K}} = \textbf{0.078 moles}$$

5.4   Gas Stoichiometry

When you finish this section you will be able to:

- Define STP conditions.
- Do a variety of calculations regarding molar mass, density, and stoichiometry of gases.

This section uses the ideal gas law to perform a variety of calculations, including molar mass, density, and volume determination. Your textbook defines the volume occupied under **standard temperature and pressure (STP)**. One mole of an ideal gas at 0°C (standard temperature) and 1.000 atm (standard pressure) will occupy **22.4 liters**.

STP MEANS 0°C and 1.000 atm.

Example 5.4 A   The Ideal Gas Law and STP

What volume will 1.18 moles of O_2 occupy at STP?

Solution

Method 1:

$$V = ? \qquad\qquad R = 0.08206 \text{ L atm/K mol} \qquad\qquad P = 1.000 \text{ atm}$$
$$n = 1.18 \text{ mol} \qquad\qquad T = 273 \text{ K}$$

$$V = \frac{nRT}{P} = \frac{(1.18 \text{ mol})(0.08206 \text{ L atm/K mol})(273 \text{ K})}{1.000 \text{ atm}} = \textbf{26.4 L}$$

Method 2:

At STP, 1 mol occupies 22.4 L, therefore,

$$V = \frac{22.4 \text{ L}}{1 \text{ mol}} = \frac{V}{1.18 \text{ mol}}$$

$$V = \frac{(22.4 \text{ L})(1.18 \text{ mol})}{1 \text{ mol}} = \textbf{26.4 L}$$

As you can see, problems are made much simpler when you are dealing with STP conditions.

Many gas law problems involve calculating the **volume** of a gas produced by the reaction of volumes of other gases. The problem-solving strategy we have used throughout your chemistry course is still the same. That is, you want to relate **moles of reactants to moles of products**. The ideal gas law will allow you to use the following strategy:

volume of reactants $\xrightarrow{\text{ideal gas law}}$ moles of reactants $\xrightarrow{\text{stoichiometry}}$ moles of products ⌐

volume of products $\xleftarrow{\text{ideal gas law}}$

In the next problem, we are given moles of reactant and asked to find the volume of product.

Example 5.4 B   Reactions and the Ideal Gas Law

A sample containing 15.0 g of dry ice, $CO_2(s)$, is put into a balloon and allowed to sublime according to the following equation:

$$CO_2(s) \rightarrow CO_2(g)$$

How big will the balloon be (i.e., what is the volume of the balloon) at 22.0°C and 1.04 atm after all of the dry ice has sublimed?

Strategy

You are given *moles of* $CO_2(s)$. You want *volume of* $CO_2(g)$. The mole ratio of the solid to the gas is 1 to 1. Given the mass of CO_2, we can go from

g CO_2 $\xrightarrow{\text{molar mass}}$ moles $CO_2(g)$ $\xrightarrow{\text{ideal gas law}}$ volume CO_2

Solution

moles CO_2 ‖ $15.0 \text{ g } CO_2 \times \dfrac{1 \text{ mol } CO_2}{44.0 \text{ g } CO_2} = \textbf{0.341 mol } CO_2$

$$V = \frac{nRT}{P} = \frac{(0.341 \text{ mol})(0.08206 \text{ L atm/K mol})(295 \text{ K})}{1.04 \text{ atm}} = \textbf{7.94 L}$$

Using dimensional analysis as a check:

L CO_2 ‖ $\dfrac{0.08206 \text{ L atm}}{\text{K mol}} \times 295 \text{ K} \times \dfrac{1}{1.04 \text{ atm}} \times 0.341 \text{ mol} = \textbf{7.94 L}$

↑
get the proper
units on top

Example 5.4 C   Practice with the Ideal Gas Law

0.500 L of $H_2(g)$ are reacted with 0.600 L of $O_2(g)$ **at STP** according to the equation:

$$2H_2(g) + O_2(g) \rightarrow 2H_2O(g)$$

What volume will the H_2O occupy at 1.00 atm and 350°C?

Strategy

Because you have two reactants, this may be a **limiting reactant** problem. As with any such problem, you must find out how many moles of each reactant you have so that you can determine the limiting reactant. Once you have done those calculations, you can determine the moles and then the volume that the product occupies.

$$\text{reactant volume} \xrightarrow{\text{ideal gas law}} \text{moles of reactants} \xrightarrow{\text{limiting reactant}} \text{moles of product}$$

$$\text{volume of product} \xleftarrow{\text{ideal gas law}}$$

Use the fact that the reactants are initially at STP to help you solve the problem.

Solution

$$\text{moles } H_2 \text{ at STP} \quad \left\| \quad \frac{1 \text{ mol}}{22.4 \text{ L}} \times 0.500 \text{ L} = \textbf{0.0223 moles } H_2 \right.$$

$$\text{moles } O_2 \text{ at STP} \quad \left\| \quad \frac{1 \text{ mol}}{22.4 \text{ L}} \times 0.600 \text{ L} = \textbf{0.0268 moles } O_2 \right.$$

To determine the limiting reactant,

$$\text{moles } H_2O \text{ from } H_2 \quad \left\| \quad 0.0223 \text{ mol } H_2 \times \frac{2 \text{ mol } H_2O}{2 \text{ mol } H_2} = 0.0223 \text{ mol } H_2O \right.$$

$$\text{moles } H_2O \text{ from } O_2 \quad \left\| \quad 0.0268 \text{ mol } O_2 \times \frac{2 \text{ mol } H_2O}{1 \text{ mol } O_2} = 0.0536 \text{ mol } H_2O \right.$$

The limiting reactant is H_2, and **0.0223 mol H_2O** will be formed.

$$\text{Volume } H_2O = \frac{nRT}{P} = \frac{(0.0223 \text{ mol})(0.08206 \text{ L atm/K mol})(623 \text{ K})}{1.00 \text{ atm}} = \textbf{1.14 L}$$

Example 5.4 D   Density and Molar Mass

A gas at 34.0°C and 1.75 atm has a density of 3.40 g/L. Calculate the molar mass of the gas.

Solution

$$\text{Molar mass} = \frac{dRT}{P}$$

$$\text{Molar mass} = \frac{(3.40 \text{ g/L})(0.08206 \text{ L atm/K mol})(307 \text{ K})}{1.75 \text{ atm}} = \textbf{48.9 g/mol}$$

As a check use dimensional analysis:

$$\frac{\text{grams}}{\text{mol}} \;\bigg\|\; \underset{\underset{\text{density}}{\uparrow}}{\frac{3.40\,\text{g}}{\text{L}}} \times \underset{\underset{R}{\uparrow}}{\frac{0.08206\,\text{L atm}}{\text{K mol}}} \times \frac{1}{1.75\,\text{atm}} \times 307\,\text{K} = \mathbf{48.9\ g/mol}$$

5.5 Dalton's Law of Partial Pressures

At the end of this section you will be able to solve a variety of problems relating to **partial pressure**, **mole fraction**, and **total pressure**.

Dalton's Law of Partial Pressures states that for a mixture of gases in a container, the **total pressure** is the **sum** of the pressures that each gas would exert if it were alone. Your textbook shows that because the ideal gas law holds,

$$P_{\text{total}} = P_1 + P_2 + \ldots + P_n = (n_1 + n_2 + \ldots + n_n)\left[\frac{RT}{V}\right]$$

because $\dfrac{RT}{V}$ will be the same for each of the different gases in the same container.

The **key problem-solving strategy** with regard to partial pressure problems is to **use the ideal gas law to interconvert between pressure and moles of each gas**. Let's try the following example.

Example 5.5 A   Partial Pressures

A volume of 2.0 L of He at 46°C and 1.2 atm pressure was added to a vessel that contained 4.5 L of N_2 at STP. What is the **total pressure** and **partial pressure of each gas at STP** after the He is added?

Solution

Find the **number of moles** of He at the original conditions. This will ultimately lead to finding the partial pressure of He at STP.

$$n = \frac{PV}{RT}$$

$$n_{\text{He}} = \frac{(1.2\ \text{atm})(2.0\ \text{L})}{(0.08206\ \text{L atm/K mol})(319\ \text{K})} = \mathbf{0.091_7\ mol\ He}$$

When the gases are combined under STP conditions, the partial pressure of He will change. That of N_2 (already at STP) will remain the same.

$$P_{\text{He}} = \frac{nRT}{V} = \frac{(0.091_7\ \text{mol})(0.08206\ \text{L atm/K mol})(273\ \text{K})}{4.5\ \text{L}} = \mathbf{0.45_7\ atm}$$

$$P_{N_2} = \mathbf{1.00\ atm}$$

$$P_{\text{total}} = 1.00\ \text{atm} + 0.46\ \text{atm} = \mathbf{1.4_6\ atm} = \mathbf{1.5\ atm}$$

The equations for **MOLE FRACTION** are derived in your textbook. Recalling,

$$\chi_i = \frac{n_i}{n_{\text{total}}} = \frac{P_i}{P_{\text{total}}}$$

The key idea here is that once you have either the number of moles **OR** the pressure of each component of your system, you can calculate the mole fraction. The next example has several parts.

Example 5.5 B   Mole Fraction and Partial Pressure

a. Calculate the number of moles of N_2 present in the previous example (5.5 A).
b. Calculate the mole fractions of N_2 and He given the following data from Example 5.5 A.

i.   mole data
ii. pressure data

Solution

a. In calculating moles of N_2 that occupy 4.5 L (as given in 5.5 A), take advantage of STP conditions.

$$\textbf{True at STP}: \quad \frac{1 \text{ mol } N_2}{22.4 \text{ L}} = \frac{x \text{ mol } N_2}{4.5 \text{ L}}$$

$$x = \textbf{0.20}_\textbf{1} \textbf{ mol } N_\textbf{2}$$

b. i.   The total number of moles $= $ mol $N_2 + $ mol He $= 0.20_1 + 0.091_7 = 0.293$ mol

$$\chi_{N_2} = \frac{0.20_1 \text{ mol}}{0.293 \text{ mol}} = \textbf{0.69} \quad \text{dimensionless, because the units cancel on top and bottom}$$

$$\chi_{He} = \frac{0.091_7 \text{ mol}}{0.293 \text{ mol}} = \textbf{0.31}$$

ii. Using partial pressure data, the total pressure was 1.46 (1.45_7) atm.

$$\chi_{N_2} = \frac{1.00 \text{ atm}}{1.45_7 \text{ atm}} = \textbf{0.69}$$

$$\chi_{He} = \frac{0.45_7 \text{ atm}}{1.45_7 \text{ atm}} = \textbf{0.31}$$

Example 5.5 C   Vapor Pressure

The vapor pressure of water in air at 28°C is 28.3 torr. Calculate the mole fraction of water in a sample of air at 28°C and 1.03 atm pressure.

Solution

$$\chi_{H_2O} = \frac{P_{H_2O}}{P_{air}} = \frac{28.3 \text{ torr}}{783 \text{ torr}} = \textbf{0.036}$$

5.6   The Kinetic Molecular Theory of Gases

When you finish this section you will be able to use the basic assumptions of the kinetic molecular theory to:

- define temperature, and
- calculate root mean square velocity.

At the beginning of this section, your textbook makes clear that the **KINETIC MOLECULAR THEORY** is simply a **model** that attempts to explain the properties of an ideal gas. The postulates of the kinetic molecular theory are listed and discussed <u>at the beginning of this section in your textbook</u>. Your textbook points out that a model is considered successful if it correctly **predicts** the behavior of the system. The postulates are:

1. The volume of the individual particles of a gas can be assumed to be negligible.
2. The particles are in constant motion. The collisions of the particles with the walls of the container are the cause of the pressure exerted by the gas.
3. The particles are assumed to exert no forces on each other.
4. The average kinetic energy of a collection of gas particles is assumed to be directly proportional to the Kelvin temperature of the gas.

Your textbook uses these postulates, along with definitions of force, momentum, and pressure, and some geometry to derive the ideal gas law.

The idea that **temperature is a measure of the average kinetic energy of a gas** is of critical importance.

$$(KE)_{average} = \frac{3}{2} RT$$

Because $KE = \frac{1}{2} mv^2$ (where m = mass and v = velocity), $\frac{1}{2} mv^2 = 3/2\, RT$, and velocity (as indicated by the random motions of the particles of a gas) increases with higher temperature. **Temperature is a measure of kinetic energy**.

Your textbook presents a discussion in which the assumptions of the kinetic molecular theory are used to derive the fundamental gas laws, the ideal gas law and Dalton's law of partial pressures. The laws themselves have not changed; Zumdahl is merely showing how looking at gases from the <u>molecular standpoint</u> results in the same relationships that were derived from experiments on "real-world-sized" samples of gases.

Root Mean Square Velocity

The expression dealing with the average velocity of gas particles is called the **root mean square velocity** and is derived in your textbook.

$$\mu_{rms} = \sqrt{\frac{3RT}{M}}$$

where $R = 8.3145$ J/K mol $= \dfrac{8.3145\ \text{kg m}^2/\text{s}^2}{\text{K mol}}$ (because 1 J = 1 kg m^2/s^2)

   T = temperature in Kelvins
   M = mass of a mole of the gas in **kilograms** ("kg/mol")

This is certainly a situation where the proper use of units will help you make sure you have the correct answer.

Example 5.6   Root Mean Square Velocity

Calculate the root mean square velocity for the atoms in a sample of oxygen gas at

   a.   0.°C
   b.   300.°C

Prediction

As the temperature is increased, we would expect the average kinetic energy of our particles to increase. The mass of the particles is constant; therefore, we would expect the rms velocity to increase when the temperature is raised from 0.°C to 300.°C.

Solution

a. at 0.°C: $R = 8.3145$ kg m^2/s^2 K mol
    $T = 273$ K

$$M = \frac{\text{kg}}{\text{mol}} \; \Big\| \; \frac{32.0 \text{ g}}{1 \text{ mol}} \times \frac{1 \text{ kg}}{1000 \text{ g}} = 0.0320 \text{ kg/mol}$$

$$\mu_{\text{rms}} = \left[\frac{3 \left[\dfrac{8.3145 \text{ kg m}^2/\text{s}^2}{\text{K mol}} \right] (273 \text{ K})}{0.0320 \text{ kg/mol}} \right]^{1/2} = \left[\frac{2.128 \times 10^5 \text{ m}^2}{\text{s}^2} \right]^{1/2} = 461 \text{ m/s}$$

b. at 300.°C: $R =$ as above
    $M =$ as above
    $T = 573$ K

$$\mu_{\text{rms}} = \left[\frac{3 \left[\dfrac{8.3145 \text{ kg m}^2/\text{s}^2}{\text{K mol}} \right] (573 \text{ K})}{0.0320 \text{ kg/mol}} \right]^{1/2} = \left[\frac{4.467 \times 10^5 \text{ m}^2}{\text{s}^2} \right]^{1/2} = 668 \text{ m/s}$$

The increase in rms velocity agreed with our prediction.

Finally, your textbook points out two very important ideas. The first is that particles collide with one another and exchange energy after traveling a very short distance. This **distance between collisions** is called the **mean free path** and is typically about 10^{-7} m. Because the exchange of energy happens at different times, particles are speeding up and slowing down. They have an average rms velocity, but rarely have precisely that velocity. **Therefore, particles have a large range of velocities, the average of which is the rms velocity**.

5.7   *Effusion and Diffusion*

When you finish this section you will be able to calculate relative rates of effusion from molar masses and vice versa.

Your textbook points out that **diffusion and effusion** are two different processes.

- **Diffusion** relates to the mixing of gases.
- **Effusion** relates to the passage of a gas through an orifice into an evacuated chamber.

Graham's Law of Effusion

If the kinetic energies of two gases, 1 and 2, in a system are the same at a given temperature, then for a mole of the particles

$$KE_{\text{avg}} = N_A \left(\tfrac{1}{2} m_1 \overline{u_1^2} \right) = N_A \left(\tfrac{1}{2} m_2 \overline{u_2^2} \right)$$

or, if M, molar mass is equal to $N_A m^*$,

$$\tfrac{1}{2} M_1 \overline{u_1^2} = \tfrac{1}{2} M_2 \overline{u_2^2}$$

---

[*]*Note that "M" has changed meanings. In Chapter 4 it meant "molarity (mol/L)." Here, it means "molar mass." Such shifts in meaning are rare, but they do happen.*

by cross multiplying and canceling, this becomes

$$\frac{M_2}{M_1} = \frac{\overline{u_1^2}}{\overline{u_2^2}}, \text{ or } \sqrt{\frac{M_2}{M_1}} = \frac{u_{rms_1}}{u_{rms_2}} = \frac{\text{rate of effusion}_1}{\text{rate of effusion}_2}$$

This is Graham's Law of Effusion. The **higher** the molar mass, the **slower** the rate of effusion through a small orifice.

Example 5.7   Graham's Law of Effusion

How many times faster than He would NO_2 gas effuse?

Solution

$$M_{NO_2} = 46.01 \text{ g/mol}$$

$$M_{He} = 4.003 \text{ g/mol}$$

$$\sqrt{\frac{M_{NO_2}}{M_{He}}} = \frac{\text{rate}_{He}}{\text{rate}_{NO_2}}$$

$$\sqrt{\frac{46.01}{4.003}} = \frac{\text{rate}_{He}}{\text{rate}_{NO_2}}$$

$$3.390 = \frac{\text{rate}_{He}}{\text{rate}_{NO_2}}$$

He would effuse 3.39 times as fast as NO_2.

Note: you could have solved the problem as

$$\sqrt{\frac{M_{He}}{M_{NO_2}}} = \frac{\text{rate}_{NO_2}}{\text{rate}_{He}}$$

in this case, $0.295 = \dfrac{\text{rate}_{NO_2}}{\text{rate}_{He}}$. The conclusion is the same.

Does the Answer Make Sense?

NO_2 has a much higher mass than He. We would expect it to effuse more slowly.

With regard to **diffusion**, the important idea is that even though gases travel very rapidly (hundreds of meters per second), their motions are in all directions, so mixing is relatively slow. In the case of diffusion, the basic structure of Graham's Law holds.

$$\frac{\text{Distance traveled}_2}{\text{Distance traveled}_1} = \sqrt{\frac{M_1}{M_2}}$$

5.8   Real Gases

When you finish this section you will be able to:

- Describe why and how gases deviate from ideal behavior.
- Solve problems relating to the extent of the deviation.

We know that no gas behaves in a truly ideal fashion. This section is devoted to **determining the deviation of a real gas from ideality**. Models of gas behavior are based on the best available data, are necessarily approximations, and when they fail, can help us learn more about our system. In this case, our "ideal gas" model fails under two important conditions: **high pressure** and **low temperature**. This forces corrections to the volume and pressure terms in the ideal gas equation.

1. *Volume Correction*: Because gas molecules **do** take up space, the free volume of the container is not as large as it would be if it were empty.

$$\text{Volume}_{\text{available}} = V_{\text{container}} - \overset{\text{correction factor}}{\underset{\text{\# of moles of gas}}{nb}}$$

Therefore, pressure can be expressed

$$P = \frac{nRT}{(V - nb)}$$

We are halfway home.

2. *Pressure Correction*: Because gas molecules can interact with each other, they do not <u>collide with the walls of the container</u> ("exert pressure") to as great an extent as when there is no intramolecular interaction (as with an ideal gas). Therefore,

$$P_{\text{observed}} = P_{\text{ideal}} - \text{a correction factor.}$$

As discussed in your textbook, the correction factor that accounts for the decrease in pressure due to intramolecular interactions equals

$$P_{\text{obs}} = P_{\text{ideal}} - a\left[\frac{n}{V}\right]^2$$

Where a = a constant that depends upon the gas
n = the number of moles of gas

Combining P_{obs} (we will call this "P") and $V_{\text{corrected}}$ into the ideal gas equation, the **CORRECTED** equation is called **van der Waal's** equation and is given by

$$\left[P + a\left(\frac{n}{V}\right)^2\right](V - nb) = nRT$$

$$\underset{\substack{\text{correction} \\ \text{for pressure}}}{\uparrow} \qquad \underset{\substack{\text{correction} \\ \text{for volume}}}{\uparrow}$$

The constants a and b have been tabulated for different gases and are given in <u>Table 5.3 in your textbook</u>. Though the numbers get a bit messy, all you are really doing is determining corrected values for P and V.

Interactions among molecules are greatest at **low temperatures (low rms velocities), high pressure, and low volumes**. Thus deviations from ideality are expected to be greatest under these conditions.

Example 5.8   Van Der Waal's Equation

Calculate the pressure exerted by 0.3000 mol of He in a 0.2000-L container at −25.0°C

    a. using the ideal gas law, and
    b. using van der Waal's equation.

Solution

a. $PV = nRT$ for this ideal situation.

b. $P = \dfrac{nRT}{V}$. But, correcting for non-ideality.

$$\left[P + a\left(\frac{n}{V}\right)^2 \right](V - nb) = nRT$$

$$\text{or }\ P + a\left(\frac{n}{V}\right)^2 = \frac{nRT}{(V - nb)}$$

$$\text{or }\ P = \frac{nRT}{(V - nb)} - a\left(\frac{n}{V}\right)^2$$

This is one of the few equations you will be working with in general chemistry that you really cannot derive. You just have to know it.

From Table 5.3 in your textbook

$$a = 0.0341 \text{ atm L}^2/\text{mol}^2 \qquad\qquad b = 0.0237 \text{ L mol}$$

a. $P = \dfrac{nRT}{V} = \dfrac{(0.3000 \text{ mol})(0.08206 \text{ L atm/K mol})(248.2 \text{ K})}{0.2000 \text{ L}}$

$P_{\text{ideal}} = \mathbf{30.55\ atm}$

b. $P = \dfrac{(0.3000 \text{ mol})(0.08206 \text{ L atm/K mol})(248.2 \text{ K})}{0.2000 \text{ L} - (0.3000 \text{ mol})(0.0237 \text{ L/mol})} - \left(\dfrac{0.0341 \text{ atm L}^2}{\text{mol}^2}\right)\left(\dfrac{0.3000 \text{ mol}}{0.2000 \text{ L}}\right)^2$

$P_{\text{nonideal}} = 31.68 \text{ atm} - 0.077 \text{ atm} = \mathbf{31.60\ atm}$

There exists a pressure difference of 1 atmosphere between the ideal equation and van der Waal's equation in this case. The error is about 3 percent.

In rereading this section in the text, you need to keep in mind the **conditions** under which gases deviate from ideality. Be able to discuss **why** certain conditions lead to non-ideal behavior.

5.9   Characteristics of Several Real Gases

The key ideas in this section are:

- The van der Waals equation contains a term, "a", which is a pressure correction.
- H_2 has a low value of "a" because it exhibits very weak intermolecular forces.
- Real gases differ from ideal gases largely due to the intermolecular forces that make gases deviate from ideal behavior.

5.10   Chemistry in the Atmosphere

When you finish this section you will be able to list and describe those reactions that are important in the study of the atmosphere.

Composition of Our Atmosphere

In the first part of this section, your textbook makes some key points with regard to the **composition of our atmosphere**.

1. The atmosphere is composed of 78% N_2, 21% O_2, 0.9% Ar, and 0.03% CO_2 along with trace gases.
2. The composition of the atmosphere varies as a function of distance from the Earth's surface. Heavier molecules tend to be near the surface due to gravity.
3. Upper atmospheric chemistry is largely affected by ultraviolet, x-ray, and cosmic radiation emanating from space. The ozone layer is especially reactive to ultraviolet radiation.
4. Manufacturing and other processes of our modern society affect the chemistry of our atmosphere. Air pollution is a direct result of such processes.

Air Pollution

The reactions that cause air pollution are extremely complex and only partially understood. Your textbook gives background reactions on **photochemical smog** and **acid rain**. This serves as but a tiny introduction to the myriad of pollution-related problems and mechanisms.

The general reactions causing smog are:

1. $N_2(g) + O_2(g) \xrightarrow[\text{temperatures}]{\text{high}} 2NO(g)$
2. $2NO(g) + O_2(g) \rightarrow 2NO_2(g)$
3. $NO_2(g) + \xrightarrow[\text{energy}]{\text{radiant}} NO(g) + O(g)$
4. $O(g) + O_2(g) \rightarrow \underset{\text{ozone}}{O_3(g)}$

Higher ozone levels that are characteristic of smog cause lung and eye irritation and can be very dangerous for people with asthma, emphysema, and other respiratory conditions. Ozone can also react with other pollutants as discussed in the text, to form the hydroxyl radical, •OH, a very reactive oxidizing agent that can react with nitrogen oxides and hydrocarbons to further increase levels of air pollutants.

Acid Rain

Your textbook points out a number of gas precursors of acid rain. One example is $SO_3(g)$, from the reaction of sulfur (from coal) and oxygen. The special section in your textbook, "Acid Rain: An Expensive Problem" gives you the background to do the next problem.

Example 5.10   Acid Rain

List the reactions that lead to acidification of rainwater.

Solution

As discussed in the special section in your textbook, there are three main acids produced: $HNO_2(aq)$, $HNO_3(aq)$, and $H_2SO_4(aq)$.

The reactions are:

1. $2NO_2(g) + H_2O(l) \rightarrow \mathbf{HNO_2(aq)} + \mathbf{HNO_3(aq)}$
2a. $2SO_2(g) + O_2(g) \rightarrow \mathbf{2SO_3(g)}$
2b. $SO_3(g) + H_2O(l) \rightarrow \mathbf{H_2SO_4(aq)}$

Exercises

Section 5.1

1. In a science demonstration, 4 to 6 plastic bags are arranged under a 2.0 m × 2.0 m piece of plywood. A volunteer stands on the plywood and others blow into the bags to "levitate" the volunteer. If four bags are used and a 140-lb person is on the plywood, what is the pressure that must be supplied by each of the four blowers (in Pa)? What if 6 bags and blowers were used? (Assume the plywood has no mass.)

2. How much force is required to inflate a high-pressure bicycle tire to 95 pounds per square inch (655 kPa) with a hand pump that has a plunger with area of 5.0 cm^2?

3. An object exerts a force of 500. N and sits on an area of 4.5 m × 1.5 m. Calculate the amount of pressure exerted by the object in torr.

4. Calculate the density of mercury. (This can be done using the fact that 760 mm Hg = 101,325 Pa.) $F = ma$ where $a = 9.81$ m/s^2. Hint: Consider a column 76.0 cm high with cross section of 1 cm^2.

5. During your travels through deep space you discover a new solar system. You land on the outermost planet and determine that the acceleration due to gravity is 2.70 m/s^2. If your mass back on Earth is 72.0 kg, what force would you exert on a scale in pounds while standing on the planet's surface?

6. As you proceed on to the next planet, some of your unbreakable equipment breaks, including that top-of-the-line machine that determines acceleration due to gravity.

   a. How do you determine the acceleration due to gravity of this planet?
   b. Calculate the acceleration due to gravity if your 72-kg mass exerts a force of 18 pounds on the planet's surface.

Section 5.2

7. A diver at a depth of 100 ft (pressure approximately 3 atm) exhales a small bubble of air with a volume equal to 100 mL. What will be the volume of the bubble (assume the same amount of air) at the surface?

8. What would the volume of gas contained in an expandable 1.0-L cylinder at 15 MPa (1 MPa = 10^6 Pa) be at 1 atm (assuming constant temperature)?

9. A sample tube containing 103.6 mL of CO gas at 20.6 torr is connected to an evacuated 1.13 liter flask. (The new volume is the sum of those of the tube and flask.) What will the pressure be when the CO is allowed into the flask?

10. A gas has a pressure of 3.2 atm and occupies a volume of 45 L. What will the pressure be if the volume is compressed to 27 L at a constant temperature?

11. The volume of a gas (held at constant pressure) is to be used "as a thermometer." If the volume at 0.0°C is 75.0 cm^3, what is the temperature when the measured volume is 56.7 cm^3?

12. If a 16.6-L sample of a gas contains 9.2 moles of F_2, how many moles of gas would there be in a 750 mL sample at the same temperature and pressure?

13. An 11.2-L sample of gas is determined to contain 0.50 moles of N_2. At the same temperature and pressure, how many moles of gas would there be in a 20.-L sample?

14. Consider a 3.57-L sample of an unknown gas at a pressure of 4.3×10^3 Pa. If the pressure is changed to 2.1×10^4 Pa at a constant temperature, what will be the new volume of the gas?

15. Calculate the volume occupied at 87.0°C and 950. torr by a quantity of gas that occupied 20.0 L at 27.0°C and 570. torr.

16. A given mass of oxygen has a volume of 100. mL at 740. mm pressure and 25.0°C. State whether the volume will be greater than 100. mL, less than 100. mL, or unchanged for each of the following new conditions (mass of oxygen remains constant).

   a. 700. mm and 25.0°C
   b. 740. mm and 50.0°C
   c. 2220. mm and 600.0 K

17. A quantity of gas at 27.0°C is heated in a closed vessel until the pressure is doubled. To what temperature is the gas heated?

Section 5.3

18. What is the volume of 16 g of sulfur dioxide at 20.0°C and 740 torr pressure?

19. A weather balloon is filled with 0.295 m^3 of helium on the ground at 18°C and 756 torr. What will the volume of the balloon be at an altitude of 10 km where the temperature is −48°C and the pressure is 0.14 atm?

20. A sample of gas occupies 3.8 L at 15°C and 1.00 atm. What does the temperature need to be for the gas to occupy 8.3 L at 1.00 atm?

21. Calculate the volume of O_2 present in a sample containing 0.89 moles of O_2 at a temperature of 40°C and a pressure of 1.00 atm.

22. Water is decomposed to $H_2(g)$ and $O_2(g)$ by electrolysis. By measuring the current it was determined that 0.365 moles of water decomposed. After the gases are dried and collected at 24.5°C and 757 torr, what are the volumes of each?

23. What pressure would be exerted by 50.0 g of He at 25.0°C in a volume of 350. L?

24. A vacuum line used in a research lab has a volume of 1.013 L. The temperature in the lab is 23.7°C and the vacuum line is evacuated to a pressure of 1×10^{-6} torr. How many gas particles remain?

25. A 10.5-g sample of CO_2 gas occupies a volume of 7.00 L at a pressure of 1.5 atm. What must be the temperature of the gas?

26. A flask that can withstand an internal pressure of 2500 torr, but no more, is filled with a gas at 21.0°C and 758. torr and heated. At what temperature will it burst?

27. Calculate the number of moles present in a quantity of gas that occupies 26,880 mL at 564°C and 380. torr.

Section 5.4

28. The density of liquid nitrogen is 0.808 g/mL at −196°C. What volume of nitrogen gas at STP must be liquefied to make 10.0 L of liquid nitrogen?

29. Calculate the volume occupied by 2.5 mol of an ideal gas at STP.

30. A hydrocarbon (compound containing only hydrogen and carbon) was analyzed to be 85.7 mass percent carbon and 14.3 mass percent hydrogen. At 26°C and 745 torr pressure a sample with a volume of 1.13 L had a mass of 1.904 g. Determine the molecular formula. (You may wish to review Chapter 3.)

31. An unknown gas has a density of 7.06 g/L at a pressure of 1.50 atm and 280 K. Calculate the molar mass of the gas.

32. Air is a mixture of about 21.0% oxygen and 79.0% nitrogen (we'll neglect the minor components and water vapor in this example). What is the density of air at 30.0°C and 1.00 atm?

33. $HCl(g)$ can be prepared by reaction of NaCl with H_2SO_4. What mass of NaCl is required to prepare enough HCl to fill a 340.-mL cylinder to a pressure of 151 atm at 20.0°C?

34. A sample of 26.81 mL of 0.1000 M HCl reacts completely with a rock containing 3.164 g $CaCO_3$. What would be the maximum theoretical volume of CO_2 collected at 30°C and 1.00 atm?

35. You are not sure whether to fill a balloon with He or hot air. To what temperature would the air have to be heated for a balloon to rise to the same height as a balloon filled with He at 25.0°C?

36. A 27.7-mL sample of $CO_2(g)$ was collected over water at 25.0°C and 1.00 atm. What is the pressure in torr due to $CO_2(g)$? (The vapor pressure of water at 25.0°C is 23.8 torr.) What will the volume of $CO_2(g)$ be at the same temperature and pressure after removing the water vapor?

Section 5.5

37. A gas-tight vessel is filled with air at 27°C and 1.00 atm. Assuming air to be 79% N_2 and 21% O_2 (by volume), calculate the following:

   a. partial pressure of N_2
   b. partial pressure of O_2

38. A gaseous mixture of O_2, H_2, and N_2 has a total pressure of 1.50 atm and contains 8.20 g of each gas. Find the partial pressure of each gas in the mixture.

39. What is the effect of adding argon gas to the sample in Exercise #38?

40. The mole fraction of argon in dry air is 0.00934. How many liters of air at STP will contain enough argon to fill a 35.4-L cylinder to a pressure of 150. atm at 20°C?

41. Assume that the mole fraction of nitrogen in the air is 0.8902. Calculate the partial pressure of N_2 in the air when the atmospheric pressure is 820 torr.

42. A flask with a volume of 1.20 L is filled with carbon dioxide at room temperature to a pressure of 650. torr. A second flask, with a volume of 900. mL, is filled at room temperature with nitrogen to a pressure of 800. torr. A stopcock connecting the two volumes is then opened and the gases allowed to mix at room temperature. What is the partial pressure of each gas in the final mixture, and what is the total pressure of the mixture?

Section 5.6

43. Calculate the temperature of a mole of oxygen molecules if the internal energy is 1.16×10^4 J. Assume ideal gas behavior.

44. Calculate the root mean square speed of O_2 gas molecules at 300. K.

45. What happens to the average kinetic energy of a mole of an ideal gas if:
    a. the volume is doubled resulting in a decrease in pressure at constant temperature?
    b. the temperature is increased at a constant pressure?
    c. absolute zero is obtained?

Section 5.7

46. Ammonia, $NH_3(g)$, and $HCl(g)$ react to form a solid precipitate, NH_4Cl. Two cotton swabs, one moistened with ammonia and the other with hydrochloric acid, are inserted into opposite ends of a 1-meter long glass tube. How far from the hydrochloric acid end of the tube would you expect to see the white NH_4Cl precipitate?

47. Calculate the rate of effusion of PH_3 molecules through a small opening if NH_3 molecules pass through the same opening at a rate of 8.02 cm³/s. Assume the same temperature and equal partial pressures of the two gases.

48. What are the relative rates of diffusion for methane, CH_4, and oxygen, O_2? If $O_2(g)$ travels 1.00 m in a certain amount of time, how far will methane be able to travel under the same conditions?

49. Which gas would effuse faster, Ne or CO_2? How much faster?

50. Which gas would effuse faster, Ar or O_2? How much faster?

Section 5.8

51. Arrange the following according to expected values for b (volume correction value) in van der Waal's equation:

    $$He, CO_2, H_2O, HF, SF_6.$$

52. Put the following gases in order from smallest to largest according to van der Waal's constant "a":

    $$H_2, N_2, CH_4, Ne, H_2O.$$

53. Put the following gases in order from smallest to largest according to van der Waal's constant "b":

    $$Kr, Cl_2, NH_3, O_2, He.$$

54. Calculate the pressure exerted by 1 mole of $Xe(g)$ using the ideal gas law and van der Waal's equation
    a. in a 100.0-L container at 23°C, and
    b. in a 1.000-L container at 23°C.

55. Calculate the pressure exerted by 100. moles of Cl_2 gas in a 20.-L container at 25.0°C using van der Waal's equation and constants in Table 5.3 of your textbook.

56. Calculate the density of a) a mixture of 21% O_2 and 79% N_2 and b) one in which He is at 760. torr and 0.0°C. Will the ratio of densities be different at 100. atm and 6.0°C?

57. Why are all gases not perfect gases?

Multiple Choice Questions

58. If a barometer were built using water instead of Hg, how high would the column of water be if the pressure were 1 atm, knowing that the density of water is 13.6 times lower than that of mercury?

    A. 10.3 m   B. 3.17 m   C. 20.0 m   D. 33.0 m

59. Calculate the pressure, in Pascals, for a column of Hg that is 2.05 m high.

    A. 2.35×10^5 Pa   B. 2.70 Pa   C. 1.56×10^6 Pa   D. 2.73×10^5 Pa

60. What is the pressure, in mm Hg, of a gas that has a pressure of 15.0 lb/in^2?

    A. 0.113 mm Hg   B. 776 mm Hg   C. 1.02 mm Hg   D. 27.6 mm Hg

61. A balloon with an internal pressure of 300. torr rises to a height of 30,000 feet, where the pressure is 15. torr. Assuming temperature remains constant, by what ratio did the volume change?

    A. 25:1   B. 1000:1   C. 20:1   D. 1.85:1

62. A 0.88-L sample of helium is heated from 68°F to 68°C. At constant pressure, what volume does this sample occupy at 68°C?

    A. 1.0 L   B. 1.6 L   C. 0.9 L   D. 2.7 L

63. A 3.00-L sample of xenon is heated from 100°F to 200°F and an initial pressure of 70.0 cm increased to 120 cm of Hg. What is the final volume, in L, of the gas?

    A. 1.80 L   B. 2.06 L   C. 3.00 L   D. 6.00 L

64. How many moles of an ideal gas are present in a sample of 1.25 L at 311 K and a pressure of 25.0 lb/in^2?

    A. 0.0833 mol   B. 0.0510 mol   C. 0.328 mol   D. 0.0102 mol

65. A 3.25-L sample of a gas at 80.0°C is heated until a final volume of 32.5 L is reached. What is the final temperature of the gas in Kelvin at constant pressure?

    A. 3.53×10^3 K   B. 151 K   C. 1.08×10^3 K   D. 1.34×10^3 K

66. Calculate the number of grams of acetylene, C_2H_2, in a 30.0 L cylinder at a temperature of 20.0°C and a pressure equal to 2500 lb/in^2.

    A. 8.47×10^3 g   B. 1000 g   C. 5.55×10^3 g   D. 2.40×10^3 g

67. A 50.0-L cylinder at a temperature of 47°C and a pressure of 50.0 atm contains how many molecules of gas per cm^3?

    A. 1.15×10^{21} B. 2.30×10^{22} C. 2.30×10^{19} D. 6.75×10^{18}

68. A 50.0-L cylinder of Cl_2 at 20.0°C and a pressure of 103,401 torr springs a leak. The following day the pressure is found to be 41,361 torr. How many moles of chlorine gas escaped during this time?

    A. 170 mol   B. 280 mol   C. 85.0 mol   D. 113 mol

69. Tin reacts with hydrochloric acid to produce hydrogen gas and tin (II) chloride. How many liters of hydrogen gas are produced at 27.0°C and a pressure of 710 torr if 2.80 g of tin reacts with excess hydrochloric acid?

$$\text{Sn }(s) + 2HCl(aq) \rightarrow SnCl_2(aq) + H_2(g)$$

    A. 0.620 L   B. 0.320 L   C. 2.00 L   D. 1.25 L

70. How many cm^3 of carbon tetrachloride are produced when 8.0 L of chlorine are allowed to react with 0.75 L of methane at STP?

$$4Cl_2(g) + CH_4(g) \rightarrow 4HCl(g) + CCl_4(g)$$

A. 1500 cm^3 B. 750 cm^3 C. 360 cm^3 D. 1080 cm^3

71. Calculate the final pressure, in atm, after 9.06 g of krypton reacts with 10.0 g of fluorine at 300 K in a 10.0-L container.

$$Kr(g) + F_2(g) \rightarrow KrF_2(s)$$

A. 0.591 atm   B. 0.382 atm   C. 0.700 atm   D. 1.90 atm

72. Calculate the density change, g/L, if 700 g of $C_2H_6(g)$ are removed from a 200. L cylinder at 200. psi (lb/in^2) and a temperature of 20°C.

A. 3.5 g/L   B. 15.0 g/L   C. 1.7 g/L   D. 16.2 g/L

73. Calculate the density, in g/L, of sulfur dioxide(g) at 37°C and a pressure of 1440 torr.

A. 6.0 g/L   B. 0.60 g/L   C. 2.38 g/L   D. 4.76 g/L

74. Calculate P_T, in atm, for three different gases at partial pressures of 144.0 cm, 800.0 mm, and 1.3 m of Hg.

A. 1.90 atm   B. 2.58 atm   C. 1.06 atm   D. 4.66 atm

75. 1.0 L of hydrogen gas is collected over water at 308 K at a pressure of 728 torr. How many grams of iron are required to react with excess HCl(aq) to produce this volume of hydrogen gas? The vapor pressure of water is 42.2 torr. The products of the reaction are iron (II) chloride and hydrogen gas.

A. 4.7 g   B. 2.35 g   C. 2.0 g   D. 1.3 g

76. Gas A diffuses twice as fast as gas B. Gas B has a molecular weight = 60.0 g/mol. What is the molar mass of gas A?

A. 15.0 g/mol   B. 120 g/mol   C. 30 g/mol   D. 90 g/mol

77. If gas B effuses four times as fast as gas A, what is the ratio of the molar masses (A/B)?

A. 2:1   B. 4:1   C. 16:1   D. 8:1

78. The rate of effusion of freon-12 to freon-11 is 1.07:1. The molar mass of freon-11 is 137.4 g/mol. Calculate the molar mass, in g/mol, of freon-12.

A. 100 g/mol   B. 182 g/mol   C. 120 g/mol   D. 118 g/mol

79. Using the van der Waals equation, calculate the pressure exerted by 10. g of methane (CH_4) in a 2.1-L container at 330 K. $a = 2.253$ L^2 atm/mol^2, $b = 0.0458$ L/mol. Calculate using the ideal gas equation, and find the difference between the ideal gas pressure and van deer Waals pressure.

A. 2.0 atm   B. 0.5 atm   C. 0.2 atm   D. 1.5 atm

Answers to Exercises

1. 39 Pa for each of 4 blowers; 26 Pa for each of 6.

2. 328 N = 73 lbs

3. 0.56 torr

4. 13.6 g/cm^3 (Steps: 1. Force = $P \times A$; 2. mass = F/a;   3. density = mass / volume, so ultimately, density = $P \times$ area / volume $\times$ acceleration)

5. 43.5 lbs

6. a. Find out what force you exert on the surface and work backwards.
   b. 1.1 m/s^2

7. 300 mL

8. 150 L

9. 1.73 torr

10. 5.3 atm

11. 206 K, −67°C

12. 0.42 moles

13. 0.89 moles

14. 0.73 L

15. 14.4 L

16. a. more than   b. more than   c. less than

17. 600. K

18. 6.18 L

19. 1.6 m^3

20. 356°C

21. 23 L

22. 8.95 L H_2, 4.47 L O_2

23. 0.873 atm

24. 3×10^{13} particles

25. 536 K or 263°C

26. 970.15 K or 697°C

27. 0.196 moles

28. 6.46×10^3 L

29. 56 L

30. C_3H_6

31. 108 g/mol

32. 1.16 g/L

33. 125 g

34. 33.4 mL = 0.0334 L of CO_2

35. 2140 K or about 1870°C

36. 736 torr, 26.8 mL

37. a. 0.79 atm
    b. 0.21 atm

38. $O_2 = 0.0832$ atm; $H_2 = 1.32$ atm; $N_2 = 0.0951$ atm

39. No effect

40. 5.30×10^5 L

41. 730 torr

42. $P_{CO_2} = 0.489$ atm; $P_{N_2} = 0.451$ atm; total = 0.940 atm

43. 930 K

44. 484 m/s

45. a. no change
    b. increase
    c. goes to zero

46. 40 cm

47. 5.67 cm^3/s

48. 1.41 CH_4 : 1 O_2; 1.41 m

49. Ne would effuse 1.48 times as fast as CO_2.

50. O_2; 1.12 times as fast

51. (smallest) He, HF, H_2O, CO_2, SF_6 (largest)

52.   Ne, H$_2$, N$_2$, CH$_4$, H$_2$O

53.   He, O$_2$, NH$_3$, Kr, Cl$_2$

54.   a.   ideal = 0.243 atm, van der Waal's = 0.243 atm
      b.   ideal = 24.3 atm, van der Waal's = 21.4 atm

55.   7.9 atm

56.   air = 1.29 g/L; He = 0.179 g/L; no

57.   Gases generally do not follow the ideal gas law, but more closely approach the behavior of an ideal gas at low pressure and high temperatures.

58.	A	59.	D	60.	B	61.	C	62.	A	63.	B
64.	A	65.	A	66.	C	67.	A	68.	A	69.	A
70.	B	71.	B	72.	A	73.	D	74.	D	75.	C
76.	A	77.	C	78.	C	79.	C				

Chapter 6

Thermochemistry

This chapter serves as an introduction to the chemistry of energy production and exchange. Additional, more advanced, material will be presented in Chapter 16.

6.1   The Nature of Energy

When you finish this section you will be able to solve problems relating to:

- work, in terms of pressure and volume, and
- the first law of thermodynamics.

Your textbook describes a number of new terms for you. The terms, along with brief definitions, are given below:

- **Thermodynamics:** The study of energy and its interconversions.
- **Energy:** The capacity to produce work or heat.
- **Kinetic Energy**: The energy of motion. Kinetic Energy = ½ mass × (velocity)^2.
- **Potential Energy**: Energy that can be converted to useful work.
- **Heat**: Involves transfer of energy between two objects.
- **Work**: Force × distance.
- **State Function**: A property that is independent of pathway. That is, it does not matter how you get there, the difference in the value is the same. For example, you can drive from New York to Los Angeles via many different routes. No matter which one you take, you are still going from New York to Los Angeles. The actual distance between the two cities is the same. **Energy** is a state function, **work** and **heat** are not.

Four more definitions will set the stage for thermodynamics. The **universe** is composed of the **system** and the **surroundings.**

- **System:** That which we are focusing on.
- **Surroundings:** Everything else in the universe.
- **Exothermic:** Energy (as heat) flows **out** of **the system.**
- **Endothermic:** Energy (as heat) flows **into the system.**

Work

We will take a different approach to this topic than your textbook does. Let's look at work from the point of view of **units**. We learned in the review chapter on <u>gas laws</u> that

$$\text{Force} = \text{mass} \times \text{acceleration} = \text{kg} \times \text{m s}^{-2}.$$

$$\text{Work} = \text{force} \times \text{distance} = \text{kg m s}^{-2} \times \text{m} = \text{kg m}^2 \text{ s}^{-2}.$$

$$\textbf{1 Joule} = \textbf{1 kg m}^2 \textbf{ s}^{-2}.$$

Example 6.1 A   *The Units of Work*

If pressure = force/area, what are the units of pressure $\times$ volume?

Solution

$$P = F/A = \text{kg m s}^{-2}/\text{m}^2 = \text{kg m}^{-1} \text{ s}^{-2} \ (= 1 \text{ Pascal})$$

$$P \times V = \text{kg m}^{-1} \text{ s}^{-2} \times \text{m}^3 = \text{kg m}^2 \text{ s}^{-2} \ (= 1 \text{ Joule})$$

Therefore, $P \times V$ has the same units as force $\times$ distance (work), and both are measures of **energy**.

Conclusion

For an ideal gas,

$$\textbf{WORK} = P\Delta V$$

This equation holds at constant pressure.

The **sign conventions** for work are as follows.

- When the **system expands,** it is doing positive work on the surroundings, therefore it is doing **negative work on the system.**
- When the **system contracts,** the surroundings have done work on the system, therefore there is **positive work done on the system.**

From the point of view of the system, then,

$$w = -P\Delta V$$

Example 6.1 B   *Work*

Calculate the work (with the proper sign) associated with the contraction of a gas from 75 L to 30 L (work is done "on the system") at a constant **external** pressure of 6.0 atm in:

   a.   L atm
   b.   Joules (1 L atm = 101.3 J)

Helpful Hint

Keep in mind that **system compression is positive work**, and **system expansion is negative work.**

Solution

a. The work is positive (compression).

$$\Delta V = \text{change in volume} = V_{\text{final}} - V_{\text{initial}} = 30\text{ L} - 75\text{ L} = \mathbf{-45\ L}$$

$$w = -P\Delta V = -6.0\text{ atm}\ (-45\text{ L}) = \mathbf{+270\ L\ atm}$$

b. $w\ (\text{in Joules}) = +270\text{ L atm} \times \dfrac{101.3\text{ J}}{1\text{ L atm}} = \mathbf{+2.7 \times 10^4\ J}$

The First Law

The law of conservation of energy, also called the **first law of thermodynamics** is described in your textbook. It states that **energy can be converted from one form to another, but can be neither created nor destroyed.** Another way of stating the first law is that

THE ENERGY OF THE UNIVERSE IS CONSTANT.

We know that energy can be changed through **work**. As chemical bonds are made and broken, energy is converted between the potential energy (stored in chemical bonds) and thermal energy (kinetic energy) as heat.

The change in the internal energy of the system, which is equal in size but opposite in sign to that of the surroundings, is equal to the **sum of the heat and work.**

$$\Delta E = q + w$$

Your textbook points out that the **SIGN** of the energy change must be viewed from the point of view of the **SYSTEM.**

$$\Delta E = -\text{ means the system loses energy.}$$
$$\Delta E = +\text{ means the system gains energy.}$$

Example 6.1 C   The First Law

Calculate the change in energy of the system if 38.9 J of work is done *by* the system with an associated heat loss of 16.2 J.

Strategy

The most important part of problem solving in thermodynamics is getting the signs correct.

$$q = \text{"}-\text{", because heat is \textbf{lost}.}$$
$$w = \text{"}-\text{", because work is done \textbf{by} the system.}$$

Solution

$$\Delta E = q + w = -16.2\text{ J} + (-38.9\text{ J}) = \mathbf{-55.1\ J}$$

The system has **lost** 55.1 J of energy.

Example 6.1 D   Practice With Heat and Work

A piston is compressed from a volume of 8.3 L to 2.8 L against a constant pressure of 1.9 atm. In the process, there is a heat gain by the system of 350 J. Calculate the change in energy of the system.

Solution

$$w = -P\Delta V = -1.9 \text{ atm} (-5.5 \text{ L}) = +10.45 \text{ L atm}$$

$$10.45 \text{ L atm} \times \frac{101.3 \text{ J}}{\text{L atm}} = \textbf{+1059 J}$$

$$q = +350 \text{ J}$$

$$\Delta E = q + w = +1059 + 350 = 1409 \text{ J} = \textbf{1400 J} \text{ (to 2 significant figures)}$$

6.2   Enthalpy and Calorimetry

When you finish this section you will be able to solve problems relating to both **enthalpy** and **calorimetry**.

Enthalpy

Your textbook derives and defines a term called **enthalpy** (H). It is a **state function,** so the change in H is independent of pathway. That is,

$$\Delta H = H_{\text{products}} - H_{\text{reactants}}$$

The change in enthalpy (ΔH) of the system is equal to the energy flow as heat **at constant pressure**.

$$\Delta H = q_p$$

If $\Delta H > 0$, the reaction is **endothermic.** (Heat is absorbed by the system.)
If $\Delta H < 0$, the reaction is **exothermic.** (Heat is given off by the system.)

Example 6.2 A   Enthalpy

When solid potassium hydroxide pellets are added to water, the following reaction takes place:

$$\text{NaOH}(s) \rightarrow \text{NaOH}(aq)$$

For this reaction at constant pressure, $\Delta H = -43$ kJ/mol. Answer the following questions regarding the addition of 14 g of NaOH to water:

    a.   Does the beaker get warmer or colder?
    b.   Is the reaction exo- or endothermic?
    c.   What is the enthalpy change for the dissolution?

Solution

    a.   If $\Delta H < 0$, then heat is given off by the system. The beaker therefore gets **warmer.**
    b.   If heat is given off by the system, the reaction is **exothermic.**
    c. $\text{kJ} \left\| \dfrac{-43 \text{ kJ}}{\text{mol}} \times \dfrac{1 \text{ mol NaOH}}{40.0 \text{ g NaOH}} \times 14 \text{ g NaOH} = -15 \text{ kJ} = \Delta H \right.$

Calorimetry

Calorimetry is the experimental technique used to determine the heat exchange (q) associated with a reaction.

At constant pressure, $q = \Delta H$.
At constant volume, $q = \Delta E$.

In both cases, however, **heat gain or loss** is being determined. The amount of heat exchanged in a reaction depends upon:

1. **The net temperature change** during the reaction.
2. **The amount of substance.** The more you have, the more heat can be exchanged.
3. The **heat capacity** (*C*) of a substance.

$$C = \frac{\text{heat absorbed}}{\text{increase in temperature}} = \text{J/}^\circ\text{C}$$

Some substances can absorb more heat than others for a given temperature change.

There are three ways of expressing heat capacity:

1. **Heat capacity** (as above) = J/°C.
2. **Specific heat capacity** = heat capacity per gram of substance (J/°C g or J/Kg).
3. **Molar heat capacity** = heat capacity per mole of substance (J/°C mol or J/K mol)

You can solve calorimetry problems very well using dimensional analysis. Before we solve numerical problems, let's do a problem involving interpretation of specific heat capacities.

Example 6.2 B   Specific Heat Capacity

Look at Table 6.1 in your textbook. Based on the values for specific heat capacity, which conducts heat better, water or aluminum? Why is this important in cooking?

Solution

The heat capacity of water is **4.18 J/°C g**. This means that it takes 4.18 J of energy to raise the temperature of one gram of water 1°C.

The heat capacity of aluminum is **0.89 J/°C g**. This means that it takes 0.89 J of energy to raise the temperature of one gram of aluminum 1°C.

In other words, it takes almost five times as much energy (4.18/0.89) to raise the temperature of an equivalent amount of water by 1°C. Therefore, **aluminum conducts heat better** because less heat causes an equal rise in temperature. This is important in cooking because pots made of aluminum transfer heat very well to food. Note that iron pots conduct heat even more readily than aluminum (but they are more difficult to take care of).

Your textbook discusses doing **constant pressure** calorimetry using a "coffee cup calorimeter." In this case, ΔH = q$_p$ in units of Joules. Remember that you may use dimensional analysis to solve calorimetry problems.

Example 6.2 C   Constant Pressure Calorimetry

Recall from Example 6.2 A that the enthalpy change (ΔH) per mole of NaOH is −43 kJ/mol when

$$\text{NaOH}(s) \;\rightarrow\; \text{NaOH}(aq)$$

If 10.0 g of solid NaOH is added to 1.00 L of water (specific heat capacity = 4.18 J/°C g) at 25.0°C in a constant pressure calorimeter, what will be the final temperature of the solution? (Assume the density of the final solution is 1.05 g/mL.)

Strategy

We need to know three things:

1. **Mass of the solution** $= 1.00 \text{ L} \times 1050 \text{ g /L} = 1050 \text{ g}$.

2. **Heat capacity of the solution** $= 4.18 \text{ J/°C g}$.

3. **The enthalpy of the reaction** $= -43 \text{ kJ/mol} \times 10.0 \text{ g NaOH} \times \dfrac{1 \text{ mol NaOH}}{40.0 \text{ g NaOH}} \times \dfrac{1000 \text{ J}}{1 \text{ kJ}}$

$$= -10{,}750 \text{ J}$$

We want to know the **change in temperature**, ΔT. We can solve using dimensional analysis. (Keep in mind the temperature will **rise** because heat is evolved.)

Solution

$$°C \; \bigg\| \; \underset{\substack{\uparrow \\ \underline{\hspace{1.5cm}} \\ \text{specific heat capacity}}}{\dfrac{1°C \text{ g}}{4.18 \text{ J}}} \; - \; \underset{\substack{\uparrow \\ \underline{\hspace{1cm}} \\ \text{soln mass}}}{\dfrac{1}{1050 \text{ g}}} \; \times \; \underset{\substack{\uparrow \\ \Delta H}}{10{,}750 \text{ J}} \; = \; 2.4°C$$

The **final temperature** will equal $25.0°C + 2.4°C = \mathbf{27.4°C}$.

Constant volume calorimetry is discussed in your textbook. The **bomb calorimeter** is used for this application. In this case, because $\Delta V = 0$, no work is done, and $\Delta E = q_V$ in units of joules. Here too, dimensional analysis works well.

Each bomb calorimeter is different. The **heat capacity (J/°C)** of the bomb and its parts must be determined using a known substance before the energy (or heat) of combustion can be determined.

Example 6.2 D   Constant Volume Calorimetry

The heat of combustion of glucose, $C_6H_{12}O_6$, is 2800 kJ/mol. A sample of glucose weighing 5.00 g was burned with excess oxygen in a bomb calorimeter. The temperature of the bomb rose 2.4°C. **What is the heat capacity of the calorimeter?**

A 4.40 g sample of propane (C_3H_8) was then burned with excess oxygen in the same bomb calorimeter. The temperature of the bomb increased 6.85°C. Calculate $\Delta E_{combustion}$ of propane.

Strategy

There are two parts to this problem. First, we must calculate the heat capacity of the bomb calorimeter using the data for glucose. Second, we can use this heat capacity to determine the energy (heat) of combustion of propane, C_3H_8. Remember that the energy of combustion is expressed in **kJ/mol**. A useful beginning is to convert grams of substance to moles of substance

Solution

A. *Heat Capacity of the Calorimeter*

$$\text{moles } C_6H_{12}O_6 = 5.00 \text{ g} \times \dfrac{1 \text{ mol}}{180.0 \text{ g}} = 2.78 \times 10^{-2} \text{ moles}$$

$$\mathbf{heat\ capacity} = \dfrac{\text{kJ}}{°C} \; \bigg\| \; \dfrac{2800 \text{ kJ}}{\text{mol}} \times 2.78 \times 10^{-2} \text{ mol} \times \dfrac{1}{2.4°C} = \mathbf{32.4\ kJ/°C}$$

B. *Energy of Combustion of Propane*

$$\text{moles } C_3H_8 = 4.40 \text{ g} \times \frac{1 \text{ mol}}{44.0 \text{ g}} = 0.100 \text{ moles}$$

$$\Delta E_{combustion} = \frac{\text{kJ}}{\text{mol}} \; \Big|\Big| \; \frac{32.4 \text{ kJ}}{°\text{C}} \times 6.85°\text{C} \times \frac{1}{0.100 \text{ mol}} = -2200 \text{ kJ/mol}$$

(Note: We add the **negative** sign because heat is **evolved**.)

6.3   Hess's Law

When you finish this section you will be able to use Hess's Law to calculate enthalpy changes for a variety of reactions.

The critical point that is made in this section is that **enthalpy changes are state functions.** The implication is that **it does not matter if ΔH for a reaction is calculated in one step or a series of steps. This idea is called Hess's law.**

By using values of ΔH of known reactions, we can use Hess's law to solve for enthalpies of reactions whose values we do not know.

Example 6.3 A   Hess's Law

Given the following reactions and ΔH values,

a. $2N_2O(g) \rightarrow O_2(g) + 2N_2(g)$ $\Delta H_a = -164 \text{ kJ}$
b. $2NH_3(g) + 3N_2O(g) \rightarrow 4N_2(g) + 3H_2O(l)$ $\Delta H_b = -1012 \text{ kJ}$

Calculate ΔH for

$$4NH_3(g) + 3O_2(g) \rightarrow 2N_2(g) + 6H_2O(l).$$

Strategy

The idea is to manipulate equations "a" and "b" so that they add up to the desired equation. <u>There are three ways that we can manipulate equations</u>:

1.   **We can reverse the entire equation.** By doing this, the products become reactants and vice versa.
2.   **We can multiply the entire equation by a factor** such as 3, 2, ½, or 1/3.
3.   **We can do both #1 and #2.**

The most important thing to keep in mind is that **WHEN YOU MANIPULATE AN EQUATION, YOU MUST MANIPULATE THE ΔH VALUE IN EXACTLY THE SAME WAY!**

If you multiply an equation by 2, you must multiply ΔH by 2. If you reverse the equation, you must multiply ΔH by −1. (An exothermic reaction becomes endothermic and vice versa.)

Solution

In my experience, the best way to solve Hess's law problems is to **find a substance that only appears once in the reactants.** Modify that reaction so the substance appears where it should be, and in the correct amount, as in the final reaction. The entire substance equation must, therefore, be correct. In our example, NH_3 appears only once in the reactants. (N_2O appears in **both** equations...STAY AWAY FROM N_2O!)

We have $2NH_3$ on the left-hand side. We want $4NH_3$ on that side. Therefore, we must multiply equation "b" and ΔH_b by +2, which gives

$$4NH_3(g) + 6N_2O(g) \rightarrow 8N_2(g) + 6H_2O(l) \qquad \Delta H_b = -2024 \text{ kJ}$$

Oxygen appears only once in the reactants. Therefore, if we modify equation "a" to get the correct amount of O_2 in the proper place, we should be done. (We have modified both equations.) We need to **reverse** equation "a" and **multiply** it by 3 to get $3O_2$ on the left side. This will agree with the desired reaction. Remember to multiply ΔH_a by −3 as well! This gives,

$$3O_2(g) + 6N_2(g) \rightarrow 6N_2O(g) \qquad \Delta H_a = +492 \text{ kJ}$$

Let's get the final ΔH by adding our new "a" and "b,"

$4NH_3(g) + 6N_2O(g) \rightarrow 8N_2(g) + 6H_2O(l)$	$\Delta H_b = -2024$ kJ
$3O_2(g) + 6N_2(g) \rightarrow 6N_2O(g)$	$\Delta H_a = +492$ kJ
$3O_2(g) + 6N_2(g) + 4NH_3(g) + 6N_2O(g) \rightarrow 6N_2O(g) + 8N_2(g) + 6H_2O(l)$	$\Delta H = -1532$ kJ

and canceling common terms,

$$3O_2(g) + 4NH_3(g) \rightarrow 2N_2(g) + 6H_2O(l) \qquad \Delta H = -1532 \text{ kJ}$$

Getting the correct final reaction serves as your check of correctness.

Example 6.3 B *Practice with Hess's Law*

Given the following reactions and ΔH values,

a.	$B_2O_3(s) + 3H_2O(g) \rightarrow B_2H_6(g) + 3O_2(g)$	$\Delta H = +2035$ kJ
b.	$2H_2O(l) \rightarrow 2H_2O(g)$	$\Delta H = +88$ kJ
c.	$H_2(g) + \frac{1}{2}O_2(g) \rightarrow H_2O(l)$	$\Delta H = -286$ kJ
d.	$2B(s) + 3H_2(g) \rightarrow B_2H_6(g)$	$\Delta H = +36$ kJ

Calculate ΔH for

$$2B(s) + {}^3\!/_2O_2(g) \rightarrow B_2O_3(s)$$

Solution

Keep the strategy in mind! Boron and B_2O_3 only appear once. Work with them first.

1. Working with $B(s)$, we see that **equation "d" is correct as is.**
2. Working with $B_2O_3(s)$, we must **multiply equation "a" by −1** (thus reversing the equation).

Now we must work with things that are present more than once.

3. Working with $O_2(g)$, we see that we **must multiply equation "c" by −3** to guarantee a total of ${}^3\!/_2O_2(g)$ on the left side when all equations are added together. (Remember, we already have $3O_2(g)$ on the left side from equation "a".)

Three of our equations are now set. Equation "b" involves $H_2O(l)$ and $H_2O(g)$. Because we multiplied equation "c" by −3, we have $3H_2O(l)$ on the left-hand side to cancel. Therefore, we must **multiply equation "b" by −3/2.**

Checking the Equations and Determining ΔH

a.	$B_2H_6(g) + 3O_2(g) \rightarrow B_2O_3(s) + 3H_2O(g)$		$\Delta H = -2035$ kJ
b.	$3H_2O(g) \rightarrow 3H_2O(l)$		$\Delta H = -132$ kJ
c.	$3H_2O(l) \rightarrow 3H_2(g) + 3/2\ O_2(g)$		$\Delta H = +858$ kJ
d.	$2B(s) + 3H_2(g) \rightarrow B_2H_6(g)$		$\Delta H = +36$ kJ

$B_2H_6(g) + 3O_2(g) + 3H_2O(g) + 3H_2O(l) + 2B(s) + 3H_2(g) \rightarrow$
$$B_2O_3(s) + 3H_2O(g) + 3H_2O(l) + 3H_2(g) + {}^3/_2O_2(g) + B_2H_6(g)$$

Canceling we get

$$\mathbf{2B(s)\ +\ {}^3/_2O_2(g)\ \rightarrow\ B_2O_3(s)} \qquad \mathbf{\Delta H = -1273\ kJ}$$

Look at Example 6.8 in your textbook. Some of the equations here were turned around and multiplied by an integer value, but you notice that the ΔH values ultimately agree with those in your textbook. Hess's law illustrates that enthalpy is a state function.

6.4 Standard Enthalpies of Formation

When you finish this section you will be able to use your knowledge of standard states and standard enthalpies of formation to calculate ΔH for a variety of reactions.

Look at your textbook where the **standard enthalpy of formation** (ΔH_f^o) of a compound is defined. There are some important points that are worth going over.

1. ΔH_f^o is always given **per mole** of compound formed.
2. ΔH_f^o involves formation of a **compound** from its **elements** with the substances in their **standard states**.
3. Your textbook lists the following standard state conditions:

   For an element:

   - It is the form which the element exists in at 25°C and 1 atmosphere.

   For a compound:

   - For a gas it is a pressure of exactly 1 atmosphere.
   - For a pure solid or liquid, it is the pure solid or liquid.
   - For a substance in solution, it is a concentration of exactly 1 M.

4. ΔH_f^o for an element in its standard state, such as $Ba(s)$ or $N_2(g)$, equals 0.

Example 6.4 A   Standard Enthalpies of Formation

By consulting Appendix 4 of your textbook, and from your knowledge of standard states, list the standard enthalpy of formation for each of the following substances.

a.	$Al_2O_3(s)$	c.	$P_4(g)$	e.	$F_2(g)$
b.	$Ti(s)$	d.	$SO_4^{2-}(aq)$		

Solution

a. -1676 kJ/mol
b. 0 kJ/mol (The solid is the standard state of titanium.)
c. 59 kJ/mol
d. -909 kJ/mol
e. 0 kJ/mol (The gaseous diatom is the standard state of fluorine.)

The key to calculating standard enthalpy changes in reactions is to remember "products minus reactants." More correctly,

$$\Delta H^{\circ}_{\text{reaction}} = \sum n_{\text{p}} \Delta H^{\circ}_{\text{f}} (\text{products}) - \sum n_{\text{r}} \Delta H^{\circ}_{\text{f}} (\text{reactants}) .$$

This reads "the sum of the $\Delta H^{\circ}_{\text{f}}$ for n moles of each of the products minus the sum of the $\Delta H^{\circ}_{\text{f}}$ for n moles of each of the reactants."

Keep in mind that, just as in Hess's law problems, when you multiply the substance by an integer coefficient in a balanced equation, you must multiply the $\Delta H^{\circ}_{\text{f}}$ value by that integer as well! Let's work the following example together.

Example 6.4 B   Calculating Standard Enthalpies of Formation

Using the data in Appendix 4 of your textbook, calculate ΔH° for the following reaction:

$$2C_3H_6(g) + 9O_2(g) \rightarrow 6CO_2(g) + 6H_2O(l)$$

Solution

The ΔH° for the reaction $= \sum n_{\text{p}} \Delta H^{\circ}_{\text{f}} (\text{products}) - \sum n_{\text{r}} \Delta H^{\circ}_{\text{f}} (\text{reactants}) .$

$$\sum n_{\text{p}} \Delta H^{\circ}_{\text{f}} (\text{products}) = [6 \times \Delta H^{\circ}_{\text{f}} \text{ for } H_2O(l)] + [6 \times \Delta H^{\circ}_{\text{f}} \text{ for } CO_2(g)]$$
$$= 6 \text{ mol} \times -286 \text{ kJ/mol} + 6 \text{ mol} \times -393.5 \text{ kJ/mol}$$
$$= \mathbf{-4077 \text{ kJ}}$$

$$\sum n_{\text{r}} \Delta H^{\circ}_{\text{f}} (\text{reactants}) = [2 \times \Delta H^{\circ}_{\text{f}} \text{ for } C_3H_6(g)] + [9 \times \Delta H^{\circ}_{\text{f}} \text{ for } O_2(g)]$$
$$= 2 \text{ mol} \times 20.9 \text{ kJ/mol} + 9 \text{ mol} \times 0 \text{ kJ/mol}$$
$$= \mathbf{+41.8 \text{ kJ}}$$

Finally,

$$\Delta H^{\circ} = (-4077 \text{ kJ}) - (+41.8 \text{ kJ}) = \mathbf{-4119 \text{ kJ}}$$

Example 6.4 C   Practice with Standard Enthalpies

The "thermite" reaction is discussed in Example 6.10 in your textbook. It is one in which molten iron is made from the reaction of aluminum powder and iron oxide. A variation on that reaction is,

$$2Al(s) + Cr_2O_3(s) \rightarrow Al_2O_3(s) + 2Cr(s)$$

a.   Calculate ΔH° for this reaction.
b.   Which reaction yields more energy **per gram of metal formed,** thermite or this one?

Solution

a. $\Delta H^{\circ} = [-1676 \text{ kJ} + 2 \times 0] - [2 \times 0 + (-1128 \text{ kJ})]$
          ↑   ↑   ↑   ↑
     $Al_2O_3(s)$ $Cr(s)$ $Al(s)$ $Cr_2O_3(s)$

   $\Delta H^{\circ} = \mathbf{-548 \text{ kJ}}$

b. $\Delta H^{\circ}_{\text{thermite}} = \mathbf{-850 \text{ kJ}}$

In the thermite reaction, 2 moles, or 111.7 g, of Fe are formed. Therefore, the heat evolved per gram of Fe formed equals:

$$\frac{-850 \text{ kJ}}{111.7 \text{ g Fe}} = -7.61 \text{ kJ/g Fe formed}$$

$\Delta H^{o}_{\text{chromium formation}} = -548$ kJ. There are 2 moles, or 104.0 g of chromium formed. The heat evolved per gram of chromium formed equals:

$$\frac{548 \text{ kJ}}{104.0 \text{ g Cr}} = -5.27 \text{ kJ/g Cr formed}$$

Therefore, the **thermite reaction yields more energy per gram of metal formed**.

6.5   *Present Sources of Energy*

Carefully read <u>Section 6.5 in your textbook</u>. This material deals with the application of thermodynamics to the "real world," and as such is perhaps the most interesting in the chapter! When you are done, use the following review questions to test your understanding of the material.

1.   What is **petroleum**?
2.   What was the most likely way in which petroleum was formed?
3.   What gases make up **natural gas**?
4.   How are hydrocarbons separated from one another?
5.   What are the major uses of the various petroleum fractions?
6.   What is **pyrolitic cracking**?
7.   There are two reasons that lead additives are a problem in gasoline. Discuss them.
8.   What are the four stages of **coal**? How do they differ?
9.   What are some of the problems with the use of coal as an energy source?
10.   What is the greenhouse effect? Why will increased amounts of CO_2 exacerbate this effect?

6.6   *New Energy Sources*

One of the impressive things about your textbook is the extent to which it deals with the application of chemistry to important issues of the day, as the following section-related questions demonstrate.

1.   What is meant by **coal gasification**?
2.   Outline the process of coal gasification.
3.   Define **synthetic gas**.
4.   Why is hydrogen a theoretically good choice for use as a fuel?
5.   Why is hydrogen a poor **practical** choice for use as a fuel? (List reactions and thermodynamic values.)
6.   What is the most common method of producing ethanol?
7.   What are some of the advantages of alcohols in fuels?

Exercises

Section 6.1

1. A system does 3 J of work on the surroundings, and 12 J of work are added to the system.

    a. What is the energy change of the system?
    b. Of the surroundings?

2. One hundred joules of work are required to compress a gas. At the same time, the gas gives off 23 J of heat to the surroundings. What is the energy change of the system?

3. A gas expands from 10 L to 20 L against a constant pressure of 5 atm. During this time it absorbs 2 kJ of heat. Calculate the work done in kJ.

4. A piston expands against 1.00 atm of pressure, from 11.2 L to 29.1 L. This is done without any transfer of heat.

    a. Calculate the change in energy of the system.
    b. Calculate the change in energy for the above change if, in addition, the system absorbs 1,037 J of heat from the surroundings.

5. If the internal energy of a thermodynamic system is decreased by 300 J when 75 J of work is done *on* the system, how much heat was transferred, and in which direction, to or from the system?

6. How much work is done by a system where pressure is kept constant but the volume changes from 20 L to 0.5 L (1.00 atm)?

Section 6.2

7. A gas is compressed against a constant pressure of 3.4 atm from 27.9 L to 16.3 L. During this process, there is a heat gain by the system of 122 J. Calculate the change in energy of the system.

8. If 596 J of heat are added to 29.6 g of water at 22.9°C in a coffee cup calorimeter, what will be the final temperature of the water?

9. A 5.037 g piece of iron heated to 100.°C is placed in a coffee cup calorimeter that initially contains 27.3 g of water at 21.2°C. If the final temperature is 22.7°C, what is the specific heat capacity of the iron (J/g°C)?

10. Calculate the heat necessary to convert 10.0 g of water (just melted) at 0.0°C to water at 20.0°C, assuming that the specific heat remains constant at 1 cal/g°C.

11. The specific heat of aluminum is 0.89 J/g°C. How much energy is required to raise the temperature of a 15.0 gram aluminum can 18°C?

12. One liter of an ideal gas at 0°C and 10 atm was allowed to expand to 1.89 L against a constant external pressure of 1 atm at a constant temperature. The enthalpy change (ΔH) for this process is -901 J. Calculate q, w, and ΔE.

13. The heat capacity of a bomb calorimeter was determined by burning 6.79 g of methane (heat of combustion = -802 kJ/mol) in the bomb. The temperature changed by 10.8°C.

    a. What is the heat capacity of the bomb?
    b. A 12.6-g sample of acetylene, C_2H_2, produced a temperature increase of 16.9°C in the same calorimeter. What is the heat of combustion of acetylene (kJ/mol)?

14. A sample of C_6H_5COOH (benzoic acid) weighing 1.221 g was placed in a bomb calorimeter and ignited in a pure O_2 atmosphere. A temperature rise from 25.24°C to 31.67°C was noted. The heat capacity of the calorimeter was 5.020 kJ/°C, and the combustion products were CO_2 and H_2O. Calculate the ΔH in kJ/mol for the reaction.

15. When 1.50 L of 1.00 M Na_2SO_4 solution at 30.0°C is added to 1.50 L of 1.00 M $Ba(NO_3)_2$ solution at 30.0°C in a calorimeter, a white solid ($BaSO_4$) forms. The temperature of the mixture increases to 42.0°C. Assuming that the specific heat capacity of the solution is 6.37 J/°C g and that the density of the final solution is 2.00 g/mL, calculate the enthalpy change per mole of $BaSO_4$ formed.

Section 6.3

16. Calculate ΔH for

$$N_2(g) + O_2(g) \rightarrow 2NO(g)$$

given:

   a. $N_2(g) + 2O_2(g) \rightarrow 2NO_2(g)$ $\Delta H_{298} = 66.4$ kJ/mol
   b. $2NO(g) + O_2(g) \rightarrow 2NO_2(g)$ $\Delta H_{298} = -114.1$ kJ/mol

17. For the reaction:

$$H_2O(l) \rightarrow H_2O(g) \qquad \Delta H = +44 \text{ kJ}$$

How much heat is <u>evolved</u> when 9.0 grams of water vapor is condensed to liquid water?

18. Given

   a. $2H_2(g) + C(s) \rightarrow CH_4(g)$ $\Delta H_{298} = -74.81$ kJ/mol
   b. $2H_2(g) + O_2(g) \rightarrow 2H_2O(l)$ $\Delta H_{298} = -571.66$ kJ/mol
   c. $C(s) + O_2(g) \rightarrow CO_2(g)$ $\Delta H_{298} = -393.52$ kJ/mol

   Calculate ΔH for

$$CH_4(g) + 2O_2(g) \rightarrow CO_2(g) + 2H_2O(l)$$

19. Given the following thermochemical data, calculate the $\Delta H°$ for:

$$Ca(s) + 2H_2O(l) \rightarrow Ca(OH)_2(s) + H_2(g)$$

   a. $H_2(g) + \frac{1}{2}O_2(g) \rightarrow H_2O(l)$ $\Delta H° = -285$ kJ
   b. $CaO(s) + H_2O(l) \rightarrow Ca(OH)_2(s)$ $\Delta H° = -64$ kJ
   c. $Ca(s) + \frac{1}{2}O_2(g) \rightarrow CaO(s)$ $\Delta H° = -635$kJ

20. Calculate the value for $\Delta H°$ for the following reaction

$$CaCO_3(s) \rightarrow CaO(s) + CO_2(g)$$

   Use the following data: the heats of formation of calcium carbonate, calcium oxide, and carbon dioxide are 288.6, 151.9, and 94.1 kcal/mol respectively.

Section 6.4

21. Using standard heats of formation (<u>Appendix 4 in your textbook</u>), calculate ΔH for the following reactions.

   a. $2H_2O_2(l) \rightarrow 2H_2O(l) + O_2(g)$ ΔH_f^o for $H_2O(l) = -187.8$ kJ/mol
   b. $HCl(g) \rightarrow H^+(aq) + Cl^-(aq)$
   c. $2NO_2(g) \rightarrow N_2O_4(g)$
   d. $C_2H_2(g) + H_2(g) \rightarrow C_2H_4(g)$
   e. $2NaOH(s) + CO_2(g) \rightarrow Na_2CO_3(s) + H_2O(g)$

22. Calculate the standard change in enthalpy for the following thermite reaction by using enthalpies of formation:

$$2Al(s) + Fe_2O_3(s) \rightarrow Al_2O_3(s) + 2Fe(s)$$

NOTE: This reaction occurs when a mixture of powdered aluminum and iron (III) oxide are ignited with a magnesium fuse.

23. a. The heat released when HNO_3 reacts with NaOH is 56 kJ/mole of water produced. How much energy is released when 400.0 mL of 0.200 M HNO_3 is mixed with 500.0 mL of 0.150 M NaOH?

    b. Now do this problem from the Additional Exercises in your textbook: The enthalpy of neutralization for the reaction of a strong acid with a strong base is −56 kJ/mol of water produced. How much energy will be released when 200.0 mL of 0.400 M HCl is mixed with 150.0 mL of 0.500 M NaOH? How does this compare with your answer in part a? Why?

Multiple Choice Questions

24. When zinc reacts with hydrochloric acid, hydrogen gas is released. In this system the release of the hydrogen gas is counteracted by an outside force that results in a smaller volume by the end of the reaction. The work done by the outside force:

    A. Is negative on the system   C. Is positive on the system
    B. Is positive on the surroundings   D. Is zero

25. A piano is brought upstairs by two workers. Due to a mistake by one of the workers, the piano rolls down the stairs and finally comes to rest by the outside door. Which sequence best describes the energy transformations for the piano from the moment it is being brought upstairs to when it stops by the door?

    A. Potential energy → Kinetic energy → Thermal energy of the ground and piano
    B. Ground energy → Potential energy → Thermal energy → Kinetic energy of the piano
    C. Potential energy → Kinetic energy → Potential energy → Thermal energy of the piano and ground
    D. Kinetic energy → Potential energy → Kinetic energy

26. While a piston performs work of 210 L atm on the surroundings, the cylinder in which it is placed expands from 10 to 25 L. At the same time, 45 J of heat is transferred from the surroundings to the system. Against what pressure was the piston working?

    A. 14 atm   B. 11 atm   C. 17 atm   D. 254 atm

27. As a system increases in volume, it releases 52.5 J of energy in the form of heat to the surroundings. The piston is working against a pressure of 10.25 atm. The final volume of the system is 58.0 L. What was the initial volume of the system if the energy of the system decreased by 102.5 J?

    A. 62.9 L   B. 53.1 L   C. 48 L   D. 68 L

28. A 500.0-g sample of an element, at 195°C is dropped into an ice-water mixture. 109.5 g of ice melts and an ice-water mixture remains. Calculate the specific heat of the element, and determine which element it is.

    A. Zn   B. Ba   C. Pb   D. Ag

29. What is the final temperature, in °C, when 20.0 g of water at 80°C is mixed with 20.0 g of water at 25°C?

    A. 12°C   B. 7.0°C   C. 8.8°C   D. 52.5°C

30. How much ozone is used if the following reaction releases 568 kJ of energy in the form of heat?

$$Pb(s) + C(s) + O_3(g) \rightarrow PbCO_3(s) \qquad \Delta H° = ?$$

    A. None   B. 192 g   C. 4 g   D. 48 g

31. Benzoic acid, $C_7H_6O_2$, is a standard used in determining the heat capacity of a calorimeter. $\Delta H°$ of combustion of benzoic acid is 3.22×10^3 kJ/mol. 0.5 g of benzoic acid was burned in a calorimeter containing 1000.0 g of water. The change in temperature of the calorimeter was 3°C. Calculate the heat capacity of the calorimeter in J/K.

   A. 450 J/K   B. 210 J/K   C. 4025 J/K   D. 2307 J/K

32. Silane, SiH_4, is highly combustible and creates a fire hazard.

$$SiH_4(g) + 2O_2(g) \rightarrow SiO_2(s) + 2H_2O(g) \quad \Delta H° = ?$$

   Calculate $\Delta H°$ for this reaction based on the following information:

   $Si(s) + 2H_2(g) \rightarrow SiH_4(g)$ $\Delta H° = 34$ kJ/mol
   $Si(s) + O_2(g) \rightarrow SiO_2(s)$ $\Delta H° = -911$ kJ/mol
   $H_2(g) + \frac{1}{2} O_2(g) \rightarrow H_2O(g)$ $\Delta H° = -242$ kJ/mol

   A. −1429 kJ   B. −733 kJ   C. 733 kJ   D. −1143 kJ

33. Ethane, C_2H_6, may be produced by using the following method

$$C(s) + CH_4(g) + H_2(g) \rightarrow C_2H_6 \qquad \Delta H° = -10 \text{ kJ/mol}$$

   Calculate the $\Delta H°$ for the following reaction given the information below:

   $C(s) + 2H_2(g) \rightarrow CH_4(g)$ $\Delta H° = ?$
   $2CH_4(g) + 2CH_4(g) \rightarrow 2C_2H_6(g) + 2H_2(g)$ $\Delta H° = 130$ kJ

   A. −150 kJ   B. −75 kJ   C. 70 kJ   D. 10 kJ

34. How much heat will be evolved if 56.08 g of calcium oxide reacts with sulfuric acid according to the following reaction?

$$CaO(s) + H_2SO_4(g) \rightarrow CaSO_4(s) + H_2O(l) \qquad \Delta H° = ?$$

   $Ca(s) + \frac{1}{2} O_2(g) \rightarrow CaO(s)$ $\Delta H° = -152$ kJ/mol
   $Ca(s) + S(s) + 2O_2(g) \rightarrow CaSO_4(s)$ $\Delta H° = -1434$ kJ/mol
   $H_2(g) + \frac{1}{2} O_2(g) \rightarrow H_2O(l)$ $\Delta H° = -286$ kJ/mol
   $H_2(g) + S(s) + 2O_2(g) \rightarrow H_2SO_4(g)$ $\Delta H° = -814$ kJ/mol

   A. −2960 kJ   B. −754 kJ   C. −10744 kJ   D. 754 kJ

35. Calculate the heat of formation of carbon monoxide based on the reaction below:

$$2CO(g) + C(s) \rightarrow C_3O_2(g) \qquad \Delta H° = 127.3 \text{ kJ}$$

   The heat of formation for carbon suboxide is −93.7 kJ/mol.

   A. 116.8 kJ/mol   B. −348.3 kJ/mol   C. −110.5 kJ/mol   D. 93.7 kJ/mol

36. Calculate the heat of reaction for the following reaction given the information below:

$$2KIO_3(s) + 12HCl(g) \rightarrow 2ICl(l) + 2KCl(s) + 6H_2O(l) + 4Cl_2(g) \qquad \Delta H° = ?$$

   The heats of formation for the reactants and products, respectively, are:

$$-501.0, -92.0, -24.0, -435.0, \text{ and } -286.0 \text{ kJ/mol}$$

   A. −2634 kJ   B. −528 kJ   C. −3227 kJ   D. −152 kJ

37. The carbon dioxide and water in the atmosphere have all of the following effects except:

   A. Increasing the temperature of the Earth
   B. Allowing visible light to escape the Earth
   C. Absorbing infrared radiation
   D. Resulting in longer daylight in summer

Answers to Exercises

1. a. 9 J
   b. −9 J

2. 77 J

3. $w = -5$ kJ

4. a. −1.81 kJ
   b. −776 J

5. −375 J was transferred *from* the system

6. 1.98 kJ

7. +4100 J (rounding of 4117 to 2 sig figs)

8. 27.7°C

9. 0.44 J/g°C

10. 837 J

11. 240 J

12. $q = -901$ J; $w = -90.1$ J; $\Delta E = -991$ J

13. a. 31.5 kJ/°C
    b. −1100 kJ/mol

14. −3230 kJ/mol

15. $\Delta H = -306$ kJ/mol

16. 180.5 kJ/mol

17. 22 kJ evolved

18. −890.37 kJ

19. −414 kJ

20. −42.5 kcal

21. a. −196 kJ   c. −58 kJ   e. −125 kJ
    b. −75 kJ   d. −175 kJ

22. −850 kJ

23. a. 4.2 kJ is released. b. 4.2 kJ is released. It is the same as part a because the total number of moles of each reagent is the same.

24. C	25. A	26. A	27. B	28. A	29. D
30. A	31. B	32. A	33. B	34. B	35. C
36. B	37. D				

Chapter 7

Atomic Structure and Periodicity

I find the material in this chapter to be among the most interesting and useful in all of chemistry. The coverage of the electromagnetic spectrum leads into the electronic structure of atoms. From this information, we can rationalize and predict such properties of atoms as size, ionization energy, and the way in which they will form bonds (a discussion that continues in Chapters 8 and 9).

7.1   Electromagnetic Radiation

When you finish this section you will be able to:

- List the regions of the electromagnetic spectrum.
- State the relationship between the wavelength and frequency of a given region.
- Convert between wavelength and frequency.

This topic is important for many reasons, as is pointed out in your textbook. From my point of view, the most fascinating aspect has to do with outer space. Electromagnetic radiation is the **only** source of information from celestial objects other than the Moon, Mars, and Venus. We can't directly touch the stars, but we can receive their radiation and learn about their composition and formation from this. Also related to this is our ability to use electromagnetic radiation to determine metal ion content in water samples.

The definitions for **electromagnetic radiation, wavelength,** and **frequency** are given in Section 7.1 of your textbook. Be able to define these terms. You should also know the different regions of the electromagnetic spectrum and the wave lengths covered by each region. The relationship between frequency and wavelength is:

$$v = c/\lambda$$
$$\uparrow \qquad \uparrow$$
$$\text{nu} \quad \text{lambda}$$

wavelength (λ) is in <u>meters</u>.
frequency (v) is in <u>sec^{-1}</u> or Hz.
speed of light (c) is in <u>meters/sec</u>.

Example 7.1 A   True - False on the Electromagnetic Spectrum

Answer the following questions as "true" or "false."

1. Blue light has a shorter wavelength than red light.
2. X-rays have lower frequencies than radio waves.
3. Microwaves have higher frequencies than gamma rays.
4. Visible radiation composes the major portion of the electromagnetic spectrum.

Solution

1. true   2. false   3. false   4. false

Example 7.1 B   Wavelength - Frequency Conversion

Photosynthesis (use of the sun's light by plants to convert CO_2 and H_2O into glucose and oxygen) uses light with a frequency of 4.54×10^{14} s^{-1}. What wavelength does this correspond to?

Solution

$$v = c/\lambda, \quad \text{so } \lambda = c/v$$

$$\lambda = \frac{2.9979 \times 10^8 \text{ m/s}}{4.54 \times 10^{14} \text{ s}^{-1}} = 6.60 \times 10^{-7} \text{ m} = 660 \text{ nm}$$

Does the Answer Make Sense?

As discussed in your textbook, the visible region goes from about 400 to 700 nm. Photosynthesis occurs at the far end (660 nm) of this, so the answer would seem to be reasonable.

7.2   The Nature of Matter

When you finish this section you will be able to solve problems relating to

- the interconversion among energy, wavelength, and frequency, and
- the de Broglie equation.

There are two critically important equations introduced at the beginning of this section of your textbook. The first one is:

$$\Delta E = hv$$

where ΔE is the change in energy for a system (in Joules **PER PHOTON**),
h is Planck's constant (6.626×10^{-34} J s), and
v is the frequency of the wave (s^{-1}).

Only certain specific amounts of energy can be gained or lost in a substance. These **quanta** have magnitudes that depend on the substance. Given the observed frequency change, the change in energy can be calculated for the absorption or emission of "**photons**" (the term used if the quanta of energy are viewed as **particles**).

From Section 7.1, you know that v and λ are related by v and c/λ. Therefore,

$$\Delta E = hv = \frac{hc}{\lambda}$$

This is the second equation you should know. These two equations are among the most important in all of chemistry, as you shall see in succeeding chapters.

Let's have some practice at interconversion among energy, frequency, and wavelength.

Example 7.2 A   Interconverting Among Energy, Frequency, and Wavelength

Sodium atoms have a characteristic yellow color when excited in a flame. The color comes from the emission of light of 589.0 nm.

    a.   What is the frequency of this radiation?
    b.   What is the change in energy associated with this photon? Per mole of photons?

Solution

a. $v = \dfrac{c}{\lambda} = \dfrac{2.9979 \times 10^8 \text{ m/s}}{5.890 \times 10^{-7} \text{ m}} = \mathbf{5.090 \times 10^{14} \text{ s}^{-1}}$

b. $\Delta E = hv = 6.626 \times 10^{-34} \text{ J s} \times 5.090 \times 10^{14} \text{ s}^{-1} = \mathbf{3.373 \times 10^{-19} \text{ J}}$

This value is *per photon*. There are $\dfrac{6.022 \times 10^{23} \text{ photons}}{\text{mole photons}}$ (Avogadro's number). To convert ΔE per photon to ΔE per mole,

$$\frac{\text{J}}{\text{mole}} = \frac{3.373 \times 10^{-19} \text{ J}}{\text{photon}} \times \frac{6.022 \times 10^{23} \text{ photons}}{\text{mole photons}} = 2.031 \times 10^5 \text{ J/mole}$$

$$\Delta E = \mathbf{203.1 \text{ kJ/mole}}$$

Example 7.2 B   Practice with Interconversions

It takes 382 kJ of energy to remove one mole of electrons from gaseous cesium. What is the wavelength associated with this energy?

Strategy

The conversion between energy and wavelength requires the value of energy to be **per photon.** This is our first conversion. We may then convert directly to wavelength.

Solution

$$\frac{\text{kJ}}{\text{photon}} = \frac{3.82 \times 10^5 \text{ J}}{\text{mol}} \times \frac{1 \text{ mol}}{6.022 \times 10^{23} \text{ photons}} = \mathbf{6.343 \times 10^{-19} \text{ J (per photon)}}$$

$$\Delta E = \frac{hc}{\lambda}, \text{ so } \lambda = \frac{hc}{\Delta E}$$

$$\lambda = \frac{(6.626 \times 10^{-34} \text{ J s})(2.9979 \times 10^8 \text{ m/s})}{6.343 \times 10^{-19} \text{ J}} = 3.13 \times 10^{-7} \text{ m} = \textbf{313 nm}$$

This lies in the ultraviolet region.

Let's review your textbook's derivation of de Broglie's equation. The goal is to relate wavelength (λ) to the mass (m) of a particle.

$$E = mc^2, \text{ and } E = \frac{hc}{\lambda}$$

This gives us $m = \frac{h}{\lambda c}$, or for a particle which is not moving at the speed of light (c), but rather at some velocity (v), **de Broglie's equation** is

$$m = \frac{h}{\lambda v} \text{ or } \lambda = \frac{h}{mv}$$

The important point is that, **at least on an atomic scale,** electromagnetic radiation has mass, and particles with mass exhibit a characteristic wavelength. This is the **dual nature of light.**

Example 7.2 C   de Broglie's Equation

What is the wavelength of an electron (mass = 9.11×10^{-31} kg) traveling at 5.31×10^6 m/s?

Solution

Before we "plug and chug," recall that $1 \text{ J} = 1 \text{ kg m}^2\text{s}^{-2}$.

$$\lambda = \frac{h}{mv} = \frac{6.626 \times 10^{-34} \text{ J s}}{(9.11 \times 10^{-31} \text{ kg})(5.31 \times 10^6 \text{ m/s})} = \frac{6.626 \times 10^{-34} \text{ kg m}^2 \text{ s}^{-2} \text{ s}}{4.83_7 \times 10^{-24} \text{ kg m s}^{-1}} = 1.37 \times 10^{-10} \text{ m}$$

$$= \textbf{0.137 nm}$$

Keep in mind that your answer is probably reasonable if the dimensions work out!

The importance of the **dual nature of light** is outlined in the paragraphs on **diffraction** at the end of Section 7.2 in your textbook.

7.3   The Atomic Spectrum of Hydrogen

The key idea presented in this section is the difference between a **continuous spectrum** and a **discrete or line spectrum.**

- A **continuous spectrum** contains all the wavelengths over which the spectrum is continuous.
- A **line spectrum** contains certain specific wavelengths that are characteristic of the substance emitting those wavelengths.

It is concluded in your textbook that the fact that hydrogen has a line spectrum shows that only certain energy transfers are allowed in hydrogen. **There are specific energy levels among which the hydrogen electron can shift.** These energy levels are said to be **quantized.**

7.4   The Bohr Model

When you finish this section you will be able to calculate the energies or associated wavelengths corresponding to electron transitions in the hydrogen atom.

At the beginning of this section, your textbook discusses Niels Bohr's reasoning for relating the **energy levels** to the observed **wavelengths** emitted by the hydrogen atom. Although Bohr's ultimate conclusions have since been enhanced, his work represented a great leap forward in 1913. The equation that is descriptive of Bohr's model is

$$E = -2.178 \times 10^{-18} \text{ J} \left(\frac{Z^2}{n^2} \right)$$

where E is the energy (in Joules)
Z is the nuclear charge (1 for hydrogen's one proton)
n is an integer related to orbital position. The farther out from the nucleus, the higher value of n. If an electron is given enough energy, it goes away from the nucleus. We say it is **ionized,** and $n = \infty$.

The lowest energy state is $n = 1$, the ground state.
The highest energy state is $n = \infty$, where an electron is ionized.

Example 7.4 A   Energy Levels in the Bohr Atom

Calculate the energy corresponding to the $n = 3$ electronic state in the Bohr hydrogen atom.

Solution

In the hydrogen atom, $Z = 1$. In this problem, $n = 3$.

$$E_3 = -2.178 \times 10^{-18} \text{ J} \left(\frac{Z^2}{n^2} \right) = -2.178 \times 10^{-18} \text{ J} \left(\frac{1^2}{3^2} \right) = \mathbf{-2.42 \times 10^{-19} \text{ J}}$$

A more challenging but **chemically meaningful** calculation is presented in the next problem.

Example 7.4 B   Changes in Energy in the Bohr Atom

Calculate the energy **change** corresponding to the excitation of an electron from the $n = 1$ to $n = 3$ electronic state in the hydrogen atom.

Solution

In this problem, we need the change in energy.

$$\Delta E = E_{\text{final}} - E_{\text{initial}} = E_3 - E_1$$

There are two possible approaches to the calculation. The first is to calculate each energy separately and take the difference. The second is to subtract the equations; then factor and calculate the change using one equation.

Approach 1:

$$\Delta E = -2.178 \times 10^{-18} \text{ J} \left(\frac{1^2}{3^2} \right) - \left(-2.178 \times 10^{-18} \text{ J} \left(\frac{1^2}{1^2} \right) \right)$$

Approach 2:

$$\Delta E = -2.178 \times 10^{-18} \text{ J} \left(\frac{1}{3^2} - \frac{1}{1^2} \right) = -2.178 \times 10^{-18} \text{ J} \left(-\frac{8}{9} \right)$$

Both approaches will yield

$$\Delta E = +1.936 \times 10^{-18} \text{ J}$$

The "+" sign is very important! It means energy was **absorbed** to excite the electron. In other words, the system has **gained energy.**

Example 7.4 C   *Wavelength from Energy in the Bohr Atom*

What wavelength of electromagnetic radiation is associated with the energy change in promoting an electron from the $n = 1$ to $n = 3$ level in the hydrogen atom? (Use the value of ΔE from the previous example.)

Solution

$$\Delta E = \frac{hc}{\lambda}, \text{ so } \lambda = \frac{hc}{\Delta E}$$

$$\lambda = \frac{(6.626 \times 10^{-34} \text{ J s})(2.9979 \times 10^8 \text{ m/s})}{1.936 \times 10^{-18} \text{ J}} = 1.026 \times 10^{-7} \text{ m} = \textbf{102.6 nm}$$

This energy corresponds to radiation in the ultraviolet region of the spectrum.

(Note: Even if ΔE = "−" (as it will be when energy is released), you **must** assign a "+" sign to λ because <u>you can't have a negative wavelength</u>!)

7.5   The Quantum Mechanical Model of the Atom

When you finish this section you will be able to:

- Describe the basic terms and ideas in the quantum mechanical model of the atom.
- Solve problems relating to the Heisenberg Uncertainty Principle.

As is pointed out in <u>Section 7.5 of your textbook</u>, there is a key assumption of the quantum mechanical model regarding electron motion around the hydrogen atom. This assumption is that **the electron is assumed to behave as a standing wave.** Only certain orbits are shaped such that the "wave" (electron) can fit. The **wave function** of an electron represents the allowed coordinates where the electron *may* reside in the atom. Each wave function is called an **orbital.** Your textbook points out that Schrödinger was not certain that treating the electron as a wave would make any sense—the key would be if the model would fit experimental atomic data.

Read the discussion on the **Heisenberg Uncertainty Principle** in your textbook. The principle says that "there is a limit to just how precisely we can know both the position and momentum of a particle at a given time." It turns out that when the radiation used to locate a particle hits that particle, it changes its momentum. Therefore, the position and momentum cannot both be measured exactly. As one is measured more precisely, the other is known less precisely. The relation is given by

$$\Delta x \cdot \Delta(mv) \geq \frac{h}{4\pi}$$

where Δx is the uncertainty in the particle's position
$\quad\quad\quad$ Δmv is the uncertainty in the particle's momentum
$\quad\quad\quad$ h is Planck's constant

The smallest possible uncertainty is **h/4π.**

The Heisenberg Uncertainty Principle tells us that we cannot know exactly where an electron is around the atom at any given time. However, we can know the **probability** of it being within a certain region. The **square of the wave function** is called the **probability distribution**, and it represents the probability (statistical likelihood) of finding the electron at a particular area around the nucleus.

This section in your textbook ends by noting that there is a probability (exceedingly small, but it does exist) of finding the electron anywhere. However, we use the (arbitrary) limit of the size of an orbital as "the radius of the sphere that encloses 90% of the total electron probability." In other words, at any time, t, there is a 90% chance of finding the electron in that orbital. It is also helpful to think of an orbital as a "three-dimensional electron density map" in which there is a finite probability of finding the electron.

7.6   Quantum Numbers

When you finish this section you will be able to:

- Describe the principle quantum numbers n, ℓ, and m_ℓ.
- Decide which values for these numbers are valid for a variety of problems.

This section in your textbook begins by describing the meaning and values of quantum numbers.

Quantum Numbers

Name	Symbol	Property of the Orbital	Range of Values
principal quantum number	n	related to **size and energy** of the orbital	integers > 0 (1,2,3,...)
angular momentum quantum number	ℓ	related to the **shape** of the orbital	integers from 0 to $n-1$*
magnetic quantum number	m_ℓ	related to the **position of the orbital in space relative to other orbitals**	integers from $-\ell$ to 0 to $+\ell$

Table 7.2 in your textbook gives the allowed values for the quantum number for the first four levels of the hydrogen atom. In examining the table, **the most important** thing you can do is **derive** the values of ℓ and m_ℓ that go with each value of n. For example, if $n = 2$, and if ℓ can go from 0 to $n-1$, then $\ell = 0$ or $\ell = 1$.

For $\ell = 0$ ("s"), m_ℓ can only equal 0 ($-\ell$ to $+\ell$).
For $\ell = 1$ ("p"), m_ℓ can equal -1, 0, or $+1$.

Therefore, there are four energy levels associated with $n = 2$, **one for 2s and three for 2p.**

---

* $\ell = 0$ is called "s"; $\ell = 1$ is "p"; $\ell = 2$ is "d"; $\ell = 3$ is "f"; $\ell = 4$ is "g"; $\ell = 5$ is "h" and so on through the alphabet. As your textbook outlines, this labeling system arose historically from early spectral line studies: s = "sharp," p = "principle," d = "diffuse," and f = "fundamental."

Example 7.6 A   Quantum Numbers

If for a given atom $\ell = 6$, how many energy levels are possible? What is the subshell designation for this orbital?

Solution

If $\ell = 6$, m_ℓ can range from -6 to $+6$ and can include 0. Therefore, there are **13 possible energy levels.** If $\ell = 3$ is called "f" and $\ell = 4$ is a "g," $\ell = 6$ **would be an "i" orbital.**

Example 7.6 B   Practice with Quantum Numbers

Which of the following sets of quantum numbers are not allowed in the hydrogen atom? For each incorrect set, state why it is incorrect.

    a. $n = 1$, $\ell = 0$, $m_\ell = 1$
    b. $n = 2$, $\ell = 2$, $m_\ell = 1$
    c. $n = 5$, $\ell = 3$, $m_\ell = 2$
    d. $n = 6$, $\ell = -2$, $m_\ell = 2$
    e. $n = 6$, $\ell = 2$, $m_\ell = -2$

Solution

    a.   Not allowed. (m_ℓ can't be greater than ℓ.)
    b.   Not allowed. (ℓ can't be greater than or equal to n.)
    c.   Allowed.
    d.   Not allowed. (ℓ can't be negative.)
    e.   Allowed.

7.7   Orbital Shapes and Energies

This section contains a discussion of orbital shapes and energies. Please use the following questions to test your understanding of the material.

    1.   What are nodes?
    2.   How many nodes are in a $5s$ orbital?
    3.   Draw a $2p$ orbital.
    4.   How do the three $2p$ orbitals differ from one another?
    5.   In what ways do d orbitals differ from p orbitals?
    6.   Why doesn't a $2d$ orbital exist?
    7.   What is the chemical meaning of a "**degenerate**?"
    8.   What is meant by the "**ground state**?" What is the ground state orbital for hydrogen?
    9.   How can a $1s$ electron be excited to a higher energy level?
  10.   How do we explain the ability of an electron to cross through a node, such as the nucleus, to occupy the region within an entire orbital?

7.8   Electron Spin and the Pauli Principle

    1.   What property of an electron does the quantum number given by m_s represent?
    2.   What is the **Pauli Exclusion Principle**? What is its consequence regarding the number of electrons that an orbital can hold?

Example 7.8 A   Practice with the Four Quantum Numbers

Which of the following sets of quantum numbers are not allowed? For each incorrect set state why it is incorrect.

a. $n = 3$, $\ell = 3$, $m_\ell = 0$, $m_s = -\frac{1}{2}$
b. $n = 4$, $\ell = 3$, $m_\ell = 2$, $m_s = -\frac{1}{2}$
c. $n = 4$, $\ell = 1$, $m_\ell = 1$, $m_s = +\frac{1}{2}$
d. $n = 2$, $\ell = 1$, $m_\ell = -1$, $m_s = -1$
e. $n = 5$, $\ell = -4$, $m_\ell = 2$, $m_s = +\frac{1}{2}$
f. $n = 3$, $\ell = 1$, $m_\ell = 2$, $m_s = -\frac{1}{2}$
g. $n = 3$, $\ell = 2$, $m_\ell = -1$, $m_s = 1$

Solution

a.   Not allowed. (ℓ cannot be equal to n.)
b.   Allowed.
c.   Allowed.
d.   Not allowed. (m_s must be either $+\frac{1}{2}$ or $-\frac{1}{2}$.)
e.   Not allowed. (ℓ must be a **positive** integer.)
f.   Not allowed. (m_ℓ must be between $-\ell$ and $+\ell$.)
g.   Not allowed. (m_s must be either $+\frac{1}{2}$ or $-\frac{1}{2}$.)

Remember the key message from Section 7.8—**an orbital can hold a maximum of two electrons,** and they must have opposite spins.

Example 7.8 B   Electrons in Orbitals

If each orbital can hold a maximum of two electrons (of opposite spin), how many electrons can each of the following subshells hold?

a. $2s$ b. $5p$ c. $4f$ d. $3d$ e. $4d$

Strategy

The key here is to figure out **how many orbitals** each contains. This is determined by the azimuthal quantum number, NOT by the principle quantum number. For example, a p orbital ($\ell = 1$) can have $m_\ell = +1, 0, \text{ or } -1$. *Each m_ℓ can have 2 electrons.* Therefore, p can have a total of 3×2, or 6 electrons. It doesn't matter if it is a $5p$, $3p$, or $2p$. Each p can have up to 6 electrons.

Solution

a. $2s \Rightarrow \ell = 0$, $m_\ell = 0$ (1 value) $\times$ 2 electrons = **2 electrons**
b. $5p \Rightarrow \ell = 1$, $m_\ell = +1, 0, -1$ (3 values) $\times$ 2 electrons = **6 electrons**
c. $4f \Rightarrow \ell = 3$, $m_\ell = +3, +2, +1, 0, -1, -2, -3$ (7 values) $\times$ 2 electrons = **14 electrons**
d. $3d \Rightarrow \ell = 2$, $m_\ell = +2, +1, 0, -1, -2$ (5 values) $\times$ 2 electrons = **10 electrons**
e. $4d \Rightarrow$ same as $3d$

7.9   Polyelectronic Atoms

1. What are the three energy contributions that must be considered when describing the helium atom?
2. What does your textbook mean by the **electron correlation problem?** How do we deal with the problem?
3. Why does it take more energy to remove an electron from Al^+ than from Al?
4. Why do electrons "prefer" to fill s, p, d, and then f within a particular quantum level?
5. What is the **penetration effect**, and why is it important?
6. Why does the **3d** orbital have a higher energy than **3p** even though it has its maximum probability closer to the nucleus than the 3p?

7.10   The History of the Periodic Table

1. What was the original basis of the construction of the periodic table?
2. What are triads?
3. List the properties that Mendeleev used to predict ekasilicon's position in the periodic table.
4. Several atoms of element 116 have been observed. Based on Table 7.4 in your textbook, what properties would you predict this element would have?
5. What is the important difference between Mendeleev's periodic table and the modern one?

7.11   The Aufbau Principle and the Periodic Table

When you finish this section you will be able to:

*   Write electron configurations using both the "conventional" (long-hand) form and the core element and valence electron (shorthand) form.
*   Determine electron configurations based on position in the periodic table.

Read the statement of the Aufbau Principle in your textbook. Your textbook presents **orbital diagrams** for the first ten elements, **hydrogen through neon.** When constructing orbital diagrams and electron configurations, please keep the following in mind:

1. Electrons fill in order from lowest to highest energy.
2. The Pauli exclusion principle holds. An orbital can hold only two electrons.
3. Two electrons in the same orbital must have opposite spins.
4. You must know how many electrons can be held by each azimuthal quantum number (i.e., s can hold 2, 6 for p, 10 for d, 14 for f).
5. **Hund's rule** applies. The lowest energy configuration for an atom is the one having the maximum number of **unpaired** electrons for a set of **degenerate** orbitals. By convention all unpaired electrons are represented as having parallel spins with the spin "up."

One excellent approach to address the order that electrons fill as the atomic number increases is shown here:

$$1s^2$$
$$2s^2\ 2p^6$$
$$3s^2\ 3p^6\ 3d^{10}$$
$$4s^2\ 4p^6\ 4d^{10}\ 4f^{14}$$
$$5s^2\ 5p^6\ 5d^{10}\ 5f^{14}$$
$$6s^2\ 6p^6\ 6d^{10}\ 6f^{14}$$
$$7s^2\ 7p^6\ 7d^{10}\ 7f^{14}$$

To determine the order of filling, draw arrows from the upper right to the lower left.

$$
\begin{array}{l}
1s^2 \\
2s^2\ 2p^6 \\
3s^2\ 3p^6\ 3d^{10} \\
4s^2\ 4p^6\ 4d^{10}\ 4f^{14} \\
5s^2\ 5p^6\ 5d^{10}\ 5f^{14} \\
6s^2\ 6p^6\ 6d^{10}\ 6f^{14} \\
7s^2\ 7p^6\ 7d^{10}\ 7f^{14}
\end{array}
$$

Get the final filling pattern by following the arrows in order:

$$1s^2\ 2s^2\ 2p^6\ 3s^2\ 3p^6\ 4s^2\ 3d^{10}\ 4p^6\ 5s^2\ 4d^{10}\ 5p^6\ 6s^2\ 4f^{14}\ 5d^{10}\ 6p^6\ 7s^2 \ldots$$

Let's apply this strategy to determining the electron configuration for oxygen (8 electrons).

- The $1s$ and $2s$ levels hold a total of 4 electrons ($1s^2 2s^2$).
- The $2p$ level can hold up to 6 electrons. However, we only have 4 remaining, which means that we will have a $2p^4$.

$$\text{O: } 1s^2 2s^2 2p^4$$

Let's translate this to an orbital diagram.

$$
\begin{array}{ccccc}
\uparrow\downarrow & \uparrow\downarrow & \uparrow\downarrow & \uparrow & \uparrow \\
1s & 2s & & 2p &
\end{array}
$$

The first three $2p$ electrons occupy their own **degenerate** $2p$ orbitals. The fourth electron shares a degenerate orbital, but does so with **opposite spin**.

Example 7.11 A   Electron Configurations

Write electron configurations for each of the following neutral atoms (**use a configuration triangle**):

    a. boron   b. sulfur   c. vanadium   d. iodine

Solution

a. B (5 electrons): $1s^2 2s^2 2p^1$
b. S (16 electrons): $1s^2 2s^2 2p^6 3s^2 3p^4$
c. V (23 electrons): $1s^2 2s^2 2p^6 3s^2 3p^6 4s^2 3d^3$
d. I (53 electrons): $1s^2 2s^2 2p^6 3s^2 3p^6 4s^2 3d^{10} 4p^6 5s^2 4d^{10} 5p^5$

Example 7.11 B   Orbital Diagrams

Draw orbital diagrams for the following:

    a. sodium   b. phosphorus   c. chlorine

Strategy

Write the electron configuration for the atom. The only sticking points will be how unpaired electrons fill degenerate orbitals (singly, if possible). The inner electrons will all be paired. The outer electrons may be unpaired. You must deal with each atom separately.

Solution

a. Na (11 electrons): $1s^2 2s^2 2p^6 3s^1$

$$\underset{1s}{\uparrow\downarrow} \quad \underset{2s}{\uparrow\downarrow} \quad \underset{2p}{\uparrow\downarrow \; \uparrow\downarrow \; \uparrow\downarrow} \quad \underset{3s}{\uparrow}$$

b. P (15 electrons): $1s^2 2s^2 2p^6 3s^2 3p^3$

$$\underset{1s}{\uparrow\downarrow} \quad \underset{2s}{\uparrow\downarrow} \quad \underset{2p}{\uparrow\downarrow \; \uparrow\downarrow \; \uparrow\downarrow} \quad \underset{3s}{\uparrow\downarrow} \quad \underset{3p}{\uparrow \; \uparrow \; \uparrow}$$

c. Cl (17 electrons): $1s^2 2s^2 2p^6 3s^2 3p^5$

$$\underset{1s}{\uparrow\downarrow} \quad \underset{2s}{\uparrow\downarrow} \quad \underset{2p}{\uparrow\downarrow \; \uparrow\downarrow \; \uparrow\downarrow} \quad \underset{3s}{\uparrow\downarrow} \quad \underset{3p}{\uparrow\downarrow \; \uparrow\downarrow \; \uparrow}$$

Look at the term "**core**" and "**valence**" electrons in your textbook. Know how to define these terms.

Notice in the problem we just completed that phosphorus and chlorine have the same core electronic structure. This **core** has the **same electron configuration as neon.** We can therefore write a shorthand version of electron configurations:

$$P = [Ne]\, 3s^2\, 3p^3 \qquad Cl = [Ne]3s^2 3p^5$$
$$\hphantom{P = [Ne]\,} \underset{\text{core}}{\uparrow} \;\; \underset{\text{valence}}{\uparrow}$$

Using the same strategy,

$$\text{yttrium (39 electrons)} = 1s^2 2s^2 2p^6 3s^2 3p^6 4s^2 3d^{10} 4p^6 5s^2 4d^1$$
$$\text{(longhand)}$$

$$= [Kr]5s^2 4d^1$$
$$\text{(shorthand)}$$

Neon and krypton are both **noble gases (Group 8A)** and are atoms that have complete inner energy levels and an outer energy level with complete s and p orbitals.

Example 7.11 C   Shorthand Configurations

Write the shorthand configuration for the atoms in Example 7.11 A, and state for parts a, b, and d how many **valence electrons** the element has.

Solution

a. B: $[He]2s^2 2p^1$ (3 valence electrons)
b. S: $[Ne]3s^2 3p^4$ (6 valence electrons)
c. V: $[Ar]4s^2 3d^3$ (transition metal)
d. I: $[Kr]5s^2 4d^{10} 5p^5$ (7 valence electrons. . . $n=4$ level is complete)

Notice that: boron is in Group 3A
sulfur is in Group 6A
iodine is in Group 7A

The group number indicates the number of valence electrons for nontransition metals. This should help you in determining electron configurations.

Also, sulfur is in Period 3. It is filling $n = 3$ electronic orbitals. Iodine is filling $n = 5$ electronic orbitals. It is in Group 5. Use the group and period locations to help do electron configurations quickly and correctly.

Two final ideas:

- For transition metals in the gaseous state, if it is possible to have a $3d^5$ (half-filled) or $3d^{10}$ (completely filled) electronic configuration at the expense of a filled $4s$, that will happen. Thus, Cu is **[Ar]$4s^1 3d^{10}$**, NOT [Ar]$4s^2 3d^9$. The same holds true for $4s/5d$-filling lanthanides and actinides.

- Know what is meant by representative, d-transition, f-transition, and noble gas elements.

Example 7.11 D   Practice with Electron Configurations

Write shorthand electron configurations for the following real and hypothetical atoms:

    a. Sr   b. Mo   c. Ge   d. Q (hypothetical, 111 electrons)

Solution

a. **Sr** is in Group 2A, Period 5. **[Kr]$5s^2$**
b. **Mo** is the fourth transition metal in Period 5, [Kr]$5s^2 4d^4$, <u>BUT</u> it can have a half-filled $4d$ if the configuration is $5s^1 4d^5$. Therefore, Mo = **[Kr]$5s^1 4d^5$**.
c. **Ge** is in Group 4A, Period 4. **[Ar]$4s^2 3d^{10} 4p^2$**.
d. **Q** would be the ninth transition metal in Period 7, [Rn]$7s^2 6d^9$, <u>BUT</u> it can have a completed $6d$ if the configuration is $7s^1 6d^{10}$. Therefore, the likely configuration is **[Rn]$7s^1 5f^{14} 6d^{10}$**.

7.12   Periodic Trends in Atomic Properties

When you finish this section you will be able to use your understanding of electronic configuration to predict **trends in ionization energy, electron affinity and atomic radius**.

Ionization Energy increases as successive electrons are removed from an atom because:

1. The value for Z_{eff} increases because there are fewer electron-electron repulsions and a higher positive to negative charge ratio than before.
2. Upon going from f to d to p to s electrons, there is a higher penetration effect. For example, the removal of a $3p$ will require more energy than the removal of a $3d$ from the same atom.
3. When you remove all of the electrons from an energy level, you begin removing **core electrons** that are more tightly bound to the nucleus than valence electrons.

Example 7.12 A   Ionization Energy

Examine <u>Table 7.5 in your textbook</u>. Justify the large increases in ionization energy at I_5 and I_7 for **sulfur**.

Solution

I_4 represents removal of the last $3p$ electron. The remaining electrons are $3s$ and core electrons. The $3s$ electron has a much greater penetration effect than the $3p$. Therefore, I_5 is much larger than I_4.

I_7 represents ionization of a core ($2p$) electron. This electron is much closer to the nucleus and thus requires more energy to ionize it.

Note that:

1. First ionization energy **increases** as we go **across a period**.
2. First ionization energy **decreases** as we go **down a group**.
3. Anomalies such as the decrease from P to S exist. (Can you explain why?)

Electron Affinity is a **change in energy** associated with the **addition of an electron** to a **gaseous atom.** In keeping with thermodynamic convention, if the addition is **exothermic,** the energy change will be **negative.** Although electron affinity *generally* increases from left to right across a period, there are several exceptions. For example, the electron affinity of phosphorus is lower than that of sulfur. That is because P is $3p^3$ (half-filled) while S is $3p^4$. If you put an extra electron on phosphorus, it must share an orbital, thus forcing electron-electron repulsion. These repulsions already exist in the $3p$ orbitals of sulfur. The trend of electron affinities is less predictable than that of ionization energies.

The **atomic radius** of an atom **decreases** from left to right **across a period**. This is because the Z_{eff} increases. Atomic radius **increases** going **down a group**. This is because of increased orbital size. (See Figure 7.35 in your textbook.) You must consider **ionic radius** in terms of the following questions: What will adding an electron do to electron-electron repulsions? What will subtracting an electron do to the effective nuclear charge?

Example 7.12 B   Trends in Atomic Radius

Order the atoms or ions in the following groups from smallest to largest radius.

a. Cs, Si, F, Ca, Ga

b. Ca^{2+}, I^-, I, Li

Solution

a. Cs is in Period 6, Group 1
    F is in Period 2, Group 7

    F, Si, Ga, Ca, Cs
    ↑   ↑
    smallest   largest

b. I is large. Adding an electron forces extreme electron-electron repulsion and makes it larger. Calcium is large, but taking its $4s$ electrons away markedly increases the Z_{eff}.

    Ca^{2+}, Li, I, I^-
    ↑   ↑
    smallest largest

7.13   The Properties of a Group: The Alkali Metals

The goal of your textbook in this section is to show you how understanding electron configurations and the structure of the periodic table can allow you to predict the properties of this group. This section begins with a review of the information contained in the periodic table.

1. Elements in a group exhibit similar properties.
2. It is primarily the number of valence electrons that determine an atom's chemistry.
3. Electron configurations can be gleaned from the periodic table.
4. Learn the names of the different groups (halogens, active metals, lanthanides, etc.). See Figure 7.36 in your textbook.
5. Metals tend to lose electrons (have low ionization energies). Nonmetals tend to gain electrons. Metalloids (semimetals) have properties of both.

The focus of the chapter is on **Group I metals.**

Properties as We Go Down the Group

A. The first ionization energy decreases.
B. The atomic radius increases.
C. The density increases.
D. The reactivity increases (they lose electrons readily).
E. The melting and boiling points decrease.

Example 7.13   The Alkali Metals

Explain the trends that the alkali metals follow in properties "A" through "D" above based on your knowledge of electronic configurations and atomic structure.

Solution

A. **First ionization energy:** As we go down the group, the valence electron falls in a higher energy level. The nuclear attraction is less, making it easier to ionize the electron.
B. **Atomic radius**: Again, the valence electron occupies a higher energy level, thereby being farther from the nucleus.
C. **Density:** Atomic mass increases faster than atomic size.
D. **Reactivity**: The electrons are easier to ionize; therefore, reactions that require less energy are possible.

Exercises

Section 7.1

1. The visible region of the spectrum ranges from 400 nm to 700 nm. What is the frequency range of the visible spectrum?

2. List the regions of the electromagnetic spectrum and the wavelengths of radiation associated with each region.

3. Calculate the frequency of blue light of wavelength 4.5×10^2 nm.

4. Calculate the wavelength of green light of frequency 5.7×10^{14} Hz.

Section 7.2

5. Red light with a wavelength of 670.8 nm is emitted when lithium is heated in a flame.

   a. What is the frequency of this radiation?
   b. What is the energy of this radiation per photon? Per mole of photons?

6. It takes 6.72×10^{-18} J of energy to remove an electron from an unknown atom. What is the maximum wavelength of light that can do this?

7. A carbon-oxygen double bond in a certain organic molecule absorbs radiation that has a frequency of 6.0×10^{13} s^{-1}.

   a. To what region of the spectrum does this radiation belong?
   b. What is the wavelength of this radiation?
   c. What is the energy of this radiation per photon? Per mole of photons?
   d. A carbon-oxygen bond in a different molecule absorbs radiation with frequency equal to 5.4×10^{13} s^{-1}. Does this radiation give more or less energy?

8. Calculate the energy of a photon that is emitted at a wavelength of 5.69×10^3 nm.

9. Calculate the energy of a mole of photons in Problem 8.

10. Many spectroscopists prefer using frequencies to wavelengths when describing electromagnetic radiation. Can you think of an advantage to the use of frequencies? (A main concern of spectroscopists is the energy of radiation that is either emitted or absorbed.)

11. Calculate the wavelength of a thoroughbred racehorse that weighs 600 pounds and is moving with a speed of 40 mi/hr.

12. What are the wavelengths associated with the following?

   a. An alpha particle (mass = 6.64×10^{-27} kg) traveling at 3.0×10^6 m/s
   b. A 1000-kg automobile traveling at 100 km/hr

13. Calculate the energy associated with a molecule of each of the following photons:

   a. Red photons of wavelength 670.0 nm
   b. Yellow photons of wavelength 580.0 nm
   c. Violet photons of wavelength 450.0 nm
   d. X-ray photons of wavelength 0.154 nm

Section 7.4

14. Make a plot of energy vs. n for the Bohr hydrogen atom for $n = 1$ to $n = 50$.

 a. What is the energy of the Bohr hydrogen atom when $n = \infty$?
 b. What is the ionization energy for the Bohr hydrogen atom (i.e., the energy required to move an electron from $n = 1$ to $n = \infty$)?

15. Calculate the wavelength of light that must be absorbed by a hydrogen atom in its ground state to reach the excited state of $\Delta E = +2.914 \times 10^{-18}$ J.

16. How much energy is required to ionize a mole of hydrogen atoms?

17. Calculate the wavelength of light emitted in the spectral transition of $n = 4$ to $n = 2$ in the hydrogen atom.

18. What region of the spectrum would you look in to find the radiation associated with the $n = 4$ to $n = 1$ transition of the Bohr hydrogen atom?

19. What region of the spectrum would you look in to find the radiation associated with the spectral transition of $n = 3$ to $n = 2$ in the hydrogen atom?

Section 7.5

20. Use the wave mechanical model to explain the quantized nature of the orbits of a hydrogen atom.

21. A chemistry book lists the radius of the hydrogen orbital as 1Å. Will the electron ever be further than 1 Å from the nucleus?

22. Even on the planet Mars, the probability of finding an electron of an atom on the nose of Mona Lisa (otherwise known as La Gioconda) is not zero. Explain.

Section 7.6

23. Which of the following sets of quantum numbers are allowed?

 a. $n = 7, \ell = 7, m_\ell = 0$
 b. $n = 7, \ell = 0, m_\ell = 1$
 c. $n = 7, \ell = 5, m_\ell = -3$
 d. $n = 3, \ell = -1, m_\ell = 0$
 e. $n = 0, \ell = 0, m_\ell = 0$

Section 7.8

24. What is the maximum number of electrons that can be accommodated in

 a. All orbits of $n = 4$
 b. All the $4f$ orbitals
 c. All the $5g$ orbitals

Section 7.11

25. Write n, ℓ, m_ℓ, and m_s quantum numbers for the 5 electrons of a boron atom.

26. Account for the fact that a p subshell containing three electrons has one in each orbital rather than two in one orbital and the third in the other.

27. What is the electron configuration for calcium?

28. How does the electron configuration of barium compare with that of calcium?

29. How many half-filled orbitals do each of the following have in the ground state?

   a. O
   b. B
   c. Ar
   d. Mn
   e. K
   f. Cf
   g. Zn

30. Indicate the higher of the two energy states in each of the following pairs:

   a. $3d$ or $4s$
   b. $4p$ or $5s$
   c. $4s$ or $4p$

31. An element, "X," combines with calcium to give the salt CaX_2. The element has its highest energy electrons in the $4p$ level. What is "X"?

32. In which orbital would an electron have a greater likelihood of being near the nucleus: $4f$ or $6s$?

33. Which group in the periodic table contains elements with the highest ionization energies? Which period in the periodic table contains elements with the highest ionization energies?

Section 7.12

34. Order the following groups from smallest to largest radius.

   a. Ar, Cl^-, K^+, S^{2-}
   b. C, Al, F, Si
   c. Na, Mg, Ar, P
   d. I^-, Ba^{2+}, Cs^+, Xe

35. Arrange the following atoms in order of increasing Z_{eff} for the highest-energy electron: Te, In, Mg, Ga, Xe, Ca.

36. Which of the following will have the most exothermic electron affinity? The least?

   a. Ge, Si, C
   b. Cl, Cl^-, Cl^+

Section 7.13

37. Properties of the alkali metals are discussed in Section 7.13. List some properties that you would expect for the alkaline earths, Be, Mg, Ca, Sr, and Ba.

38. Which elements are metalloids, and why are they called metalloids?

Multiple Choice Questions

39. The frequency of an electromagnetic wave is 1.5×10^{14} Hz. Calculate its wavelength in meters.

   A. 2.0×10^{-6} m
   B. 6.6×10^{-9} m
   C. 5.0×10^5 m
   D. 5.0×10^{-5} m

40. Calculate the wavelength of an electromagnetic wave with a frequency of 1.7×10^{14} Hz.

   A. 5.9×10^6 m
   B. 0.67×10^{-15} m
   C. 0.33×10^8 m
   D. 1.8×10^{-6} m

41. Carbon absorbs energy at a wavelength of 150 nm. The total amount of energy emitted by a carbon sample is 1.98×10^5 J. Calculate the number of carbon atoms present in the sample, assuming that each atom emits one photon.

   A. 1.50×10^{23}
   B. 2.50×10^{19}
   C. 1.48×10^{20}
   D. 1.65×10^5

42. Calculate the wavelength of a photon traveling at a velocity of 0.01 c (speed of light). $c = 3.0 \times 10^8$ m/s.

    A. 3.4×10^{-55} m
    B. 1.3×10^{-15} m
    C. 1.9×10^{-28} m
    D. 1.9×10^{-25} m

43. An electron has an associated wavelength of 1.0×10^{-6} m. Calculate the velocity of the electron in m/s.

    A. 45 m/s
    B. 91.3 m/s
    C. 730 m/s
    D. 458 m/s

44. A particle has a velocity equal to 0.25 c and a wavelength of 1.3×10^{-16} m. Calculate the mass of the particle in kilograms. $c = 3.0 \times 10^8$ m/s.

    A. 1.7×10^{-20} kg
    B. 6.8×10^{-26} kg
    C. 8.5×10^{-19} kg
    D. 3.3×10^{-28} kg

45. A photon with an energy of 5.0×10^{-20} J strikes an electron. All of the photon's energy is converted into kinetic energy of the electron. Calculate the velocity of the electron after the strike, assuming it was at rest before the energy transfer.

    A. 1.2×10^5 m/s
    B. 2.4×10^4 m/s
    C. 1.2×10^3 m/s
    D. 3.3×10^5 m/s

46. The Heisenberg Uncertainty Principle states that:

    A. Both position and momentum of a particle cannot be known precisely at the same time.
    B. The position and the momentum of the particle cannot be known precisely at any time.
    C. The probability of finding an electron near a nucleus is related to the square of its wave function.
    D. The probability of finding an electron near an atom is at least 95%.

47. The depiction of an "s" electron orbital as a three-dimensional sphere is not completely correct because:

    A. The electron orbitals are not spheres, but vary in shape from spheres to octagons.
    B. The electron orbitals are not three dimensional, rather they are n-dimensional depending on the probability of an electron being near a nucleus.
    C. The depiction is correct, and is the best fit to experimental data.
    D. The electron orbitals are waves rather than spheres.

48. How many quantum numbers are required to describe the electrons of any system?

    A. 1
    B. 3
    C. 4
    D. 2

49. How many distinct magnetic quantum numbers are possible if the angular momentum quantum number is 6?

    A. 13
    B. 7
    C. 12
    D. 3

50. If the principal quantum number is 3, how many values of the angular momentum quantum numberare possible?

    A. 3
    B. 4
    C. 5
    D. 2

51. Which of the following quantum number sets is unacceptable?

    A. 1, 0, 0
    B. 6, 2, 0
    C. 4, 3, 3
    D. 4, 2, 3

52. Arrange the following numbers to obtain an acceptable set of quantum numbers for an electron: 4, 5, −2

    A. 4, 5, −2
    B. −2, 5, 4
    C. 5, −2, 4
    D. 5, 4, −2

53. A g-orbital has an angular momentum value equal to 4. How many total electrons can the g-orbitals hold?

    A. 18
    B. 10
    C. 6
    D. 2

54. The Pauli Exclusion Principle states:

    A. The position and velocity of an electron can never be known precisely at the same time.
    B. No two electrons in the system can have the same quantum numbers.
    C. The spin number must be $-\frac{1}{2}$ for the first electron in an orbital, and $+\frac{1}{2}$ for the second one.
    D. The azimuthal number can never be the same for two electrons in two identical orbitals of two different atoms.

55. The Pauli Exclusion Principle is violated by one of the following electron systems

    A. 1, 0, 0, $\frac{1}{2}$ 1, 0, 0, $\frac{1}{2}$ C. 4, 3, -3, $\frac{1}{2}$ 4, 3, -3, $-\frac{1}{2}$
    B. 5, 4, -2, $\frac{1}{2}$ 5, 4, -2, $-\frac{1}{2}$ D. 3, 2, -2, $\frac{1}{2}$ 3, 2, -2, $-\frac{1}{2}$

56. Which of the following sets is unacceptable for any electron system?

    A. 3, 2, 3, $-\frac{1}{2}$ B. 2, 0, 0, $\frac{1}{2}$ C. 2, 1, 1, $-\frac{1}{2}$ D. 2, 1, -1, $\frac{1}{2}$

57. We first encounter a d-electron in which row of the periodic table?

    A. 3   B. 6   C. 4   D. 2

58. Which one of the following elements does the following configuration describe: $1s^2 2s^2 2p^6 3s^2 3p^5$

    A. Cl   B. Ar   C. K   D. S

59. Which one of the following elements has a $5s^2 5p^5$ valence shell configuration?

    A. Te   B. I   C. Sb   D. As

60. Which one of the following elements has a $6s^2 4f^{14} 5d^3$ valence shell configuration?

    A. Re   B. Ta   C. Mo   D. Hf

61. What elements in the periodic table have the following electron configuration: [Noble gas] $ns^2 nd^5$

    A. Fe, Ru, Os, Uno   B. Mn, Tc, Re   C. F, Cl, Br, I, At   D. Co, Rh, Ir, Une

62. Place the following atoms, P, Kr, Mg, Li, in order of increasing first ionization energies.

    A. P < Kr < Mg < Li   C. Kr < P < Mg < Li
    B. Mg < Li < P < Kr   D. Li < Mg < P < Kr

63. Place the following atoms, Cl, F, Na, C, in order of decreasing electron affinity values.

    A. C > Cl > F > Na   C. F > Na > Cl > C
    B. Cl > F > C > Na   D. F > Cl > C > Na

64. At what step of ionization does astatine exhibit a sudden marked increase in its ionization energy?

    A. 6th   B. 5th   C. 4th   D. 3rd

65. Place the following elements, Br, Kr, C, Se, Te, in order of increasing atomic size.

    A. Br < Te < Kr < Se < C   C. Te < Se < Br < Kr < C
    B. C < Kr < Br < Se < Te   D. Br < Kr < C < Se < Te

66. Which of the following elements has the lowest reducing ability?

    A. Li   B. Cs   C. Na   D. K

Answers to Exercises

1. $7.5 \times 10^{14}\, s^{-1}$ to $4.3 \times 10^{14}\, s^{-1}$

2. See <u>Figure 7.2 in your textbook</u>.

3. $6.7 \times 10^{14}\, Hz$

4. $5.3 \times 10^{-7}\, m$

5. a. $4.469 \times 10^{14}\, s^{-1}$ b. $2.961 \times 10^{-19}\, J$, $178.3\, kJ$

6. $29.6\, nm$

7. a. infrared   c. $4.0 \times 10^{-20}\, J$, $24\, kJ$
   b. $5 \times 10^{-6}\, m$ d. less

8. $3.49 \times 10^{-20}\, J$

9. $21.0\, kJ$

10. Direct proportionality between frequency and energy

11. $1.4 \times 10^{-37}\, m$

12. a. $3.3 \times 10^{-14}\, m$ b. $2.4 \times 10^{-38}\, m$

13. a. $2.965 \times 10^{-19}\, J$ c. $4.414 \times 10^{-19}\, J$
    b. $3.424 \times 10^{-19}\, J$ d. $1.29 \times 10^{-15}\, J$

14. a. 0   b. $2.18 \times 10^{-18}\, J$

15. $6.817 \times 10^{-8}\, m$

16. $1.31 \times 10^{6}\, J$

17. $486\, nm$

18. Ultraviolet

19. Visible

20. The key point is that the Schrödinger equation shows that electrons can reside in certain mathematically allowed locations.

21. Yes

22. The probability of finding the electron on Mars mathematically approaches zero with increasing distance from the nucleus, but it never actually reaches zero.

23. C

24. a. 32 electrons   b. 14 electrons   c. 18 electrons

25. $1, 0, 0, -\frac{1}{2}$; $1, 0, 0, +\frac{1}{2}$; $2, 0, 0, -\frac{1}{2}$; $2, 0, 0, +\frac{1}{2}$; $2, 1, 0, -\frac{1}{2}$

26. Mutually repulsive electrons will occupy separate p orbitals. This behavior is summarized by Hund's rule, which states that the lowest energy configuration for an atom is the one having the maximum number of unpaired electrons allowed by the Pauli principle in a particular set of degenerate orbitals.

27. Ca: $1s^2\ 2s^2\ 2p^6\ 3s^2\ 3p^6\ 4s^2$

28. Barium has the same number of valence electrons and therefore similar chemical properties as calcium. Ba = [Xe] $6s^2$; Ca = [Ar] $4s^2$.

29. a. 2   d. 5   f. 4
    b. 1   e. 1   g. 0
    c. 0

30. a. $3d$ b. $5s$ c. $4p$

31. X = bromine

32. $6s$

33. VIII A (noble gases); Periods I and II

34. a. K^+, Ar, Cl^-, S^{2-} c. Ar, P, Mg, Na
    b. F, C, Si, Al   d. Ba^{2+}, Cs^+, Xe, I^-

35. Mg < Ca < Ga < In < Te < Xe

36. a. C, Ge   b. Cl^+, Cl^-

37. Examples are reaction with water to give bases and metallic character, among others.

38. Si, Ge, As, Sb, Te, Po, and At are called metalloids because they exhibit both metallic and nonmetallic properties under certain circumstances.

39. A	40. D	41. A	42. B	43. C	44. B
45. D	46. A	47. C	48. C	49. A	50. A
51. D	52. D	53. A	54. B	55. A	56. A
57. C	58. A	59. B	60. B	61. B	62. D
63. B	64. A	65. B	66. A		

Chapter 8

Bonding: General Concepts

In this chapter, you will use many of the concepts you learned in Chapter 7, especially electronic configurations. You will learn why different types of bonds form, the nature of those bonds, and a model to predict the three-dimensional structure of molecules formed from covalent bonds.

8.1   Types of Chemical Bonds

When you finish this section you will be able to list and define **three types of bonding** and solve for **the energy of interaction of an ionic solid**.

Your textbook says that bonds are *"forces that hold groups of atoms together and make them function as a unit."* Bonds form because the energy of the system is lower than if bonds did not form.

Ionic bonding is due to electrostatic attraction. It results from the loss of an electron from a metal and its gain by a nonmetal.

Your textbook gives a formula for the energy of an ion pair called the **energy of interaction**. Let's use that formula to calculate the stability gain when Mg^{2+} and O^{2-} interact.

Example 8.1   Coulomb's Law

Calculate the energy of interaction (in kJ/mol) between Mg^{2+} and O^{2-} if the distance between the centers of Mg^{2+} and O^{2-} is 0.205 nm (2.05 Å).

Solution

$$E = 2.31 \times 10^{-19} \text{ J nm} \left[\frac{Q_1 Q_2}{r} \right] = 2.31 \times 10^{-19} \text{ J nm} \left[\frac{(+2)(-2)}{0.205 \text{ nm}} \right]$$

$$= -4.51 \times 10^{-18} \text{ J/ion pair} = \mathbf{-4.51 \times 10^{-21} \text{ kJ/ion pair}}$$

For a mole of ion pairs,

$$E = \frac{-4.51 \times 10^{-21} \text{ kJ}}{\text{ion pair}} \times \frac{6.022 \times 10^{23} \text{ ion pairs}}{\text{mole}} = -2710 \text{ kJ/mol}$$

Commentary

Note how much higher this value is than the value for the NaCl example in your textbook! Part of the reason for this is the **smaller radius** owing to the higher charge on Mg and O. The more important contribution is from the **higher charge** on each ion. This allows for more powerful electrostatic interaction, which gives a more stable bond.

Covalent Bonding occurs when bonds form between similar kinds of atoms for the same reason as between dissimilar atoms: the energy of the system is lowered as a result of the bond formation.

Look at Figure 8.1 in your textbook. At a particular distance apart, the combination of repulsive and attractive forces allows the system to have a minimum energy. In this case, we have a **covalent bond** in which **electrons are shared by both nuclei approximately equally.** Examples are S_8, graphite, diamond, and fullerenes.

In the case where there is **unequal sharing of electrons, a polar covalent** bond exists. Charges are not distributed equally in such a molecule. Positive and negative poles exist. Examples of polar covalent bonds are C–Cl, H–Cl, and O–H.

In summary, the nature of the bond will depend upon the ability of each atom in the bond to attract electrons to itself. This is called **electronegativity**.

8.2   Electronegativity

When you finish this section you will be able to use the positions of atoms in the periodic table and their electronegativities to predict relative bond polarities.

Recall that unequal sharing of electrons in a bond results in a polar covalent bond. Ionic bonds result from the transfer of electrons between atoms.

Your textbook says that we can get a measure of the degree of ionic character of a bond by **measuring bond energies**. The more the electrostatic interaction that occurs between two atoms, the greater will be the difference in bond energies when compared to the average of the perfectly covalent bonds involving the atoms. This difference in bond energy is called Δ and is the relative **electronegativity** difference between the bonding atoms.

For the representative elements, **electronegativity decreases going down a group** and **increases going across a period**. Thus francium has the lowest electronegativity, and fluorine has the highest.

The greater the Δ, the more ionic character the bond has. If $\Delta = 0$, the bond is said to be perfectly covalent. But there are no precise cut-offs. All bonds have some ionic and some covalent character.

Example 8.2 A   Electronegativities of Atoms

Look at the periodic table at the front of your textbook. Based only on their positions in the table, place the following atoms in order of increasing electronegativity.

Sr, Cs, Se, O, Ba

Solution

Remember the trend: electronegativity increases from the lower left to the upper right of the periodic table. Based on their positions, the order is:

Cs < Ba < Sr < Se < O
least electronegative   most electronegative

Try the next example, keeping in mind that the determinant of bond polarity is the value of Δ.

Example 8.2 B   Bond Polarity

Using <u>Figure 8.3 in your textbook</u>, calculate Δ for each of the following bonds, and order the set from most covalent to most ionic character.

a. Na–Cl   b. Li–H   c. H–C   d. H–F   e. Rb–O

Solution

a.   Na = 0.9, Cl = 3.0 Δ = 3.0 – 0.9 = **2.1**
b.   Li = 1.0, H = 2.1 Δ = 2.1 – 1.0 = **1.1**
c.   H = 2.1, C = 2.5 Δ = 2.5 – 2.1 = **0.4**
d.   H = 2.1, F = 4.0 Δ = 4.0 – 2.1 = **1.9**
e.   Rb = 0.8, O = 3.5 Δ = 3.5 – 0.8 = **2.7**

In order of increasing ionic character:

H–C, Li–H, H–F, Na–Cl, Rb–O
most covalent   most ionic

The information gained from electronegativities will be used in the next section to figure out the polarities of molecules.

8.3   Bond Polarity and Dipole Moments

When you finish this section you will be able to determine whether or not simple molecules have a dipole moment.

Recall from Section 8.2 that it is possible to determine the **polarity of a bond** by the **size of Δ**. If a molecule is diatomic (two atoms), there is often only one bond, and that will determine whether the molecule is polar.

For instance, in Example 8.2 B, we determined that H–F was polar, with fluorine being the more electronegative atom. A **partial negative charge ($\delta-$)** resides on the fluorine atom, and a **partial positive charge ($\delta+$)** resides on the hydrogen atom.

+——→
H———F
$\delta+$ $\delta-$

The **arrow** points to the **center of negative charge** while the **tail** is at the **center of positive charge**. A **dipole moment** means that the molecule has **two poles**.

The situation is clear-cut with HF. It becomes more difficult with 3 or more atoms in a molecule because the **individual dipoles can cancel each other out**. Look at Table 8.2 in your textbook. This shows how individual bond polarities can cancel each other out to yield a molecule with no dipole moment. Although you will be able to derive the geometries in Table 8.2 later on, you should memorize them for now.

Example 8.3 A   Dipole Moment

Does $CHCl_3$ (a tetrahedral molecule with carbon at the center) have a dipole moment? If so, show the orientation of the dipole moment.

Strategy

Perform the following steps:

1. Look up the electronegativity of each atom.
2. Draw the molecule in three-dimensional space.
3. Determine the polarity of each bond and the net polarity on each atom.
4. Draw the dipoles, and determine the direction (if any) of the molecule dipole moment.

Solution

$C = 2.5$, $H = 2.1$, $Cl = 3.0$

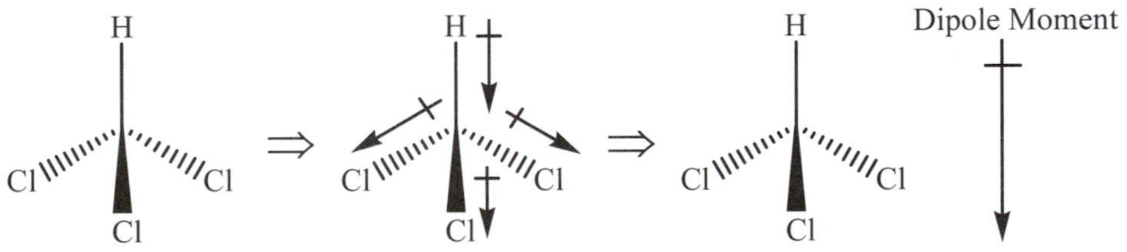

Example 8.3 B   Practice with Dipole Moments

For each of the following, determine the orientation of the dipole moment (if any).

    a. HI   b. N_2 c. CCl_2F_2 (carbon is the central atom)

Solution

a.   H———I   ⟹   H———I   ⟹   H———I
      2.1   2.5 $\delta+$ $\delta-$

b.   N≡≡≡N   ⟹   No dipole moment ($\Delta = 0$)
      3.0   3.0

c. $C = 2.5$, $Cl = 3.0$, $F = 4.0$ ⟹

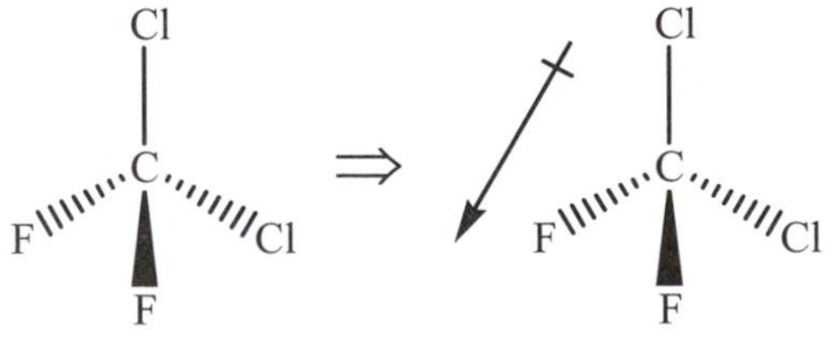

(Fluorine is more electronegative than chlorine; therefore the dipole moment on this molecule is tipped slightly toward the fluorine atoms.)

182 Chapter 8

8.4   Ions: Electron Configurations and Sizes

When you finish this section you will be able to predict

- The formulas of simple ionic compounds.
- The relative sizes of ions.

Your textbook deals only with **nonmetals, representative metals,** and ionic bonds in this discussion. It has been observed that atoms that form bonds in stable compounds have a **noble gas electronic configuration**. (Each is isoelectronic with a noble gas.)

Example 8.4 A   Ionic Electron Configuration

Write the electronic configuration, and determine the charge on each of the following atoms when it forms its most stable ion (noble gas electronic configuration).

    a. Mg   b. P   c. Br   d. Rb

Solution

Determine how many electrons the atom must gain or lose (metals will lose; nonmetals will gain) to obtain the electronic structure of the **nearest noble gas**.

a. Magnesium is $1s^2 2s^2 2p^6 3s^2$. It will lose the two $3s$ electrons to have the electronic configuration of neon. Magnesium will therefore ionize to $\mathbf{Mg^{2+}}$.

$$\mathbf{Mg^{2+} = 1s^2 2s^2 2p^6} \text{ (isoelectronic with neon)}$$

b. P (nonmetal) will gain 3 electrons to be isoelectronic with argon.

$$\mathbf{P^{3-} = 1s^2 2s^2 2p^6 3s^2 3p^6}$$

c. Br (nonmetal) will gain one electron to be isoelectronic with krypton.

$$\mathbf{Br^- = 1s^2 2s^2 2p^6 3s^2 3p^6 4s^2 3d^{10} 4p^6}$$

d. Rb (metal) will lose one electron to be isoelectronic with krypton.

$$\mathbf{Rb^+ = 1s^2 2s^2 2p^6 3s^2 3p^6 4s^2 3d^{10} 4p^6}$$

Note that Br^-, Rb^+, and Kr are all **isoelectronic**.

To determine the formula of binary ionic compounds remember that **chemical compounds are electrically neutral**. (The sum of the cation charges must equal the anion charges.)

For example, the formula of the ionic compound formed from combining magnesium and chlorine can be determined by:

1. assessing the charge on the ions, and
2. determining how many of each ion is required to combine to make the compound electrically neutral.

The ions are $\mathbf{Mg^{2+}}$ and $\mathbf{Cl^-}$. In order to maintain electronic neutrality,

$$Mg^{2+} + 2Cl^- \rightarrow MgCl_2.$$

Example 8.4 B   Formulas of Binary Ionic Compounds

Determine the formula for each of the following sets of atoms when they combine to form a binary ionic compound.

    a.   K and Br   b.   Sr and F   c.   Al and Se   d.   Ba and O

Solution

a. $K^+ + Br^- \rightarrow$ **KBr**
b. $Sr^{2+} + 2F^- \rightarrow$ **SrF$_2$**
c. $2Al^{3+} + 3Se^{2-} \rightarrow$ **Al$_2$Se$_3$**
d. $Ba^{2+} + O^{2-} \rightarrow$ **BaO**

We have maintained electronic neutrality throughout.

We discussed ion size in our review of Chapter 7. As we go down a group, ion size increases (higher energy levels have a larger average distance). Because the ratio of protons to electrons becomes greater, **cations are smaller than their neutral atoms**. Because of electron-electron repulsion and less effective shielding, **anions are always larger than their neutral atoms**. *The larger the charge, the more pronounced the effect*. Thus,

S^{2-} is larger than S.
Ca^{2+} is smaller than Ca.
S^{2-} is **isoelectronic** with Ca^{2+}. Because Ca^{2+} has two more protons than electrons, while S^{2-} has more
    electrons than protons, S^{2-} is much larger than Ca^{2+}. Therefore, we can conclude that for an isoelectronic
    series, **the more positive the nuclear charge (Z), the smaller the ion**.

Example 8.4 C   Ion Size

Order the following ions from smallest to largest.

    a. O^{2-}, Na^+, Mg^{2+}, F^-
    b. Se^{2-}, Te^{2-}, Rb^+, Mg^{2+}

Solution

a.   These ions are all **isoelectronic with neon**. Therefore the smallest will have the highest positive
    charge. The ions will get larger until we have the highest negative charge.

$$Mg^{2+} < Na^+ < F^- < O^{2-}$$
     smallest   largest

b.   Because Se^{2-} and Te^{2-} are anions, they will be larger than Rb^+ and Mg^{2+}. Because Te^{2-} is farther
    down the periodic table than Se^{2-}, it will be larger. Magnesium is higher on the periodic table and
    has a greater charge than rubidium; therefore, it will have the smallest ion.

$$Mg^{2+} < Rb^+ < Se^{2-} < Te^{2-}$$
     smallest   largest

8.5   Energy Effects in Binary Ionic Compounds

When you finish this section you will be able to explain the different steps involved when elements form ionic solids. We will pay special attention to the idea of lattice energy.

The theme of this section is that there are **many separate processes** which go into **forming** an **ionic solid**. As we have said before, *the ionic compound forms because its energy is lower than if its elements remain separated.* However, not every part of the process is energetically favorable. Your textbook points out that it is the **lattice energy** (the energy released when an ionic solid is formed from its gaseous ions) that is the most favorable and that more than makes up for some parts of the process that are energetically unfavorable. Let's examine the formation of the **KCl ionic solid** ("salt") from **K(s)** and **Cl₂(g)**.

Processes That Must Occur

1. $K(s)$ must form $K(g)$; **(Energy of Sublimation = +64 kJ)**.
2. $K(g)$ must form $K^+(g)$; **(First Ionization Energy = +419 kJ)**.
3. $\frac{1}{2}Cl_2(g)$ must form $Cl(g)$; **(Bond Energy for ½ mole = 120 kJ)**.
4. $Cl(g)$ must form $Cl^-(g)$; **(Electron Affinity = −349 kJ)**.
5. $K^+(g)$ must combine with $Cl^-(g)$ to form $KCl(s)$; **(Lattice Energy = −690 kJ)**.

The net energy of formation (ΔH_f^o) equals the sum of the energy changes, −436 kJ. So you see that the value for any of the processes that make up salt formation can be obtained if you understand the processes involved and are given suitable data.

Example 8.5   Calculation of Lattice Energy

Given the following data, determine ΔH_f^o for LiBr.

$$Li(s) + \tfrac{1}{2}Br_2(g) \rightarrow LiBr(s)$$

ionization energy for Li =	+ 520 kJ/mol
electron affinity for Br =	− 324 kJ/mol
energy of sublimation for Li =	+ 161 kJ/mol
lattice energy =	−787 kJ/mol
bond energy of Br₂ =	+ 193 kJ/mol

Solution

While ΔH_f^o will be the sum of all energy changes, be careful to **multiply the bond energy by ½** because we have ½ of a mole of $Br_2(g)$.

$$520 + (-324) + (+161) + (-787) + (+96) = -334\,kJ = \Delta H_f^o$$

Note that here, as in the previous example, the lattice energy is the most significant contributor to salt formation.

The remainder of the section makes the point that the **higher the charge** on each ion, the greater the lattice energy will be. This value counteracts the higher endothermic ionization energies, thus resulting in a **more energetically stable crystal**.

8.6   Partial Ionic Character of Covalent Bonds

If you can answer the questions posed in this review section, then you understand the material that your textbook is trying to get across.

1. Why do we say that there are **no totally ionic** bonds?
2. How do we determine the **percent ionic character** of a bond?
3. How do we define (as an operating definition) an ionic compound (salt)?

8.7   The Covalent Chemical Bond: A Model

If you can answer the questions posed in this review section, then you understand the material that your textbook is trying to get across.

1. (Review) What is a chemical bond?
2. (Review) Why do chemical bonds occur?
3. Why is it useful to think of each bond in a molecule as being **environment independent**?
4. Knowing the energy of stabilization of CCl_4, discuss how we would determine the bond energy of a C–H bond in $CHCl_3$.
5. What is **a model**?
6. Why do we develop models?
7. List the fundamental properties of a model.
8. What are the **limitations** of models?
9. Is a wrong model useless? Why or why not?

8.8   Covalent Bond Energies and Chemical Reactions

When you finish this section you will be able to calculate heats of reaction from bond energies.

Recall from the last section that your textbook calculated the average bond energy for a C–H bond for methane (CH_4). The assumptions made were:

1. that the energy needed to break each C–H bond was the same (413 kJ), and
2. that each of the bonds was not sensitive to its environment.

Assumption #2 is not really correct, as can be shown by the inset in which hydrogens and ethane (C_2H_6) are replaced by bromine, chlorine, and fluorine (see your textbook). In spite of this, average bond energies can help give us fair estimates of heats of reaction, and as such they are useful tools.

Look at Table 8.4 in your textbook. This table gives **average bond energies** for many covalent bonds. Notice that **multiple bonds** (bonds that involve **sharing more than 2 electrons** between atoms) require more energy to break than single bonds.

Your textbook uses the example of the combination of hydrogen and fluorine gas to make hydrogen fluoride. It states that

$$\Delta H = \Sigma D \text{ (bonds broken)} - \Sigma D \text{ (bonds formed)}$$

where ΔH = heat of reaction and D = bond energy per mole of bonds.

The key to calculating the heat of reaction from the average bond energy is to **carefully list all bonds broken and all bonds formed**. Also, recognize that *it takes energy to break bonds* (endothermic, $\Delta H = +$) while *energy is released when bonds are formed* (exothermic, $\Delta H = -$).

Example 8.8   *Heat of Reaction from Bond Energy*

Using data from Table 8.4 in your textbook, calculate ΔH for the following reaction. Compare this to ΔH_f° calculated from the thermodynamic values in the Appendix. (See Chapter 6 if you need a review of this.) (ΔH_f° [for CF_4] = -680 kJ/mol)

$$CH_4(g) \ + \ 4F_2(g) \ \rightarrow \ CF_4(g) \ + \ 4HF(g)$$

Solution

Let's make a list of bonds broken and bonds formed.

Bonds Broken	Energy per Bond (kJ)		Total Energy (kJ)
4 C–H	413×4 bonds	=	1652
4 F–F	154	=	616 (154×4 moles)
Total Energy to Break Bonds		=	**2268 kJ**

Bonds Formed	Energy per Bond (kJ)		Total Energy (kJ)
4 C–F	485×4 bonds	=	1940
4 H–F	565	=	2260 (565×4 moles)
Total Energy to Form Bonds			**4200 kJ**

ΔH = (Energy to break bonds) − (Energy released from forming the bonds)
     = 2268 kJ − 4200 kJ
     = **−1932 kJ**

This reaction is exothermic.

The value for ΔH_f° = (ΔH_f° [for CF_4] + 4 ΔH_f° [for HF]) − (ΔH_f° [for CH_4] + 4 ΔH_f° [for F_2])
           = (−680 + 4 (−271)) − (−75 + 4 (0))
           = **−1689 kJ**

8.9   *The Localized Electron Bonding Model*

If you can answer the following review questions, you understand the important ideas in this brief section.

1. What is the **basic assumption** about electron position in the LE model?
2. What are **lone pairs** and **bonding pairs**?
3. What are the parts of the LE model?

8.10   *Lewis Structures*

When you finish this section you will be able to draw **Lewis structures** for a variety of **covalent molecules**.

Lewis structures are often used to depict **bonding pairs** and **lone pairs** in molecules. We are concerned only with **valence** electrons because these are the ones (for Period 1 and 2 atoms) that are involved in bond making and breaking.

Individual atoms are represented with Lewis structures by putting valence electrons (as dots or circles) around the atomic symbol. For example, magnesium ([Ne] $3s^2$) would be represented as

$\cdot$ **Mg** $\cdot$

Mg^{2+} ([Ne]) would be Mg. It no longer has its valence electrons. Chlorine ([Ne] $3s^2\,3p^5$) would be

:$\overset{\bullet\bullet}{\underset{\bullet\bullet}{Cl}}$ $\cdot$

and Cl^- ([Ne] $3s^2\,3p^6$) would be

:$\overset{\bullet\bullet}{\underset{\bullet\bullet}{Cl}}$:

Example 8.10 A   Lewis Dot Structures

Draw Lewis dot structures for the following atoms or ions.

 a. N   b. N^{3-} c. I   d. Ba   e. Ba^{2+}

Solution

 a. N = [He] $2s^2\,2p^3$ (5 valence electrons) $\cdot\overset{\bullet\bullet}{\underset{\bullet}{N}}\cdot$

 (I drew the $2s^2$ on the top and the three 2p electrons singly. You can draw a maximum of 2 electrons on any side unless there is a triple bond involved).

 b. N^{3-} = [He] $2s^2\,2p^6$ (8 valence electrons) :$\overset{\bullet\bullet}{\underset{\bullet\bullet}{N}}$:

 c. I = [Kr] $5s^2\,4d^{10}\,5p^5$ (7 valence electrons) :$\overset{\bullet\bullet}{\underset{\bullet\bullet}{I}}\cdot$

 d. Ba = [Xe] $6s^2$ (2 valence electrons) $\cdot$ **Ba** $\cdot$

 e. Ba^{2+} = [Xe] (0 valence electrons) **Ba**

EVERY PERIOD 1 AND 2 ELEMENT (with the exception of H, He, B, and Be) CAN FORM COMPOUNDS OF LOWEST ENERGY IF THEIR HIGHEST ENERGY LEVELS ARE FILLED (s^2p^6). THIS IS CALLED THE <u>OCTET RULE</u>. If an ion or atom observes the octet rule, we will say that it is "happy" (using terminology first coined by Nobel Prize winner in Chemistry Roald Hoffmann). Hydrogen is "happy" if its 1s orbital is filled. We will discuss boron and beryllium later on.

Your textbook discusses a strategy for drawing Lewis structures. I will propose a different strategy. Use the one you feel more comfortable with.

Kelter Strategy for Writing Lewis Structures

1. **Total number of valence electrons in the system:** Sum the number of valence electrons on all of the atoms. Add the total negative charge if you have an anion. Subtract the charge if you have a cation.

 e.g., CO_3^{2-};   C has 4 valence electrons   = 4
                O has 6 valence electrons $\times$ 3 atoms   = 18
                <u>charge on the ion is -2 so add</u>   = <u>2</u>
                electrons in the system =   24

2. **Number of electrons if each atom is to be happy:** Atoms in our examples will need 8 electrons (octet rule) or 2 electrons (hydrogen). Using CO_3^{2-} as an example:

                C needs 8 electrons   = 8
                <u>O needs 8 electrons $\times$ 3 atoms</u>   = <u>24</u>
                electrons for happiness=   32

The -2 charge comes as a result of the electrons in the system. The charge is **never** counted toward happiness.

3. **The number of bonds in the system:** Covalent bonds are made by sharing electrons. You need 32 electrons and you have 24. You are 8 electrons deficient. If you make 4 bonds (with 2 electrons per bond), you will make up the deficiency. Therefore,

$$\# \textbf{ bonds } = \frac{\#2 - \#1}{2} = \frac{32 - 24}{2} = \textbf{4 bonds}$$

4. **Draw the structure.** The central atom is carbon. The oxygens surround it. Because there are 4 bonds, there will be two single bonds and one double bond. Each bond accounts for two electrons. Then complete the octets by putting electrons around each atom. Double-check your results by counting total electrons in the system.

$$\left[:\!\ddot{O}\!=\!\!\underset{\underset{\displaystyle :\ddot{O}:}{|}}{C}\!\!-\!\!\ddot{O}\!: \right]^{2-} = 24 \text{ electrons}$$

We will discuss resonance in this system later on.

Example 8.10 B   More Lewis Structures

Using the steps outlined above (or in your textbook), write Lewis structures that obey the octet rule for the following:

        a. Cl_2 b. CH_2Cl_2 c. H_2CO d. NH_3

Solution

a. 1. Total of valence electrons $= 7$ per Cl $\times 2 = $ **14**
    2. Total if happy $= $ per Cl $\times 2 = $ **16**
    3. # bonds $= (16 - 14)/2 = $ **1 bond**

$$:\!\ddot{Cl}\!-\!\ddot{Cl}\!: = 14 \text{ electrons}$$

b. 1. Total of valence electrons $=$ 4 for the C $= $ 4
                                 7 per Cl $\times 2 = $ 14
                                 <u>1 per H $\times 2 = $ 2</u>
                        valence electrons $= $ 20

    2. Total if happy $= $ 8 for the C $= $ 8
                             8 per Cl $\times 2 = $ 16
                        <u>2 per H $\times 2 = $ 4</u>
                electrons if happy $= $ 28

    3. # bonds $= (28 - 20)/2 = $ **4 bonds**

$$:\!\ddot{Cl}^{\backslash\backslash\backslash\backslash}\!\!\underset{\underset{\displaystyle H}{|}}{\overset{\displaystyle H}{C}}\!\!^{////}\ddot{Cl}\!: = 20 \text{ electrons (Hydrogen does NOT get a complete octet!)}$$

c. 1. Total of valence electrons $= $ 4 for the C $= $ 4
                                 6 for the O $= $ 6
                               <u>1 per H $\times 2 = $ 2</u>
                        valence electrons $= $ 12

2. Total if happy = 8 for the C = 8
                        8 for the O = 8
                        2 per H × 2 = 4
            electrons if happy = 20

3. # bonds = (20 − 12)/2 = **4 bonds**

$$\underset{H}{\overset{H}{>}}C\!=\!\ddot{O}\!: \quad = 12 \text{ electrons}$$

(Only C, O, N, S, and P commonly have double bonds. Only C and N commonly have <u>stable</u> triple bonds.)

d. 1. Total of valence electrons =   5 for the N = 5
                                        1 per H × 3 = 3
                valence electrons = 8

2. Total if happy = 8 for the N = 8
                        2 per H × 3 = 6
            electrons if happy = 14

3. # bonds = (14 − 8)/2 = **3 bonds**

$$\underset{\underset{H}{|}}{\overset{H\diagdown \overset{..}{N} \diagup H}{}} \quad = 8 \text{ electrons}$$

8.11   Exceptions to the Octet Rule

When you finish this study section you will be able to draw Lewis structures for covalent molecules whose central atom does not obey the octet rule.

Although your textbook deals with boron in this section, we will focus on the more general case of **central atoms that can exceed the octet rule**. This can happen because there are **empty *d* orbitals** that shared electrons can occupy. This can happen with third period (row) elements.

Writing Lewis Structures for Exceptions to the Octet Rule

To determine if you have an exception to the rule, proceed as if your molecule obeys the octet rule. Let's use ICl_3 as an example.

1. Total of valence electrons =   7 for the I   =   7
                                     7 per Cl × 3   = 21
                valence electrons = 28

2. Total if happy =   8 for the I   = 8
                        8 per Cl × 3   = 24
            electrons if happy = 32

3. # bonds = (32 − 28)/2 = **2 bonds**

CAUTION! We have 2 bonds for 3 chlorine atoms! **There are not enough bonds to account for all the atoms!** This is how you know that we have an exception to the octet rule.

To write the Lewis structure of exceptions, draw the structure with **one bond to each ligand, complete the octets, and add any extra electrons to the central atom**. With ICl_3,

$$\ddot{:}\underset{\cdot\cdot}{Cl}\diagdown\quad\diagup\underset{\cdot\cdot}{Cl}\ddot{:}$$
$$\overset{\cdot}{\underset{\cdot}{\underset{I}{\cdots}}}$$
$$\ddot{:}\underset{\cdot\cdot}{Cl}\ddot{:}$$

= 28 valence electrons (I has <u>10 electrons</u> represented here.)

Example 8.11   Exceptions to the Octet Rule

Write Lewis structures for the following molecules:

    a. IF_2^{-1} b. CO_2 c. XeF_4

Solution

a. 1. Total of valence electrons =   7 for the I =   7
                                        7 per F $\times$ 2   = 14
                                 <u>add 1 for "–"   =   1</u>
                            valence electrons   = 22

    2. Total if happy =   8 for the I =   8
                              <u>8 per F $\times$ 2 = 16</u>
                  electrons if happy = 24

    3. # bonds = (24 – 22)/2 = **1 bond** <u>NOT ENOUGH! EXCEPTION!</u>

$$\left[\ddot{:}\underset{\cdot\cdot}{F}-\overset{\cdot\cdot}{\underset{\cdot\cdot}{I}}-\underset{\cdot\cdot}{F}\ddot{:}\right]^{-}$$

b. 1. Total valence electrons = **16**
    2. Total if happy = **24**
    3. # bonds = **4 bonds** Not an exception!

$$\ddot{O}=C=\ddot{O}$$

c. 1. Total of valence electrons = **36** (Xe = 8!)
    2. Total if happy = **40**
    3. # bonds = **2 bonds** <u>Exception!</u>

8.12   Resonance

Recall the example of CO_3^{2-} that we did at the beginning of Section 8.10. We determined the Lewis structure to be

However, the double bond could have been placed on any of the three oxygens:

Measurements in bond lengths suggest that **all three C—O bond lengths are equivalent**. Electrons move around the entire molecule. Therefore, the actual structure is a **time-average** of all these structures. These structures are called **resonance** structures. The Lewis structure of the molecule can be drawn any of three ways. The double bond seems to resonate between the carbon and oxygen atoms.

Formal Charge and Resonance

Let's review your textbook's discussion on formal charge and how it is used to draw resonance structures. Formal charge is, "the difference between the number of valence electrons on the free atom and the number of valence electrons assigned to the atom in a molecule." Formal charge is a **computational device** based on a **localized electron (LE) model** and as such is not perfectly correct. To determine formal charge (a somewhat more realistic estimate of charge distribution in a molecule), we need to know:

1.   How many electrons each atom "owns."

> **Electrons Owned** =   [# valence electrons around the atom +
>
> # bonds (which equals ½ # shared electrons)]

2.   The **formal charge** on each atom.

> **Formal Charge** =   # valence electrons on the neutral atom
>
> − # electrons owned by the atom based on the resonance structure you drew

Let's look at CO_3^{2-} again.

- Carbon owns 4 electrons (4 bonds). The formal charge on carbon = **4 valence electrons** on the neutral atom minus **4 electrons owned = 0**.

- Oxygen$_a$ has 6 valence electrons and 1 bond. It owns **7 total valence electrons**. The formal charge = 6 valence electrons on the neutral atom minus 7 electrons owned = **−1**.

- Oxygen$_b$ has the same formal charge as oxygen$_a$ = **−1**.

- Oxygen$_c$ has 4 valence electrons and 2 bonds. It owns 6 total valence electrons. The formal charge = 6 valence electrons on the neutral atom minus 6 electrons owned = **0**.

The sum of the formal charges, $0 + (-1) + (-1) + 0 = -2$, **must always equal the charge on the ion** (or molecule, if that's what you are dealing with). Your textbook says that if you can write **nonequivalent** Lewis structures (different numbers of single and double bonds) for a molecule or ion, those with formal charges closest to zero and with any negative formal charges on the most electronegative atoms are considered to best describe the bonding.

Example 8.12 A   Formal Charges

Assign formal charges to each atom in the following resonance structures of CO_2.

1. $\overset{\cdot\cdot}{\underset{\cdot\cdot}{O}} = C = \overset{\cdot\cdot}{\underset{\cdot\cdot}{O}}$
      a   b

2. $:O \equiv C - \overset{\cdot\cdot}{\underset{\cdot\cdot}{O}}:$
       a   b

Which structure is more likely to be correct?

Solution

Let's establish formal charges for each atom.

Structure 1:
    C owns 4 electrons. Formal charge $= 0$
    O_a owns 6 electrons. Formal charge $= 0$
    O_b owns 6 electrons. Formal charge $= 0$

Structure 2:
    C owns 4 electrons. Formal charge $= 0$
    O_a owns 5 electrons. Formal charge $= +1$
    O_b owns 7 electrons. Formal charge $= -1$

Structure 1 is more likely because all formal charges are zero.

Note: Structure 2 can be represented with its formal charges:

$$:O \equiv C - \overset{\cdot\cdot}{\underset{\cdot\cdot}{O}}:$$
$$\overset{}{+}\qquad\qquad\underset{-}{}$$

Example 8.12 B   Resonance

Draw all resonance structures and select the most stable one for SCN^-.

Solution

First determine the number of bonds for the Lewis structures.

    There are **16 valence electrons** in the system.
    The total for happiness is **24 electrons**.
    There are $(24-16)/2 = $ **4 bonds** in this ion.

Now let's draw some possible resonance structures based on 4 bonds.

    a. $\left[:\overset{\cdot\cdot}{\underset{\cdot\cdot}{S}} - C \equiv N: \right]^-$
         b. $\left[:S \equiv C - \overset{\cdot\cdot}{\underset{\cdot\cdot}{N}}: \right]^-$
         c. $\left[\overset{\cdot\cdot}{\underset{\cdot\cdot}{S}} = C = \overset{\cdot\cdot}{\underset{\cdot\cdot}{N}} \right]^-$

Now evaluate the formal charges on each atom.

a. $\left[:\ddot{S}-C\equiv N: \right]^{-}$
 $-1\quad\ 0\quad\ 0$

b. $\left[:S\equiv C-\ddot{N}: \right]^{-}$
 $+1\quad\ 0\quad\ -2$

c. $\left[\ddot{S}=C=\ddot{N} \right]^{-}$
 $0\quad\ 0\quad\ -1$

Nitrogen is more electronegative than sulfur. Therefore structure "c" is the most likely.

8.13   Molecular Structure: The VSEPR Model

When you finish this section you will be able to predict molecular geometries by using the VSEPR model.

The Valence Shell Electron Pair Repulsion (VSEPR) model assumes that atoms will orient themselves so as to minimize electron pair repulsions around the central atom.

Memorize the information in Table 8.6 in your textbook. Each lone pair or bond around the central atom occupies a position in space. The effect of lone pairs around the central atom is to squeeze bonded pairs closer together (see your textbook's discussion regarding bond angles in CH_4, NH_3, and H_2O). Multiple bonds are counted as "one bonding pair" in the VSEPR model because the double bonds are constrained in space. Let's determine the VSEPR structure of formaldehyde, H_2CO.

Step 1:   Determine the Lewis structure.

Total valence electrons = **12**
Total if happy = **20**
bonds = **4**

$$\begin{array}{c} H \\ \diagdown \\ C=\ddot{O}: \\ \diagup \\ H \end{array}$$

Step 2: Count the number of bonds and lone pairs on the central atom.

2 C–H + 1 C=O = **3 bonds**

Step 3: Determine the geometry based on Table 8.6 in your textbook.

3 "electron pairs" = **trigonal planar**

Now let's try **IBr$_2$**.

Step 1:   Total valence electrons = 22
Total if happy = 24
bonds = 1 **Exception!**

$$:\ddot{B}r—\ddot{I}—\ddot{B}r:$$

Step 2:   There are **2 bonds** and **3 lone pairs** around the central atom. The total is **5 electron pairs**.

Step 3: The molecule will orient itself with a trigonal bipyramid geometry. The electron pairs will orient in the equatorial positions first (see discussion in the text), and the bonding pairs will make up the remaining positions:

The molecule will be **linear.**

(Note that when we draw VSEPR structures, we are only concerned with the central atom. We often omit valence electrons around the ligands.)

Example 8.13   VSEPR Structures

Determine the geometry of each of the following molecules or ions.

   a. CO_2 b. SO_4^{2-} c. BrF_3 d. XeO_4 e. ICl_2^+

Solution

a.   The Lewis structure for CO_2 (**16 valence electrons**) is

The double bonds each count in the VSEPR model as 1 restricted bond, so CO_2 acts as if it has **2 electron pairs** around it. Geometry = **linear.**

b. SO_4^{2-} = **32 valence electrons**

The double-bonded resonance structure seems to be better because of the lower formal charges on the oxygens, although even experts in inorganic chemistry still debate which is correct. It turns out that for purposes of the VSEPR model, both structures will give the same 4 electron pairs (2 double bonds "count" as 2 electron pairs). Geometry = **tetrahedron.**

c. BrF_3 = **28 valence electrons**

There are **2 lone pairs** and **3 bonding pairs = 5 electron pairs**. It is a **trigonal pyramid** basis. The electrons take up two equatorial spots, leaving two F's axial and one equatorial. Geometry = **T-shaped**.

d. XeO_4 = **32 valence electrons**

$$:\overset{\displaystyle ..}{O}:$$
$$|$$
$$:\overset{..}{\underset{..}{O}}-Xe-\overset{..}{\underset{..}{O}}:$$
$$|$$
$$:\overset{}{\underset{..}{O}}:$$

There are 4 bonds, thus **4 electron pairs**. Geometry = **tetrahedral**.

e. $ICl_2{}^+$ = **20 valence electrons**

$$\left[:\overset{..}{\underset{..}{Cl}}-\overset{..}{\underset{..}{I}}-\overset{..}{\underset{..}{Cl}}:\right]^+$$

There are 2 bonding pairs and 2 lone pairs, thus 4 electron pairs. It is a **tetrahedral basis**.
Geometry = **V-shaped**.

Exercises

Section 8.1

1. Indicate whether the bonds between the following would be primarily covalent, polar covalent, or ionic:

    a. O–H
    b. Cs–Cl

    c. H–Cl
    d. Br–Br

2. Calculate the energy of interaction for KCl if the internuclear distance is 0.314 nm.

3. Calculate the energy of interaction between Ag^+ and Br^- if the internuclear distance of AgBr is 0.120 nm (in kJ/mole).

Section 8.2

4. Using a periodic table, order the following from lowest to highest electronegativity:

    a. Fr, Mg, Rb
    b. B, Al, C, N

    c. P, As, Ga, O
    d. Cl, S, P

5. Using the periodic chart of elements, place the following in order from the lowest to the highest electronegativity:

    F, Nb, N, Si, Rb, Ca, Pt

6. Using Figure 8.3 in your textbook, calculate the difference in electronegativity (Δ) for each of the following bonds:

    a. Cl–Cl
    b. K–Br

    c. Fe–O
    d. H–O

    e. S–H

7. Place the following in order of increasing polarity:

    NaBr, I_2, H_2O, MnO_2, CN^-

8. Which of the following molecules contain polar covalent bonds? List in order of increasing bond polarity. (Use Fig. 8.3 in your text).

    O_3, P_8, NO, CO_2, CH_4, H_2S

9. How will the charge be distributed on each of the following molecules: HF, NO, CO, and HCl?

10. Why is it that BeF_2 is ionic, and $BeCl_2$ is covalent?

Section 8.3

11. Determine the orientation of the dipole of the following, if any.

    a. $AlCl_3$ (planar with aluminum atom at the center)
    b. CH_3F (tetrahedral with carbon at the center)
    c. N_2O (linear with N–N–O structure)
    d. $AgCl_4$ (planar molecule, silver atom at center, Ag–Cl bonds 90° apart)

12. Which of the molecules in problem 11 contain one or more polar bonds?

13. Which of the following molecules would you expect to have a dipole moment of zero? Describe the dipole orientation of the other two molecules.

    a. KI
    b. CF_4 (tetrahedral structure)
    c. H_2Se (bent structure)

Section 8.4

14. Determine the most stable ion for each of the following atoms, and indicate which element they would be isoelectronic with if they lost or gained electrons:

    a. O
    b. Be
    c. I
    d. Te
    e. Na

15. List four ions that are isoelectronic with argon and have charges from −2 to +2. Arrange these in order of increasing ionic radius.

16. List four ions that are isoelectronic to Kr. Arrange these in order of increasing radius.

17. Determine the formula of the binary compound formed from the following sets of atoms.

    a. Ca and O
    b. K and Cl
    c. Rb and S
    d. Ba and P

18. Predict formulas for the following binary ionic compounds.

    a. Mg and N
    b. Na and F
    c. Ca and S
    d. Sr and Te

19. Using shorthand notation, list the core electron configurations for the ions in the compounds in problem 18.

20. Place the following in an order of increasing ionic size. Use the shorthand notation to list the core electron configuration for each of the ions.

    a. Ba^{2+}, Te^{2-}, Cs^+, I^-
    b. Rb^+, S^{2-}, O^{2-}, K^+

21. Place the following in an order of increasing ionic size. Use shorthand notation to list the core electron configuration for each of the ions.

    a. Cl^-, F^-, Sr^{2+}, Ca^{2+}
    b. Na^+, Mg^{2+}, Li^+, Be^{2+}

Section 8.8

22. Using the bond energy values listed in <u>Table 8.4 of your text</u>, calculate the ΔH for the following reactions:

    a. $2H_2(g) + O_2(g) \rightarrow 2H_2O(g)$
    b. $2C_2H_6(g) + 7O_2(g) \rightarrow 4CO_2(g) + 6H_2O(g)$
    c. $HCN(g) + 2H_2(g) \rightarrow CH_3NH_2(g)$

23. Use bond energy values from Table 8.4 in your textbook to calculate ΔH for the following reactions:

    a. $H-C\equiv C-H(g) + H_2(g) \rightarrow CH_2=CH_2(g)$
    b. $2CH_4(g) + 3O_2(g) \rightarrow 2CO(g) + 4H_2O(g)$
    c. $N_2(g) + 2H_2(g) \rightarrow N_2H_4(g)$ (NH_2-NH_2)

24. Compare the values obtained in parts "a" and "b" of problem 23 to ΔH values calculated from ΔH_f° data in Appendix 4 in your textbook.

25. Calculate the enthalpy of reaction ΔH_f° for the following reaction. Use the enthalpies of formation found in Appendix 4 of your textbook.

$$H_2(g) + C_2H_4(g) \rightarrow C_2H_6(g)$$

Section 8.10

26. Draw Lewis dot structures for the following atoms, ions, or molecules:

    a. Sr
    b. Br^-
    c. ICN

    d. Ga
    e. $GaCl_4^-$
    f. P^{3-}

    g. NH_2^-
    h. CSe_2

27. Draw Lewis structures for the following:

    a. H^+
    b. C

    c. P
    d. P^{5+}

    e. Cl^-

28. Draw Lewis structures for the following:

    a. AsF_3
    b. O_3

    c. H_3O^+
    d. BH_4^-

    e. NH_4^+
    f. O_2

Section 8.11

29. Draw Lewis dot structures for the following:

    a. BCl_3
    b. AsF_5

    c. BrO_3^-
    d. S_2F_{10} (contains a S–S bond)

30. Draw Lewis dot structures for the following:

    a. $SbCl_3$
    b. AlF_6^{3-}
    c. PCl_5

Section 8.12

31. Assign formal charges to each of the labeled atoms, a-e.

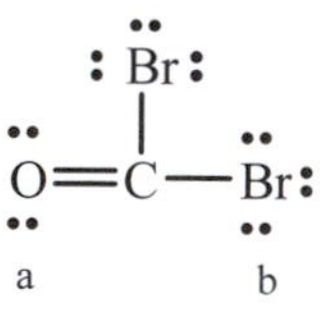

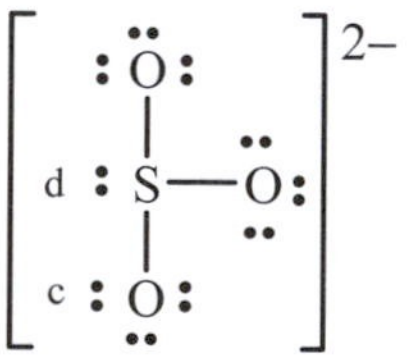

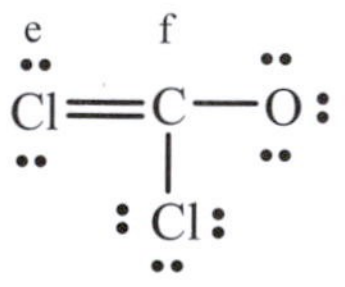

32. Draw the remaining resonance forms for N_2O_4.

33. How many reasonable resonance structures are there for carbon monoxide, CO? What are the formal charges?

34. Hydrazine, N_2H_4, is used as a propellant on the Space Shuttle. Draw all reasonable structures for N_2H_4, and assign formal charges.

35. Draw a Lewis structure and any resonance forms of benzene, C_6H_6. (Benzene consists of a ring of six carbon atoms with one hydrogen bonded to each carbon.)

Section 8.13

36. Predict the structure of each of the following molecules or ions:

   a. SeF_6 c. ClF_4^+ e. CF_3Cl (carbon is central atom)
   b. N_2O d. ClO^-

37. Predict the structure of HCN.

38. Using the VSEPR model, determine the molecular geometry for each of the following molecules:

   a. SCl_4 c. IF_4^- e. $TlCl_2^+$
   b. H_2Se d. $SnCl_5^-$

39. Which of the molecules or ions in problem 36 contain polar covalent bonds? Are polar?

Multiple Choice Questions

40. The bond in RbF is:

   A. Covalent   B. Molecular   C. Polar covalent   D. Ionic

41. Which of the following bonds do you expect to be polar covalent?

   A. H–N   B. H–H   C. Cs–F   D. H–O

42. In a polar bond, electrons:

   A. spend equal time around both nuclei
   B. are localized between both nuclei
   C. spend more time around the bigger nucleus
   D. spend more time around one of the nuclei than the other one

43. What is the electronegativity difference between At and H?

   A. 0.1   B. −0.1   C. 4.3   D. 0.0

44. Which of the following bonds is the most polar one?

   A. H–O   B. Cs–Cl   C. N–O   D. C–H

45. Order the following bonds in order of increasing bond polarity: H–F, Se–Cl, C–O, C–At

A. C–At < Se–Cl < C–O < H–F
B. C–O < Se–Cl < H–F < C–At

C. H–F < C–O < Se–Cl < C–At
D. C–At < C–O < Se–Cl < H–F

46. Order the following bonds in order of decreasing bond polarity:

$$Ca–O, Ca–Cl, P–Cl, Fe–O, B–O, N–O$$

A. N–O > P–Cl > B–O > Fe–O > Ca–Cl > Ca–O
B. Ca–Cl > P–Cl > Ca–O > Fe–O > B–O > N–O

C. Ca–O > Ca–Cl > Fe–O > B–O > P–Cl > N–O
D. Fe–O > Ca–O > B–O > N–O > Ca–Cl > P–Cl

47. Which of the following molecules would exhibit the greatest polarity? All molecules are tetrahedral in shape.

A. $CHCl_3$ B. CH_4 C. CCl_4 D. CH_3Cl

48. Which of the following molecules has a dipole moment equal to zero?

A. SiO_4 (tetrahedral)   B. H_2O (bent)   C. $C_2H_2F_2$ (tetrahedral)   D. $CBrCl_2F$

49. Which ion could the following electron configuration describe? $1s^2 2s^2 2p^6 3s^2 3p^4$

A. K^+ B. Cl^+ C. S^{2-} D. Ca^{2+}

50. Which of the following ions does not have the following configuration? $1s^2 2s^2 2p^6 3s^2 3p^6 3d^2$

A. V^+ B. Nb^{3+} C. Mn^{2+} D. all of them

51. Place the following species in order of increasing size: Ne, B^{3+}, O^{2-}, and Be^{2+}

A. $B^{3+} < Be^{2+} < Ne < O^{2-}$
B. $Ne < B^{3+} < Be^{2+} < O^{2-}$

C. $O^{2-} < Ne < Be^{2+} < B^{3+}$
D. $Ne < O^{2-} < B^{3+} < Be^{2+}$

52. Determine the formula for the following sets of atoms when they combine to form binary compounds: Cs and F, Al and O, B and F, Ag and Cl

A. Cs_2F, Al_2O_3, BF, AgCl
B. CsF, Al_2O_3, BF_3, AgCl

C. CsF_2, AlO, B_3F, $AgCl_2$
D. Cs_2F_2, Al_2O_5, B_2F_2, Ag_2Cl_2

53. Select the crystal that would have the largest lattice energy. Assume that the internuclear distance is the same in all these crystals.

A. NaCl   B. KCl   C. K_2S D. CaO

54. Chemical bonds between two atoms result because:

A. The atoms can thus achieve a state of higher energy
B. The atoms can thus achieve a state of lower energy
C. The atoms fit together nicely
D. The atoms can react better when bonded

55. Two bonded atoms:

A. React more readily with other substances
B. Are less reactive compared to when free

C. Share all their electrons
D. Behave in unpredictable ways

56. The reaction of hydrogen with fluorine gas is highly exothermic (releases a high degree of energy). Calculate the F–F bond energy knowing that: H–H = 432 kJ/mol, H–F = 565 kJ/mol, and $\Delta H = -543$ kJ.

$$H_2(g) + F_2(g) \rightarrow 2HF(g)$$

A. 155 kJ/mol   B. 543 kJ/mol   C. 698 kJ/mol   D. 1019 kJ/mol

57. A truck uses propane (C_3H_8) to power its engine. Calculate how much heat will be released when 5 moles of propane are burned, knowing that the reaction of propane with oxygen gas produces carbon dioxide and water.

   A. 7330 kJ   B. 25 kJ   C. 10,000 kJ   D. 4784 kJ

58. Chlorine trifluoride is prepared by reacting chlorine gas with fluorine gas. The heat of the reaction is −803 kJ/mol of chlorine reacted. Calculate the Cl–Cl bond energy.

   A. 1091 kJ/mol   B. 155 kJ/mol   C. 238 kJ/mol   D. 50 kJ/mol

59. How many valence electrons does selenium have?

   A. 6   B. 4   C. 3   D. 5

60. How many of the six valence electrons in sulfur are used in covalent bonding in sulfur tetrachloride and disulfur difluoride?

   A. 4 and 2   B. 3 and 2   C. 6 and 1   D. 2 and 2

61. How many of the six valence electrons in oxygen are usually used in covalent bonding?

   A. 4   B. 3   C. 6   D. 2

62. In the $POCl_3$ molecule, how many double bonds are there? How about single bonds?

   A. 1 and 3   B. 4 and 1   C. 2 and 1   D. 1 and 2

63. Which one of the following molecules possesses a triple bond?

   A. SF_4 B. PCl_5 C. C_2H_2 D. C_2H_6

64. Which one of the following molecules does not possess a double bond?

   A. C_2F_4 B. $C_2H_4F_2$ C. OCH_2 D. $HOCOCl$

65. Which one of the following molecules contains a central atom that violates the octet rule?

   A. SF_4 B. COF_2 C. $Si(OH)_4$ D. PBr_3

66. Calculate the formal charge on chlorine in ClO_4^-

   A. −1   B. +3   C. +6   D. +4

Answers to Exercises

1. a. polar covalent   c. polar covalent
    b. ionic   d. covalent

2. -7.36×10^{-19} J

3. E = -1160 kJ/mole

4. a. Fr, Rb, Mg   c. Ga, As, P, O
    b. Al, B, C, N   d. P, S, Cl

5. Rb < Ca < Nb < Si < Pt < N < F
    0.8   1.0   1.6   1.8   2.2   3.0   4.0

6. a. 0   c. 1.7   e. 0.4
    b. 2.0   d. 1.4

7. I_2 < CN^- < H_2O < NaBr < MnO_2
    0   0.5   1.4   1.9   2.0

8. O_3, P_8 < H_2S, CH_4 < NO < CO_2

9. HF; NO; CO; HCl
    + − + − + − + −

10. The difference in the electronegativity between Be and F is higher than the one between Be and Cl.

    $BeCl_2$: 3.0 − 1.5 = 1.5 (covalent bond)
    BeF_2:   4.0 − 1.5 = 2.5 (ionic)

11. a. no dipole   c. negative toward O
    b. negative toward F   d. no dipole

12. all

13. b. The opposing bond polarities in a tetrahedral structure cancel out. Thus CF_4 has no dipole moment. KI is a binary ionic compound that has a negative dipole toward I. Selenium will have a partial negative charge as its electronegativity is greater than that of hydrogen. Thus the resulting dipole moment of H_2Se would be orientated as shown:

14. a. O^{2-}, isoelectronic with neon
    b. Be^{2+}, isoelectronic with helium
    c. I^-, isoelectronic with xenon
    d. Te^{2-}, isoelectronic with xenon
    e. Na^+, isoelectronic with neon

15. Ca^{2+}, K^+, Cl^-, S^{2-}

16. Sr^{2+} < Rb^+ < Br^- < Se^{2-}

17. a. CaO   c. Rb_2S
    b. KCl   d. Ba_3P_2

18.   a. Mg_3N_2 c.   CaS
       b.   NaF   d.   SrTe

19.   a.   [Ne], [Ne]   c.   [Ar], [Ar]
       b.   [Ne], [Ne]   d.   [Kr], [Xe]

20.   a. $Ba^{2+} < Cs^+ < I^- < Te^{2-}$;   all can be written as [Xe]
       b. $K^+ < O^{2-} < Rb^+ < S^{2-}$; $K^+ = [Ar]$, $O^{2-} = [Ne]$, $Rb^+ = [Kr]$, $S^{2-} = [Ar]$

21.   a. $Ca^{2+} < Sr^{2+} < F^- < Cl^-$ $Ca^{2+} = [Ar]$, $Sr^{2+} = [Kr]$, $F^- = [Ne]$, $Cl^- = [Ar]$
       b. $Be^{2+} < Li^+ < Mg^{2+} < Na^+$ $Be^{2+} = [He]$, $Li^+ = [He]$, $Mg^{2+} = [Ne]$, $Na^+ = [Ne]$

22.   a.   −509 kJ   b.   −2881 kJ   c.   −158 kJ

23.   a.   −169 kJ   b.   −1091 kJ   c.   +81 kJ

24.   a.   6 kJ difference   b.   7 kJ difference

25. $\Delta H_f^\circ = -136.7$ kJ/mole

26.   a.   • Sr •   d.   • Ga •   g. $\left[H-\overset{..}{\underset{..}{N}}-H \right]^-$

       b. $\left[:\overset{..}{\underset{..}{Br}}: \right]^-$ e. $\left[\begin{array}{c} :\overset{..}{\underset{..}{Cl}}: \\ | \\ :\overset{..}{\underset{..}{Cl}}-Ga-\overset{..}{\underset{..}{Cl}}: \\ | \\ :\overset{..}{\underset{..}{Cl}}: \end{array} \right]^-$ h. $\overset{..}{\underset{..}{Se}}=C=\overset{..}{\underset{..}{Se}}$

       c. $:\overset{..}{\underset{..}{I}}-C\equiv N:$ f. $\left[:\overset{..}{\underset{..}{P}}: \right]^{3-}$

27.   a. $[H]^+$ c. $:\overset{.}{P}\,•$ e. $\left[:\overset{..}{\underset{..}{Cl}}: \right]^-$

       b.   • $\overset{.}{\underset{.}{C}}$ •   d. $[P]^{5+}$

28.   a. $\begin{array}{c} :\overset{..}{F}-\overset{..}{As}-\overset{..}{F}: \\ | \\ :\overset{..}{\underset{..}{F}}: \end{array}$ c. $\left[\begin{array}{c} H-\overset{..}{O}-H \\ | \\ H \end{array} \right]^+$ e. $\left[\begin{array}{c} H \\ | \\ H-N-H \\ | \\ H \end{array} \right]^+$

       b. $\overset{..}{O}=\overset{..}{O}-\overset{..}{\underset{..}{O}}:$ d. $\left[\begin{array}{c} H \\ | \\ H-B-H \\ | \\ H \end{array} \right]^-$ f. $\overset{..}{\underset{..}{O}}=\overset{..}{\underset{..}{O}}$

29. a. 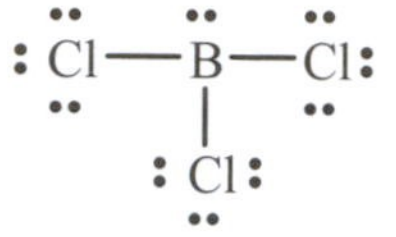 :Cl—B—Cl: c. 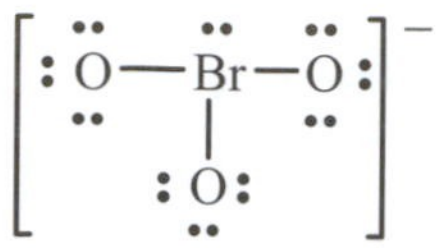 [:O—Br—O:]⁻
         | |
         :Cl: :O:

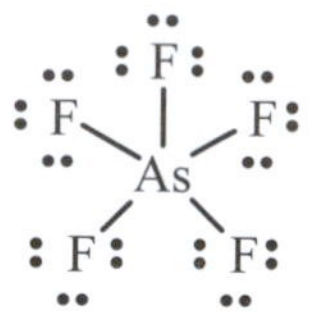

     b. :F\ :F: /F: d. 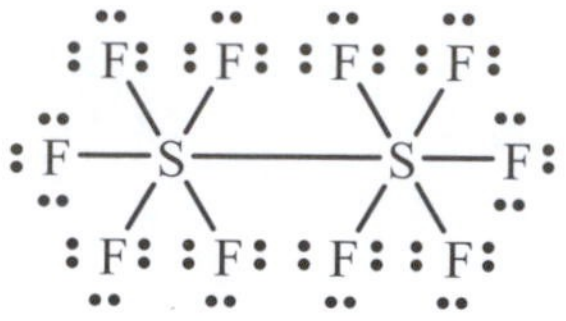 :F: :F: :F: :F:
          \ | /   \ | | /
           As   :F—S———S—F:
          /   \   / | | \
       :F: :F: :F: :F: :F: :F:

30. a. 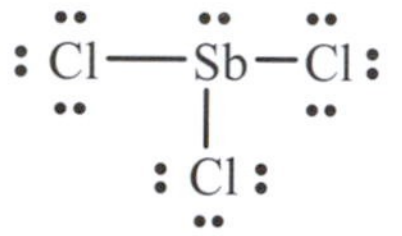 :Cl—Sb—Cl: b. 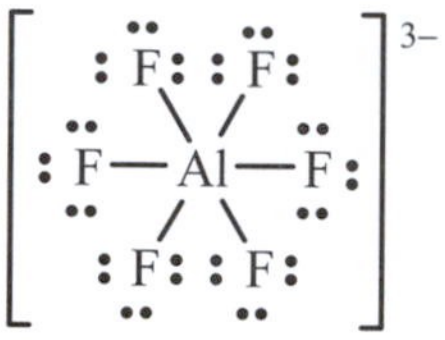 [:F: :F:]³⁻   c. 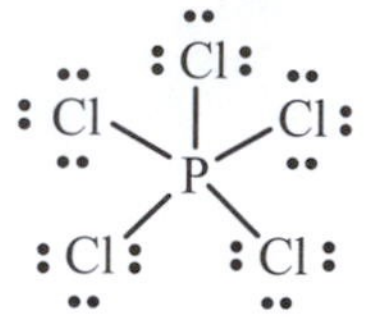   :Cl:
              | [\ /   ] |
             :Cl: [:F—Al—F:]   :Cl—P—Cl:
                                  [/ \   ] |
                                  [:F: :F:]   :Cl: :Cl:

31. a. 0   c. −1   e. +1
     b. 0   d. +1   f. 0

32.

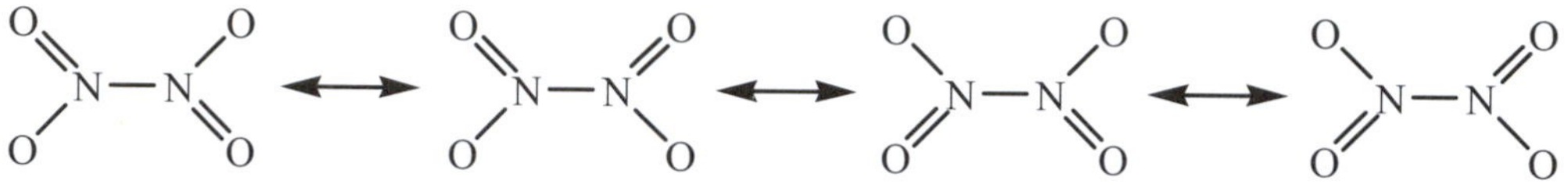

33. :O≡C: is the only reasonable structure. The carbon has a formal charge of −1, and oxygen has a +1
     formal charge. It is possible to draw a structure such as :O=C: in which both C and O have a zero
     formal charge, but the carbon would be electron deficient (lacks an octet).

34. The only reasonable structure has an N—N single bond. All formal charges = 0.

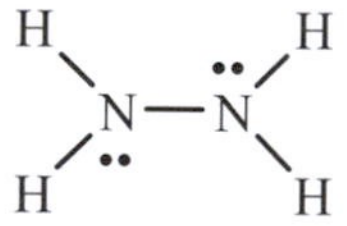

     Each nitrogen has a formal charge of +1 and the two hydrogens that are single have formal charges of −1.

35.

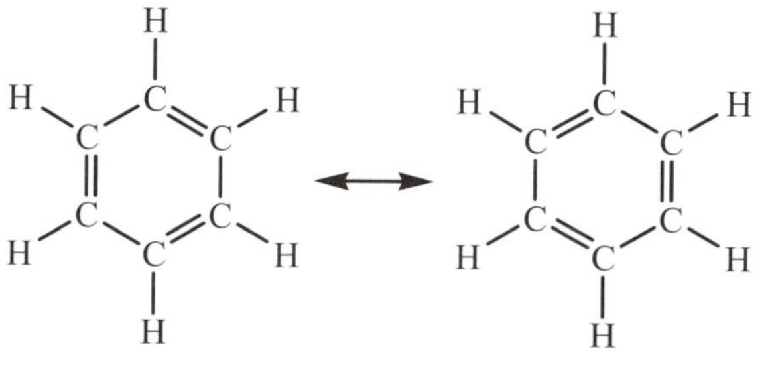

36. a. octahedral   c. see-saw   e. tetrahedral
     b. linear   d. linear

37. HCN is linear (H—C≡N:)

38. a.   see-saw   c.   square planar   e.   linear
     b.   bent   d.   trigonal bipyramidal

39. All contain polar covalent bonds; b, c, d, and e are polar. Negative is toward: O in b; equatorial fluorine in
     c; O in d; fluorine in e.

40. D   41. D   42. D   43. A   44. B   45. A
46. C   47. A   48. A   49. B   50. C   51. A
52. B   53. D   54. B   55. B   56. A   57. C
58. C   59. A   60. A   61. D   62. A   63. C
64. B   65. A   66. B

Chapter 9

Covalent Bonding: Orbitals

Chapter 8 dealt with ways of representing structures on paper and with predicting the three-dimensional structure that covalent molecules should have. This chapter reviews the two most important models that attempt to explain covalent molecular and ionic structure and shape—the **Localized Electron (LE) Model** and the **Molecular Orbital (MO) Model**.

In the **LE model**, electron pairs are still localized around specific atoms, but the orbitals around the central atom are modified.

In the **MO model**, all of the electrons in the molecule are combined into a set of molecular orbitals that describe bonding in the entire molecule. As we shall review, each model has its special strengths.

9.1   Hybridization and the Localized Electron Model

When you finish this section you will be able to describe the bonding in a variety of covalent molecules and ions using the concept of hybridization.

Your textbook points out the difficulty with assuming that the 4 C–H bonds in methane (CH_4) are made by interacting the $2s$ and $2p$ orbitals of carbon (valence orbitals) with the $1s$ orbitals of hydrogen. It says that the three $2p$-$1s$ C–H interactions would be expected to be located at 90° to one another. This does not agree with the observed bond angles of 109.5°. Recall that these bond angles **minimize electron-electron repulsions** in methane.

To work around this difficulty, the concept of **hybrid orbitals** is suggested. By combining the **one 2s** and the **three 2p** orbitals of carbon, **four hybrid sp^3** orbitals are formed. These orbitals are equivalent, and that is consistent with the observed presence of 4 equivalent C–H bonds. Figure 9.6 in your textbook displays the interaction between the sp^3 hybrid orbitals of carbon and the $1s$ orbitals of the hydrogens.

The bottom line is this: When the VSEPR model indicates that you have a **tetrahedron** as the basis of your structure (4 effective electron pairs), the atom is sp^3 **hybridized**.

Example 9.1 A   sp³ Hybridization

Describe the bonding in the water molecule using the localized electron model.

Strategy

We must establish the VSEPR structure. This is done by drawing the Lewis structure and determining the number of effective electron pairs around the central atom.

Solution

The Lewis structure of H_2O is (see Sections 8.10 and 8.13)

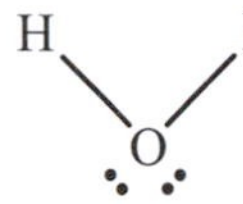

Four effective electron pairs around the oxygen atom indicate a **tetrahedral** basis leading to a V-shaped structure. Therefore the oxygen is sp^3 hybridized. Two sp^3 orbitals are occupied with $1s$ electrons from hydrogen. The other two sp^3 orbitals are occupied by lone pairs.

In molecules or ions with a **trigonal planar** configuration (3 effective electron pairs around an atom), sp^2 **hybridization** occurs. With this hybridization, an s orbital and two p orbitals are used. That leaves an **unhybridized** (unchanged) p **orbital perpendicular** to the sp^2 plane.

Look at Figure 9.12 in your textbook.

- Bonds formed from the overlap of orbitals in the plane between two atoms are called **sigma (σ) bonds**.
- Bonds formed by the overlap of unhybridized p orbitals (above and below the center plane) are called **pi (π) bonds**.
- A **single bond** is a σ bond.
- A **double bond** consists of **one σ and one π bond**.
- A **triple bond** consists of **one σ bond and two π bonds**.

Example 9.1 B   Sigma and Pi Bonds

How many σ bonds are there in the commercial insecticide, "Sevin," shown below? How many π bonds?

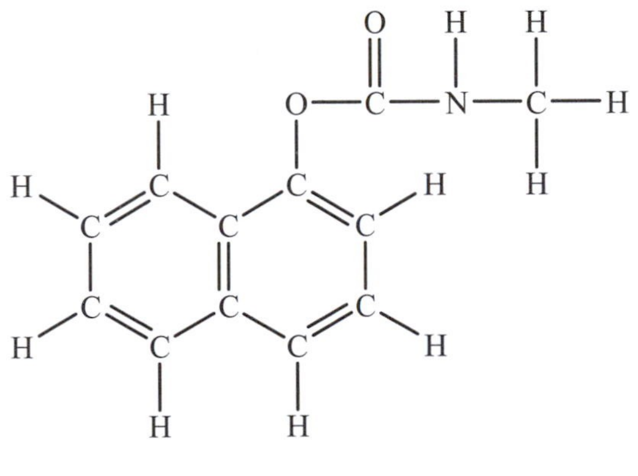

Solution

There are **27 σ bonds** (single bonds and one of the bonds in a double bond).
There are **6 π bonds** (the other bond in a double bond).

Your textbook goes over a variety of different hybridization schemes, all of which center on the idea that **the hybridization of the orbitals of an atom depends on the total number of effective electron pairs around it**. In order to determine the hybridization of an atom, it is essential that you can figure out its VSEPR structure.

The following table summarizes effective electron pairs around an atom with hybridization. Remember that, for purposes of the VSEPR model, **double and triple bonds count as only one effective electron pair**.

Effective Electron Pairs Around an Atom	Arrangement	Hybridization
2	linear	sp
3	trigonal planar	sp^2
4	tetrahedral	sp^3
5	trigonal bipyramid	dsp^3
6	octahedral	d^2sp^3

Keep in mind that a ligand can have a hybridization different from that of the central atom. Each atom in a molecule must be considered separately based on the Lewis and VSEPR structures of that molecule.

Example 9.1 C   Practice with Hybrid Orbitals

Give the hybridization, and predict the geometry of each of the central atoms in the following molecules or ions.

    a. IF_2^+
    b. OSF_4 (sulfur is the central atom)
    c. SiF_6^{2-}
    d.   HCCH (work with both carbons)

Solution

a. Lewis structure:

$$:\!\overset{\cdot\cdot}{\underset{\cdot\cdot}{F}}\!-\!\overset{\cdot\cdot}{\underset{\cdot\cdot}{I}}\!-\!\overset{\cdot\cdot}{\underset{\cdot\cdot}{F}}\!:$$

All three atoms have 4 effective electron pairs. They are **all sp^3 hybridized**. The VSEPR structure has a tetrahedral basis. Because the central atom (iodine) has two bonding pairs, it will take on a **V-shape**, but the bond angle will be smaller than 109.5°.

b. Lewis structure:

Sulfur has 5 effective electron pairs. The VSEPR structure is a trigonal bipyramid. **Sulfur is dsp^3 hybridized**. Each of the fluorines has 4 effective electron pairs.

c. Lewis structure:

Silicon has 6 effective electron pairs. The VSEPR structure is octahedral. **Silicon is d^2sp^3 hybridized**. Each of the fluorines has 4 effective electron pairs.

d. Lewis structure: $H-C\equiv C-H$

Each carbon has 2 effective electron pairs. The VSEPR structure is linear. **Each carbon is *sp* hybridized**. Hydrogen atoms bond using $1s$ orbitals. The orbitals are unhybridized.

Example 9.1 D   Summing it All up

Answer the following questions regarding aspartame (NutraSweet™).

a. How many σ bonds are in the molecule?
b. How many π bonds?
c. What is the hybridization on carbon "a"? Carbon "b"?
d. What is the hybridization on nitrogen "a"? Oxygen "a"?
e. What is the $C_b-O_b-H_a$ bond angle?

Solution

You must remember to complete the octets in this "shorthand" Lewis structure. Lone pairs are often "assumed," so always be on the lookout.

a. 39 σ bonds
b. 6 π bonds
c. sp^3, sp^2 (3 effective electron pairs)
d. sp^3 (remember the "assumed" lone pair to complete the octet), sp^2 (3 effective electron pairs, including the two to complete the octet)
e. O_b is sp^3 hybridized (complete the octet!); therefore, the angle is based on a tetrahedron, with two lone pairs compressing the C–O–H bond angle to 104.5°

9.2   The Molecular Orbital Model

When you finish this section you will be able to:

- State the major ideas of the molecular orbital (MO) model.
- Compute bond orders for homonuclear diatoms of hydrogen and helium.

The localized electron model does an excellent job of predicting and justifying molecular shapes. It does not deal with molecules with **unpaired electrons**. It also neglects **bond energies**. The molecular orbital (MO) model also gives a view of electrons in a molecule that allows us to get a clearer understanding of what we had called resonance.

Key Ideas of the MO Model

1. All the valence electrons in a molecule exist in a set of **molecular orbitals** of a given energy. The valence electrons of each atom are not acting independently, but rather act with other valence electrons to form a set of MO's.
2. Figure 9.27 in your textbook illustrates that there are **bonding** and **antibonding** MO's. Bonding results in **lower energy** than if no interaction occurred. Antibonding results in higher energy.
3. A molecule will be stable if there is more bonding than antibonding interaction.
4. **Bond order (BO)** is a measure of net bonding interactions.

$$BO = \frac{\text{\# bonding electrons} - \text{\# antibonding electrons}}{2}$$

5. BO must be greater than zero for a stable molecule to form.
6. The higher the BO, the stronger the bond.

Your text goes over the bonding in H_2, H_2^-, and He_2. Let's try two more.

Example 9.2   MO Theory

Use MO theory to describe the bonding and stability of

    a. H_2^{2-} b. H_2^+

Solution

We must fill in the molecular orbital energy-level diagram for each of the species.

a. H_2^{2-} has 4 electrons. They will fill pairwise, and with opposite spins, the σ_{1s} and $\sigma_{1s}*$ orbitals.

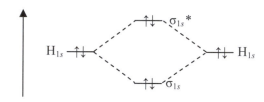

Bond order $= \dfrac{\textbf{2 bonding electrons} - \textbf{2 antibonding electrons}}{\textbf{2}} = \textbf{0}$

We would not expect H_2^{2-} to be a stable ion.

b. H_2^+ has 1 electron. We must again fill in our MO energy-level diagram.

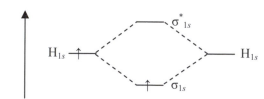

Bonding order $= \dfrac{\textbf{1 bonding electron} - \textbf{0 antibonding electrons}}{\textbf{2}} = \textbf{½}$

We would expect H_2^+ to form, but it would not be much more stable than two separate hydrogens.

9.3 Bonding in Homonuclear Diatomic Molecules

When you finish this section you will be able to describe the bonding in homonuclear diatomic molecules with regard to:

- Stability.
- Bond order, relative bond energy, and bond length.
- Paramagnetism or diamagnetism.

In the last section we saw how to use the MO model to assess the **stability** of diatomic molecules with 1s orbitals. In this section we will expand the discussion to **diatoms containing 2s and 2p orbitals**. Bonds can be formed when atomic orbitals overlap in space. This is not possible with 1s orbitals in Period 2 elements because each of the orbitals is too close to its own nucleus. Only the 2s and 2p orbitals (containing the valence electrons) can participate in bond formation.

We know that **σ bonds** form between nuclear centers, and these MO's can be formed using 2s orbitals or **$2p_x$ orbitals**. In addition, **π orbitals** can be formed by the overlap of both **$2p_y$ or $2p_z$ orbitals**. Also σ and π antibonding orbitals exist.

Your textbook uses **paramagnetism** and **diamagnetism** to prove the relative positions of MO energy levels. **Paramagnetism** indicates **unpaired electrons** in a substance. This causes the substance to be **attracted to** a magnetic field. **Diamagnetism** indicates **paired electrons** in a substance. This causes the substance to be **repelled from** the magnetic field. Paramagnetism is a much stronger effect than diamagnetism. It will dominate if both effects are present. (The substance will be attracted to a magnetic field.)

Your textbook describes how para- and diamagnetism help explain the order of molecular-orbital energy levels. This order, for valence electrons in, for example, carbon and nitrogen is

For oxygen and fluorine, the energy level diagram is slightly different as shown:

Filling molecular orbitals from atoms containing 2s and 2p electronic orbitals works the same as when filling from 1s orbitals. Just fill from lowest to highest energy, and remember to fill degenerate orbitals separately; then pairwise with the electrons having opposite spins. Let's try the following example together.

Example 9.3 A   *σ and π Molecular Orbitals*

Determine the following regarding F_2^-: electron configuration, bond order, para- or diamagnetism and the bond energy relative to F_2.

Solution

To get the **electron configuration** we must fill in the molecular orbital energy-level diagram. We know that F_2^- has **15 valence electrons** (7 on F, 8 on F^-).

$$
\begin{array}{lll}
\sigma_{2p}^* & \uparrow & \\
\pi_{2p}^* & \uparrow\downarrow & \uparrow\downarrow \\
\pi_{2p} & \uparrow\downarrow & \uparrow\downarrow \\
\sigma_{2p} & \uparrow\downarrow & \\
\sigma_{2s}^* & \uparrow\downarrow & \\
\sigma_{2s} & \uparrow\downarrow & \\
\end{array}
$$

E

Electron Configuration (valence only) $F_2^- = (\sigma_{2s})^2(\sigma_{2s}^*)^2(\sigma_{2p})^2(\pi_{2p})^4(\pi_{2p}^*)^4(\sigma_{2p}^*)^1$

Bond Order $= \dfrac{\textbf{8 bonding electrons} - \textbf{7 antibonding electrons}}{\textbf{2}} = \frac{1}{2}$

F_2^- has one unpaired electron. It would be expected to be paramagnetic. Molecular fluorine, F_2, has a bond order of 1 and is therefore more stable than F_2^-. Its bond energy would be expected to be higher.

Example 9.3 B   *Practice with σ and π Orbitals*

Determine the electron configuration and bond orders for S_2^{2-} and Cl_2^{2+}. If they can exist, discuss their magnetism.

Solution

S_2^{2-} has **14 valence electrons** (6 + 6 from the sulfur atoms and a −2 charge).
Cl_2^{2+} has **12 valence electrons** (7 + 7 from the chlorine atoms and a +2 charge).

$$
\begin{array}{llll}
\sigma_{3p}^* & \text{——} & & \text{——} \\
\pi_{3p}^* & \uparrow\downarrow \quad \uparrow\downarrow & & \uparrow \quad \uparrow \\
\pi_{3p} & \uparrow\downarrow \quad \uparrow\downarrow & & \uparrow\downarrow \quad \uparrow\downarrow \\
\sigma_{3p} & \uparrow\downarrow & & \uparrow\downarrow \\
\sigma_{3s}^* & \uparrow\downarrow & & \uparrow\downarrow \\
\sigma_{3s} & \uparrow\downarrow & & \uparrow\downarrow \\
& S_2^{2-} & & Cl_2^{2+} \\
\end{array}
$$

E

Electron Configuration of S_2^{2-} $= (\sigma_{3s})^2(\sigma_{3s}^*)^2(\sigma_{3p})^2(\pi_{3p})^4(\pi_{3p}^*)^4$
Electron Configuration of Cl_2^{2+} $= (\sigma_{3s})^2(\sigma_{3s}^*)^2(\sigma_{3p})^2(\pi_{3p})^4(\pi_{3p}^*)^2$

Bond Order of S_2^{2-} $= (8 − 6)/2 = \mathbf{1}$
Bond Order of Cl_2^{2+} $= (8 − 4)/2 = \mathbf{2}$

S_2^{2-} would be diamagnetic.
Cl_2^{2+} would be paramagnetic.

9.4 Bonding in Heteronuclear Diatomic Molecules

When you finish this section you will be able to describe the bonding in some heteronuclear diatomic molecules.

The **MO model** for homonuclear molecules works well for describing the bonding in atoms **adjacent to one another** on the periodic table. The model breaks down with two **very different** atoms.

Let's reinforce our understanding with practice on some more homonuclear species.

Example 9.4   Practice with the MO Model

Using the MO model, describe the bonding, magnetism, and relative bond energies in the following species:

$$O_2, \ O_2^-, \text{ and } O_2^{2-}$$

Solution

Valence Electrons: O_2 has 12, O_2^- has 13, O_2^{2-} has 14

Bond Orders: O_2 $= (8 - 4)/2 = \textbf{2}, \textbf{O}_2^- = (8 - 5)/2 = \textbf{1.5}, \textbf{O}_2^{2-} = (8 - 6)/2 = \textbf{1}$

Both O_2 and O_2^- are expected to be paramagnetic. We expect O_2^{2-} to be diamagnetic. The bond energy of O_2 is expected to be the highest, followed by O_2^-, with O_2^{2-} having the lowest.

In the heteronuclear MO description of HF, your textbook notes that the MO model accounts for the polarity in that molecule. This is a great strength of the model.

9.5 Combining the Localized Electron and Molecular Orbital Models

When you finish this section you should be able to answer the following review questions:

- What is the primary problem with the localized electron (LE) model?
- Why is the use of resonance unsatisfactory for describing molecular bonding?
- What is the major disadvantage of the molecular orbital (MO) model?
- What bond is depicted as shifting in LE resonance structures? Why is that important in our combination model?
- In summary, what are the advantages and disadvantages of the LE model? The MO model?

Exercises

Section 9.1

1. What geometry do the following hybrid bonds possess?

    a. sp
    b. sp^2
    c. sp^3
    d. dsp^3
    e. d^2sp^3

2. Predict the type of hybrid orbital that the central atoms of each of the following compounds display:

    a. SiH_4
    b. H_3O^+
    c. PCl_5
    d. NCl_3
    e. AsH_3
    f. SiF_6
    g. CH_3^+

3. Predict the geometries of the following compounds:

    a. SF_2 b. SF_4 c. XeF_2 d. XeF_4 e. IF_5 f. ClF_3

4. Predict the geometry about the indicated atom, and identify the hybridization of each atom.

    a. the two carbon atoms and the nitrogen atom of glycine

    b. the carbon atom in CF_2Cl_2
    c. the phosphorous atom in PCl_5
    d. the nitrogen atom in NH_2^-

Section 9.2

5. Determine the number of sigma and pi bonds in each of the following:

6. The structure of urea is

    a. How many σ bonds are there?
    b. How many π bonds?
    c. What is the hybridization at the carbon?
    d. How are the nitrogen atoms hybridized?
    e. What is the N−C−N bond angle expected to be?
    f. How many lone pairs of electrons are there?

Section 9.3

7. Determine the hybridization of each numbered atom of the following molecule:

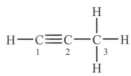

8. Draw an energy level diagram for He_2^+. What is the bond order?

9. What is the bond order for H_2^-?

10. The elements N, O, and F exist as diatomic molecules. Use MO theory to explain why Ne, the next element on the periodic table, does not exist as a diatomic molecule.

Section 9.4

11. Which of the following are paramagnetic species: Li_2, N_2, He_2^+, H_2, O_2, Be_2?

12. For N_2, N_2^+, N_2^-, and N_2^{2-}, use molecular orbitals to predict bond order and if the species are diamagnetic or paramagnetic.

13. If liquid nitrogen were poured between the poles of a powerful magnet, would it "stick" to the poles or simply pour through the space between the poles? Why?

14. What is the bond order and expected magnetism of carbon monoxide? Of CO^-?

15. Summarize the relative bond order for the diatomic molecules of Period 2 from boron to fluorine.

16. What is the bond order and expected magnetism of IBr?

17. Which of the following are diamagnetic: CN^-, CN, CN^+?

Multiple Choice Questions

18. What hybridization describes square planar geometry?
    A. sp^3 B. dsp^3 C. d^2sp^3 D. sp^2

19. Formaldehyde is used as a preservative. In the presence of air, formaldehyde is oxidized to formic acid, HCOOH. What hybridization does the carbon atom have in formic acid?
    A. sp^2 B. sp^3 C. sp D. dsp

20. Phosphorus pentachloride is produced upon reaction of phosphorus trichloride with chlorine. What hybridization is present in the phosphorus atom of PCl_3 and PCl_5 molecules, respectively?
    A. dsp, d^2sp^3 B. sp^3, dsp^3 C. sp^3, d^2sp^3 D. d^2sp^3, dsp^3

21. How many σ, and how many π bonds, respectively, are there in the following molecule:

    $$CH_3-CH_2-CH_2-CH_2-CH=C=C=CH_2$$

    Remember that carbon needs to have four bonds to be satisfied.
    A. 19, 3   B. 16, 7   C. 16, 3   D. 20, 4

22. The following molecule, CH_3CH_2CHO, is reduced to $CH_3CH_2CH_2OH$. What orbital is most probably used in the reduction process?

   A. π orbital of one of the sp^3 carbons
   B. σ orbital of one of the sp^2 carbons
   C. σ orbital of one of the sp^3 carbons
   D. π orbital of the sp^2 carbon

23. What is the hybridization of phosphorus in PCl_6^-?

   A. d^2sp^3
   B. dsp^3
   C. sp^3
   D. sp^2

24. How many π bonds are in the following molecule?

$$CH_3-CH=CH-CH=C=CH-CH_3$$

   A. 4
   B. 3
   C. 0
   D. 1

25. Which of the following bond orders would show the greatest bond strength?

   A. ½
   B. 1
   C. 0
   D. 0.3

26. According to the molecular orbital model, a bonding orbital:

   A. Is unstable
   B. Is as stable as an antibonding orbital
   C. Is more stable than an antibonding orbital
   D. Is less stable than a nonbonding orbital

27. Find out how many bonding electrons, and calculate the bond order of the molecular ion $(O_2)^{2+}$, respectively.

   A. 10, 3
   B. 4, 3
   C. 6, 1.5
   D. 8, 1

28. According to molecular orbital theory, $(C_2)^{2-}$ should be

   A. Paramagnetic
   B. Ferromagnetic
   C. Diamagnetic
   D. Antimagnetic

29. The bond order for C_2 is

   A. 0
   B. 2
   C. 2.5
   D. 3

30. Which of the following species is paramagnetic?

   A. O_2
   B. C_2
   C. Be_2
   D. N_2

31. Which of the following species has a bond order that differs from the others?

   A. O_2
   B. NO
   C. BN
   D. NO^-

32. Which of the following species is paramagnetic?

   A. NO
   B. CO
   C. BN
   D. CN^-

Answers to Exercises

1.   a.   linear   c.   tetrahedral   e.   octahedral
     b.   trigonal planar   d.   trigonal bipyramidal

2.   a. sp^3 c. dsp^3 e. sp^3 g. sp^2
     b. sp^3 d. sp^3 f. d^2sp^3

3.   a.   angular (like water)   c.   linear   e.   square pyramidal
     b.   see-saw   d.   square planar   f.   T-shaped

4.   a.   carbon a-sp^3, tetrahedral; carbon b-sp^2, trigonal planar; nitrogen-sp^3, based on tetrahedron-trigonal pyramid
     b. sp^3-tetrahedral
     c. dsp^3-trigonal bipyramid
     d. sp^3-based on tetrahedron-bent

5.   a.   8 sigma, 2 pi
     b.   5 sigma, 2 pi
     c.   7 sigma, 3 pi

6.   a.   7   c. sp^2 e.   120°
     b.   1   d. sp^3 f.   4

7.   1. sp 2. sp 3. sp^3

8.   ½

9.   ½

10.   Ne has an equal number of bonding and antibonding electrons.

11. He_2^+ and O_2

12. N_2 : bond order = 3, diamagnetic
       N_2^+ : bond order = 2.5, paramagnetic
       N_2^- : bond order = 2.5, paramagnetic
       N_2^{2-}: bond order = 2, paramagnetic

13.   Liquid nitrogen would simply pour through the space between the poles because it is diamagnetic as opposed to oxygen, which would be held between the poles because it is paramagnetic.

14.   CO: bond order = 3, diamagnetic; CO^-: bond order = 2.5, paramagnetic

15. $B_2 < C_2 < N_2 > O_2 > F_2$

16.   IBr: bond order = 1, diamagnetic

17. CN^- and CN^+

18.   C   19.   A   20.   B   21.   A   22.   D   23.   A
24.   B   25.   B   26.   C   27.   A   28.   C   29.   B
30.   A   31.   B   32.   A

Chapter 10

Liquids and Solids

Your textbook justifies **grouping solids and liquids** together, separate from gases, by **examining common properties**. The **densities**, **compressibilities** and **heats of phase change** all indicate that the forces that hold solids and liquids together are similar. Gases have no such forces. In this chapter, your textbook discusses the **bonding models**, **structure,** and **other properties** of liquids and solids.

10.1 Intermolecular Forces

When you finish this section you will be able to list intermolecular forces and describe some of the effects on liquids and solids.

Your textbook differentiates between **intra**molecular and **inter**molecular forces.

- **Intra**molecular forces mean **forces within a molecule**. We dealt with these in Chapter 8.
- **Inter**molecular forces mean **forces between molecules**. These are the forces that hold molecules together as liquids and solids.

There are three kinds of forces that are discussed in this section.

- **Dipole-dipole forces** result when the partial positive and negative charges of neighboring polar covalent molecules attract. These forces are about 1% as strong as intramolecular covalent bonds.
- **Hydrogen bonds** are a special case where dipoles in **small, highly electronegative atoms** (such as fluorine) form a **surprisingly strong** interaction with the small hydrogen, which has a highly positive charge per unit size. Hydrogen bonding leads to substances with unusually high intermolecular bond energies. This is shown in <u>Figure 10.3 in your textbook</u>.
- **London dispersion forces** are caused by the instantaneous dipoles that arise in a molecule as a result of momentary imbalances in electron distribution. These are very weak forces that become more important as the size of the atom of interest increases. (See the discussion on **polarizability** in your textbook.)

Example 10.1 A   *London Dispersion Forces*

The boiling point of argon is −189.4°C.

- a.   Why is it so low?
- b.   How does this boiling point help prove that London dispersion forces exist?
- c.   The boiling point of xenon is −119.9°C. Why is it higher than that of argon?

Solution

- a.   Argon does not interact with other substances because it is so small and has a complete octet of valence electrons. Argon must be made quite cool to allow liquefication via London dispersion forces.

- b.   If these forces did not exist, argon would never liquefy.

- c.   Xenon is bigger and has more electrons than argon. The likelihood of momentary dipoles is thus greater. (It has a greater polarizability than argon.)

Example 10.1 B   *The Effect of Intermolecular Forces*

Put the following substances in order from lowest to highest boiling point.

$$C_2H_6, \ NH_3, \ F_2$$

Solution

$$\mathbf{F_2, \ C_2H_6, \ NH_3}$$

F_2 can exhibit only intermolecular London dispersion forces. **C_2H_6** is not especially polar, but it does have a very slight electronegativity difference between the carbons and the hydrogens. **NH_3** exhibits hydrogen bonding, thus giving it a relatively high boiling point.

10.2   The Liquid State

The following review questions will serve to test your understanding of the material in this section.

1.   Why do liquids tend to bead up when on solid surfaces?
2.   What are **cohesive forces**? **Adhesive forces**? What causes these forces?
3.   What is surface tension? Why does it arise?
4.   Why does water form a concave meniscus when in a thin tube? Why does mercury form a convex meniscus?
5.   What is **viscosity**? What is a requirement for a liquid to be viscous?
6.   Why do models of liquids tend to be more complex than those for either solids or gases?

Example 10.2   *Properties of Liquids*

Which would have a higher surface tension, H_2O or C_6H_{14}? Why? Would the shape of the H_2O meniscus in a glass tube be the same or different than C_6H_{14}?

Solution

Water, having a large dipole moment, has relatively **large cohesive forces**. **Hexane**, C_6H_{14}, is essentially **nonpolar**. It has low cohesive forces. **Water** would therefore have the **higher surface tension**.

The water meniscus is **concave** because the adhesive forces of water to polar constituents on the surface of the glass are stronger than its cohesive forces. **Hexane** would have a **convex meniscus**. It has very small adhesive forces, and the slightly larger cohesive forces would dominate.

10.3   An Introduction to Structures and Types of Solids

When you finish this study section you will be able to:

- Define the basic terms relating to the structure of solids.
- Solve simple problems relating to X-ray diffraction by crystals.

This section in your textbook begins by contrasting **crystalline solids** (highly regular atomic arrangements) and **amorphous solids** (disordered atomic arrangements). The focus of the section (and the bulk of the chapter) is on **crystalline solids**.

Example 10.3 A   Basic Terms

Answer the following questions regarding the terms introduced in this section of your textbook.

   a.   Define **lattice**.
   b.   Define **unit cell**.
   c.   List three types of **cubic unit cells**.
   d.   What is diffraction?
   e.   Why are X-rays used in diffraction analyses of solids?
   f.   What is the difference among **ionic**, **molecular**, and **atomic solids**?

Solution

   a.   **Lattice**: A three-dimensional system of points designating the centers of the constituent building units in a crystalline solid.

   b.   **Unit cell**: The smallest repeating unit of a lattice.

   c.   As shown in <u>Figure 10.9 of your textbook</u>, the three types of cubic unit cells are **simple cubic, body-centered cubic** and **face-centered cubic**.

   d.   **Diffraction** is the **scattering of light** from a **regular array** of points or lines, where the spacings between the points or lines (in our case planes of atoms) are related to the wavelength of the light.

   e.   X-rays are used because their wavelengths are similar to the distances between atomic nuclei.

   f.   **Ionic solids** form electrolytes when dissolved in water. **Molecular solids** do not. An **atomic solid** contains atoms of only one element. These atoms are covalently bonded to each other.

There are three types of atomic solids described in your textbook: metallic, network, and Group VIIIA solids. Note the description and examples of these (as given in <u>Table 10.3 in your textbook</u>) just before the start of Section 10.4. Review the meaning of each of the terms of the **Bragg equation** (<u>Equation 10.3 in your textbook</u>), which is used to determine the structures of crystalline solids. When you are comfortable with the terms, try the next example.

Example 10.3 B   The Bragg Equation

A topaz crystal has a lattice spacing (d) of 1.36 Å (1 Å = 1×10^{-10} m). Calculate the wavelength of X-ray that should be used if $\theta = 15.0°$ (assume n = 1).

Solution

The Bragg equation is in the form

$$n\lambda = 2d \sin \theta$$

where n = the order of diffraction
λ = the wavelength of the "incident energy" (the beam hitting the sample)
d = the spacing between planes of atoms
$\sin \theta$ = the sine of the angle of reflection of the beam. $\sin(15.0°) = 0.259$

$$\lambda = \frac{2d \sin \theta}{n} = \frac{2\,(1.36\ \text{Å})\,(0.259)}{1} = \textbf{0.704 Å} = 70.4\ \text{pm}$$

10.4 Structure and Bonding in Metals

When you finish this section you will be able to:

- Discuss bonding in metals.
- Calculate the density of metals based on the structure of their unit cell.
- Define and give examples of substitutional and interstitial alloys.

This section of your textbook begins by listing properties of metals. These are:

- high thermal and electrical conductivity,
- malleability (can form thin sheets), and
- ductility (can be pulled into a wire).

A model that describes the structure of metals must be able to explain these properties.

The proposed model assumes that the atoms of metals are **uniform hard spheres** that are packed to best utilize available space **(closepacking)**. Figures 10.13 - 10.15 in your textbook give examples of packing in layers.

- **Hexagonal closest packing** (hcp) has every **other** layer being spatially equivalent ("*ababab*..."). Examples include magnesium, zinc, cadmium, cobalt, and lithium.
- **Cubic closest packing** (ccp), or a **face-centered cubic** structure has every **third** layer being spatially equivalent ("*abcabcabc*..."). Examples include silver, aluminum, nickel, lead, and platinum.

Each sphere in each structure has **12 equivalent nearest neighbors (the coordination number is 12)**: 6 in its layer, 3 in the layer above, and 3 in the layer below (see Figure 10.16 in your textbook). The exception to the above list is

- **Body-centered cubic structure** (bcc). In this structure, spheres are **not** closest packed. The **coordination number** for such a sphere is **8** (see Figure 10.9(b) in your textbook). Iron and alkali metals have bcc structures.

The calculation of properties such as **density** from crystal structures is possible if you know the type of unit cell that a metal (or ionic solid, as we shall see later) forms. Your textbook deals only with the **ccp** structure in this regard. The following example shows how density is calculated using the atomic radius and the ccp structure.

Example 10.4 A   Crystal Structure And Density

The radius of a nickel atom is 1.24 Å ($1\text{Å} = 1 \times 10^{-8}$ cm). Nickel crystallizes with a cubic closest packed structure (face-centered cubic). Calculate the **density** of solid nickel.

Solution

The goal is to calculate density, which is mass/volume. The volume in this case can be represented by the unit cell. As shown in your textbook (see Figure 10.17 in your textbook), the ccp unit cell contains 4 atoms of nickel.

Use your understanding of stoichiometry to calculate the mass of nickel in the unit cell.

The volume for the unit cell can be calculated using the Pythagorean theorem, as illustrated in Example 10.2 in your textbook. The volume of the unit cell, a cube, is

$$V = d^3$$

where d = the length on a side.

According to the theorem, the square of the hypotenuse is equal to the sum of the squares of the sides of a right triangle, or in our case,

$$h^2 = d^2 + d^2$$

$h = 4r$ (r = the radius of nickel), as illustrated in the example.

$$(4r)^2 = 2d^2$$
$$16r^2 = 2d^2$$
$$8r^2 = d^2$$
$$d = r\sqrt{8}$$

Therefore,

$$V = d^3 = r^3 8^{1.5} = 22.63r^3$$

We can now solve for density.

$$\textbf{mass} = 4 \text{ atoms} \times \frac{58.70 \text{ g Ni}}{6.022 \times 10^{23} \text{ atoms}} = \textbf{3.900} \times \textbf{10}^{-22} \textbf{ g}$$

$$\textbf{volume} = 22.63 \times (1.24 \text{ Å} \times 1 \times 10^{-8} \text{ cm/Å})^3 = \textbf{4.315} \times \textbf{10}^{-23}$$

$$\textbf{density} = 3.900 \times 10^{-22} \text{ g} / 4.315 \times 10^{-23} \text{ cm}^3 = \textbf{9.04 g/cm}^3$$

This is in good agreement with the actual value of 8.90 g/cm^3.

Your textbook justifies **conduction** of **electricity** and **heat** by metals by using the **band model**. The discussion says that if you have a **few** molecular orbitals (MOs) when a couple of atoms covalently bond; then, a covalently bonded system with **many** atoms should lead to MOs with many energies (a "continuum"). Electrons in filled MOs can be excited into empty MOs and thus travel throughout the metal. Notice how your knowledge of MO theory (Chapter 9) helps you understand bonding in metals.

The section ends with a discussion of **metal alloys**. When you think you are comfortable with the material, try the following example.

Example 10.4 B   Metal Alloys

Why are high-carbon steels **interstitial alloys** while brass is a **substitutional alloy**?

Solution

An **interstitial alloy** is formed when holes in the closest packed metal structure are occupied by **small atoms** (in high carbon steels the iron holes are filled by carbon).

A **substitutional alloy** contains **similar-sized** atoms of more than one element (the holes are not occupied). An example is the combination of copper and zinc to form the brass alloy.

Note that the makeup of the alloy greatly affects its properties.

10.5 Carbon and Silicon: Network Atomic Solids

The questions listed below will help you to review the material presented in this section.

1. Define **network** solid.
2. List the **properties** of network solids.
3. Why does diamond have carbons with sp^3 hybridizations while those in graphite are sp^2 hybridized?
4. Use MO theory to explain why graphite conducts electricity while diamond does not.
5. Why can carbon form π-bonds while silicon cannot?
6. What is the difference between **silica** and **silicate**?
7. What property does glass share with ceramics? How are they different?
8. What can be done to make ceramics somewhat flexible?
9. What is the purpose of adding arsenic to a semiconductor?
10. What is a **p-type** semiconductor? How does it work?
11. Describe how a p-n junction acts as a rectifier.

10.6 Molecular Solids

1. Define **molecular solid**.
2. Give some **examples** of molecular solids.
3. Describe the relative bonding strength and bond distances **within** and **between** molecules of a molecular solid.
4. Why are some larger nonpolar molecular solids solid at 25°C?

10.7 Ionic Solids

When you finish this section you will be able to:

- Discuss the rationale for the packing that is observed in closest packed structures.
- Classify substances according to the types and properties of the solids they form.

Your textbook says that the structure of most binary ionic solids can be explained by the **closest packing of spheres**. Anions, which are usually larger than the cations with which they combine (see Section 8.4 in your textbook), are packed in either an hcp or ccp arrangement. Cations fill the holes within the packed anions.

Key Idea: The packing arrangement is done in such a way as to minimize anion-anion and cation-cation repulsions.

The nature of the holes depends on the ratio of the anion to cation size. Trigonal holes are smallest, followed by tetrahedral, and octahedral are the largest.

Example 10.7 A Packing in Ionic Structures

Would AlP have a closest packed structure that is more like NaCl or ZnS?

Ionic Radii are: $Al^{3+} = 50$ pm, $P^{3-} = 212$ pm
$Zn^{2+} = 74$ pm, $S^{2-} = 184$ pm
$Na^+ = 95$ pm, $Cl^- = 181$ pm

Solution

We need to take the anion to cation radius ratio in each of the three cases.

$$S^{2-} / Zn^{2+} = 2.49 \text{ (tetrahedral holes)}$$
$$Cl^- / Na^+ = 1.91 \text{ (octahedral holes)}$$
$$P^{3-} / Al^{3+} = 4.24 \text{ (?)}$$

The aluminum ion is very small compared to the phosphorus ion. Therefore, not much room is needed by the Al^{3+} cations. **Tetrahedral holes** are adequate. **AlP** is **more like ZnS** than NaCl.

Study Table 10.7 in your textbook. Based on the information given in the table, as well as your overall knowledge of chemistry, try the next example, which deals with classifying solids.

Example 10.7 B   Types of Solids

Based on their properties, classify each of the following substances as to the **type of solid** it forms:

    a. Fe   b. C_2H_6 c. $CaCl_2$ d. graphite   e. F_2

Solution

a. Solid iron is an **atomic solid** with metallic properties.
b. Solid C_2H_6, ethane, contains nonpolar molecules and is a **molecular solid**.
c. Solid $CaCl_2$ contains Ca^{2+} and Cl^- ions and is an **ionic solid**.
d. Graphite is made up of nonpolar carbon atoms covalently bonded in directional planes. It is a **network solid**.
e. Solid fluorine is made up of nonpolar fluorine molecules. It is a **molecular solid**.

10.8   Vapor Pressure and Changes of State

When you finish this section you will be able to:

- Interconvert among **vapor pressure**, **temperature**, and **enthalpy of vaporization** of a liquid.
- Perform calculations regarding the energy of phase changes.

Your textbook introduces some very useful terms in this section. You need to be able to define **vaporization, enthalpy of vaporization, condensation, sublimation, enthalpy of fusion, melting point,** and **boiling point.**

Dynamic equilibrium is a concept that you will be using a great deal in the latter half of your chemistry studies. It means that *two opposing processes are occurring at the same rate.* The net effect is *no observable change.* But the system is not static. In this section, **the equilibrium vapor pressure** means that evaporation and condensation by a liquid are occurring at the same rate. The net effect is to have a **constant** vapor pressure exerted by the liquid.

The vapor pressure of a liquid varies with the molecular weight of the liquid and other molecular properties such as polarity and hydrogen bonding.

A heavier substance will have a lower vapor pressure than a lighter substance, all other things being equal, because the atoms are more polarizable, leading to larger intermolecular forces. A substance with hydrogen bonding interactions will have a lower vapor pressure (will be less volatile) than a nonpolar substance. Your textbook introduces the Clausius-Clapeyron equation (Equation 10.5 in your textbook), which interrelates the **vapor pressure, temperature,** and **enthalpy** of a liquid.

$$\ln\left[\frac{P_{vap, T_1}}{P_{vap, T_2}}\right] = \frac{\Delta H_{vap}}{R}\left[\frac{1}{T_2} - \frac{1}{T_1}\right]$$

Where T_1 and T_2 are **temperatures** in Kelvins,
        ΔH_{vap} is the **enthalpy of vaporization** of the liquid,
        P_{vap, T_1} and P_{vap, T_2} are **vapor pressures** of the liquid temperatures T_1 and T_2

Example 10.8 A   Clausius-Clapeyron Equation

The vapor pressure of 1-propanol at 14.7°C is 10.0 torr. The heat of vaporization is 47.2 kJ/mol.
Calculate the vapor pressure of 1-propanol at 52.8°C.

Solution

Let's list what we are given.

ΔH_{vap} = 47.2 kJ/mol
R = 8.314 J/K mol = 0.008314 kJ/K mol
T_1 = 14.7°C = 287.9 K
T_2 = 52.8°C = 326.0 K
P_{vap, T_1} = 10.0 torr
P_{vap, T_2} = x

Substituting,

$$\ln\left[\frac{10.0}{x}\right] = \frac{47.2 \text{ kJ/mol}}{0.008314 \text{ kJ/mol}}\left[\frac{1}{326.0 \text{ K}} - \frac{1}{287.9 \text{ K}}\right]$$

$$\ln\left[\frac{10.0}{x}\right] = 5677 \,(-4.06 \times 10^{-4})$$

$$\ln\left[\frac{10.0}{x}\right] = -2.305$$

Taking the antilog of both sides,

$x/10 = 10.02$

$x = P_{vap, T_2} = 100.2 = 100 \text{ torr}$

Look at <u>Figure 10.44 in your textbook</u>. This illustrates the **heating curve** for water. Note two important observations:

*   The **temperature** of a substance **remains constant** during a **phase change**.
*   The **temperature rises** when heat is input while a substance is in **one phase**.

You can find the amount of **energy required** to convert water from ice at T_1 to steam at T_2 by using the following information:

*   specific heat capacity of ice (**2.1 J/g °C**)
* ΔH_{fusion} of water (**6.0 kJ/mol**)
*   specific heat capacity of liquid water (**4.2 J/g °C**)
* ΔH_{vap} of water (**43.9 kJ/mol**)
*   specific heat capacity of steam (**1.8 J/g °C**)

How much of the information will be used will depend on the problem you have to solve.

Example 10.8 B   Heating Curve

How much energy does it take to convert 130. g of ice at −40°C to steam at 160°C?

Strategy

There are **5 steps** involved in this conversion from ice to steam.

1. Heating ice from −40°C to the melting point.
2. Melting ice to form liquid water.
3. Heating liquid water to its boiling point.
4. Boiling liquid water to form steam.
5. Heating to 160°C.

The **total energy** required is the **sum** of the energy required in each of the 5 steps. The appropriate constants for each step are given in the discussion preceding this problem. The units of heat capacity contain °C because the temperature is rising in each of these steps. The units "enthalpy of fusion and vaporization" do not because the temperature is constant at a phase change.

Solution

Energy used = sum of energies from individual steps. There are 7.22 mol of water in 130 g.

$$
\begin{aligned}
\text{Step 1} &= 40°C \times 130\ g \times 2.1\ J/g\ °C &&= 10.92\ \text{kJ} \\
\text{Step 2} &= 7.22\ mol \times 6.0\ kJ/mol &&= 43.3\ \text{kJ} \\
\text{Step 3} &= 100°C \times 130\ g \times 4.2\ J/g\ °C &&= 54.6\ \text{kJ} \\
\text{Step 4} &= 7.22\ mol \times 43.9\ kJ/mol &&= 317.0\ \text{kJ} \\
\text{Step 5} &= 60°C \times 130\ g \times 1.8\ J/g\ °C &&= \underline{14.04\ \text{kJ}} \\
&\textbf{Total Energy} &&= \textbf{440.}\ \textbf{kJ}
\end{aligned}
$$

10.9   Phase Diagrams

When you finish this section you will be able to extract information from simple phase diagrams.

The beauty of this section is that it helps explain a large number of real-world phenomena. (See the Chemical Impact on diamonds in this section of your textbook.)

You should be able to define the following terms: **phase diagram, critical temperature, critical pressure, critical point,** and **triple point**. You should also be able to answer the following general questions regarding material presented in this section.

1. Why does the solid/liquid line in the phase diagram of water have a negative slope? Why is it positive for carbon dioxide?
2. Why does it take longer to cook an egg in the Rocky Mountains than at sea level?
3. How does the phase diagram for water help explain why your blades glides on a liquid layer when you ice skate?
4. How does the phase diagram for carbon dioxide help explain how a CO_2 fire extinguisher works?
5. Snow sometimes sublimes. How can this be so in spite of the phase diagram?

Example 10.9   Phase Diagrams

What phase changes does water undergo (see Figure 10.49 in your textbook) as the pressure changes while the temperature is held constant at −12°C?

Solution

At very low pressures, water exists as a gas at −12°C. As the pressure is increased, it turns into a solid. At very high pressures the water will liquefy.

Exercises

Section 10.1

1. Define the following terms in your own words:

    a. dipole-dipole forces
    b. hydrogen bonding
    c. London dispersion forces

2. Define the following terms in your own words:

    a. crystalline solids
    b. amorphous solids
    c. lattice
    d. unit cell

3. Of HF, HCl, and HBr, which has the highest boiling point? Why? Which has the lowest boiling point?

4. Propane, C_3H_8, is a gas at room temperature; hexane, C_6H_{14}, is a liquid; and dodecane, $C_{12}H_{26}$, is a solid. Explain.

5. Which would you expect to have a lower melting point, C_3H_8 or CH_3OH? Why?

6. Would you expect methane (CH_4) and water to have about the same boiling points? Why or why not?

7. Arrange the following in order of increasing boiling points, and justify your assignments: CH_3F, CH_3Cl, CH_3Br, CH_3I.

Section 10.2

8. Why does water "bead up" more on a car that is waxed than isn't?

9. Would mercury bead up more on a waxed or unwaxed car?

10. Which would have greater surface tension $N_2(l)$ or $Br_2(l)$?

Section 10.3

11. X-rays of wavelength 0.1541 nm produce a reflection angle $\theta = 15.55°$. What is the spacing between crystal planes ($n = 1$)? What would be the angle of reflection for $n = 2$?

12. X-rays of wavelength 263 pm were used to analyze a crystal. A reflection was produced at $\theta = 13.9°$. Assuming $n = 1$, calculate the spacing between the planes of atoms producing this reflection.

13. The second-order reflection ($n = 2$) for a gold crystal is an angle of 22.20° for X-rays of 154 pm. What is the spacing between these crystal planes?

14. Classify the following as ionic, molecular, or atomic crystalline solids.

    a. dry ice, $CO_2(s)$
    b. graphite
    c. $CaF_2(s)$
    d. $MnO_2(s)$
    e. $C_{10}H_8(s)$ (naphthalene)

Section 10.4

15. Tungsten crystallizes in a body-centered cubic structure with a unit cell edge length of 315.83 pm. The density of tungsten metal is 19.3 g/cm^3, and its atomic weight is 183.85 g/mol. Calculate the value of Avogadro's number by this method.

16. Silver crystallizes in a face-centered cubic structure with a unit cell edge length of 407.76 pm. The density is determined to be 10.5 g/cm^3. Calculate the atomic weight of silver.

17. The density of gold is 19.3 g/cm^3. Gold crystallizes in a face-centered cubic unit cell. What is the radius of a gold atom?

18. Potassium crystallizes in a body-centered cubic structure. Calculate the atomic radius of potassium if the unit cell edge length is 533.3 pm.

19. Why is iron relatively soft, ductile, and malleable while high carbon steels are much harder, stronger, and less malleable?

20. Alloys, in general, are harder, lower melting, and poorer conductors of electricity than the pure metals of which they are composed. Why?

Section 10.6

21. What accounts for the lubricating ability of graphite?

22. Which electrons account for the conductivity of graphite?

23. Explain why graphite conducts electricity parallel to the layers much better than it conducts electricity perpendicular to the layers.

24. How many bonds would each sulfur atom in an S_8 molecule have to make to have the nearest noble gas electron configuration?

Section 10.7

25. Cobalt fluoride crystallizes in a closest packed array of fluoride ions with cobalt ions filling one half of the octahedral holes. What is the formula of this compound?

26. Explain why anions usually have larger radii than cations.

27. What is the formula for the compound that crystallizes with a closest packed array of sulfur ions, that contains zinc ions in 1/8 of the tetrahedral holes and aluminum ions in 1/2 of the octahedral holes?

28. Manganese and fluoride ions crystallize such that each cubic unit cell has manganese ions at the corners and fluoride ions at the center of each edge. What is the formula of this compound?

Section 10.8

29. Using the following vapor pressure data for CCl_4, make a graph, and determine the normal boiling point of the liquid.

Temp (°C)	20.0	40.0	60.0	70.0	80.0	90.0	100.0
V.P. (torr)	91.0	213.0	444.3	617.4	836.0	1110	1459

30. The vapor pressure of water at 25°C is 23.8 torr. Confirm the value of 43.9 kJ/mol for the heat of vaporization of water. (Use data for the normal boiling point as well as the vapor pressure given.)

31. On top of one of the peaks in Rocky Mountain National Park the pressure of the atmosphere is 550 torr. Determine the boiling point of water at this location.

32. In Denver, Colorado, the pressure of the atmosphere is 697 torr. Determine the boiling point of water at this location.

33. Calculate the vapor pressure of water at $0°C$ (in kilopascals, kPa).

34. Isopropanol, C_3H_8O, is also known as rubbing alcohol. The heat of vaporization is 42.1 kJ/mol. How much heat is needed to evaporate 25 g of isopropanol?

35. How much isopropanol must evaporate to cool 1.00 kg from $25°C$ to $20°C$? (specific heat of isopropanol is 2.59 J/g°C)

36. What quantity of heat is required to melt 1.0 kg of ice at its melting point?

37. What quantity of heat is required to vaporize ($100°C$) 1.0 kg of ice at $0°C$?

38. What is the final temperature when 10 g of water at $0°C$ is added to 100 g of water at $75°C$?

39. If 10 g of ice at $0°C$ comes into contact with 10 g of water at $50°C$, calculate the final temperature reached by the system at equilibrium.

40. What is the final temperature when 10 g of ice at $0°C$ is added to 100 g of water at $75°C$? ($\Delta H_f =$ 6.0 kJ/mol, heat capacity of water is 4.2 J/g°C)

41. If 10 g of ice at $0°C$ comes in contact with 50 g of water at $10°C$, calculate the final temperature reached by the system at equilibrium.

42. Calculate the amount of energy in Joules required to change 10 g of solid mercury at its melting point to mercury vapor at the boiling point. The m_p, b_p, and specific heat of mercury are $-39°C$, $375°C$, and 0.140 J/g°C, respectively. Compare with the amount of heat needed to change 10 g of ice at $0°C$ to steam at $100°C$. (heat of fusion Hg = 11.4 J/g, heat of vaporization Hg = 5.91×10^4 J/mol)

Section 10.9

43. How is the change in density for a solid-to-liquid phase related to the slope of the liquid-solid line of a phase diagram?

Multiple Choice Questions

44. If water and carbon dioxide molecules did interact, what major intermolecular force would exist between these molecules?

    A. Hydrogen bonding   B. London dispersion   C. Ion-dipole forces   D. Dipole-dipole forces

45. Which of the following molecule pairs are not involved in hydrogen bonding?

    A. $HCOOH$, H_2O B. H_2O, NH_3 C. CH_3OH, CH_3COOH D. H_2 and I_2

46. Arrange the following molecules in order of decreasing intermolecular interaction: SO_2, Cl_2, CH_3OH, CH_3NH_2

    A. $CH_3OH > CH_3NH_2 > SO_2 > Cl_2$ C. $SO_2 > CH_3NH_2 > CH_3OH > Cl_2$
    B. $Cl_2 > SO_2 > CH_3OH > CH_3NH_2$ D. $CH_3NH_2 > CH_3OH > SO_2 > Cl_2$

47. Knowing that solutes with a certain polarity (or absence of it) are best dissolved in solutions with similar polarity, which of the following solvents would be optimal for the solvation of CH_3COOH?

    A. CH_4 B. CH_3OH C. C_2H_6 D. C_6H_6

48. Which of the following molecules interact primarily through London dispersion forces?

    A. SO_2 B. CCl_4 C. CH_2Cl_2 D. H_2S

49. Which of the following has the highest boiling point?

    A. H_2O B. HF   C. HI   D. HBr

50. Surface tension is due to:

    A. The liquid molecules being more attracted to surrounding surface molecules than other liquid molecules
    B. The liquid molecules being as attracted to surrounding molecules as to other liquid molecules
    C. The liquid molecules being more attracted to other liquid molecules than surrounding molecules
    D. The liquid molecules being attracted only to surrounding molecules

51. Which of the following liquids will be the most viscous?

    A. C_3H_8 B. C_6H_6 C. CH_4 D. C_2H_6

52. In the liquid state, molecules:

    A. Are 100% hydrogen bonded to their neighbors   C. Are frozen in space
    B. Are part of a crystal lattice   D. Are considerably restricted in movement

53. The smallest repeating unit of the lattice is called:

    A. unit cell   B. unit lattice   C. cell   D. unit crystal

54. Which of the following types of solids exist as crystals?

    A. ionic solids   B. atomic solids   C. molecular solids   D. metallic solids

55. X-rays with a particular wavelength were used to analyze a crystalline solid. Assuming $n = 1$, and knowing that the distance between the planes of atoms producing a diffraction with an angle $= 20.0°$, is 200 pm, calculate the frequency (in Hz) of the X-rays.

    A. 365   B. 6.22×10^{15} C. 0.110   D. 2.19×10^{18}

56. The successive packing pattern for an hcp cell is which one of the following?

    A. *ABABAB*   B. *ABAABA*   C. *ABCABC*   D. *ABCCBA*

57. Conducting electrons in metals are situated in:

    A. Localized orbitals   B. *s*-orbitals   C. Metallic orbitals   D. Conduction bands

58. If the conduction band of metallic chromium were a pure 3d band with 10,000 chromium ions in it, how many conduction electrons would this cluster have?

    A. 10   B. 10,000   C. 40,000   D. 50,000

59. The difference between an interstitial alloy and a substitutional alloy is that in substitutional alloys:

    A. Some atoms of one element are replaced by atoms of another element
    B. Atoms of one element are inserted in the spaces between the atoms of the other element
    C. Atoms of one element fuse with the atoms of the other element
    D. Three elements, rather than two, are combined to form the alloy

60. An n-type semiconductor is produced by:

    A. Increasing the number of atoms of the semiconductor
    B. Increasing the number of valence electrons of the semiconductor by the introduction of different atoms
    C. Decreasing the number of valence electrons of the semiconductor by the introduction of different atoms
    D. Combining two different semiconductors, which allows the new alloy to be stronger, but less conductive

61. If a p-n type semiconductor is connected to a battery so that the negative pole of the battery is in the n-type region and the positive pole is in the p-type region, which of the following statements is true?

    A. Electrons will move from the p to the n-type region
    B. Electrons will move from the n to the p-type region
    C. Electrons will move in the opposite direction to the original movement
    D. Electron movement will be opposed by the battery

62. The slope of the solid-liquid equilibrium line for water is −99.4 atm/°C. Calculate the pressure, in atm, if the melting point is −4.30°C.

    A. 327 atm   B. 428 atm   C. 214 atm   D. 99.4 atm

63. Using the Clausius-Clapeyron equation calculate P_1, in Torr, at 300 K for ether knowing that: T_2 (normal boiling point) = 319 K and ΔH_{vap} = 29.69 kJ/mol.

    A. 302 torr   B. 403 torr   C. 69.2 torr   D. 579 torr

64. Calculate the normal boiling point (T_2), in K, of a liquid knowing that ΔH_{vap} = 29.09 kJ/mol, and P_1 = 69.2 torr at 300 K.

    A. 305 K   B. 378 K   C. 250 K   D. 325 K

65. The change from a solid to a gas state is known as:

    A. Sublimation   B. Evaporation   C. Condensation   D. Gas-state melting

66. Dichlorodifluoromethane (CCl_2F_2) is a liquid that cools by evaporating. How many kilograms of CCl_2F_2 must be evaporated to freeze a tray of water (1050 g of water) at 273.15 K to ice at the same temperature? Heat of evaporation of CCl_2F_2 is 17.4 kJ/mol.

    A. 2.44 kg   B. 1.22 kg   C. 10.3 kg   D. 12.2 kg

67. How much heat is necessary to melt 175.32 g of NaCl at 801°C? (heat of fusion NaCl = 28.16 kJ/mol)

    A. 22.5 kJ   B. 9.39 kJ   C. 30.2 kJ   D. 84.5 kJ

68. The heat of crystallization for substance A = −65.0 J/g. The heat of fusion of water is 335 J/g. If 2000.0 g of liquid A is added to an excess of ice, how many grams of ice will melt, assuming no temperature change?

    A. 388 g   B. 200.0 g   C. 10.3 g   D. 10300 g

69. Arrange the following liquids, A, B, C, with vapor pressures at room temperature of 88, 680, and 155, respectively, in order of decreasing boiling points.

    A. $B > C > A$ B. $A > B > C$ C. $A > C > B$ D. $C > A > B$

70. The melting point of ice will change in what direction as pressure decreases?

    A. No change
    B. Decreases
    C. Increases
    D. Depends on the pressure

Answers to Exercises

1. Student's own answers.

2. Student's own answers.

3. HF has the highest boiling point because of hydrogen bonding. HBr has the lowest boiling point.

4. All other things being equal, the higher the molecular weight of nonpolar compounds, the greater the intermolecular London forces.

5. C_3H_8 has a lower melting point because CH_3OH exhibits hydrogen bonding.

6. No. Water would have a higher boiling point than methane because water is polar due to its angular arrangement.

7. $CH_3F < CH_3Cl < CH_3Br < CH_3I$

8. There are fewer adhesive forces between the nonpolar wax and water molecules than among water molecules.

9. an unwaxed car

10. $Br_2(l)$

11. 0.2874 nm; $\theta_{n=2} = 32.42°$

12. 547 pm

13. 408 pm

14. a. molecular   c. ionic   e. molecular
    b. atomic   d. ionic

15. 6.05×10^{23}

16. 107 g/mol

17. 144 pm

18. 230.9 pm

19. Iron forms new structures with carbon that exhibit stronger attractive forces involving directional bonding.

20. Alloys are harder, lower melting, and poorer conductors of electricity than the pure metals of which they are composed due to the destruction of slippage planes brought about by displacing slightly larger or slightly smaller atoms throughout the crystal.

21. weak attractive forces between layers of bonded carbon atoms

22. the (delocalized) electrons in the π bonds

23. Parallel conduction is favored because π electrons are localized above and below the layers.

24. 2 bonds

25. CoF_2

26. Extra electrons repel each other and have a lower Z_{eff}.

27. $ZnAl_2S_4$

28. MnF_3

29. 77°C

30. The calculated value is 42.7 kJ/mol.

31. 91.6°C

32. 97.7°C

33. 0.61 kPa

34. 18 kJ

35. 18.5 g

36. 330 kJ

37. 3200 kJ

38. 68°C

39. 0°C

40. 61°C

41. 0°C

42. 30,096 J for water vs. 3630 J for Hg.

43. If solid is more dense - positive slope; if solid is less dense - negative slope.

44. A	45. D	46. A	47. B	48. B	49. A
50. C	51. B	52. D	53. A	54. A	55. D
56. A	57. D	58. D	59. A	60. B	61. B
62. B	63. B	64. B	65. A	66. A	67. D
68. A	69. C	70. C			

Chapter 11

Properties of Solutions

This chapter deals with **solutions - homogeneous mixtures** of solids, liquids, or gases, as shown in Table 11.1 of your textbook.

11.1   Solution Composition

When you finish this section you will be able to solve problems relating to the **mass percent**, **mole fraction**, **molality**, and **normality** of a solution.

As you read the textbook material in this section, pay particular attention to the **terms and definitions** given. You will use these for as long as you are involved with the study of chemistry. Let's review, and add to, the definitions of the key terms of this section.

solute: 1.   If it and the solvent are present in the *same phase*, it is the one in *lesser amount*.
           2.   If it and the solvent are present in *different phases*, it is the one that *changes phase*.
           3.   Your book puts it in more general terms by saying that it is the one that *dissolves into the solvent*.

solvent: 1.   If it and the solute are present in the *same phase*, it is the one in *greater amount*.
           2.   If it and the solute are present in *different phases*, it is the one that *retains its phase*.
           3.   Your book says that it is the one *into which the solute dissolves*.

This section in the textbook introduces you to four new concentration terms. **Mass percent, mole fraction,** and **molality** are useful because they are **temperature-independent**. As the temperature of a solution changes, the **molarity changes slightly** because the **solution volume changes**. As your textbook points out, these new terms depend only on mass, which is temperature-independent.

The formulas for the new concentration units are:

$$\textbf{mass percent} = \frac{\text{g solute}}{\text{g solution}} \times 100\%$$

$$\text{mole fraction of A } (\chi_A) = \frac{n_A}{n_{total}} = \frac{\text{number of moles of A}}{\text{total number of moles}}$$

$$\text{molality } (m) = \frac{n \text{ solute}}{\text{kg solvent}}$$

Also, recall the definition of **molarity** (moles of solute/liter of solution) that we have used throughout the course.

Example 11.1 A Concentration Units

A solution was prepared by adding 5.84 g of formaldehyde, H_2CO, to 100.0 g of water. The final volume of the solution was 104.0 mL. Calculate the molarity, molality, mass percent, and mole fraction of the formaldehyde in the solution.

Solution

Molarity: You have $5.84 \text{ g } H_2CO \times \dfrac{1 \text{ mole } H_2CO}{30.03 \text{ g } H_2CO} = 0.194 \text{ mol } H_2CO$

$$M = \frac{0.194 \text{ moles } H_2CO}{0.104 \text{ L solution}} = \mathbf{1.87\ M\ H_2CO}$$

Molality: $m = \dfrac{0.194 \text{ mol } H_2CO}{0.100 \text{ kg } H_2O \text{ (solvent)}} = \mathbf{1.94\ m\ H_2CO}$

Mass percent $= \dfrac{5.84 \text{ g } H_2CO}{105.84 \text{ g } (H_2CO + H_2O)} \times 100\% = \mathbf{5.52\%\ H_2CO}$

(The mass percent of H_2O must be 94.48%. Can you see why?)

Mole fraction: We know that there is 0.194 mole of H_2CO.

$$n_{H_2O} = 100.0 \text{ g } H_2O \times \frac{1 \text{ mol } H_2O}{18.02 \text{ g } H_2O} = 5.55 \text{ mol } H_2O$$

$$\chi_{H_2CO} = \frac{0.194 \text{ mol } H_2CO}{0.194 \text{ mol } H_2CO + 5.55 \text{ mol } H_2O} = \mathbf{0.0338}$$

(The mole fraction of H_2O must be 0.9662. Can you see why?)

The molarity will be equal to the molality only in very dilute aqueous solutions. The final concentration term in this section is **normality (N)**.

$$N = \frac{\text{equivalents of solute}}{\text{liter of solution}}$$

An **equivalent** is related to a mole. The number of equivalents of a given solute will always be **greater than or equal to** the number of moles of the solute. **An equivalent is the mass of the solute that can furnish or accept one mole of protons (H^+ in acid/base) or one mole of electrons (redox).**

The **equivalent mass** of a substance is the **mass of one equivalent** of the substance. The number of equivalents per mole of solute (hence the equivalent mass and the normality) depends upon the reaction. This is why it is such an awkward term. We can only hope the unit will eventually fall into disuse.

If we look at the reaction involving H_3PO_4,

$$H_3PO_4 + 3NaOH \rightarrow PO_4^{3-} + 3H_2O + 3Na^+$$

Each mole of H_3PO_4 (97.99 g/mol) can supply 3 moles of protons (there are 3 equivalents per mole). The equivalent mass (the mass that can supply one mole of protons) is

$$\frac{97.99 \text{ g/mol}}{3 \text{ eq/mol}} = \textbf{32.66 g/eq}$$

Example 11.1 B   Normality

Given the reaction listed above, if we have 28.42 g H_3PO_4 in 800 mL of water, what are the normality and molarity of the solution?

Solution

We know from the discussion preceding this example that phosphoric acid has a mass of **32.66 g/eq**.

$$\text{equivalents of } H_3PO_4 = \frac{1 \text{ eq}}{32.66 \text{ g}} \times 28.42 \text{ g} = \textbf{0.870 eq}$$

$$N = \frac{0.870 \text{ eq } H_3PO_4}{0.800 \text{ L solution}} = \textbf{1.09 } \textit{N} \textbf{ } H_3PO_4$$

$$M = \frac{28.42 \text{ g}}{0.800 \text{ L}} \times \frac{1 \text{ mol}}{97.99 \text{ g}} = \textbf{0.363 } \textit{M} \textbf{ } H_3PO_4$$

Note the normality, as expected, is 3 times the molarity.

11.2   The Energies of Solution Formation

When you finish this section you will be able to determine the miscibility of a variety of substances.

This section concerns **how**, and **under what conditions**, a solution will be formed from the interaction of two substances. Your textbook lists **three steps** that must occur for solution formation to occur.

1. Expanding the solute (endothermic, $\Delta H = +$)
2. Expanding the solvent (endothermic, $\Delta H = +$)
3. Allowing the solute and solvent to interact (often exothermic, $\Delta H = -$)

In solution formation the enthalpy of solution (the sum of steps 1, 2, and 3) is often "−" (exothermic). However, this is not the determinant of whether a reaction will occur. It is just one outcome of solution formation. We will learn how to specifically determine if a reaction will occur in Chapter 16.

Although we cannot yet determine explicitly if solution formation will occur, we can make the generalization that **like dissolves like**. This is illustrated in Table 11.3 in your textbook.

The key to predicting whether two substances will mix (are "miscible") is to establish the polarity of each. If they are similar, you can probably form a solution. If they are very different, a solution is not likely to form. A more extensive explanation is given in the paragraph just preceding Example 11.3 in your textbook.

Example 11.2   *Predicting Solubility*

The following substances are slowly added to a 250 mL graduated cylinder: 50 mL of carbon tetrachloride (CCl_4, density = 1.4 g/mL), 50 mL of water (density = 1.0 g/mL), and 50 mL of cyclohexane (C_6H_{12}, density = 0.8 g/mL). After cyclohexane has been added, **how would the liquids appear in the cylinder** (i.e., would there be solution formation)? If solid I_2 flakes are added to the system, **in which layers (if any) will they dissolve?**

Solution

CCl_4 has no dipole moment (the molecule is **nonpolar**).
H_2O is a **polar** molecule.
C_6H_{12} is essentially **nonpolar**.

Because of the different densities of the liquids, CCl_4 would be the bottom layer. Water would sit on top of CCl_4. C_6H_{12} would be the top layer. If shaken, the CCl_4 and C_6H_{12} layers would mix.

Iodine is nonpolar. When solid I_2 flakes are added to the layers, they will dissolve in our nonpolar layers, CCl_4 and C_6H_{12}. There would ultimately be purple $CCl_4 + I_2$ and $C_6H_{12} + I_2$ layers surrounding a clear water layer.

11.3   *Factors Affecting Solubility*

When you finish this section you will be able to:

- Predict relative solubilities of simple molecules based on structure.
- Solve problems relating to Henry's Law.

In considering **solubility effects** it is important to **separate the behavior of solids and liquids from gases**. The solubility of gases is for the most part independent of structure. Solid and liquid solubility is highly structure dependent.

Reread the discussion in your textbook on "structure effects" as it relates to the solubilities of vitamins A and C, then work the next example.

Example 11.3 A   *Structure and Solubility*

Determine whether or not each of the following compounds is likely to be **water soluble**.

a. H_3C—$\underset{\underset{\displaystyle OH}{|}}{CHNH_2}$

d. $CH_3(CH_2)_4CH_2NH_2$

b. H_2C—$\underset{\underset{\displaystyle NC_6H_5}{\|}}{\overset{\overset{\displaystyle H}{|}}{N}}$—$C_6H_5$

e. $\underset{\underset{\displaystyle OH}{|}}{H_2C}$—$\underset{\underset{\displaystyle OH}{|}}{CH_2}$

c. C_4H_9—$\underset{\underset{\displaystyle H}{|}}{C}$=$CH_2$

Strategy

An important factor in water solubility is the **ratio of polar to nonpolar groups** in the covalent molecule. The higher the ratio (with hydrogen-bonding groups being especially good), the more likely the covalent substance is to be water soluble unless there is no dipole moment (such as in CCl_4).

However, there are no absolutes. There are other factors involved in solubility that you will learn more about when you are introduced to **organic chemistry** (Chapter 22 in your textbook).

Solution

a. **water soluble** (small molecule with 3 polar bonds)
b. **insoluble** (large molecule with only 1 polar bond)
c. **insoluble** (nonpolar molecule)
d. **insoluble** (large molecule)
e. **water soluble** (small molecule with 2 polar bonds)

Note how the answers here are consistent with Example 11.2 in this study guide.

Example 11.3 B   Polarity and Carotenoids

Carotenoids are compounds found in photosynthetic microorganisms and plants, but not in animals. They give rise to the red color of tomatoes (lycopene), the orange color of carrots (α-carotene and β-carotene), the yellow in sweet corn (lutein), and the pink color of flamingos and salmon (astaxanthin). Carotenoids are converted by the body to vitamin A, critical for vision and the development of embryos, among other uses. The structures of four carotenoids are given below. Notice how similar they are to vitamin A (Figure 11.4 in your textbook)! Chemists have struggled to differentiate among these in instrumental analysis. Some recent techniques have proven successful. Many of these techniques use *liquid chromatography* as their basis. The substances travel along a column **filled with a nonpolar substance**. The polar carotenoids travel through most quickly and are separated from the nonpolar compounds which travel through more slowly. Please sort the following compounds in order of speed traveling through the liquid chromatograph column. (Which would come out first, second, etc.)

a.

Lutein

b.

Lycopene

c.

Astaxanthin

d.

α-Carotene

Solution

Here we have four substances that have very similar structures except for just a couple of groups on either end of each molecule. This small difference, however, is enough to separate the molecules using liquid chromatography. Going from most polar to least polar we have,

astaxanthin > lutein > α-carotene > lycopene
most polar   least polar

Your textbook points out that **pressure has little effect on the solubilities of liquids and solids.** The solubility of gases is, for the most part, **independent of structure** and is given by **Henry's law,**

$$C = kP$$

where C = concentration of the dissolved gas (in mol/L)
P = partial pressure of the gaseous solute above the solution (in atm)
k = a constant for a particular solution (in L atm/mol)

Keep in mind that Henry's law applies for **low gas concentrations** and gases that **do not react with the solution.**

Example 11.3 C   Henry's Law

The solubility of O_2 is 2.2×10^{-4} M at 0°C and 0.10 atm. Calculate the solubility of O_2 at 0°C and 0.35 atm.

Solution

Because the temperature is the same in both solutions, the Henry's law constant is the same. Therefore, if $k = C/P$ then

$$\frac{C_1}{P_1} = k = \frac{C_2}{P_2}$$

$C_1 = 2.2 \times 10^{-4} M$ $C_2 = ?$
$P_1 = 0.10$ atm $P_2 = 0.35$ atm

$$\frac{2.2 \times 10^{-4} M}{0.10 \text{ atm}} = \frac{C_2}{0.35 \text{ atm}}$$

$$C_2 = 7.7 \times 10^{-4} M O_2$$

With regard to temperature effects, the solubility of most, but not all, solids increases with temperature. Your textbook points out that the solid dissolves **more rapidly** at higher temperatures but the **amount** that dissolves may not increase. Also, review the discussion on **thermal pollution** and **boiler scale.**

11.4   The Vapor Pressures of Solutions

When you finish this section you will be able to calculate

- The vapor pressure of a variety of solutions.
- The molar mass from vapor pressure information.

This section deals with the vapor pressure relationships of solutions containing both volatile and nonvolatile solutes. Let's deal with each of these separately.

The behavior of **nonvolatile solutes** is described by **Raoult's law,**

$$P_{soln} = \chi_{solvent} P^0_{solvent}$$

where P_{soln} = vapor pressure of the solution

$\chi_{solvent}$ =   mole fraction of the solvent (see <u>Section 5.5 in your textbook</u> for a review of mole fractions)

$P^0_{solvent}$ = vapor pressure of the pure solvent

This equation says that the addition of a nonvolatile solute will cause the vapor pressure of the solution to fall in direct proportion to the mole fraction of the solute.

Example 11.4 A   Raoult's Law

Glycerin, $C_3H_8O_3$, is a nonvolatile liquid. What is the vapor pressure of a solution made by adding 164 g of glycerin to 338 mL of H_2O at 39.8°C? The vapor pressure of pure water at 39.8°C is 54.74 torr and its density is 0.992 g/mL.

Strategy

We know that the vapor pressure of the solution will be lowered upon addition of (nonvolatile) glycerine. To determine the new vapor pressure, we must calculate the **mole fraction of water**, then multiply by the vapor pressure of pure water.

Solution

$$\text{moles of glycerin} = 164 \text{ g} \times \frac{1 \text{ mol}}{92.1 \text{ g}} = 1.78 \text{ mol}$$

$$\text{moles of water} = 338 \text{ mL} \times \frac{0.992 \text{ g}}{\text{mL}} \times \frac{1 \text{ mol}}{18.0 \text{ g}} = 18.63 \text{ mol}$$

$$\chi_{water} = \frac{n_{water}}{n_{water} + n_{gly}} = \frac{18.63 \text{ mol}}{18.63 \text{ mol} + 1.78 \text{ mol}} = \mathbf{0.913 \text{ mol}}$$

$$P_{water} = \chi_{water} P^0_{solvent} = 0.913 \times 54.74 \text{ torr} = \mathbf{50.0 \text{ torr}}$$

Let's extend the concept further using a **nonvolatile electrolyte** that **dissociates completely** in water. When solving the problem, we are interested in **moles of solute** actually present **in the solution** (after dissociation), not moles added.

Example 11.4 B   Raoult's Law with Electrolytes

What is the vapor pressure of a solution made by adding 52.9 g of $CuCl_2$, a strong electrolyte, to 800.0 mL of water at 52.0°C? The vapor pressure of water is 102.1 torr, and its density is 0.987 g/mL.

Solution

We know that the vapor pressure of the solution will be lowered upon the addition of $CuCl_2$. Keep in mind however that $CuCl_2$ dissociates in aqueous solution:

$$CuCl_2(s) \rightarrow Cu^{2+}(aq) + 2Cl^-(aq)$$

As a result, we have **3 moles of solute formed for every mole we add.** To determine the new vapor pressure, we must calculate the **mole fraction** of water, then multiply by the vapor pressure of pure water.

$$\text{moles of } CuCl_2 = 52.9 \text{ g} \times \frac{1 \text{ mol}}{134.5 \text{ g}} = 0.393 \text{ mol}$$

$$\text{total moles of solute} = 0.393 \text{ mol} \times 3 = \textbf{1.18 mol solute}$$

$$\text{moles of water} = 800.0 \text{ mL} \times \frac{0.987 \text{ g}}{\text{mL}} \times \frac{1 \text{ mol}}{18.0 \text{ g}} = \textbf{43.8}_7 \textbf{ mol } H_2O$$

$$\chi_{water} = \frac{n_{water}}{n_{water} + n_{solute}} = \frac{43.87 \text{ mol}}{43.87 \text{ mol} + 1.18 \text{ mol}} = \textbf{0.974 mol}$$

$$P_{water} = \chi_{water} P^0_{water} = 0.974 \times 102.1 \text{ torr} = \textbf{99.4 torr}$$

The discussion before <u>Example 11.6 in your textbook</u> considers how it is possible to use Raoult's law to determine the number of moles of a solute present. Since the number of grams is known, molar mass can be determined. Consider the following to be a challenge problem.

Example 11.4 C   Molar Mass via Raoult's Law

At 29.6°C pure water has a vapor pressure of 31.1 torr. A solution is prepared by adding 86.7 g of "Y," a nonvolatile nonelectrolyte to 350.0 g of water. The vapor pressure of the resulting solution is 28.6 torr. Calculate the molar mass of Y.

Strategy

In this problem, we know the vapor pressure drops due to the addition of Y.

If $P_{water} = \chi_{water} P^0_{water}$, then $\chi_{water} = \dfrac{P_{water}}{P^0_{water}}$. You can thus determine χ_{water}.

$$\chi_{water} = \frac{n_{water}}{n_Y + n_{water}}$$

You know n_{water} and χ_{water}. You can therefore solve for n_Y, which leads to the molar mass of Y.

Solution

$$\chi_{water} = \frac{28.6 \text{ torr}}{31.1 \text{ torr}} = \textbf{0.920}$$

$$n_{water} = 350.0 \text{ g} \times \frac{1 \text{ mol}}{18.0 \text{ g}} = \textbf{19.4 mol}$$

$$\chi_{water} = \frac{n_{water}}{n_Y + n_{water}}$$

$$0.920 = \frac{19.4}{n_Y + 19.4}$$

$$0.920(n_Y) + 17.8 = 19.4$$

$$0.920(n_Y) = 1.60$$

$$n_Y = 1.74 \text{ mol}$$

$$\textbf{molar mass of Y} = \frac{86.7 \text{ g}}{1.74 \text{ mol}} = \textbf{49.9 g/mol}$$

We have thus far worked with nonvolatile solutes. **Volatile solutes** contribute to the vapor pressure such that

$$\boldsymbol{P_{total} = P_{solute} + P_{solvent}}$$

The vapor pressure of each component can be expressed by Raoult's law:

$$P_{solute} = \chi_{solute} P^0_{solute}$$

$$P_{solvent} = \chi_{solvent} P^0_{solvent}$$

therefore,

$$P_{total} = \chi_{solute} P^0_{solute} + \chi_{solvent} P^0_{solvent}$$

By extension, for a solution that contains n components,

$$P_{total} = \chi_1 P^0_1 + \chi_2 P^0_2 + ... + \chi_n P^0_n$$

Let's apply this to the following example.

Example 11.4 D   Volatile Solutes

The vapor pressure of pure hexane (C_6H_{14}) at 60.0°C is 573 torr. That of pure benzene (C_6H_6) at 60.0°C is 391 torr. What is the expected vapor pressure of a solution prepared by mixing 58.9 g of hexane with 44.0 g of benzene (assuming an ideal situation)?

Solution

$$n_{hexane} = 58.9 \text{ g} \times \frac{1 \text{ mole}}{86.0 \text{ g}} = 0.685 \text{ mol}$$

$$n_{benzene} = 44.0 \text{ g} \times \frac{1 \text{ mole}}{78.0 \text{ g}} = 0.564 \text{ mol}$$

$$\chi_{hexane} = \frac{n_{hexane}}{n_{hexane} + n_{benzene}} = \frac{0.685 \text{ mol}}{0.685 \text{ mol} + 0.564 \text{ mol}} = \textbf{0.548}$$

$$\chi_{benzene} = 1 - 0.548 = \textbf{0.452}$$

$$P_{total} = \chi_{hexane} P^0_{hexane} + \chi_{benzene} P^0_{benzene} = 0.548 (573) + 0.452 (391) = \textbf{491 torr}$$

11.5 Boiling Point Elevation and Freezing Point Depression

When you finish this section you will be able to solve problems relating to the **colligative properties** of **boiling point elevation** and **freezing point depression**.

This section in your textbook introduces us to **colligative properties**. These properties depend **only on the concentration of solute particles**, not on the nature of the particles.

Your textbook deals first with **boiling point elevation**, that is, **the increase in boiling point of a liquid due to the addition of a nonvolatile solute**. The boiling point elevation, ΔT, is given by

$$\Delta T = K_b m_{solute}$$

where ΔT is the boiling point elevation (in °C)

          K_b is the molal boiling point elevation constant (in °C kg/mol)

          m is the molality of the solute (in mol solute/kg solvent)

Table 11.5 in your textbook gives values of K_b for some common solvents. One thing to keep in mind is that much useful information can be obtained from colligative property experiments, including molar mass of the solute, molality, and mass percent.

Example 11.5 A   Boiling Point Elevation

A solution is prepared by adding 31.65 g of sodium chloride to 220.0 mL of water at 34.0°C (density = 0.994 g/mL, K_b for water is 0.51°C kg/mol). Calculate the boiling point of the solution, assuming complete dissociation of NaCl in solution (which is not true, strictly speaking - see Section 11.7).

Solution

In order to calculate ΔT, we need to calculate the **molality of NaCl**. In this case,

$$\text{molality} = \frac{\text{mol NaCl}}{\text{kg water}}$$

$$\text{mol NaCl} = 31.65 \text{ g} \times \frac{1 \text{ mol}}{58.44 \text{ g}} = \textbf{0.542 mol}$$

$$\text{kg water} = 220.0 \text{ mL} \times \frac{0.994 \text{ g}}{\text{mL}} \times \frac{1 \text{ kg}}{1000 \text{ g}} = \textbf{0.219 kg}$$

$$m_{NaCl} = \frac{0.542 \text{ mol}}{0.219 \text{ kg}} = \textbf{2.47}_7 \textit{ m}$$

But NaCl **dissociates** into Na^+ and Cl^-, giving 2 moles of solute per 1 mole dissolved. Therefore,

$$m_{total} = 2.477 \times 2 = \textbf{4.95} \textit{ m}$$

We can now calculate ΔT directly.

$$\Delta T = K_b m = 0.51°C \text{ kg/mol} \times 4.95 \text{ m} = 2.5°C$$

The boiling point $= 100 + 2.5 = \textbf{102.5°C}$

As discussed in your textbook, the effect of adding a nonvolatile solute to a liquid is to lower the freezing point by an amount, ΔT, given by

$$\Delta T = K_f m_{solute}$$

where ΔT is the **freezing point depression** (in°C)
   K_f is the **molal freezing point depression** constant (in °C kg/mol)
   m is the **molality of the solute** (in mol solute/kg solvent)

The overall effect of adding a nonvolatile solute to a liquid is to **extend the liquid range of the solvent**. The same information can be found from both freezing point depression and boiling point elevation problems.

Example 11.5 B   Freezing Point Depression

How many grams of glycerin ($C_3H_8O_3$) must be added to 350.0 g water in order to lower the freezing point to −3.84°C (K_f for water = 1.86°C kg/mol)?

Solution

$\Delta T = 3.84°C$
$K_f = 1.86°C$ kg/mol

$$\text{molality} = \frac{\text{moles of glycerin}}{\text{kg of water}}$$

The bottom line then is to solve for the **molality of glycerin**, which leads to **moles** and finally **grams of glycerin.**

$$m_{glycerin} = \frac{\Delta T}{K_f} = \frac{3.84°C}{1.86°C \text{ kg/mol}} = \mathbf{2.06_4 \ m}$$

$$\textbf{moles of glycerin} = \frac{2.064 \text{ mol glycerin}}{\text{kg water}} \times 0.350 \text{ kg water} = \mathbf{0.722_5 \ mol}$$

$$\textbf{g of glycerin} = 0.7225 \text{ mol} \times 92.1 \text{ g/mol} = \textbf{66.5 g of glycerin}$$

11.6   Osmotic Pressure

When you finish this section you will be able to solve problems regarding osmosis and molar mass.

This section in your textbook begins with the definition of **osmosis** as **the flow of solvent into a solution through a semipermeable membrane. Osmotic pressure** is **the pressure that just stops the osmosis.** Osmotic pressure is a **colligative property** because its value depends on the **concentration of the solute,** not its nature. Your textbook points out that osmotic pressure is especially useful for determining molar masses experimentally because **small solute concentrations produce large osmotic pressures.**

The relationship between osmotic pressure and solution concentration is given by

$$\Pi = MRT$$

where Π is the **osmotic pressure** (in atm)
   M is the **molarity of the solute** (in mol/L)
   R is the **universal gas constant** (0.08206 L/K mol)
   T is the **temperature** (in K)

Let's try an example that uses osmotic pressure to determine the molar mass of a substance.

Example 11.6   Osmotic Pressure

The osmotic pressure of a solution containing 26.5 mg of aspartame per liter is 1.70 torr at 30°C. Calculate the molar mass of aspartame.

Solution

$$\Pi = MRT$$

We want to solve for molar mass. This can be done through **molarity**.

$$\Pi = 1.70 \text{ torr} \times \frac{1 \text{ atm}}{760 \text{ torr}} = 2.24 \times 10^{-3} \text{ atm}$$

$R = 0.08206$ L atm/K mol
$T = 303$ K

$$M = \frac{\Pi}{RT} = \frac{2.24 \times 10^{-3} \text{ atm}}{(0.08206 \text{ L atm/K mol}) (303 \text{ K})} = \mathbf{9.01 \times 10^{-5} \text{ mol/L}}$$

$$\textbf{molar mass} = \frac{\text{g aspartame}}{\text{mol}} = \frac{1 \text{ L}}{9.01 \times 10^{-5} \text{ mol}} \times \frac{0.0265 \text{ g}}{\text{L}} = \mathbf{294 \text{ g/mol}}$$

Several new terms are introduced in the discussion following Example 11.11 in your textbook. You should be able to define **dialysis, hypertonic, hypotonic, isotonic solutions, crenation,** and **hemolysis.** You should also be able to describe the concept of **reverse osmosis** and how this can be used to **desalinate water.**

11.7   *Colligative Properties of Electrolyte Solutions*

At the end of this section you will be able to solve colligative property problems requiring the van't Hoff factor.

Recall Example 11.5 A in this book, where we calculated the boiling point elevation of a solution to which NaCl had been added. We made an **assumption** that **NaCl completely dissociates.** Thus, our 2.48 *m* solution dissociated into 2.48 *m* Na$^+$ and 2.48 *m* Cl$^-$.

This section of your textbook points out that this assumption is **not valid, especially at high solute concentrations.** The reason seems to be **ion pairing,** where some sodium and chloride ions encounter one another, pair, and are counted as a single particle.

The equation that takes experimentally observed dissociation into account is

$$\Delta T = imK$$

where *i* is the **van't Hoff factor** representing electrolyte dissociation in solution. Keep in mind that for a given electrolyte, *i* is **concentration dependent.** Also, *i* = **1 for all nonelectrolytes.** The expected and observed values for several electrolytes are given in Table 11.6 of your textbook.

Example 11.7   Van't Hoff Factor

Use data from Table 11.6 in your textbook to calculate the **freezing point** and **expected osmotic pressure** of a 0.050 m FeCl$_3$ solution at 25.0°C. (Assume the density of the final solution equals 1.0 g/mL and that the liquid volume is unchanged by the addition of FeCl$_3$.)

Solution

a. *Freezing point*

From <u>Table 11.6 in your textbook</u>, $i_{observed} = 3.4$ for this 0.05 *m* solution.

$$\Delta T = imK_f = 3.4 \times 0.050\ m \times 1.86°C\ kg/mol = 0.32°C$$
$$\uparrow\uparrow\uparrow$$
$$imK_f$$

The **freezing point** would be $-0.32°C$.

b. *Osmotic pressure*

Analogous to the formula for freezing point depression, the formula for osmotic pressure (itself a colligative property) is

$$\Pi = iMRT$$

We assumed that the density of the solution is 1.0 g/mL, and the volume of the liquid was unchanged by the addition of $FeCl_3$. In this **unusual** case, the molarity = 0.05 *M*.

$$\Pi = 3.4 \times 0.050\ M \times 0.08206\ L\ atm/K\ mol \times 298.2\ K = \textbf{4.16 atm}$$
$$\uparrow\uparrow\uparrow\uparrow$$
$$iMRT$$

Osmotic pressure (π) = 4.16 atm

11.8   Colloids

The following review questions will help guide your study of this section.

1. What is the **Tyndall effect?**
2. How can the Tyndall effect be used to distinguish between a suspension and a true solution?
3. Define **colloid.**
4. Give some examples of **colloids.**
5. How can we prove that **electrostatic repulsion** helps stabilize a colloid?
6. Define **coagulation.**
7. How does heating destroy a colloid?
8. How does the addition of an electrolyte destroy a colloid?
9. What is the relationship between proteins in fish and the manufacture of ice cream?

Exercises

Section 11.1

1. Rubbing alcohol contains 585 g of isopropanol (C_3H_7OH) per liter (aqueous solution). Calculate the molarity.

2. The density of a 10.0% (by mass) solution of NaOH is 1.109 g/cm^3. Calculate the concentration of this solution in molarity, molality, and mole fraction.

3. Toluene ($C_6H_5CH_3$), an organic compound often used as a solvent in paints, is mixed with a similar organic compound, benzene(C_6H_6). Calculate the molarity, molality, mass percent, and mole fraction of toluene in 200. mL of solution that contains 75.8 g of toluene and 95.6 g of benzene. The density of the solution is 0.857 g/cm^3.

4. A hydrochloric acid solution was made by adding 59.26 g HCl to 100. g H_2O. The density of the solution was 1.19 g/cm^3. Calculate the concentration of HCl in molarity, molality, mass percent, and mole fraction.

5. What is the molarity of a solution that is 5.00% $Pb(NO_3)_2$ by weight? The density is 1.05 g/mL.

6. What mass of H_2SO_4 is required to prepare 250 mL of a 6.00 *M* solution?

7. In what circumstance is molality a more useful unit than molarity?

8. How many grams of NaOH and H_2O are required to prepare 100.0 g of a 28.7% NaOH by weight solution?

9. Find the concentration of NaOH in molarity, molality, and mole fraction for the solution prepared in the previous problem, assuming a solution density of 1.10 g/mL . Determine the mole fraction of H_2O.

10. Seawater contains 1.94% chlorine (by mass). How many grams of chlorine are there in 400 mL of seawater if the density of seawater is 1.025 g/cm^3?

11. The molal concentration of a solution of salicylic acid ($C_7H_6O_3$), the primary organic molecule used in aspirin production, and ethanol ($C_2H_5O_3$) is 17.5 *m*. If 42.0 g of ethanol were used to make the solution, how many grams of salicylic acid were needed?

12. The concentration of glucose, $C_6H_{12}O_6$, in a biological fluid is 75 mg/100 g. What is the molal concentration?

13. A solution of phosphoric acid was made by dissolving 10.0 g of hydrogen phosphate, H_3PO_4, in 100.0 mL of water. The resulting volume was 104 mL. Calculate the density, mole fraction, molarity, and molality of the solution.

Section 11.2

14. Rationalize the water solubilities for the gases listed below (units are g/100 cm^3).

Gas	Solubility
H_2	0.09191
CO_2	0.141
HCl	82.3
NH_3	89.9

    What would you expect for relative solubilities in hexane (C_6H_{14})?

15.   Would boric acid, $B(OH)_3$, be more soluble in ethanol, C_2H_5OH, or in benzene, C_6H_6?

16.   Predict which substance would be more soluble in water.

   a. $NH_2CH_2CH_2OCH_3$ or NH_2CH_2COOH

   b. CH_3CH-CH_2 or $CH_3CH_2CHCH_2CH_3$
         $\quad\ \ |\quad\ |$ $\qquad\qquad\ \ |$
         $\quad\ \ OH\ OH$ $\qquad\qquad\ \ CH_2OH$

   c. C_4H_9OH or C_4H_9SH

Section 11.3

17.   Ammonium salts can be used to make chemical cold packs. When these salts are dissolved in water the solution gets quite cold. What can we say about the heat of solution?

18.   The solubility of oxygen in water at 0°C and 1 atm is 0.073 g per liter. What is k in Henry's law for this temperature?

19.   The solubility of nitrogen at 0°C in water is 8.21×10^{-4} mol/L if the N_2 pressure above the water is 0.790 atm. What is the solubility at 1 atm of N_2 and 0°C?

20.   What is the Henry's law k for nitrogen in Problem 19?

21.   The solubility of nitrogen in blood at 37°C and 0.80 atm is 5.6×10^{-4} M. If a deep sea diver breathes compressed air from a tank at a partial pressure of 3 atm, what would the solubility of nitrogen be in the diver's blood at 37°C?

22.   Why doesn't Henry's law work for ammonia?

23.   The solubility of CO_2 in water at 25°C and 1 atm is 0.034 M. What is the Henry's law constant (k)? What would the solubility of CO_2 in water be at 0.038 atm and 25°C?

Section 11.4

24.   The vapor pressure of ethanol at 40°C is 135.3 torr. Calculate the vapor pressure of a solution of 53.6 g of glycerin, $C_3H_8O_3$, in 133.7 g of ethanol, C_2H_5OH, at 40°C.

25.   Antifreeze solutions are mainly ethylene glycol, $C_2H_6O_2$, in water. Calculate the vapor pressure of a solution made by adding 101.6 g of ethylene glycol to 139.6 g of water at 50.0°C. At this temperature ethylene glycol is essentially nonvolatile, and the vapor pressure of water is 92.51 torr.

26.   Sucrose, $C_{12}H_{22}O_{11}$ (a nonvolatile substance), is a sweetener. A solution was made with 35.2 g of sucrose and 78.0 g of water at 30°C. Calculate the vapor pressure of the solution if the vapor pressure of water at 30°C is 31.824 torr.

27.   A solution was prepared by dissolving 16.8 g of camphor ($C_{10}H_{16}O$) in 86.1 g of benzene (C_6H_6). At 23°C, the vapor pressure of benzene is 86.0 torr. If the vapor pressure of the resulting solution was 78.2 torr, calculate the vapor pressure of camphor at 23°C (note: solid camphor has a low volatility).

28.   How many grams of a nonvolatile compound B (molar mass = 97.8 g/mol) would need to be added to 250 g of water to produce a solution with a vapor pressure of 23.756 torr? The vapor pressure of water at this temperature is 42.362 torr.

29. Calculate the vapor pressure of a solution of 40.27 g $MgCl_2$ in 500.3 mL of water at 25.0°C. The density of water is 0.9971 g/mL, and the vapor pressure is 23.756 torr.

30. A 12.97-g sample of a metal chloride, general formula MCl, is added to 100.0 g of water at 30.0°C. The vapor pressure of the solution is 30.60 torr. The vapor pressure of pure water at 30.0°C is 31.824 torr. What is the formula of the metal chloride?

31. Calculate the vapor pressure of a solution made with 321 g of toluene (C_7H_8) and 398 g of benzene (C_6H_6). At 60°C, the vapor pressures of the two volatile components are 140. torr and 396 torr, respectively.

32. What is the vapor pressure of a solution made by adding 26.93 g potassium sodium tartrate, $KNaC_4H_4O_6$ to 676.5 g of water at 25°C? The vapor pressure of pure water is 23.756 torr at 25°C. Assume that potassium sodium tartrate is nonelectrolyte (a poor assumption).

Section 11.5

33. An aqueous solution of sucrose, $C_{12}H_{22}O_{11}$, boils at 112°C. What is the molality?

34. 1.51 g of white phosphorus (P_4) was dissolved in 18.0 g of carbon disulfide (CS_2). The boiling point elevation of the carbon disulfide solution was found to be 1.36°C. Calculate the molality of the white phosphorus. What is K_b for carbon disulfide?

35. What is the freezing point of a solution that contains 15.0 g of ethylene glycol, $C_2H_6O_2$, in 250 g of water?

36. A solution of carbon tetrachloride (CCl_4) has a molality of 1.35 m. The freezing point depression constant for CCl_4 is 30°C kg/mol, and its freezing point is −22.99°C. At what temperature would the solution freeze?

37. What are the boiling and freezing points of a solution of 50.3 g of I_2 in 350 g of chloroform? (See Table 11.5 in your textbook for constants.)

38. A 4.367-g sample of an unknown hydrocarbon is dissolved in 21.35 g benzene. The freezing point of the solution is observed to be −0.51°C. Calculate the molar mass of the unknown.

39. Calculate the molecular weight of a compound if 4.00 g of it plus 50.0 g of water give a solution with a boiling point of 100.41°C.

40. During a Wisconsin winter, the temperature can reach −25°C (or colder!). How many grams of antifreeze (ethylene glycol, $C_2H_6O_2$) would you need to add to your radiator to keep 7.5 liters of water from freezing? Assume that the density is 1.0 g/mL (K_f for water = 1.86°C kg/mol).

Section 11.6

41. Calculate the osmotic pressure of a solution made by adding 13.65 g of sucrose, $C_{12}H_{22}O_{11}$, to enough water to make 250. mL of solution at 25°C.

42. What is the osmotic pressure of 1.38×10^{-2} M KBr at 25°C?

43. The osmotic pressure of blood is 7.65 atm at 37°C. How much glucose, $C_6H_{12}O_6$, should be used per liter for an intravenous injection having the same osmotic pressure as blood?

44. The osmotic pressure of an aqueous solution of a polypeptide solution was determined to be 4.80 torr at 25°C. How many grams of polypeptide would be in 2.00 L of solution? (The molar mass of the peptide is 269 g/mol.)

45. The osmotic pressure of a tryptophan solution containing 1.136 g per liter is 103.5 torr at 25°C. What is the molar mass of tryptophan?

46. What is the minimum pressure required to desalinate 1.0 M salt (NaCl) solution at 25°C?

47. Calculate the osmotic pressure (in atm) of a 0.225 M aqueous solution of urea that is isotonic with sea water at 10°C.

Section 11.7

48. Table 11.6 in your textbook lists values of observed van't Hoff factors for 0.05 m solutions. Would you expect the van't Hoff factors for 0.5 m solutions to be greater or less than the values for 0.05 m?

49. Using the observed van't Hoff factor from Table 11.6 in your textbook, calculate the freezing point, boiling point, and osmotic pressure (at 25°C) of a 0.050 m MgCl$_2$ solution. (Assume that 0.050 m = 0.050 M.)

50. Calculate the osmotic pressure of a 0.05 M solution of NaCl at 30°C. The van't Hoff factor for NaCl is 2.0.

Section 11.8

51. Indicate the type of the colloid that each of the following represents (i.e., aerosol, foam, emulsion, sol, or gel):

    a. Milk of Magnesia
    b. salad dressing
    c. meringue
    d. rain cloud

52. The black ink of this printed page could have been prepared by burning natural gas in a limited amount of air, so that hydrogen burned away leaving colloid carbon (carbon black) that was then mixed with a liquid to make printer's ink. Is this an example of the condensation or dispersion method for preparing colloids? Explain.

Multiple Choice Questions

53. Which of the following terms is not a quantitative description of a solution?

    A. Molarity   B. Molality   C. Mole fraction   D. Supersaturation

54. The solubility of a certain compound is 29.3 g/100 g of water. How many grams of this solute will dissolve in 87.9 g of water?

    A. 300 g   B. 19.4 g   C. 29.3 g   D. 25.8 g

55. How many grams of nitric acid, 70.5% by weight, are present in 1500.0 mL of solution with density of 1.42 g/cm^3?

    A. 352 g   B. 718 g   C. 2130 g   D. 1.50×10^3 g

56. The mole fraction of calcium chloride in water is 0.326. How many grams of $CaCl_2$ are present in 90.0 g of solution?

    A. 32.1 g   B. 38.1 g   C. 60.0 g   D. 67.5 g

57. A solution contains 1300. g of solvent, 40.0 g of solute, and is known to be 0.170 molal. What is the molar mass of the solute in g/mol?

    A. 90 g/mol   B. 181 g/mol   C. 133 g/mol   D. 72.0 g/mol

58. Calculate the molarity of a solution prepared by mixing 50.0 cm^3 of 0.82 M NaCl with 30.0 cm^3 0.52 M NaCl solution, assuming volumes are additive.

    A. 0.71 M B. 1.02 M C. 0.80 M D. 0.66 M

59. What volume of solution is required to prepare a 0.014 M solution containing 1.40 g of $FeCl_2$?

    A. 0.100 L   B. 1.80 L   C. 0.79 L   D. 0.010 L

60. Which of the following molecules would be the most soluble in water?

    A. CCl_4 B. CH_3NH_2 C. HI   D. CH_4

61. Which of the following molecules would be the most soluble in CH_3CH_3?

    A. CH_4 B. KI   C. HI   D. CH_3NH_2

62. Oil and water cannot mix together, because:

    A. The water molecules cannot break the London forces between oil molecules
    B. The hydrogen bonds between the water molecules are stronger than the hydrogen bonds between water and oil molecules
    C. The London forces between the water molecules are not as strong as the ones between oil molecules
    D. The hydrogen bonds between water molecules are hard to break by oil molecules

63. How will increasing the pressure of a gas affect the solubility of the gas in a solvent?

    A. increase it   B. decrease it   C. no effect   D. depends on pressure

64. Which of the following is not a hydrophobic vitamin?

    A. A   B. K   C. C   D. D

65. 4.3 g of Vitamin B_{12} is dissolved in 74 g of ether $[(C_2H_5)_2O]$ at 25°C. The change in vapor pressure amounts to 1.00 torr. Calculate the molar mass of this vitamin. Pressure of ether at 25°C is 315 torr.

    A. 422 g/mol   B. 106 g/mol   C. 675 g/mol   D. 1.36×10^3 g/mol

66. A 10.0-g sample of unknown material is dissolved in 80.0 g of water. Calculate the molar mass of the solute if water begins to freeze at −0.44°C.

    A. 250 g/mol   B. 528 g/mol   C. 265 g/mol   D. 800 g/mol

67. 0.015 moles of a compound are dissolved in 18.0 fl. oz. of water. At what temperature will the solution begin to freeze? The density of water = 1.00 g/mL.

    A. −0.055°C   B. −0.210°C   C. −1.86°C   D. −2.00°C

68. What is the boiling point, in degrees Celsius, of a solution that is prepared by mixing 500.0 mL of water with 5.844 g of sodium chloride?

    A. 100.1°C   B. 102.0°C   C. 99.80°C   D. 100.05°C

69. Four distinct solvents, A, B, C, and D have the following K_b values $= 0.51, 4.59, 5.07,$ and $40.0°C/m$. Their respective molar masses are $= 60, 88, 98,$ and 152 g/mol. 3.0 g of a solute with molar mass $= 138.0$ g/mol is dissolved in 200.0 g of one of the solvents. The boiling point is changed by $4.34°C$. Which solvent is it?

   A. A   B. B   C. C   D. D

70. What is the osmotic pressure created by 20.0 mg of insulin dissolved in 10.0 mL of solution at $27°C$? The molar mass of insulin is 6.02×10^3 g/mol.

   A. 3.11 torr   B. 6.22 torr   C. 0.0493 torr   D. 0.0818 torr

71. What molar glucose solution would be isotonic with blood? The osmotic pressure of blood at $37°C$ is 7.65 atm. The molar mass of glucose is 180 g/mol.

   A. 0.30 M B. 0.15 M C. 2.5 M D. 1.5 M

72. How many grams of NaCl would be required to lower the freezing point of 2.00 kg of water to $0.0°F$?

   A. 280 g   B. 560 g   C. 200 g   D. 117 g

Answers to Exercises

1.   9.75 M

2.   molarity = 2.77 M; molality = 2.78 m; mole fraction = 0.0476

3.   molality = 8.62 m $C_6H_5CH_3$; molarity = 4.11 M $C_6H_5CH_3$; mass percent = 44.2%; mole fraction = 0.402

4.   molarity = 12.1 M; molality = 16.3 m; mass percent = 37.2%; mole fraction = 0.226

5. 1.59×10^{-1} mol/L

6.   147 g

7.   Molality does not change when temperature changes. Molarity does change.

8.   28.7 g NaOH, 71.3 g H_2O

9.   molarity = 7.89 M; molality = 7.2 m; mole fraction of NaOH = 0.15; mole fraction of H_2O = 0.85

10.   7.95 g Cl

11.   101 g $C_7H_6O_3$

12. 4.2×10^{-3} m

13.   Density = 1.057 g/mL   Mole fraction: H_3PO_4 = 0.018   Molarity = 0.981 mol/L
                                                  H_2O = 0.982   Molality = 1.02 mol/kg

14. H_2 and CO_2 are nonpolar and therefore not very soluble in water. HCl and NH_3 are polar and therefore more soluble in water. In hexane it is expected that H_2 and CO_2 will be more soluble than HCl and NH_3.

15.   Ethanol. Ethanol is more polar than benzene. Because boric acid is also a polar substance, it would be more soluble in ethanol.

16.   a. NH_2CH_2COOH c. C_4H_9OH
      b. $CH_3CH{-}CH_2$
               | |
              OH   OH

17.   It is an endothermic reaction, therefore, $\Delta H_{sol} > 0$.

18. 2.3×10^{-3} mol/L atm

19. 1.04×10^{-3} M

20.   962 L atm/mol

21. 2.1×10^{-3} mol/L

22. $NH_3 + H_2O \rightarrow NH_4^+ + OH^-$, ammonia reacts with water

23. k = 29 L atm/mol; 1.3×10^{-3} M

24.   113 torr

25.   76.39 torr

26.   31.1 torr

27. 0.05 torr

28. 1.063×10^3 g compound B

29. 23.40 torr

30. NaCl

31. 291 torr

32. 23.66 torr

33. 23.5 m

34. 6.77×10^{-1} m; 2.01°C kg/mol

35. −1.8°C

36. −63.49°C

37. 63.3°C, −66.2°C

38. 174 g/mol

39. 101.46 g/mol

40. 6.3×10^3 g (or 6.3 kg)

41. 3.90 atm

42. 0.675 atm, 513 torr

43. 54.3 g/L

44. 0.138 g

45. 204 g/mol

46. 49 atm

47. 5.23 atm

48. 0.5 m factors would be less than 0.05 m factors. The more concentrated the solute, the more chances for interactions.

49. Freezing pt. = −0.25°C; Boiling pt. = 100°C; osmotic pressure 3.3 atm.

50. 2.49 atm

51. a. sol   b. emulsion   c. foam   d. aerosol

52. Condensation method: The formation of insoluble substances from solutions is another illustration of this method.

53. D	54. D	55. D	56. D	57. B	58. A
59. C	60. C	61. A	62. D	63. A	64. C
65. D	66. B	67. A	68. A	69. D	70. B
71. A	72. B				

Chapter 12

Chemical Kinetics

In the introductory remarks to this chapter, your textbook makes the critical distinction between kinetics and thermodynamics. Thermodynamics can be used to **predict if a reaction will occur**. It says nothing about how fast it occurs. **Kinetics** describes the **rate and mechanism** of a reaction **given that it occurs**. Thermodynamics says "if." Kinetics says "how" and "how fast."

12.1   Reaction Rates

When you finish this section you will be able to use concentration vs. time data to determine relative reaction rates.

Your textbook defines the **rate of a reaction** as

$$\text{rate} = \frac{\text{change in concentration of a substance}}{\text{change in time}}$$

Let's look at data for the reaction of hydrogen and oxygen to give water,

$$2H_2(g) + O_2(g) \rightarrow 2H_2O(l)$$

Time (s)	$[H_2]$	$[O_2]$	$[H_2O]$
0.0020	0.050	0.080	0
0.0040	0.025	0.0675	0.0250
0.0060	0.018	0.064	0.032
0.0080	0.0125	0.0612	0.0375

The rate of disappearance of H_2 must be twice the rate for O_2, and equal to the appearance of H_2O. This is because the *coefficients of the chemical equation tell you the relative rates.* That is

$$\frac{-\Delta[H_2]}{\Delta t} = \frac{-2\Delta[O_2]}{\Delta t} = \frac{\Delta[H_2O]}{\Delta t}$$

where "[ ]" means **concentration in moles/liter.** This says that for **every mole of O$_2$ that reacts,** per unit time, 2 moles of H$_2$ react, and 2 moles of H$_2$O are produced. Note that the *rate of disappearance* of the reactants is given a negative ("−") sign. That implies that we are *losing reactants* while gaining products ("+" rate). Note also that your textbook uses the standard that all reaction rates should be *positive.*

Using the data from our reaction, the rate of reaction of <u>hydrogen</u> between 0.004 and 0.008 seconds is

$$\text{rate} = \frac{-\Delta[H_2]}{\Delta t} = \frac{-(0.0125 - 0.025)}{(0.008 - 0.004)} = \frac{-(-0.0125)}{0.004}$$

$$\textbf{rate} = \textbf{3.1 mol/L s}$$

We are *losing* hydrogen at a rate of 3.1 mol/L s.

Example 12.1 A   Reaction Rates

Answer the following questions using the hydrogen and oxygen rate data just presented.

    a.   Based on the *coefficients of the chemical equation alone*, what is the rate of oxygen reaction between 0.004 and 0.008 seconds?

    b.   What is the rate of water production during the same time period?

    c.   Use the concentration vs. time data to prove your answers.

Solution

    a.   One mole of oxygen reacts for every two moles of hydrogen. That is, *oxygen reacts at ½ the rate of hydrogen.* The rate of reaction of hydrogen is **3.1 mol/L s.** The **rate of reaction of oxygen** is therefore (3.1)/2 = **1.6 mol/L s** (rounded off to two significant figures).

    b.   The balanced chemical equation says that one mole of hydrogen reacts to produce one mole of water. Therefore, the *rate of consumption of hydrogen equals the rate of production of water.*

$$\textbf{rate of water production} = \textbf{3.1 mol/L s}$$

    c.   For oxygen,

$$\frac{-\Delta[O_2]}{\Delta t} = \frac{-(0.0612 - 0.0675)}{(0.008 - 0.004)} = \textbf{1.6 mol/L s}$$

        For water,

$$\frac{\Delta[H_2O]}{\Delta t} = \frac{(0.0375 - 0.0250)}{(0.008 - 0.004)} = \textbf{3.1 mol/L s}$$

Remember the order of operations of your calculator. After subtracting 0.0250 from 0.0375, press the "=" key before dividing by Δt; otherwise, you will divide 0.0250 by Δt.

Look at <u>Table 12.2 in your textbook</u>. Notice that the *rate of NO$_2$ decomposition decreases with time* until equilibrium is reached. **The rate of a reaction is not constant.**

Example 12.1 B   Practice With Reaction Rates

Given the hypothetical reaction

$$2A + nB \rightarrow qC + rD$$

If $\quad \dfrac{-\Delta[A]}{\Delta t} = 0.050 \text{ mol/L s}$

$\quad\quad \dfrac{-\Delta[B]}{\Delta t} = 0.150 \text{ mol/L s}$

$\quad\quad \dfrac{-\Delta[C]}{\Delta t} = 0.075 \text{ mol/L s}$

$\quad\quad \dfrac{-\Delta[D]}{\Delta t} = 0.025 \text{ mol/L s}$

What are the coefficients n, q, and r?

Solution

The rate of disappearance of B is 3 times that of A. Therefore, n must be 3×2 or **6**. The rate of formation of C is 3/2 that of the decomposition of A. So $q = 3/2 \, (2) = $ **3**. D appears at ½ the rate of disappearance of A. Therefore, $r = ½ \, (2) = $ **1**. The equation is, therefore,

$$2A + 6B \rightarrow 3C + D$$

12.2   Rate Laws: An Introduction

Answer the following review questions after you have read the entire section.

1. What **simplifying assumption** do we make regarding the rates of forward and reverse reactions in our study of rate laws?
2. Define **rate law.**
3. What are "**k**" and "**n**"?
4. Why don't **products** appear in a rate law?
5. Why is it important to **specify the component** whose rate we are describing with the rate law?
6. What is a **differential rate law?**
7. What is an **integrated rate law?**
8. Give a couple of **practical examples** of the importance of determining rate laws.

12.3   Determining the Form of the Rate Law

When you finish this study section you will be able to use the **method of initial rates** to determine **differential rate laws and rate constants**.

Recall from the last section that the **differential rate law** deals with the **dependence of rate on concentration.** The important question is, "how will the rate of reaction change if we change the concentration of our reactants?"

Let's take the hypothetical example of reactant "A" going to product "P."

$$A \rightarrow P$$

A plot of the concentration of A, "[A]", vs. time is similar in shape to the one shown in Figure 12.3 in your textbook. Examine the following data where the **initial rate** = $\Delta[A]/\Delta t$ and is taken as close to $t = 0$ as possible.

$[A]_o$	Initial Rate of Reaction (mol/L s)
0.35	7.2×10^{-4}
0.70	2.90×10^{-3}
1.05	6.45×10^{-3}

("$[A]_o$" means the **initial concentration** of A. The subscript "o" means "initial.")

How does the rate of decomposition of A change as we change [A]?

As [A] doubles from 0.35 to 0.70, the rate goes from 7.20×10^{-4} mol/L s to 29.0×10^{-4} mol/L s. The rate has increased by a factor of 4.0. As [A] triples, the rate increases by a factor of 9. This 2:4, 3:9 relationship suggests that the rate depends on the square of the concentration, or

$$\text{rate} = \frac{-\Delta[A]}{\Delta t} = k[A]^2$$

where k = the rate constant for the reaction.

Your textbook points out that such differential rate laws can only be found by experiments, not theoretically. To solve for k, substitute several rate values with the corresponding values for [A], and average the resultant values for k. Using [A] = 0.70 M as an example,

$$\text{rate} = k[A]^2$$

$$2.90 \times 10^{-3} \text{ mol/L s} = k (0.70 \text{ } M)^2$$

$$k = \frac{2.90 \times 10^{-3} \text{ mol/L s}}{(0.70 \text{ } M)^2} = 5.92 \times 10^{-3} \text{ L/mol s}$$

while the rate is not constant, *the rate constant is constant* for a particular reaction at a particular temperature.

Example 12.3 A   Differential Rate Law

Using the data given above, what would the rate of reaction for A be (rate = $-\Delta[A]/\Delta t$) if [A] = 0.16 M?

Strategy

You know the rate law and the value for k (which you would check by doing further experiments). You can therefore solve for the rate by substitution.

Solution

$$\text{rate} = k[A]^2 = (5.92 \times 10^{-3} \text{ L/mol s})(0.16 \text{ } M)^2 = 1.52 \times 10^{-4} \text{ mol/L s}$$

Note that the rate of reaction is *considerably slower* with the *lower concentration* than with [A] = 0.70 M. That makes sense because the rate equation says that *the rate will increase as the square of the concentration increases*. If the concentration decreases, the rate does as well.

Let's look at the more complex example of

$$A + B \rightarrow \textbf{products}$$

We want to determine a rate law for the decomposition of "A" that takes into account both [A] and [B]. We can hypothesize that the rate of reaction is proportional to the concentrations of A and B. The more of each, the faster the reaction.

$$\text{rate} = \frac{-\Delta[A]}{\Delta t} = k[A]^n[B]^p$$

Gathering experimental data will allow us to determine the rate law. Our goal is to determine *n*, *p*, and *k*. The sum of the exponents, *n + p*, is called the reaction order. Let's say the data are:

Reaction	$[A]_o$	$[B]_o$	Initial Rate of Reaction (mol/L s)
1	0.100	0.100	1.53×10^{-4}
2	0.100	0.300	4.59×10^{-4}
3	0.200	0.100	6.12×10^{-4}
4	0.100	0.200	3.06×10^{-4}
5	0.300	0.600	8.26×10^{-3}

The key process here is to **determine how the rate varies with varying concentration of one component while the concentration of the other is held constant.**

What happens to the rate when you vary [B] while holding [A]$_o$ constant?

In reactions #1, #2, and #4, [A]$_o$ is constant. Therefore, any increase in rate must be related to the increase in [B]$_o$. Using data from reactions #1 and #2,

$$\text{increase in [B]}_o = \frac{0.300\ M}{0.100\ M} = \textbf{factor of 3}$$

$$\text{increase in rate} = \frac{4.59 \times 10^{-4}\ \text{mol/L s}}{1.53 \times 10^{-4}\ \text{mol/L s}} = \textbf{factor of 3}$$

The rate increases *linearly* with [B]$_o$.

$$\text{rate} = k[A]^n[B]^1$$

What happens to the rate when you vary [A]$_o$ while holding [B]$_o$ constant?

In reactions #1 and #3, [B]$_o$ is constant. Therefore, any increase in rate must be related to the increase in [A]$_o$.

$$\text{increase in [A]}_o = \frac{0.200\ M}{0.100\ M} = \textbf{factor of 2}$$

$$\text{increase in rate} = \frac{6.12 \times 10^{-4}\ \text{mol/L s}}{1.53 \times 10^{-4}\ \text{mol/L s}} = \textbf{factor of 4}$$

The rate increases with [A]$_o$ *squared*.

$$\text{rate} = k[A]^2[B] \quad \text{(if no exponential is given, "1" is assumed)}$$

We now have a complete rate law. However, we need to **solve for *k*** and use other data to verify our solution. Using data set #4 (we really should average the values obtained from all 5 reactions),

$$k = \frac{\text{rate}}{[A]^2[B]} = \frac{3.06 \times 10^{-4}\ \text{mol/L s}}{(0.100\ M)^2(0.200\ M)} = \textbf{0.153 L}^2\textbf{/mol}^2\textbf{ s}$$

(We will address the unusual units in the next section.)

As a **double check**, let's solve for the rate in reaction #5.

$$\text{rate} = (0.153 \text{ L}^2/\text{mol}^2 \text{ s})(0.300 \text{ } M)^2 (0.600 \text{ } M) = 8.26 \times 10^{-3} \text{ mol/L s}$$

Our rate law properly reflects the data.

Example 12.3 B Rate Law Determination

Determine the rate law, and solve for the order and value of the rate constant for the reaction

$$\text{C} + \text{D} + \text{E} \rightarrow \text{Products}$$

given the following data:

Reaction #	$[C]_o$	$[D]_o$	$[E]_o$	Initial rate (mol/L s)
1	0.400	0.300	0.560	7.14×10^{-4}
2	0.100	0.500	0.200	4.55×10^{-5}
3	0.100	0.200	0.200	4.55×10^{-5}
4	0.400	0.300	0.750	1.28×10^{-3}
5	0.100	0.300	0.560	3.57×10^{-4}

Solution

$$\text{rate} = k[C]^n[D]^p[E]^q$$

We need to see how each component varies while the others remain constant.

1. *Regarding dependence on [C]:* $[D]_o$ and $[E]_o$ remain constant in reactions #1 and #5. In going from #5 to #1, $[C]_o$ increases.

$$[C]_o = \frac{0.400 \text{ } M}{0.100 \text{ } M} = \textbf{factor of 4}$$

$$\text{rate} = \frac{7.14 \times 10^{-4} \text{ mol/L s}}{3.57 \times 10^{-4} \text{ mol/L s}} = \textbf{factor of 2}$$

Therefore, rate increases as $[C]_o^{1/2}$!

$$\text{rate} = k[C]_o^{1/2}[D]_o^p[E]_o^q$$

2. *Regarding dependence on [D]:* $[C]_o$ and $[E]_o$ remain constant in reactions #2 and #3. In going from #3 to #2, $[D]_o$ increases.

$$[D]_o = \frac{0.500 \text{ } M}{0.200 \text{ } M} = \textbf{factor of 2.5}$$

$$\text{rate} = \frac{4.55 \times 10^{-5} \text{ mol/L s}}{4.55 \times 10^{-5} \text{ mol/L s}} = \textbf{NO CHANGE!}$$

Therefore, the rate is **not dependent** on $[D]_o$.

$$\text{rate} = k[C]_o^{1/2}[D]_o^0[E]_o^q$$

3. *Regarding dependence on [E]_o:* $[C]_o$ and $[D]_o$ remain constant in reactions #1 and #4. In going from #1 to #4, $[E]_o$ increases.

$$[E]_o = \frac{0.750 \text{ } M}{0.560 \text{ } M} = \textbf{factor of 1.34}$$

$$rate = \frac{1.28 \times 10^{-3} \text{ mol/L s}}{7.14 \times 10^{-4} \text{ mol/L s}} = \textbf{factor of 1.79}$$

The square of 1.34 is 1.79, or $(1.34)^2 = 1.79$. Therefore, rate increases as $[\text{E}]_o^2$.

$$rate = k[\text{C}]_o^{1/2}[\text{D}]_o^0[\text{E}]_o^2$$

or $\textbf{rate} = \boldsymbol{k[\text{C}]_o^{1/2}[\text{E}]_o^2}$, because $[\text{D}]_o^0 = 1$

4. *Solving for the rate constant, k:* Pick any of the equations (though it is best to average all five). Using equation #3,

$$k = \frac{rate}{[\text{C}]_o^{1/2}[\text{E}]_o^2} = \frac{4.55 \times 10^{-5} \text{ mol/L s}}{(0.100 \ M)^{1/2}(0.200 \ M)^2} = \textbf{3.60} \times \textbf{10}^{-3} \ \textbf{L}^{3/2}/\textbf{mol}^{3/2}$$

The order of k equals the sum of the exponents $n + p + q = \frac{1}{2} + 0 + 2 = \textbf{2}\frac{1}{2}$ **order**.

Checking Your Work

Use **data set #5** and solve for the rate.

$$rate = 3.6 \times 10^{-3}(0.100)^{1/2}(0.560)^2 = \textbf{3.57} \times \textbf{10}^{-4} \ \textbf{mol/L s}$$

This agrees with the observed rate.

12.4 The Integrated Rate Law

When you finish this section you will be able to:

- Determine the proper units of a rate constant for any order rate law.
- Determine the reaction order from concentration v. time data.
- Calculate the half-life of a first- or second-order reaction.

Units of k

In the last section, we reviewed the development of a differential rate law from initial rate data. The general form of the rate equation was

$$rate = \frac{-\Delta[\text{A}]}{\Delta t} = k[\text{A}]^n$$

where A is a reactant.

If $n = 1$ (a first-order reaction), then the units of k are

$$k = \frac{rate}{[\text{A}]} = \frac{\text{mol/L s}}{\text{mol/L}} = \textbf{1/s}$$

If $n = 2$ (a second-order reaction), then the units of k are

$$k = \frac{rate}{[\text{A}]^2} = \frac{\text{mol/L s}}{\text{mol}^2/\text{L}^2} = \textbf{L/mol s}$$

If $n = 3$ (a third order reaction), then

$$k = \frac{rate}{[\text{A}]^3} = \frac{\text{mol/L s}}{\text{mol}^3/\text{L}^3} = \textbf{L}^2/\textbf{mol}^2 \ \textbf{s}$$

Do you see a pattern emerging? For an **nth order reaction** the units of k will be $\mathbf{L^{(n-1)}/mol^{(n-1)}}$ **s**. Thus, for our 2 ½-order reaction in Example 12.3 B, k is in units of $M^{1-(5/2)}\,s^{-1} = \mathbf{M^{-3/2}\,s^{-1}} \equiv \mathbf{L^{3/2}/mol^{3/2}}$ **s**.

Example 12.4 A   The Units of the Rate Constant

Determine the units of k for the following differential rate law.

$$rate = k[A]^4[B][C]^2$$

Solution

The order of this reaction is $4 + 1 + 2 = 7$. The units of k are $mol^{(1-7)}/L^{(1-7)}\,s = \mathbf{L^6/mol^6}$ **s**.

First-Order Rate Law

Recall that the difference between a **differential rate law** and an **integrated rate law** is that the latter relates **concentration of a substance to reaction.** Your textbook presents two *mathematically identical* forms of the first-order integrated rate law (rate = $k[A]$).

- $\ln[A] = -kt + \ln[A]_0$

- $\ln\left(\dfrac{[A]_0}{[A]}\right) = kt$

where $[A]_0$ = initial concentration of A
$[A]$ = concentration of A at time t
k = rate constant

I have added a third equation (by exponentiating both sides of the previous equations), which you may find useful.

- $[A] = [A]_0\,e^{-kt}$

Your textbook makes several points regarding this concentration vs. time relationship. To reemphasize some of them,

- The equation can be expressed in the form $y = mx + b$, where $y = \ln[A]$, $m = -k$, $x = t$, and $b = \ln[A]_0$.
- As a result, if you plot $\ln[A]$ vs. time and get a straight line, your data is likely to be first-order. If you do not get a straight line, some other order may predominate.
- The form of the equation that you should use will depend upon what you are given. You generally will avoid equations where the unknown is part of a log term.

Example 12.4 B   Practice With First-Order Kinetics

All radioactive elements have nuclei that follow a first-order rate law when decaying. (We will learn more about this in Chapter 18.) Radon decays to polonium according to the following equation,

$$^{222}Rn \rightarrow {}^{218}Po + {}^4He$$

The first-order rate constant for decay is **0.181 days^{-1}**. If you begin with a 5.28-g sample of pure ^{222}Rn, how much will be left after **1.96 days**? **3.82 days**?

Strategy

Your goal is to solve for [A]. You are given $[A]_0$, k, and t. Although any of the three forms of the integrated rate law is usable, the most useful one solves for [A] directly,

$$[A] = [A]_0 e^{-kt}$$

Solution

Given: $[A]_o = 5.28$ g, $k = 0.181$ d^{-1}, $t = 1.96$ d

$$[A] = 5.28e^{-(0.181 \times 1.96)} = 5.28 \times 0.701 = \textbf{3.70 g }^{222}\textbf{Rn}$$

When we want to **increase** t to 3.82 days,

$$[A] = 5.28e^{-(0.181 \times 3.82)} = 5.28 \times 0.501 = \textbf{2.64 g }^{222}\textbf{Rn}$$

Does the Answer Make Sense?

In each case, the amount we ended with was *less* than the amount we started with. That makes sense as we are *decomposing* ^{222}Rn. The longer time period in the second part of the problem allowed even more to decompose. This answer should be less than in the first part. That, too, makes sense.

Note that in the problem you just finished, **3.82 days** were required to **decompose ½ of the reactant** (5.28 g → 2.64 g). The time it takes to lose half of your reactant is called the **half-life**, or $t_{1/2}$. For ^{222}Rn, $t_{1/2} = 3.82$ days. For a first-order reaction (see your textbook for the proof),

$$t_{1/2} = \frac{0.693}{k}$$

In a first-order reaction, the half-life is *independent* of the initial concentration of the reactant.

Look at Figure 12.4 in your textbook. This says that you *lose 1/2 of the remaining substance for each half-life* that occurs. In effect,

$$[A] = \frac{[A]_o}{2^n}$$

where n = the number of half-lives that have occurred.

Example 12.4 C   First-Order Half-Life

Using the same ^{222}Rn nuclear decay data as in the previous example, how much ^{222}Rn would remain from a 5.82-g sample after **11.46 days? 21.01 days?** (Solve for 21.01 days using the half-life method; then, for practice, by using the integrated rate law.)

Solution

The key is to calculate the number of half-lives ("n") in the given time period. For $t = 11.46$ days,

$$n_{(11.46)} = \frac{1 \text{ half-life}}{3.82 \text{ days}} \times 11.46 \text{ days} = 3.0 \text{ half-lives}$$

$$[A] = \frac{[A]_o}{2^n} = \frac{5.82 \text{ g}}{2^3} = \frac{5.82 \text{ g}}{8} = \textbf{0.728 g }^{222}\textbf{Rn}$$

For $t = 21.01$ days (half-life method),

$$n_{(21.01)} = \frac{1 \text{ half-life}}{3.82 \text{ days}} \times 21.01 \text{ days} = 5.50 \text{ half-lives}$$

$$[A] = \frac{5.82 \text{ g}}{2^{5.5}} = \frac{5.82 \text{ g}}{45.3} = \textbf{0.129 g }^{222}\textbf{Rn}$$

For $t = 21.01$ days (integrated rate law, $k = 0.181$ d^{-1})

$$[A] = [A]_o e^{-kt} = 5.82 e^{-(0.181 \times 21.01)} = 0.130 \text{ g } ^{222}\text{Rn}^*$$

*The *slight difference* in the last two answers came as a result of rounding off when *converting $t_{1/2}$ to k*. Without rounding off, $k = 0.1814$, and $[A] = 0.129$ g.

Second-Order Rate Law

Your textbook gives the following equation relating to the **integrated rate law for second-order data.**

$$a\text{A} \rightarrow \text{Products}$$

$$\frac{1}{[A]} = kt + \frac{1}{[A]_o}$$

The equation is in the form $y = mx + b$ where $y = 1/[A]$, $m = k$, $x = t$, and $a = 1/[A]_o$. Thus, if your data fit a second-order model, a plot of $1/[A]$ vs. t should give a **straight line** with a **slope = k** and **intercept = $1/[A]_o$**. (See Figure 12.5 in your textbook.)

The equation relating half-life to k is

$$t_{1/2} = \frac{1}{k[A]_o}$$

Notice that for a second-order system, the half-life depends on the initial concentration!

Example 12.4 D   Second-Order Rate Law

The differential rate law for the association of iodine atoms to molecular iodine,

$$2\text{I}(g) \rightarrow \text{I}_2(g)$$

is given by

$$\text{rate} = k[\text{I}]^2, \quad \text{where } k = 7.0 \times 10^9$$

    a.   What are the proper units for k?
    b.   If $[\text{I}]_o = 0.40$ M, calculate $[\text{I}]$ at $t = 2.5 \times 10^{-7}$ seconds.
    c.   If $[\text{I}]_o = 0.40$ M, calculate $t_{1/2}$.
    d.   If $[\text{I}]_o = 0.80$ M, how much time would it take for 75% of the iodine atoms to react?

Solution

a.   The units for a second-order rate constant are L/mol s.

b. $\dfrac{1}{[\text{I}]} = kt + \dfrac{1}{[\text{I}]_o}$

$$\frac{1}{[\text{I}]} = 7.0 \times 10^9 \text{ L/mol s } (2.5 \times 10^{-7}) + \frac{1}{0.40 \text{ mol/L}}$$

$$\frac{1}{[\text{I}]} = 1.75 \times 10^3$$

$$[\text{I}] = 5.7 \times 10^{-4} \text{ s}$$

c. $t_{1/2} = \dfrac{1}{k[I]_o} = \dfrac{1}{7.0 \times 10^9 \,(0.40)} = \mathbf{3.6 \times 10^{-10} \ s}$

Note that $t_{1/2}$ is so short that if we let the reaction proceed for as few as 2.5×10^{-7} seconds, most of the iodine atoms have already reacted.

d. $[I]_o = 0.80 \ M$. After 75% has reacted, 25% is left. Therefore, $[I] = [I]_o/4 = 0.20 \ M$.

$$\dfrac{1}{0.20 \ \text{mol/L}} = 7.0 \times 10^9 \ \text{L/mol s} \ (t) + \dfrac{1}{0.80 \ \text{mol/L}}$$

$$5 = 7.0 \times 10^9 \ (t) + 1.25$$

$$3.75 = 7.0 \times 10^9 \ (t)$$

$$t = \mathbf{5.4 \times 10^{-10} \ s}$$

You might think that an alternative method is to recognize that 75% reacted is 2 half-lives. If $[I]_o = 0.80 \ M$

$$t_{1/2} = \dfrac{1}{7.0 \times 10^9 \,(0.80)} = 1.8 \times 10^{-10} \ s$$

But this method is not valid here! The first half-lives would occur in 1.8×10^{-10} s. However, the **initial concentration** for the second half-life **is now 0.40 *M*** (½ of 0.80 *M*!). As a result, $t_{1/2}$ for the second half-life is **twice as long** (3.6×10^{-10} s). The sum of $t_{1/2}$ for both half-lives equals 1.8×10^{-10} s + 3.6×10^{-10} s = **5.4×10^{-10} s.** This agrees with our results, but is the difficult way of arriving at an answer.

The remainder of this section in your textbook discusses zero, multicomponent, and pseudo-first-order rate laws. The last one is especially important for you to read about because so many multicomponent kinetic analyses are done using the strategy where one component is in much lower concentration than all of the others.

This section ends with a presentation of kinetics from two different standpoints. The first regards experimentation and is dealt with in points 4 and 5 in the Rate Law Summary. In Table 12.6, a more mathematical approach is taken. Please pay special attention to the plot needed to give a straight line for each rate law.

12.5   *Reaction Mechanisms*

When you finish this section you will be able to evaluate the validity of simple reaction mechanisms.

Your textbook begins this chapter by defining several terms. You should know the definitions of:

- **reaction mechanisms** - the series of steps by which a chemical reaction occurs.
- **intermediate** - a species that is neither a reactant nor product. It is formed and consumed in the reaction sequence.
- **molecularity** - the number of species that collide to produce the reaction indicated by an elementary step. Uni- and bimolecular are two important molecularities. (Why isn't a termolecular molecularity likely?)
- **rate-determining step** - the slowest step. The reaction can only go as fast as the rate-determining step.

One idea in this section is that a **chemical equation** only tells us **what** reactants become what products. It does not tell us **how** these changes occur. This is the goal of reaction mechanism studies.

Another key idea is that the differential rate law is determined by the **rate-determining step.** This step may be in one direction or reversible (both directions). Your goal is to be able to determine whether mechanisms are consistent with experimentally determined rate laws.

Review Example 12.6 in your textbook, then try the following examples.

Example 12.5 A   Reaction Mechanisms

The balanced equation for the reaction of nitric oxide with hydrogen is

$$2NO + 2H_2 \rightarrow 2H_2O + N_2$$

The experimentally-determined rate law is

$$\text{rate} = k[NO]^2[H_2]$$

The following mechanism has been proposed.

$$NO + H_2 \xrightarrow{\ k_1\ } N + H_2O \qquad\qquad \text{(slow)}$$
$$N + NO \xrightarrow{\ k_2\ } N_2O \qquad\qquad \text{(fast)}$$
$$N_2O + H_2 \xrightarrow{\ k_3\ } N_2 + H_2O \qquad\qquad \text{(fast)}$$

Is this mechanism consistent with the observed rate law?

Solution

Your textbook says that there are two criteria that must be met if a mechanism is to be considered acceptable.

1. **The sum of the steps must give the balanced equation.** If you add up the three steps, you will find that this is in fact true here.

2. **The mechanism must agree with the observed rate law.** According to the proposed mechanism, the first step is rate-determining. This means that the overall mechanism must be given by the first step.

$$\text{rate} = k_1[NO][H_2]$$

This does not agree with the experimentally-determined rate law; therefore **our observed mechanism is incorrect.**

Example 12.5 B   More Reaction Mechanisms

Consider the same balanced equation as in the previous example. Is the following mechanism consistent with the experimentally determined rate law?

$$NO + H_2 \underset{k_{-1}}{\overset{k_1}{\rightleftharpoons}} N + H_2O \qquad\qquad \text{(fast, with equal rates)}$$
$$N + NO \xrightarrow{\ k_2\ } N_2O \qquad\qquad \text{(slow)}$$
$$N_2O + H_2 \xrightarrow{\ k_3\ } N_2 + H_2O \qquad\qquad \text{(fast)}$$

Solution

If you add up the individual steps, you will find that they equal the balanced chemical equation. So far, so good. We must now **evaluate the mechanism.** The rate-determining step is the second one,

$$N + NO \xrightarrow{\ k_2\ } N_2O$$

Based on this

$$\text{rate} = k_2[N][NO]$$

However, **N is an intermediate.** You can't control the initial amount of nitrogen you add because there is none present at the beginning. We always want to express rate laws in terms of reactants and/or products. The **first step** allows us to solve for N in terms of reactants and products.

$$\text{forward rate} = k_1[NO][H_2]$$
$$\text{reverse rate} = k_{-1}[N][H_2O]$$

The rates are equal. Therefore,

$$k_1[NO][H_2] = k_{-1}[N][H_2O]$$

Solving for [N],

$$[N] = \frac{k_1[NO][H_2]}{k_{-1}[H_2O]}$$

Substituting for [N] in our rate-determining equation,

$$\text{rate} = \frac{k_2 k_1[NO][H_2][NO]}{k_{-1}[H_2O]} = \frac{k'[NO]^2[H_2]}{[H_2O]}$$

This is close but **not quite there.** If you got this far, you have done excellent work. For an extra pat on the back, try the next challenging problem.

Example 12.5 C   Even More Reaction Mechanisms

Same problem as before, with the same data. Test this mechanism:

$$NO + H_2 \xrightarrow{k_1} N + H_2O \qquad \text{(fast)}$$
$$N + NO \xrightarrow{k_2} N_2O \qquad \text{(fast)}$$
$$N_2O + H_2 \xrightarrow{k_3} N_2 + H_2O \qquad \text{(slow)}$$

Solution

Again, our mechanisms add up to give the balanced chemical equation. The rate determining step is the third.

$$\text{rate} = k_3[N_2O][H_2]$$

However, **N₂O is an intermediate.** There are no reverse reactions here. We can't explicitly substitute for [N₂O] with rate equations. **However,** N_2O is formed **very quickly,** and irreversibly, from the first two reactions. Using **mole-ratios,**

- **1 mole of N₂O comes from the combination of 2 moles of NO** (2 nitrogens from 2 nitrogens).

Therefore, we can say

$$[N_2O] = 2[NO] \quad \text{(from stoichiometry)}$$

In the rate-determining step, we can then substitute

$$2NO + H_2 \rightarrow \text{products}$$

because [N₂O] is directly related to [NO], and the formation of N_2O is fast. Our rate equation becomes

$$\text{rate} = k_3[NO]^2[H_2]$$

And *this agrees with the observed rate law.*

(Reference for 12.5 A, B, and C is T.W. Lippencott, A.B. Garret and F.H. Verhock, Chemistry, J.W. Wiley, New York, (1977))

12.6   A Model for Chemical Kinetics

When you finish this section you will be able to:

- State the criteria for reaction according to the **collision model**.
- Solve problems relating to **activation energy**.

Your textbook points out two empirical facts about the rate of chemical reactions. All other things being equal,

1. The more **concentrated** your reactants, the **faster** the reaction.
2. The **higher the temperature,** the **faster** the reaction.

These trends can be explained using **collision theory.** The theory says that there are two steps involved in having a successful reaction.

1. Molecules must hit with sufficiently high energy.
2. Molecules must hit with the proper orientation (see Figure 12.12 in your textbook).

The combination of steps 2 and 3 (orientation and energy) makes it unlikely that reaction will occur even in the best of circumstances. However, the odds of collision are increased with higher concentrations. The odds of collision with a sufficiently high energy (kinetic energy is proportional to temperature) are increased with higher temperature.

The rate constant, k, is a measure of the **fraction** of collisions with **sufficient energy** to produce a reaction,

$$k = Ae^{-E_a/RT}$$

where k is the **rate constant.**
A is the **frequency factor** (related to the collision frequency and orientation of collisions).
E_a is the **activation energy** (in J/mol).
R is the **universal gas constant** (8.3145 J/K mol).
T is the **temperature** (in Kelvins).

Your textbook puts the equation in "$y = mx + b$" form by taking the ln of both sides,

$$\ln(k) = -\frac{E_a}{R}\left(\frac{1}{T}\right) + \ln(A)$$

$$\underset{y}{\uparrow} \qquad \underset{m}{\uparrow}\,\underset{x}{\uparrow} \qquad \underset{b}{\uparrow}$$

$\ln(k)$ can be plotted graphically vs. $1/T$ for a series of data to determine the slope and intercept, or for quick work, two sets of data can be used mathematically, yielding Equation 12.11 in your textbook.

$$\ln\left(\frac{k_2}{k_1}\right) = \frac{E_a}{R}\left(\frac{1}{T_1} - \frac{1}{T_2}\right)$$

Example 12.6 A   Activation Energy

Your textbook gives the activation energy for the reaction

$$H_2(g) + I_2(g) \rightarrow 2HI(g)$$

as being **167 kJ/mol**. It also gives the rate constant at 302°C as being **2.45×10^{-4} L/mol.** What is the rate constant for the reaction at 205°C?

Solution

You are given data at two temperatures. Your goal is to solve for k_2, the rate constant at the second temperature.

$k_1 = 2.45 \times 10^{-4}$ L/mol s $T_1 = 302 + 273.2 = 575.2$ K
$k_2 = ?$ $T_2 = 205 + 273.2 = 478.2$ K
$E_a = 167$ kJ/mol $= 167,000$ J/mol (the units (J) must match that in R)

$$\ln\left(\frac{k_2}{k_1}\right) = \frac{E_a}{R}\left(\frac{1}{T_1} - \frac{1}{T_2}\right)$$

$$\ln\left(\frac{k_2}{2.45 \times 10^{-4} \text{ L/mol s}}\right) = \frac{1.67 \times 10^5 \text{ J/mol}}{8.3145 \text{ J/K mol}}\left(\frac{1}{575.2 \text{ K}} - \frac{1}{478.2 \text{ K}}\right)$$

$$\ln\left(\frac{k_2}{2.45 \times 10^{-4} \text{ L/mol s}}\right) = -7.083$$

Taking the **inverse ln** of both sides,

$$\frac{k_2}{2.45 \times 10^{-4} \text{ L/mol s}} = e^{-7.083} = 8.394 \times 10^{-4}$$

$$k_2 = \mathbf{2.06 \times 10^{-7} \text{ L/mol s}}$$

Does the Answer Make Sense?

T_2 was lower than T_1. We would therefore expect k_2 to be lower than k_1. Also, for every 10°C, the rate should be cut in half. The temperature was reduced here by 100°C. The rate constant should be reduced substantially, as it was.

Example 12.6 B   Practice with Activation Energy

A second-order reaction has rate constants of **8.9×10^{-3} L/mol** and **7.1×10^{-2} L/mol** at 3°C and 35°C respectively. Calculate the value of the activation energy for the reaction.

Solution

$k_1 = 8.9 \times 10^{-3}$ L/mol s $T_1 = 276.2$ K
$k_2 = 7.1 \times 10^{-2}$ L/mol s $T_2 = 308.2$ K

Using Equation 12.11 from your textbook,

$$\ln\left(\frac{7.1 \times 10^{-2} \text{ L/mol s}}{8.9 \times 10^{-3} \text{ L/mol s}}\right) = \frac{E_a}{8.3145 \text{ J/K mol}}\left(\frac{1}{276.2 \text{ K}} - \frac{1}{308.2 \text{ K}}\right)$$

$$2.077 = (4.52 \times 10^{-5}) E_a$$

$$E_a = 4.6 \times 10^4 \text{ J/mol} = \mathbf{46 \text{ kJ/mol}}$$

12.7   *Catalysis*

The following questions will help you review the material in this section.

1. Define **catalysis.**
2. What are **enzymes?**
3. Why are catalysts more useful for increasing the reaction rate than raising the temperature?
4. How do catalysts work? How does this reflect the collision theory model?
5. What is a **heterogeneous catalyst?** Give some examples.
6. Differentiate between **absorption** and **adsorption.**
7. Describe the involvement of a heterogeneous catalyst in hydrogenation.
8. Discuss the role of catalysis in the conversion of sulfur dioxide to acid rain.
9. What is a **homogeneous catalyst?** Give an example.
10. How does nitric oxide act as a catalyst in the production of ozone?
11. How can chlorine catalyze the decomposition of ozone?

Exercises

Section 12.1

1. At 40°C $H_2O_2(aq)$ will decompose according to the following reaction:

    $$2H_2O_2(aq) \rightarrow 2H_2O(l) + O_2(g)$$

    The following data were collected for the concentration of H_2O_2 at various times.

times(s)	$[H_2O_2](M)$
0	1.000
2.16×10^4	0.500
4.32×10^4	0.250

    a. Calculate the average rate of decomposition of H_2O_2 between 0 and 2.16×10^4 s. Use this rate to calculate the rate of production of $O_2(g)$.
    b. What are these rates for the time period 2.16×10^4 s to 4.32×10^4 s?

2. Given the following hypothetical equation and data:

    $$fA \rightarrow gB + 3C$$

    $-\Delta[A]/t = 0.156$ mol/L s
    $+\Delta[B]/t = 0.026$ mol/L s
    $+\Delta[C]/t = 0.078$ mol/L s

    find the coefficients f and g.

Section 12.3

3. Use the given data for the hypothetical reaction:

    $$2A + B \rightarrow \text{products}$$

    to determine the rate law and to evaluate the rate constant at 30°C.

Reaction #	[A]	[B]	Initial rate (mol/L s)
1	0.1	0.1	3×10^{-2}
2	0.1	0.3	3×10^{-2}
3	0.2	0.3	6×10^{-2}

4. Derive the rate law expression and calculate the rate constant for the reaction:

    $$A + B + 3C \rightarrow \text{products}$$

    given the following data for 15°C.

Reaction #	[A]	[B]	[C]	Initial rate (mol/L s)
1	0.4	0.1	0.1	6.0×10^{-3}
2	0.4	0.2	0.1	6.0×10^{-3}
3	0.4	0.3	0.2	1.2×10^{-2}
4	1.2	0.4	0.2	0.11

5. Indicate the overall order of reaction for each of the following rate laws.

    a. $R = k[NO_2][F_2]$
    b. $R = k[I]^2[H_2]$
    c. $R = k[H_2][Cl_2]^{1/2}$

6. The following hypothetical reaction was performed:

$$A + 2B + C + 4D \rightarrow products$$

Determine the rate law, and calculate the rate constant for the following data collected at 20°C.

Reaction #	[A]	[B]	[C]	[D]	Initial rate (mol/L s)
1	0.25	0.30	0.60	0.15	7.20×10^{-5}
2	0.75	0.30	0.60	0.15	2.17×10^{-4}
3	0.25	0.30	0.20	0.15	7.20×10^{-5}
4	0.75	0.30	0.60	0.45	6.51×10^{-4}
5	0.75	0.44	0.60	0.15	4.67×10^{-4}

Section 12.4

7. The decomposition of $H_2O_2(aq)$ into $H_2O(l)$ and $O_2(g)$ is first order. From the data in Problem 1, determine the rate constant and the half-life.

8. Using the answer from the previous problem, calculate the concentration of H_2O_2 remaining after 8 hours if the initial concentration was 5.0 M.

9. Determine the rate constant (k) and $t_{1/2}$ for the data pertaining to the decomposition of phosphine (PH_3).

$$4PH_3(g) \rightarrow P(g) + 6H_2O(g)$$

Reaction #	[PH₃]	Rate (mol/L s)
1	0.18	2.4×10^{-3}
2	0.54	7.2×10^{-3}
3	1.08	1.4×10^{-2}

10. Using the data and answers from the previous problem, calculate the amount of PH_3 remaining after 2 minutes if the initial concentration of phosphine is 1.00 M. How much time would be needed for 90% of the phosphine to decompose?

11. Data for the decomposition of compound AB to give A and B is given below. Determine the rate expression, the rate constant, and $t_{1/2}$ for a 1 M solution.

Reaction #	[AB]	Rate (mol/L s)
1	0.2	3.2×10^{-3}
2	0.4	12.8×10^{-3}
3	0.6	28.8×10^{-3}

12. The rate law for the following second-order reaction at 10°C can be written as:

$$2NOBr(g) \rightarrow 2NO(g) + Br_2(g)$$

$$rate = k[NOBr]^2, \quad where \ k = 0.80 \ L/mol \ s$$

a. Determine $t_{1/2}$ when $[NOBr]_o = 0.650 \ M$.
b. Calculate $[NOBr]$ at $t = 5.80 \times 10^{-3}$ s if $[NOBr]_o = 0.650 \ M$.
c. If $[NOBr]_o = 1.00 \ M$, how long would it take for 50% NOBr to react?

13. Using the information in the previous problem, calculate $[NOBr]_o$ if it took 4.31 s for 75% to react.

14. The ketone acid, $(CH_2COOH)_2CO$, undergoes a first-order decomposition in aqueous solution to yield acetone and carbon dioxide:

$$(CH_2COOH)_2CO \ \rightarrow \ (CH_3)_2CO + 2CO_2$$

    a. Write the expression rate for the reaction.
    b. The rate constant, k, has been determined experimentally as 5.48×10^{-2}. Calculate $t_{1/2}$ at 60°C.
    c. The rate constant at 0°C has been determined as 2.46×10^{-5}. Calculate $t_{1/2}$ at this temperature.

15. The decomposition of NOCl is a second-order reaction with $k = 4.0 \times 10^{-8}$ L/mol s. Given an initial concentration of 0.50 M, what is the half-life? How much is left after 1×10^8 s? What is the half-life for an initial concentration of 0.25 M?

16. IBr(g) decomposes to form $I_2(g)$ and $Br_2(g)$. A plot of 1/[Br] v. time gave a straight line. Write the general rate law for the reaction.

17. The rate law for the reaction:

$$2NO(g) \ + \ Cl_2(g) \ \rightarrow \ 2NOCl(g)$$

is $R = k[NO]^2[Cl]$. If an experiment was performed in which the partial pressure of NO(g) initially was 0.1 atm and the initial partial pressure of $Cl_2(g)$ was 10 atm, what experimental data would give a straight line plot?

18. For the reaction A + B $\rightarrow$ C the following data were obtained:

- In experiment I, with 0.10 M as the initial concentrations for both A and B, the observed formation of C was 1.0×10^{-4} M/min.
- In experiment II, A and B were, respectively, 0.10 M and 0.30 M. The initial rate was 9.0×10^{-4} M/min.
- In experiment III, with the initial concentrations of both A and B at 0.30 M, the initial rate of formation was 2.7×10^{-3} M/min.

    a. Write the rate law for the reaction.
    b. Calculate the value of the specific rate constant.
    c. If the initial concentrations of both A and B were 0.40 M, what was the initial rate of formation of C?

Section 12.5

19. Write the rate laws for the following proposed mechanisms for the decomposition of IBr to I_2 and Br_2.

    a.
      $IBr(g) \ \rightarrow \ I(g) \ + \ Br(g)$ (fast)
      $IBr(g) \ + \ Br(g) \ \rightarrow \ I(g) \ + \ Br_2(g)$ (slow)
      $I(g) \ + \ I(g) \ \rightarrow \ I_2(g)$ (fast)

    b.
      $IBr(g) \ \rightarrow \ I(g) \ + \ Br(g)$ (slow)
      $I(g) \ + \ IBr(g) \ \rightarrow \ I_2(g) \ + \ Br(g)$ (fast)
      $Br(g) \ + \ Br(g) \ \rightarrow \ Br_2(g)$ (fast)

    c.
      $IBr(g) \ + \ IBr(g) \ \rightarrow \ I_2Br^+(g) \ + \ Br^-(g)$ (fast)
      $I_2Br^+(g) \ \rightarrow \ Br^+(g) \ + \ I_2(g)$ (slow)
      $Br^-(g) \ + \ Br^+(g) \ \rightarrow \ Br_2(g)$ (fast)

    d. $IBr(g) \ + \ IBr(g) \ \rightarrow \ I_2(g) \ + \ Br_2(g)$ (one step)

20. Given:

$$R = [H_2O_2][I^-][H_3O^+],$$

for the reaction

$$H_2O_2 + 3I^- + 2H^+ \rightarrow 2H_2O + I_3^-$$

Which of the following mechanisms are consistent with the observed rate law? If any are not, write a rate equation that is consistent with the mechanism.

a. $H_3O^+ + I^- \rightarrow HI + H_2O$ (fast)
    $H_2O_2 + HI \rightarrow H_2O + HOI$ (slow)
    $HOI + H_3O^+ + I^- \rightarrow 2H_2O + I_2$ (fast)
    $I^- + I_2 \rightarrow I_3^-$ (fast)

b. $H_3O^+ + H_2O_2 \rightarrow H_3O_2^+ + H_2O$ (fast)
    $H_3O_2^+ + I^- \rightarrow H_2O + HOI$ (slow)
    $HOI + H_3O^+ + I^- \rightarrow 2H_2O + I_2$ (fast)
    $I_2 + I^- \rightarrow I_3^-$ (fast)

c. $H_2O_2 + I^- \rightarrow H_2O + OH^-$ (slow)
    $H_3O^+ + I^- \rightarrow H_2O + HI$ (fast)
    $H_3O^+ + OI^- + HI \rightarrow 2H_2O + I_2$ (fast)
    $I_2 + I^- \rightarrow I_3$ (fast)

21. Write the rate law for the following predicted mechanism for the production of nitrogen dioxide (NO_2):

$$NO + O_2 \rightarrow NO_3 \quad \text{(fast)}$$
$$NO_3 + NO \rightarrow 2NO_2 \quad \text{(slow)}$$

22. Which of the following rate laws is consistent with the proposed mechanism for the reaction below?

$$3ClO^- \rightarrow ClO_3^- + 2Cl^-$$

$$ClO^- + ClO^- \rightarrow ClO_2^- + Cl^- \quad \text{(slow)}$$
$$ClO^- + ClO_2^- \rightarrow ClO_3^- + Cl^- \quad \text{(fast)}$$

a. $R = k[ClO^-][ClO_2^-]$
b. $R = k[ClO^-]^3$
c. $R = k[ClO^-]^2[ClO_2^-]$
d. $R = k[ClO^-]^2$

Section 12.6

23. The activation energy for the decomposition of $HI(g)$ to $H_2(g)$ and $I_2(g)$ is 186 kJ/mol. The rate constant at 555 K is 3.52×10^{-7} L/mol s. What is the rate constant at 645 K?

24. The rate constant for the decomposition of acetone to carbonic acid was determined to be 6.42×10^{-5} L/mol s at 10°C and 2.03×10^{-1} at 78°C. Calculate the activation energy.

25. The rate constant for the reaction:

$$C_4H_8 \rightarrow 2C_2H_4$$

at 325°C is 6.1×10^{-8} s^{-1}. At 525°C, the rate constant is 3.16×10^{-2} s^{-1}. Calculate the activation energy.

26. The rate of a reaction increases 2.21 times as the temperature changes from 70°C to 80°C. Calculate the activation energy.

27. The activation energy for the hypothetical reaction $2A + B \rightarrow C$ is 8.9×10^4 J/mol. If the original rate of the reaction is tripled at a temperature of 325 K, at what temperature did the reaction begin?

Multiple Choice Questions

28. The rate of decomposition of ammonia to hydrogen gas and nitrogen gas

$$2NH_3(g) \rightarrow N_2(g) + 3H_2(g)$$

is expressed as $-\Delta[NH_3]/\Delta t$. Express the rate of reaction in terms of $\Delta[H_2]/\Delta t$

A. Rate = $^2/_3\Delta[H_2]/\Delta t$ B. Rate = $\Delta[H_2]/\Delta t$ C. Rate = $3\Delta[H_2]/\Delta t$ D. Rate = $2\Delta[H_2]/\Delta t$

29. Given the following information, calculate the average rate, $\Delta[SO_2]/\Delta t$, between 10 and 40 minutes for the production of SO_2.

$$2SO_3(g) \rightarrow 2SO_2(g) + O_2(g)$$

t (min)	$[SO_2]$
0.0	0.0
10.0	0.032
20.0	0.056
30.0	0.074
40.0	0.087
50.0	0.096

A. 1.8×10^{-3} M/min   B. 1.5×10^{-3} M/min   C. 3.0×10^{-3} M/min   D. 3.2×10^{-3} M/min

30. Given the following information, calculate the average rate, $\Delta[O_2]/\Delta t$, between 20 and 40 minutes for the production of O_2.

$$2SO_3(g) \rightarrow 2SO_2(g) + O_2(g)$$

t (min)	$[O_2]$
0.0	0.0
10.0	0.01
20.0	0.028
30.0	0.037
40.0	0.044
50.0	0.048

A. 8.0×10^{-3} M/min   B. 1.6×10^{-3} M/min   C. 6.7×10^{-4} M/min   D. 8.0×10^{-4} M/min

31. Given the following information, calculate the average rate, $-\Delta[NH_3]/\Delta t$, between 10 and 30 minutes for the production of NH_3.

$$2NH_3(g) \rightarrow N_2(g) + 3H_2(g)$$

t (min)	$[NH_3]$
0.0	0.130
10.0	0.083
20.0	0.063
30.0	0.045
40.0	0.033
50.0	0.025

A. 8.0×10^{-3} M/min   B. 1.9×10^{-3} M/min   C. 1.5×10^{-2} M/min   D. 2.3×10^{-3} M/min

32. Based on the following equation, which one of the following compounds would you expect to undergo the most change in concentration in a certain amount of time?

$$2NH_3\,(g) \rightarrow N_2(g) + 3H_2(g)$$

A. Nitrogen   B. Ammonia   C. Hydrogen   D. None of the molecules

33. For the following reaction, under circumstances where the reverse reaction can be neglected, the reaction rate will depend on which of the options below?

$$AlCl_3\,(g) + PH_3(g) \rightarrow Cl_3AlPH_3(g)$$

A. $AlCl_3$ B. $AlCl_3$ and PH_3 C. PH_3 D. Cl_3AlPH_3

34. For the following rate law:

$$R = k[X]^3$$

What is the order on X?

A. 2   B. 4   C. 9   D. 3

35. Based on the following data, determine the rate law of this reaction:

$$SO_2\,(g) + Cl_2(g) \rightarrow SOCl_2 + Cl_2O(g)$$

Experiment	[SO₂] (M)	[Cl₂] (M)	Initial rate (M/s)
1	0.400	0.400	0.2918
2	0.400	0.200	0.0730
3	0.400	0.800	1.1674
4	0.200	0.800	0.5837

A. $R = k[SO_2]$ B. $R = k[SO_2][Cl_2]$ C. $R = k[Cl_2]$ D. $R = k[SO_2][Cl_2]^2$

36. Based on the following data, determine the rate constant for the reaction:

$$SO_2\,(g) + Cl_2(g) \rightarrow SOCl_2 + Cl_2O(g)$$

Experiment	[SO₂] (M)	[Cl₂] (M)	Initial rate (M/s)
1	0.400	0.400	0.2918
2	0.400	0.200	0.0730
3	0.400	0.800	1.1674
4	0.200	0.800	0.5837

A. 1.82   B. 9.12   C. 4.56   D. 0.0351

37. Based on the following data, determine the overall order of this reaction:

$$2PO\,(g) + Cl_2(g) \rightarrow 2POCl(g)$$

Experiment	[PO] (M)	[Cl₂] (M)	Initial rate (M/s)
1	0.20	0.20	0.40
2	0.20	0.40	0.80
3	0.60	0.20	3.2
4	0.60	0.60	9.6

A. 3   B. 2   C. 5/2   D. 4

38. Fill in the missing data item in this table:

Experiment	[AB] (M)	[CV] (M)	Initial rate (M/hr)
1	1.00	0.500	Z
2	1.00	0.750	?
3	2.00	0.750	3.00Z
4	1.00	0.250	0.500Z

   A. 1.00Z B. 1.50Z C. 2.00Z D. 1.75Z

39. In a zero-order rate expression, what units must the specific rate constant possess?

   A. t/M B. $1/t$ C. M/t D. $1/M \times t$

40. The rate of decomposition of a substance is first-order. If $k = 2.46 \times 10^{-3}$ s^{-1}, what concentration of this substance remains after 2 minutes, knowing that [Substance]$_o$ = 0.550 M?

   A. 0.409 M B. 0.547 M C. 0.553 M D. 0.739 M

41. A particular drug can be sold until 20% of the original drug has undergone change. Knowing that $k = 1.25 \times 10^{-2}$/day and the change is first-order, how long, in days, will it take before the drug can no longer be sold?

   A. 17 days   B. 18 days   C. 18 hours   D. 35 hours

42. A gas phase reaction in which substance A reacts with substance B to produce AB, is found to be second-order on A. Knowing that $k = 0.0368$ $M^{-1} \times$ hr^{-1}, and that [A]$_o$ = 2.25 M, what percent of A remains after 177 hours of reaction?

   A. 3.31%   B. 85.6%   C. 14.4%   D. 6.39%

43. The decomposition of N_2O gas obeys zero-order kinetics. Given a rate constant = 2.46×10^{-3} M/s and [N_2O]$_{120}$ = 0.155 M, calculate [N_2O]$_o$.

   A. 0.445 M B. 0.450 M C. 0.550 M D. 0.225 M

44. The decomposition of N_2O gas obeys zero-order kinetics. Given a rate constant = 2.46×10^{-3} M/s and [N_2O]$_o$ = 0.450 M, calculate [N_2O] at the end of 0.0445 hours.

   A. 0.0558 M B. 0.221 M C. 0.225 M D. 0.450 M

45. To determine whether data from different experiments correspond to a zero-order rate expression, a plot of what variables will yield a straight line?

   A. [X] vs. $1/t$ B. [X] vs. t C. $1/$[X] vs. t D. [X]^3 vs $1/t$

46. A specific reaction is known to have a first-order rate expression. If $k = 1.52 \times 10^{-2}$/min, what is the half life, in minutes, of this reaction?

   A. 91.2   B. 66.7   C. 33.3   D. 45.6

47. The following is a proposed mechanism for this reaction:

$$H_2(g) + Br_2(g) \rightarrow 2HBr(g)$$

   a. Br_2 + light $\rightarrow$ 2Br
   b. 2Br + 2H_2 $\rightarrow$ 2HBr + 2H
   c. H + Br_2 $\rightarrow$ HBr + Br
   d. H + HBr $\rightarrow$ H_2 + Br
   e. Br + Br $\rightarrow$ Br_2

In this proposed mechanism, which steps are bimolecular?
   A. a   B. b, c, d, e   C. All   D. b, c, d

48. The following is a proposed mechanism for this reaction:

$$H_2(g) + Br_2(g) \rightarrow 2HBr(g)$$

   a. $Br_2 + light \rightarrow 2Br$
   b. $2Br + 2H_2 \rightarrow 2HBr + 2H$
   c. $H + Br_2 \rightarrow HBr + Br$
   d. $H + HBr \rightarrow H_2 + Br$
   e. $Br + Br \rightarrow Br_2$

   In this proposed mechanism, which step consists only of intermediates as reactants?

   A. e   B. a   C. c, d   D. b

49. For a reaction to take place, the molecules that are reacting:

   A. Must have more energy than the products
   B. Must have less energy than the products
   C. Must be able to reach the activation energy
   D. Must be in considerable numbers

50. Which of the following is not a factor determining the energy of activation according to the Arrhenius equation?

   A. Orientation of molecules   C. Frequency factor
   B. Temperature   D. None of these choices

51. Calculate E_a when $k_1 = 2.00$ and $k_2 = 10.0$, $T_1 = 318$ K and $T_2 = 371$ K.

   A. 9.0 J   B. 30.0 kJ   C. 85.0 kJ   D. 3.1 J

52. Calculate T_2 when $E_a = 30.0$ kJ, $T_1 = 285$ K, $k_1 = 3.00$, and $k_2 = 15.0$.

   A. 327 K   B. 253 K   C. 158 K   D. 2.53 K

53. In a "reaction progress" graph, reacting molecules are most unstable at:

   A. Their initial position   C. Right after they collide
   B. When they are about to collide   D. At the transition state

54. A catalyst:

   A. Is consumed during a reaction while effectively increasing the number of reacting molecules that can reach the energy of activation
   B. Changes an endothermic reaction into an exothermic reaction
   C. Increases the energy of the products
   D. Provides an alternate pathway to the reaction, effectively lowering E_a

55. In which of the following examples is a heterogeneous catalyst NOT used?

   A. Hydrogenation of fats
   B. Oxidation of sulfur dioxide
   C. Decomposition of ozone
   D. Catalytic converters of automobile exhaust systems

Answers to Exercises

1. a. Average rate of decomposition of $H_2O_2 = 2.31 \times 10^{-5}$ mol/L s
        Rate of production of $O_2 = 1.16 \times 10^{-5}$ mol/L s
    b. Average rate of decomposition of $H_2O_2 = 1.16 \times 10^{-5}$ mol/L s
        Rate of production of $O_2 = 5.79 \times 10^{-6}$ mol/L s

2. $f = 6$; $g = 1$

3. $R = k[A]$; $k = 0.3$ s^{-1}

4. $R = k[A]^2[C]$; $k = 0.38$ L^2/mol^2 s

5. a. 2
    b. 3
    c. $1^1/_2$

6. $R = k[A][B]^2[D]$; $k = 2.14 \times 10^{-2}$ L^3/mol^3 s

7. $k = 3.21 \times 10^{-5}$s^{-1}; $t_{1/2} = 2.16 \times 10^4$ s

8. 2.0 *M*

9. $k = 1.3 \times 10^{-2}$ s^{-1}; $t_{1/2} = 53$ s

10. $[PH_3] = 0.21M$; 177 s = 2 min. 57 sec.

11. $R = k[AB]^2$; $k = 0.080$ L/mol s; $t_{1/2} = 12.5$ s

12. a. $t_{1/2} = 1.92$ s   b. $[NOBr] = 0.648$ *M*   c. $t = 1.25$ s

13. $[NOBr]_o = 0.87$ *M*

14. a. $R = -k[(CH_2COOH)_2CO]$
    b. $t_{1/2} = 12.64$ seconds
    c. $t_{1/2} = 2.81 \times 10^4$ seconds

15. $t_{1/2}(0.50\ M) = 5.0 \times 10^7$ s; $[NOCl]_{10^8\ s} = 0.17$ *M*; $t_{1/2}(0.25\ M) = 1 \times 10^8$ s

16. $R = k[IBr]^2$

17. 1/[NO] vs. *t* (pseudo second-order)

18. a. $R = k[A][B]^2$
    b. $k = 0.1$ L^2/mol^{2}s
    c. $R = 0.064$ mol/Ls

19. a. $R = k[IBr]^2$ c. $R = k[IBr]^2$
    b. $R = k[IBr]$ d. $R = k[IBr]^2$

20. a and b; for c the rate law would be $R = k[H_2O_2][I^-]$

21. $R = k[NO]^2[O_2]$

22. d

23. 9.76×10^{-5} L/mol s

24. 98.0 kJ/mol

25. 2.61×10^5 J/mol = 261 kJ/mol

26. 79.9 kJ/mol

27. 315 K

28. A	29. A	30. D	31. B	32. C	33. B
34. D	35. D	36. C	37. D	38. B	39. C
40. A	41. B	42. D	43. B	44. A	45. B
46. D	47. B	48. A	49. C	50. D	51. B
52. A	53. D	54. D	55. C		

Chapter 13

Chemical Equilibrium

Your textbook introduces the concept of **equilibrium** by noting that **no reaction goes fully to completion**. Some reverse reaction, however small, always exists. The direction of equilibrium may lie "**far to the right**" (reaction almost complete), "**far to the left**' (negligible reaction), or somewhere in between. Whatever the extent of reaction, the key point is that equilibrium is reached when **the rates of the forward and reverse reactions are equal**.

13.1   The Equilibrium Condition

When you finish this section you will be able to list some characteristics of reactions at equilibrium.

Try the following questions to test your general understanding of equilibrium.

Example 13.1   Background Questions

1. What is **dynamic equilibrium**?
2. What is true about the initial rate of forward and reverse reactions in a system where only reactants are present?
3. What is true about the rates of forward and reverse reactions **at equilibrium?**
4. Why does equilibrium occur?
5. What are some of the factors that determine the equilibrium position of a reaction?

Solution

1. Dynamic equilibrium is a process in which the rates of the forward and reverse reactions are equal. As a result, the overall reaction appears static, but is not.
2. Initially, there is only forward reaction. After some product builds up, the reverse reaction begins.
3. The rates will be **equal** at equilibrium.

4. Equilibrium occurs because products, once they are produced, can combine to form reactants at a rate that will eventually **equal** that of product formation by reactants.
5. Among the factors listed in your textbook are initial concentrations, relative energies of the reactants and products, and the energy and organization of the products.

13.2 The Equilibrium Constant

When you finish this section you will be able to write a mass action expression for a given balanced chemical equation.

Section 13.2 opens with the statement of the **Law of Mass Action**. For a reaction

$$jA + kB \rightleftharpoons lC + mD$$

the equilibrium constant K is given by

$$K = \frac{[C]^l [D]^m}{[A]^j [B]^k}$$

One way to determine the significance of K is to assume that the forward reaction involves jA and kB in its rate-determining step, and the reverse reaction involves lC and mD in its rate-determining step.

$$\text{rate}_{forward} = k_f [A]^j [B]^k$$
$$\text{rate}_{reverse} = k_r [C]^l [D]^m$$

At equilibrium, $\text{rate}_{forward} = \text{rate}_{reverse}$, or $k_f [A]^j [B]^k = k_r [C]^l [D]^m$. Rearranging,

$$\frac{k_f}{k_r} = K = \frac{[C]^l [D]^m}{[A]^j [B]^k}$$

K is thus the **ratio** of the forward to reverse **rate constants** (not the rate)! As with the kinetic rate-determining step, reactant or product **coefficients become exponents** when put in the mass action expression, or "equilibrium expression."

Example 13.2 A   Equilibrium Expressions

Write the equilibrium expression for each of the following reactions:

a. $PCl_5(g) \rightleftharpoons PCl_3(g) + Cl_2(g)$
b. $S_8(g) \rightleftharpoons 8S(g)$
c. $Cl_2O_7(g) + 8H_2(g) \rightleftharpoons 2HCl(g) + 7H_2O(g)$

Solution

a. $K = \dfrac{[PCl_3][Cl_2]}{[PCl_5]}$ (products ↓, reactant ↑)

b. $K = \dfrac{[S]^8}{[S_8]}$ (coefficient of S ↓)

c. $K = \dfrac{[H_2O]^7[HCl]^2}{[Cl_2O_7][H_2]^8}$

where the first $[H_2O]^7$ has coefficient of H_2O and $[HCl]^2$ has coefficient of HCl, and $[H_2]^8$ has coefficient of H_2.

There are several types of equilibrium problems that you will need to solve in this chapter:

- Solving for K (knowing equilibrium concentrations).
- Solving for equilibrium concentrations (knowing K).
- Solving for equilibrium concentrations (knowing K and initial concentrations).

We will work with the first two types of problems below.

Example 13.2 B   Calculating an Equilibrium Constant

Calculate the equilibrium constant, K, for the following reaction at 25°C,

$$H_2(g) + I_2(g) \rightleftharpoons 2HI(g)$$

if the equilibrium concentrations are $[H_2] = 0.106\ M$, $[I_2] = 0.022\ M$, and $[HI] = 1.29\ M$.

Solution

The mass action (or equilibrium) expression is

$$K = \frac{[HI]^2}{[H_2][I_2]}$$

We are given the **equilibrium concentrations.** We can substitute to get

$$K = \frac{(1.29)^2}{(0.106)(0.022)} = \mathbf{7.1 \times 10^2} \qquad \text{(dimensionless - the units cancel)}$$

Example 13.2 C   Practice with Equilibrium Expressions

Using the same reaction as in our previous problem (with $K = 7.1 \times 10^2$ at 25°C), if the **equilibrium concentrations** of H_2 and I_2 are 0.81 M and 0.035 M respectively, calculate the equilibrium concentration of HI.

Solution

As before, the equilibrium expression is valid.

$$K = \frac{[HI]^2}{[H_2][I_2]}$$

Substituting,

$$7.1 \times 10^2 = \frac{[HI]^2}{(0.81)(0.035)}$$

$$[HI]^2 = 20.13, \quad \mathbf{[HI] = 4.49\ M = 4.5\ M} \text{ (2 sig figs)}$$

Checking Your Answer

In the majority of equilibrium problems, you will be *given K* and asked to find the concentration of a substance. After finding the solution, the best way of checking your answer is to *substitute concentration values into the equilibrium expression to make sure you get the known value of K*. If you do (with a reasonable round-off error) your solution is correct. If you don't, there is probably an error. Checking by solving for *K*,

$$K = \frac{(4.49)^2}{(0.81)(0.035)} = 7.1 \times 10^2$$

Our [HI] is therefore correct.

Example 13.2 D   Changing K When Changing the Reaction Coefficients or Direction

Your textbook discusses certain conclusions regarding the value of *K* when you modify a balanced chemical equation. Reread that discussion. Given our balanced chemical equation and the value for *K* from the past two problems, calculate the value of *K* for the following reactions.

   a. $2HI \rightleftharpoons H_2 + I_2$
   b. $\frac{1}{3}H_2 + \frac{1}{3}I_2 \rightleftharpoons \frac{2}{3}HI$
   c. $8HI \rightleftharpoons 4H_2 + 4I_2$

Solution

a.   The equation is the **reverse** of the original.

$$K' = 1/K = 1/710 = \mathbf{1.4 \times 10^{-3}}$$

b.   The equation is **1/3** of the original.

$$K' = K^{1/3} = (710)^{1/3} = \mathbf{8.9}$$

c.   The equation is **reversed** and is **four times** the original.

$$K' = 1/K^4 = K^{-4} = (710)^{-4} = \mathbf{3.9 \times 10^{-12}}$$

13.3   Equilibrium Expressions Involving Pressures

When you finish this section you will be able to solve for *K* or K_p, given appropriate data.

Your textbook points out that equilibria for gases can be expressed in either **concentration** or in **pressure terms**. We use K for the equilibrium constant gotten by using **concentrations**, and K_p for that using **pressures**.

K and K_p can be related. Before we explicitly show this relationship, let's solve a problem that uses pressures of gases rather than concentrations.

Example 13.3 A   Calculating K_p

Calculate K_p for the following reaction

$$CH_3OH(g) \rightleftharpoons CO(g) + 2H_2(g)$$

given the equilibrium pressures as follows:

$$P_{CH_3OH} = 6.10 \times 10^{-4} \text{ atm} \qquad P_{CO} = 0.387 \text{ atm} \qquad P_{H_2} = 1.34 \text{ atm}$$

Solution

The **form** of the equilibrium expression is the **same** whether it involves concentration **or** pressure data. That is,

$$K = \frac{\text{products}}{\text{reactants}}$$

For this example,

$$K_p = \frac{P_{H_2}^{\ 2}\, P_{CO}}{P_{CH_3OH}} = \frac{(1.34 \text{ atm})^2\, (0.387 \text{ atm})}{(6.10 \times 10^{-4} \text{ atm})} = 1.14 \times 10^3 \text{ atm}^2$$

Example 13.3 B   Relating K to K$_p$

Derive an expression that relates K to K_p for the reaction given in the previous problem. Then **calculate** K for the reaction at 25°C using the value for K_p.

Solution

According to the ideal gas equation, $PV = nRT$. Concentration units can be expressed as moles/liter, or n/V which is the same as C, the molar concentration of the gas. Relating this to the ideal gas equation,

$$\frac{n}{V} = \frac{P}{RT} = C$$

$$K = \frac{[H_2]^2 [CO]}{[CH_3OH]} = \frac{C_{H_2}^{\ 2} C_{CO}}{C_{CH_3OH}}$$

Substituting P/RT for C (or n/V),

$$K = \frac{\left(\dfrac{P_{H_2}}{RT}\right)^2 \left(\dfrac{P_{CO}}{RT}\right)}{\left(\dfrac{P_{CH_3OH}}{RT}\right)}$$

Factoring out P's,

$$K = \frac{P_{H_2}^{\ 2} P_{CO}}{P_{CH_3OH}} \left(\frac{(RT)^{-2}\ (RT)^{-1}}{(RT)^{-1}}\right)$$

$$K_p = \frac{P_{H_2}^{\ 2} P_{CO}}{P_{CH_3OH}}, \text{ therefore } K = K_p(RT)^{-2}$$

Note how this agrees with the formula given in your textbook.

$$K_p = K(RT)^{\Delta n} \quad \text{or} \quad K = K_p(RT)^{-\Delta n}$$

Using $R = 0.08206$ L atm K^{-1} mol^{-1}
$T = 298.2$ K,

$$K = 1.14 \times 10^3 \text{ atm}^2\ [(0.08206 \text{ L atm K}^{-1} \text{ mol}^{-1})\ (298.2 \text{ K})]^{-2}$$
$$= 1.14 \times 10^3 \text{ atm}^2\ (1.67 \times 10^{-3} \text{ L}^{-2} \text{ mol}^2 \text{ atm}^{-2})$$
$$K = 1.90 \text{ mol}^2 \text{ L}^{-2}$$

13.4 Heterogeneous Equilibria

When you finish this section you will be able to write equilibrium expressions for reactions involving pure solids and liquids.

Equilibrium expressions involve concentrations (or pressures) of substances that **change** from initial to equilibrium conditions. A pure substance such as water changes **amount**, but not **concentration.**

The concentrations of pure solids and liquids remain constant. That means that pure solids and liquids **can be incorporated into the equilibrium constant.** For example, the equilibrium expression for

$$CaCO_3(s) \rightleftharpoons CaO(s) + CO_2(g)$$

$$\textbf{IS NOT } K = \frac{[CaO][CO_2]}{[CaCO_3]}$$

Rather, $[CaCO_3]$ and $[CaO]$ are constant. The **amounts** of each will change, but their **concentrations** (a <u>derivative</u> term) are constant. Thus we can incorporate them into our equilibrium expression.

$$K = [CO_2]$$

Example 13.4   Equilibrium Expressions Involving Pure Solids and Liquids

Please write equilibrium expressions for each of the following reactions:

a. $Ba(OH)_2(s) \rightleftharpoons Ba^{+2}(aq) + 2OH^-(aq)$
b. $NH_4NO_2(s) \rightleftharpoons N_2(g) + 2H_2O(g)$
c. $HCl(g) + NH_3(g) \rightleftharpoons NH_4Cl(s)$
d. $Zn(OH)_2(s) + 2OH^-(aq) \rightleftharpoons Zn(OH)_4^{-2}(aq)$
e. $CH_3CO_2H(aq) \rightleftharpoons H^+(aq) + CH_3CO_2^-(aq)$
f. $Al(NO_3)_3(s) + 6H_2O(l) \rightleftharpoons [Al(H_2O)_6]^{3+}(aq) + 3NO_3^-(aq)$

Solution

Remember that concentrations of pure solids and liquids are **not** put into the equilibrium expression. They are incorporated into the equilibrium **constant**.

a. $K = [Ba^{2+}][OH^-]^2$

b. $K = [N_2][H_2O]^2$, or $K_p = P_{N_2} P_{H_2O}^2$

c. $K = \dfrac{1}{[HCl][NH_3]}$ or $K_p = \dfrac{1}{P_{HCl} P_{NH_3}}$

d. $K = \dfrac{[Zn(OH)_4^{2-}]}{[OH^-]^2}$

e. $K = \dfrac{[H^+][CH_3CO_2^-]}{[CH_3CO_2H]}$

f. $K = [NO_3^-]^3[Al(H_2O)_6^{3+}]$

13.5 Applications of the Equilibrium Constant

When you finish this section you will be able to:

- Use the **reaction quotient (Q)** to predict the direction of chemical reactions toward equilibrium.
- Use the reaction quotient to aid in solving simple equilibrium problems.

Your textbook begins this section by reviewing those things that K can and cannot indicate:

- K **does not** reflect **how fast** a reaction goes.
- A **large** value of K means that mostly **products** will be present at equilibrium.
- A **small** value of K means that mostly **reactants** will be present at equilibrium.

The **theme** of this section is the use of the **reaction quotient, Q,** which predicts the direction that the reaction will go to reach equilibrium. You calculate Q by using the law of mass action on the **initial, not equilibrium,** concentrations (or pressures) of the reaction substances.

- If Q **is equal to K,** the system is **at equilibrium.**
- If Q **is greater than K,** mathematically, there is too much product present. The system will **shift to the left** to reach equilibrium.
- If Q **is less than K,** there is too much reactant present. The system will **shift to the right** to reach equilibrium.

Example 13.5 A   Predicting the Direction of Equilibrium

Let's reexamine the reaction between hydrogen gas and iodine gas that we used in examples 13.2 B and C.

$$H_2(g) + I_2(g) \rightleftharpoons 2HI(g)$$
$$K = 7.1 \times 10^2 \text{ at } 25°C$$

Predict the direction that the system will shift in order to reach equilibrium given each of the following initial conditions:

a. $Q = 427$
b. $Q = 1522$
c. $[H_2]_o = 0.81\ M$ $[I_2]_o = 0.44\ M$ $[HI]_o = 0.58\ M$
d. $[H_2]_o = 0.078\ M$ $[I_2]_o = 0.033\ M$ $[HI]_o = 1.35\ M$
e. $[H_2]_o = 0.034\ M$ $[I_2]_o = 0.035\ M$ $[HI]_o = 1.50\ M$

Solution

a. $Q < K$, so the reaction will form more products and **shift to the right.**
b. $Q > K$, so the reaction will form more reactants and **shift to the left.**

c. $Q = \dfrac{[HI]_o^2}{[H_2]_o[I_2]_o} = \dfrac{(0.58)^2}{(0.81)(0.44)} = 0.94$

$Q < K$, so the reaction will **shift to the right.**

d. $Q = \dfrac{(1.35)^2}{(0.078)(0.033)} = 7.1 \times 10^2$

$Q = K$, so the system is **at equilibrium.**

e. $Q = \dfrac{(1.50)^2}{(0.034)(0.035)} = 1.9 \times 10^3$

$Q > K$, so the reaction will **shift to the left.**

Earlier in our discussion (see Section 13.2) we mentioned three types of equilibrium problems. The third, and by far most interesting, is the problem where we are *given initial concentrations and K* and are asked to *find equilibrium concentrations*. In the remainder of this section, we will introduce this type of problem. We will examine it in some detail in the next section.

Example 13.5 B   Equilibrium Concentrations - Large Value For K

Let's continue to use our data from our hydrogen and iodine reaction ($K = 7.1 \times 10^2$ at 25°C). Calculate the **equilibrium concentrations** if a 5.00-L vessel initially contains 15.7 g of H_2 and 294 g of I_2.

Solution

The key is to determine **the direction of the reaction**. Thus we must **compare Q to K.** (Note that our data must be converted from g/L to mol/L.)

$$[H_2]_0 = \frac{15.7 \text{ g}}{5.00 \text{ L}} \times \frac{1 \text{ mol}}{2.016 \text{ g}} = 1.56 \ M$$

$$[I_2]_0 = \frac{294 \text{ g}}{5.00 \text{ L}} \times \frac{1 \text{ mol}}{253.8 \text{ g}} = 0.232 \ M$$

$$[HI]_0 = 0.00 \ M$$

$$Q = \frac{[HI]_0^{\ 2}}{[H_2]_0[I_2]_0} = \frac{0^2}{(1.56)(0.232)} = 0$$

$Q < K$, so the reaction will **go to the right.** (which makes sense, as there is no product!)

Our next step is critical - setting up the **table of initial and final conditions**. In order to do this well, we must **take K into account.** You **cannot** memorize the equilibrium table set-up. You must **evaluate** each problem separately.

The value of K is 710. We can **assume** that the reaction will go **essentially to completion**, although there will be some small amount that will be unreacted. If this is true, the **stoichiometry** of the reaction dictates that I_2 **is the limiting reactant** and that H_2 will be in excess.

	H_2	+	I_2	$\rightleftharpoons$	2HI
initial	1.56 *M*		0.232 *M*		0 *M*
final	1.328 *M*		0 *M*		0.464 *M*

However, the reaction does not go all the way. There will be some I_2 left over. Call the amount "+X." An identical amount of H_2 will be left over (+X). Twice the amount of HI will **not be formed** (2 moles of HI to 1 mole of I_2 or H_2), so the amount that will not be formed is "−2X." The "−" indicates that some small amount will remain on the reactant side. Summarizing in table form,

	initial	change	final
H_2	1.56 *M*	−0.232 + *X*	1.328 + *X*
I_2	0.232 *M*	−0.232 + *X*	+ *X*
HI	0 *M*	+0.464 − 2*X*	0.464 − 2*X*

Assumption

Since the reaction is fairly complete, neglect *X* **relative to** 1.328 and 0.464. If the assumption is valid, *X* will be <5% of 1.328 and 0.464. Therefore, $[H_2] \approx 1.328 \ M$, $[I_2] = X$, and $[HI] \approx 0.464 \ M$. Solving,

$$K = \frac{[HI]^2}{[H_2][I_2]} \quad \Rightarrow \quad 710 = \frac{(0.464)^2}{(1.328)(X)}$$

$$X = [I_2] = 2.3 \times 10^{-4} M$$

$$[H_2] = 1.33 M, \quad [I_2] = 2.3 \times 10^{-4} M, \quad [HI] = 0.464 M$$

Testing our Assumption

Is 2.28×10^{-4} less than 5% of 1.328? Yes. Is it less than 5% of 0.464? Yes. If it were not, we would solve the equation by the quadratic formula (see the next section).

Double Check

Solve for K.

$$K = \frac{(0.464)^2}{(1.328)(2.28 \times 10^{-4})} = 7.1 \times 10^2$$

Remember that the key to solving equilibrium problems lies with **making and testing valid assumptions.**

13.6   Solving Equilibrium Problems

When you finish this section you will be able to solve many equilibrium problems and test any assumptions you make.

At the beginning of this section, your textbook presents a **"Procedure for Solving Equilibrium Problems."** When following that procedure you must always think about what assumptions can make your problem solving easier and **test the validity** of your assumptions.

We have already dealt with **large values of K.** Now let's do an example which involves a **small value of K.** (What assumptions might you make?)

Example 13.6 A   Equilibrium Calculations - Small Value For K

The reaction between nitrogen and oxygen to form nitric oxide has a value for the equilibrium constant at 2000 K of $K = 4.1 \times 10^{-4}$. If 0.50 moles of N_2 and 0.86 mole of O_2 are put into a 2.0-L container at 2000 K, what would be the equilibrium concentrations of all species?

$$N_2(g) + O_2(g) \rightleftharpoons 2NO(g)$$

Solution

The **most important question** you must raise in doing equilibrium problems is *"What does the value of K tell me about the extent of the reaction?"*

With **very small values of K** such as this one, the reaction will **stay far to the left.** Only a small amount of N_2 and O_2 (call the amount "X") will react. The small amount of NO that will be formed will be **2X,** because of the 2:1 stoichiometry between NO and N_2 or O_2. We may now set up our equilibrium table.

$$[N_2]_0 = \frac{0.50 \text{ moles}}{2.0 \text{ L}} = 0.25 M \qquad\qquad [O_2]_0 = \frac{0.86 \text{ moles}}{2.0 \text{ L}} = 0.43 M$$

	initial (M)	change	final (M)
N_2	0.25	$-X$	$0.25 - X$
O_2	0.43	$-X$	$0.43 - X$
NO	0	$+2X$	$2X$

Because K is small, **let us assume that X is negligible** compared to 0.25 and 0.43. That is, $0.25 - X \approx$ 0.25 and $0.43 - X \approx 0.43$. We can now solve for X (and test the validity of our assumption).

$$K = \frac{[NO]^2}{[N_2][O_2]}$$

$$4.1 \times 10^{-4} = \frac{(2X)^2}{(0.25)(0.43)}$$

$$4X^2 = 4.41 \times 10^{-5}$$

$$X = 3.3 \times 10^{-3} \, M$$

$$[NO] = 2X = 6.6 \times 10^{-3} \, M, \qquad [N_2] = 0.25 \, M, \qquad [O_2] = 0.43 \, M$$

We are not finished yet. We must **test our assumption** and **check the answer.** Is X less than 5% of 0.25 M? Yes (it is about 1.3%). Is X less than 5% of 0.43 M? Yes (it is about 0.8%). Our assumption was O.K.

We can check our math by **solving for K** using our equilibrium concentrations,

$$K = \frac{(6.6 \times 10^{-3})^2}{(0.25)(0.43)} = 4.0_5 \times 10^{-4}$$

Acceptably close, considering round-off.

We have solved problems with very large K's and very small K's. The majority of our problems will have these kinds of equilibrium constants. Occasionally you will have to solve a problem where the **value for K is intermediate** (roughly between 0.01 and 100). But remember, there are **no absolutes** in solving these kinds of problems. Assumptions that you make must **always be tested**.

In problems with intermediate K's, you **cannot (in general) make assumptions** based on the extent of reaction. It is too uncertain. The trade-off is that you must explicitly solve for X, often using the **quadratic formula**.

Example 13.6 B   Equilibrium Calculations - Intermediate Value For K

Sulfurous acid dissociates in water as follows:

$$H_2SO_3(aq) \rightleftharpoons H^+(aq) + HSO_3^-(aq)$$

If $[H_2SO_3]_0 = 1.50 \, M$ and $[H^+] = [HSO_3^-] = 0 \, M$, calculate the equilibrium concentrations of all species at 25°C if $K = 1.20 \times 10^{-2}$ for this reaction.

Solution with Assumptions

The value of K is small. Let us **assume** we can **neglect dissociation of H_2SO_3** (i.e., X is negligible compared to 1.5 M).

	initial (M)	change	final (M)
H_2SO_3	1.50	$-X$	$1.50-X$ (≈ 1.50??)
H^+	0	$+X$	X
HSO_3^-	0	$+X$	X

$$K = \frac{[H^+][HSO_3^-]}{[H_2SO_3]} \qquad \Rightarrow \qquad 1.20 \times 10^{-2} = \frac{X^2}{1.50}$$

$$X = 0.134 \, M = [H^+] = [HSO_3^-]$$

Let's test our assumption that X is negligible compared to 1.5 M. In fact, **0.134 is 9% of 1.5! Our assumption is not valid!** We must solve the problem explicitly.

Explicit Solution

$$1.20 \times 10^{-2} = \frac{X^2}{1.50 - X}$$

You may use the quadratic equation (see <u>Section 13.6 in your textbook</u>). Multiplying out,

$$0.0180 - 0.0120X = X^2$$

Setting our equation equal to zero,

$$X^2 + 0.0120X - 0.0180 = 0$$

The quadratic formula is

$$X = \frac{-b \pm \sqrt{b^2 - 4ac}}{2a}$$

Because "$-b$" is negative (-0.012), the "$\pm$" in the quadratic must be "+," or one of our solutions will be a negative concentration, a physical impossibility. That is,

$$X = \frac{-b + \sqrt{b^2 - 4ac}}{2a}$$

$a = 1$
$b = 0.0120$
$c = -0.0180$

$$X = \frac{-0.0120 + \sqrt{(0.0120)^2 - (4)(1)(-0.0180)}}{(2)(1)}$$

$$X = \frac{-0.0120 + 0.2686}{2} = 0.128 \, M$$

$$[H_2SO_3] = 1.50 - 0.128 = \mathbf{1.37 \, M}, \qquad\qquad [H^+] = [HSO_3^-] = \mathbf{0.128 \, M}$$

$$\text{Checking, } K = \frac{(0.128)^2}{1.37} = 1.20 \times 10^{-2}$$

The key to doing equilibrium problems is to **make assumptions when you can** to simplify the math. But you must **test all assumptions that you make.**

13.7   Le Châtelier's Principle

When you finish this section you will be able to predict the response of a system to stresses placed on it.

Your textbook defines **Le Châtelier's Principle:** "If a change is imposed on a system **at equilibrium** the position of the equilibrium will shift in a direction that tends to reduce that change."

Your textbook uses the **Haber process**,

$$N_2 + 3H_2 \rightleftharpoons 2NH_3$$

to conceptualize the following effects on the position of equilibrium:

- concentration
- pressure
- temperature

Look over the effects of each on the direction of equilibrium; then try the following problem.

Example 13.7 A   Le Châtelier's Principle

Nitrogen gas and oxygen gas combine at 25°C in a closed container to form nitric oxide as follows:

$$N_2(g) + O_2(g) \rightleftharpoons 2NO(g) \qquad \Delta H = +1.81 \text{ kJ} \qquad K_p = 3.3 \times 10^{30}$$

What would be the effect on the direction of equilibrium (i.e., would it shift to the **left, right**, or **not at all**) if the following changes were made to the system?

    a. N_2 is added
    b.   He is added
    c.   the container is made larger
    d.   the system is cooled

Solution

    a.   From a mathematical point of view, if P_{N_2} is larger, Q_p will be smaller than K_p. To compensate, more product must be formed. The reaction will shift **to the right.**
    b.   The addition of an **inert gas** such as He does not alter the partial pressure of any of the gases. Therefore, Q_p is still equal to K_p, and the position of equilibrium **remains the same.**
    c.   **Enlarging** the container will favor the side with **more gas molecules** because they can exist with fewer collisions per unit time. In this case, both sides have the same number of molecules. Therefore, there will be **no change** in the position of equilibrium.
    d.   This is an **endothermic reaction** ($\Delta H =$ "+"). It requires heat to go. Cooling the system will thus shift the equilibrium **to the left.**

Example 13.7 B   Practice with Le Châtelier's Principle

The combination of hydrogen gas and oxygen gas to give water vapor can be expressed by

$$2H_2(g) + O_2(g) \rightleftharpoons 2H_2O(g) \qquad \Delta H = -484 \text{ kJ}$$

Predict the effect of each of the following changes to the system on the direction of equilibrium.

    a. H_2O is removed as it is being generated
    b. H_2 is added
    c.   the system is cooled

Solution

    a.   moves **to the right** (removal of product forces Q to be less than K and system compensates by making more product).
    b.   moves **to the right** (same as above, except Q is less than K because $[H_2]_o$ is larger than at previous equilibrium).
    c.   moves **to the right** because an **exothermic** reaction gives off heat. Cooling the system increases the temperature gradient between the system and surroundings, thus allowing a continued flow of heat to the surroundings.

Exercises

Section 13.2

1. Write the equilibrium expression for each of the following reactions:

    a. $2H_2(g) + O_2(g) \rightleftharpoons 2H_2O(g)$
    b. $Cl_2(g) + 2Fe^{2+}(aq) \rightleftharpoons 2Fe^{3+}(aq) + 2Cl^-(aq)$
    c. $Cu^{2+}(aq) + 4NH_3(aq) \rightleftharpoons Cu(NH_3)_4^{2+}(aq)$

2. Write the equilibrium expression for each of the following reactions:

    a. $2NO(g) + O_2(g) \rightleftharpoons 2NO_2(g)$
    b. $Ag^+(aq) + I^-(aq) \rightleftharpoons AgI(s)$
    c. $Fe^{3+}(aq) + 3OH^-(aq) \rightleftharpoons Fe(OH)_3(s)$

3. Write the equilibrium expression for each of the following reactions:

    a. $2H_2S(g) \rightleftharpoons 2H_2(g) + S_2(g)$
    b. $4NO_2(g) + 6H_2O(l) \rightleftharpoons 4NH_3(g) + 7O_2(g)$
    c. $2NH_3(g) + 4H_2O(g) \rightleftharpoons 2NO_2(g) + 7H_2(g)$
    d. $PCl_5(g) \rightleftharpoons PCl_3(g) + Cl_2(g)$

4. Write the equilibrium expression for each of the following reactions:

    a. $Zn_2Fe(CN)_6(s) \rightleftharpoons 2Zn^{2+}(aq) + Fe(CN)_6^{-4}(aq)$
    b. $H^+(aq) + OH^-(aq) \rightleftharpoons H_2O(l)$

5. Calculate the equilibrium constant, K, at 25°C for the reaction:

    $$2NO(g) + O_2(g) \rightleftharpoons 2NO_2(g)$$

    If the equilibrium concentrations are $NO_2 = 0.55$ atm, $NO = 6.5 \times 10^{-5}$ atm, $O_2 = 4.5 \times 10^{-5}$ atm.

6. Calculate the equilibrium, K, at 25°C for the Haber process:

    $$3H_2(g) + N_2(g) \rightleftharpoons 2NH_3(g)$$

    If the equilibrium concentrations are $[H_2] = 0.85\ M$, $[N_2] = 1.33\ M$, $[NH_3] = 0.22\ M$.

7. Calculate the value of K for the reaction: $2NO(g) + Cl_2(g) \rightleftharpoons 2NOCl(g)$ if the equilibrium concentrations are:

    $[NOCl] = 4.2 \times 10^{-2}$ mol/L, $[NO] = 6.7 \times 10^{-1}$ mol/L, $[Cl] = 2.9 \times 10^{-3}$ mol/L

8. If the equilibrium constant at 444°C for

    $$2HI(g) \rightleftharpoons H_2(g) + I_2(g)$$

    is 1.39×10^{-2}, calculate the equilibrium constant for the reverse reaction at 444°C.

9. Write the equilibrium expression for each of the following reactions:

    a. $N_2(g) + 3H_2(g) \rightleftharpoons 2NH_3(g)$
    b. $I_2(s) + Cl_2(g) \rightleftharpoons 2ICl(g)$
    c. $2B(s) + 3F_2(g) \rightleftharpoons 2BF_3(g)$

10. Given your answer from Problem #5, calculate the value for K at 25°C for each of the following reactions:

   a. $\frac{1}{2}NO(g) + \frac{1}{4}O_2(g) \rightleftharpoons \frac{1}{2}NO_2(g)$
   b. $2NO_2(g) \rightleftharpoons 2NO(g) + O_2(g)$
   c. $NO_2(g) \rightleftharpoons NO(g) + \frac{1}{2}O_2(g)$

11. Given your answer from Problem #6, calculate the value for K at 25°C for each of the following reactions:

   a. $2NH_3(g) \rightleftharpoons 3H_2(g) + N_2(g)$
   b. $NH_3(g) \rightleftharpoons {}^3/_2H_2(g) + {}^1/_2N_2(g)$
   c. $6NH_3(g) \rightleftharpoons 9H_2(g) + 3N_2(g)$

12. Write equilibrium expressions for each of the following reactions:

   a. $CaCO_3(s) \rightleftharpoons CaCO(s) + CO_2(g)$
   b. $2HCN(aq) + Zn(s) \rightleftharpoons H_2(g) + 2CN^-(aq) + Zn^{2+}(aq)$
   c. $2NaHCO_3(s) + 2CaHPO_4(s) \rightleftharpoons 2H_2O(g) + 2CO_2(g) + 2CaNaPO_4(s)$

13. For which of the following cases does the reaction go farthest to completion: $K = 1$, $K = 10^{10}$, $K = 10^{-10}$?

14. The dissociation of acetic acid, CH_3COOH, has an equilibrium constant at 25°C of 1.8×10^{-5}. The reaction is

$$CH_3COOH(aq) \rightleftharpoons CH_3COO^-(aq) + H^+(aq)$$

If the equilibrium concentration of CH_3COOH is 0.46 moles in 0.500 L of water and that of CH_3COO^- is 8.1×10^{-3} moles in the same 0.500 L, calculate $[H^+]$ for the reaction.

15. For the system:

$$2HI(g) \rightleftharpoons H_2(g) + I_2(g)$$

the specific rate constant of the forward reaction is 0.018 at 490°C. Calculate the specific rate constant for the backward reaction.

16. Calculate the equilibrium constant, K, for the following reaction at 25°C if the equilibrium concentrations are $[Cl_2] = 0.371\ M$, $[F_2] = 0.194\ M$, and $[ClF] = 1.02\ M$.

$$Cl_2(g) + F_2(g) \rightleftharpoons 2ClF(g)$$

17. Write the equation for the reaction in Problem #14 if the value of K is 5.6×10^4 ($1/K_{original}$).

18. Indicate the effect of each of the following on: a. The speed of reaction, b. The position of equilibrium.

   1. Catalyst   2. Pressure   3. Temperature   4. Concentration

Section 13.3

19. Derive an expression that relates K to K_p, and calculate the value of K at 25°C for the reaction given in Problem #5.

20. At 700 K, the measured values for the partial pressures of ammonia, hydrogen, and nitrogen are 0.400 atm, 7.20 atm and 2.40 atm, respectively. Calculate K_p and K_c at 700 K for the ammonia synthesis:

$$N_2(g) + 3H_2(g) \rightleftharpoons 2NH_3(g).$$

21. For the following process at 700°C, what is the partial pressure of each gas at equilibrium if the total pressure is 0.750 atm?

$$C(s) + CO_2(g) \rightleftharpoons 2CO(g) \qquad K_p = 1.50$$

22. Calculate the value for K_p at 25°C if the value for K is 3.7×10^9 L mol^{-1} for the reaction:

$$CO(g) + Cl_2(g) \rightleftharpoons COCl_2(g)$$

23. Given the initial partial pressures of $P_{PCl_5} = 0.0500$ atm, $P_{PCl_3} = 0.150$ atm, and $P_{Cl_2} = 0.250$ atm at 250°C for the following reaction, what must each equilibrium partial pressure be?

$$PCl_5(g) \rightleftharpoons PCl_3(g) + Cl_2(g) \qquad K_p = 2.15$$

24. Consider the dimerization of nitrogen dioxide:

$$2NO_2(g) \rightleftharpoons N_2O_4(g) \qquad K_p = 8.8$$

If the temperature is 298 K and the total pressure is 0.220 atm, what are the equilibrium partial pressures?

Section 13.5

25. The reaction of methane with water is given by the following equation:

$$CH_4(g) + H_2O(l) \rightleftharpoons CO(g) + 3H_2(g) \qquad K = 5.67$$

Predict the direction that the system will shift in order to reach equilibrium given the following initial values of Q.

a. $Q = 11.85$
b. $Q = 3.8 \times 10^{-4}$
c. $Q = 5.67$

26. Determine what the system will do to reach equilibrium given the following values:

a. $K = 2.9 \times 10^2$; $Q = 3.1 \times 10^1$
b. $K = 0.621$; $Q = 6.21 \times 10^{-1}$
c. $K = 7.3 \times 10^2$; $Q = 8.2 \times 10^2$

27. $K_p = 0.133$ atm at a particular temperature for the reaction:

$$N_2O_4(g) \rightleftharpoons 2NO_2(g)$$

Calculate the reaction quotient, Q, if $P_{N_2O_4} = 0.048$ atm and $P_{NO_2} = 0.056$ atm.

28. Using the same reaction as in Problem #25, determine the direction the system will shift in order to reach equilibrium given the **initial concentrations.**

	[CH$_4$]	[H$_2$O]	[CO]	[H$_2$]
a.	4.6×10^{-3}	0.800	0.200	1.00
b.	0.500	0.300	0.620	0.100
c.	0.818	0.750	0.650	2.00

29. The reaction

$$2NO(g) \rightleftharpoons N_2(g) + O_2(g)$$

has a value of $K = 2.4 \times 10^3$ at 2000 K. If 0.61 g of NO are put in a previously empty 3.00-L vessel, calculate the equilibrium concentrations of NO, N_2, and O_2.

30. At 250°C, the equilibrium constant for the reaction

$$PCl_5(g) \rightleftharpoons PCl_3(g) + Cl_2(g)$$

is 2.15. If PCl_5 was initially the only gas present in the reaction vessel at 0.012500 atm, calculate the partial pressures of all the gases after equilibrium has been reached.

31. Using the same reaction at 2000 K as in Problem #29, calculate the equilibrium concentrations of NO, N_2, and O_2 if the initial concentrations of each species are: $[NO] = 0 \, M$, $[N_2] = 0.850 \, M$, $[O_2] = 0.560 \, M$.

32. Using the same reaction as in Problem #23, calculate the equilibrium partial pressures if the initial partial pressures are $P_{PCl_5} = 0.850$ atm, $P_{PCl_3} = 0.440$ atm, and $P_{Cl_2} = 0.935$ atm.

Section 13.6

33. Hypobromous acid, HOBr, dissociates in water according to the following reaction:

$$HOBr(aq) \rightleftharpoons OBr^-(aq) + H^+(aq) \qquad K = 2.06 \times 10^{-9} \text{ at } 25°C$$

Calculate $[H^+]$ of a solution originally 1.25 M in HOBr.

34. The following reaction has an equilibrium constant of 6.2×10^2 at a certain temperature. Calculate the equilibrium concentrations of all species if 4.5 mol of each component were added to a 3.0-L flask.

$$H_2(g) + F_2(g) \rightleftharpoons 2HF(g)$$

35. Ammonia undergoes hydrolysis according to the following reaction:

$$NH_3(aq) + H_2O(l) \rightleftharpoons NH_4^+(aq) + OH^-(aq) \qquad K = 1.8 \times 10^{-5} \text{ at } 25°C$$

Calculate $[NH_3]$, $[NH_4^+]$, and $[OH^-]$ in a solution originally 0.200 M in NH_3.

36. Using the same reaction and value for K as in Problem #35, determine $[OH^-]$ if $[NH_3] = 0.500 \, M$ and $[NH_4^+] = 0.750 \, M$.

37. Given the following reaction at 25°C:

$$2SO_2Cl(g) \rightleftharpoons 2SO_2(g) + Cl_2(g)$$

Calculate the equilibrium constant if the equilibrium concentrations are $[SO_2Cl] = 0.037 \, M$, $[SO_2] = 0.591 \, M$, and $[Cl_2] = 1.24 \, M$.

38. Calculate the equilibrium concentration of Cl_2 for the following reaction at 25°C. The equilibrium constant is 2.3×10^2, and the equilibrium concentrations of PCl_3 and PCl_5 are 2.00 M and 0.04 M, respectively.

$$PCl_5(g) \rightleftharpoons PCl_3(g) + Cl_2(g)$$

39. The equilibrium constant for the reaction:

$$SbCl_3(g) + Cl_2(g) \rightleftharpoons SbCl_5(g)$$

at 448°C is 40. What are the equilibrium concentrations of $SbCl_3$, Cl_2, and $SbCl_5$ if $[Cl_2]_o = 0.620$ M and $[SbCl_5]_o = 0.180$ M?

40. Using the same reaction and K as in Problem #39, calculate the equilibrium concentrations of all species if $[SbCl_5]_o = 1.25$ M, and $[Cl_2]_o = [SbCl_3]_o = 0$.

Section 13.7

41. The following chemical process is at equilibrium:

$$2H_2(g) + O_2(g) \rightleftharpoons 2H_2O(g)$$

How would the process respond if the pressure were increased at a constant temperature?

42. The reaction of carbon disulfide with chloride is as follows:

$$CS_2(g) + 3Cl_2(g) \rightleftharpoons CCl_4(g) + S_2Cl_2(g) \qquad \Delta H° = -238 \text{ kJ}$$

Predict the effect of the following changes to the system on the direction of equilibrium:

a. The pressure on the system is doubled by halving the volume.
b. CCl_4 is removed as it is generated.
c. Heat is added to the system.

43. Given the following reaction at equilibrium,

$$Cl_2(g) + 3F_2(g) \rightleftharpoons 2ClF_3(g)$$

a. Predict the effect if the pressure were reduced at constant temperature.
b. Predict the effect if the volume were reduced by increasing the pressure at constant temperature.

44. The reaction of nitrogen gas with hydrogen chloride is as follows:

$$N_2(g) + 6HCl(g) \rightleftharpoons 2NH_3(g) + 3Cl_2(g) \qquad \Delta H = +461 \text{ kJ}$$

Predict the effect of each of the following changes to the system on the direction of equilibrium:

a. Triple the volume of the system.
b. The amount of nitrogen is doubled.
c. Heat is added to the system.

45. Using the following equation, what is the effect on the equilibrium when the partial pressure of ammonia is increased?

$$NH_4Cl(s) \rightleftharpoons NH_3(g) + HCl(g)$$

46. In the reaction

$$2H_2(g) + O_2(g) \rightarrow 2H_2O(l)$$

list two ways of maximizing water recovery.

Multiple Choice Questions

47. When a reaction has reached equilibrium:

    A. The molecules are in a passive state; therefore, no more products are formed.
    B. The products are reacting while the reactants are passive.
    C. The reactants are reacting while the products are passive.
    D. Both reactants and products are formed continuously.

48. Which of the following changes will change the position of equilibrium?

    A. Allow more time to pass.
    B. Remove some products.
    C. Add a catalyst.
    D. All of these.

49. Which of the reactions below has the following equilibrium expression:

$$K = \frac{[A]^2[B]^2}{[D][C]^3}$$

    A. $2A + 2B \rightleftharpoons D + 3C$
    B. $3C + D \rightleftharpoons 2A + 2B$
    C. $A_2 + B_2 \rightleftharpoons D + C_3$
    D. $D + C_3 \rightleftharpoons A_2 + B_2$

50. Calculate K from the following information: $K_f = 1.00 \times 10^2 /(M \times t)$, $K_r = 1.60 \times 10^{-2}/(M \times t)$.

    A. 6.25×10^3
    B. 1.60
    C. 4.00×10^{-5}
    D. 1.00×10^2

51. Given the following information, calculate K:

$$COCl_2(g) \rightleftharpoons CO(g) + Cl_2(g)$$
$$0.21\ M \qquad 0.020M \quad 0.020\ M \qquad \text{at equilibrium}$$

    A. 1.9×10^{-3}
    B. 2.6×10^2
    C. 0.19
    D. 3.8×10^{-3}

52. At equilibrium, $[PSCl_3] = 1.00\ M$, $[PCl_3] = 7.8 \times 10^{-4}\ M$, while $K = 1.3 \times 10^{-29}$. Calculate $[S_8]$.

$$8PSCl_3(g) \rightleftharpoons 8PCl_3(g) + S_8(g)$$

    A. $1.00\ M$
    B. $1.6 \times 10^{-26}\ M$
    C. $9.5 \times 10^{-5}\ M$
    D. $94.88\ M$

53. What is the equilibrium constant for the reverse reaction of the previous problem?

    A. 1.3×10^{-29}
    B. 7.7×10^{28}
    C. 4.1×10^{-15}
    D. 1.3

54. Calculate the equilibrium constant for the following reaction:

$$PCl_5(g) \rightleftharpoons PCl_3(g) + Cl_2(g)$$

knowing that: $[PCl_5] = 0.00325\ M$, $[PCl_3] = 2.52\ M$, and $[Cl_2] = 0.02175\ M$ at equilibrium

    A. 16.9
    B. 0.0296
    C. 33.7
    D. 7.82

55. What is the equilibrium constant of the reverse reaction for the previous problem?

    A. 0.059
    B. 3.82
    C. 5.81
    D. 0.128

56. The equation that relates K_p to K is:

    A. $K_p + K = (RT)^{\Delta n}$
    B. $K_p = K$
    C. $K_p = K(RT)^{\Delta n}$
    D. $K = K_p(RT)^{\Delta n}$

57.   Calculate K_p at 160°C for the following reaction:

$$4NO_2(g) + 6H_2O(g) \rightleftharpoons 4NH_3(g) + 5O_2(g) \qquad K = 0.455$$

   A. 0.0128   B. 16.16   C. 5.97   D. 2.20

58.   Calculate K_p given the following information:

$$N_2(g) + O_2(g) \rightleftharpoons 2NO(g)$$

A 1.00-L vessel at 273 K contains 0.0290 moles of nitrogen gas, 0.00290 moles of oxygen gas, and 1.92 micromoles of nitrogen oxide at equilibrium.

   A. 4.38×10^{-8} B. 762   C. 250   D. 7.97×10^{-3}

59.   A 0.250 L closed vessel at 487°C contains 0.500 g of PCl_5, 19.55 g of PCl_3, and 10.1 g of Cl_2 at equilibrium. Calculate K_p based on the following equation:

$$PCl_5(g) \rightleftharpoons PCl_3(g) + Cl_2(g)$$

   A. 781   B. 0.0292   C. 33.65   D. 2.10×10^3

60.   Which of the compounds of the following reaction would not appear in an equilibrium expression?

$$AgNO_3(aq) + HCl(aq) \rightleftharpoons AgCl(s) + HNO_3(aq)$$

   A. HCl   B. $AgNO_3$ C. AgCl   D. HNO_3

61.   What is the correct equilibrium expression for the following reaction:

$$HCl(aq) + H_2O(l) \rightleftharpoons H_3O^+(aq) + Cl^-(aq)$$

   A. $\dfrac{[H_3O^+][Cl^-]}{[HCl]}$

   B. $\dfrac{[H_3O^+][Cl^-]}{[H_2O][HCl]}$

   C. $\dfrac{[H_2O][HCl]}{[H_3O^+]}$

   D. $\dfrac{[Cl^-]}{[HCl][H_2O]}$

62.   0.0500 moles of PCl_5 and 5.00 moles of PCl_3 are introduced into an evacuated 1.00-L chamber. Calculate the equilibrium concentration of PCl_3, knowing that $K = 33.3$.

$$PCl_5(g) \rightleftharpoons PCl_3(g) + Cl_2(g)$$

   A. 0.0063   B. 5.04   C. 0.0435   D. 1.50

63.   0.125 mole of oxygen gas is added to carbon in a 0.250-L container. The mixture equilibrates at 500 K. Calculate the equilibrium concentration of carbon monoxide, knowing that $K = 0.086$ at 500 K.

$$C(s) + O_2(g) \rightleftharpoons 2CO(g)$$

   A. 0.19   B. 0.100   C. 1.00   D. 0.041

64.   For a certain reaction, $Q = 2.33$, while $K = 3.54$. What do you expect to happen?

   A. The reaction will proceed forward.
   B. The reaction will proceed backward.
   C. The reaction will proceed away from equilibrium.
   D. The direction cannot be determined.

65.   3.0 moles of each of the reactants and products for the following reaction are placed in a 2.00-L chamber. Predict the direction of the reaction, knowing that $K = 33.3$.

$$PCl_5(g) \rightleftharpoons PCl_3(g) + Cl_2(g)$$

A. The reaction will proceed to the right.   C. The reaction is at equilibrium.
B. The reaction will proceed to the left.   D. The direction is unpredictable.

66.   For the following reaction, what must the minimum pressure of NO be to allow the reaction to proceed to the left? $K_p = 4.38 \times 10^{-8}$.

$$N_2(g) + O_2(g) \rightleftharpoons 2NO(g)$$
0.65 atm 0.065 atm   ? atm

A. 0.065 atm   B. 0.0085 atm   C. 0.000022 atm   D. 0.000044 atm

67.   Calculate the equilibrium pressure of nitrogen dioxide when the equilibrium pressures of the nitric oxide and oxygen are 0.0100 and 0.500 atm, respectively. $K_p = 1.00 \times 10^4$ at 200 K.

$$2NO(g) + O_2(g) \rightleftharpoons 2NO_2(g)$$

A. 22.4 atm   B. 0.707 atm   C. 500 atm   D. 3.98 atm

68.   Calculate the amount of $COCl_2$ heated to 575 K in a 0.220-L vessel when the equilibrium concentration of CO and Cl_2 is 0.0367 M. $K = 2.50x\ 10^{-3}$.

$$COCl_2(g) \rightleftharpoons CO(g) + Cl_2(g)$$

A. 11.2 g   B. 0.128 g   C. 0.580 g   D. 1.36 g

69.   For the following reaction, calculate the concentration of C(aq) at equilibrium when 0.135 M B solution is allowed to react with A.

$$A(l) + B(aq) \rightleftharpoons D(aq) + E(aq) \qquad K = 1.5 \times 10^{-10}$$

A. 0.01353 M   B. 4.5×10^{-6} M   C. 2.03×10^{-11} M   D. 2.03 M

70.   Which of the following changes has no effect on the position of the equilibrium?

A. Change in concentration.   C. Passing of time.
B. Change in pressure.   D. Volume change.

71.   Which of the following changes will not affect the equilibrium position of the following equation?

$$A(g) + 4B(s) \rightleftharpoons 2D(g) + E(g) + G(s)$$

A. Removal of A.   B. Increase in pressure. C. Addition of G.   D. Addition of D.

72.   What would you change to increase the yield of the following reaction?

$$A(g) + 4B(s) \rightleftharpoons 2D(g) + E(g) + G(s)$$

A. Increase in pressure.   C. Increase in temperature.
B. Decrease in pressure.   D. Decrease in temperature.

73.   What would you change to increase the yield of the following reaction?

$$A(g) + 4B(s) \rightleftharpoons 2D(g) + E(g) + G(s) \quad \Delta H = -258\ kJ$$

A. Add B.   C. Increase temperature.
B. Decrease temperature.   D. Remove G.

Answers to Exercises

1. a. $K = \dfrac{[H_2O]^2}{[H_2]^2[O_2]}$ b. $K = \dfrac{[Cl^-]^2[Fe^{3+}]^2}{[Cl_2][Fe^{2+}]^2}$ c. $K = \dfrac{[Cu(NH_3)_4^{2+}]}{[Cu^{2+}][NH_3]^4}$

2. a. $K = \dfrac{[NO_2]^2}{[NO]^2[O_2]}$ b. $K = \dfrac{1}{[Ag^+][I^-]}$ c. $K = \dfrac{1}{[Fe^{3+}][OH^-]^3}$

3. a. $K = \dfrac{[H_2]^2[S_2]}{[H_2S]^2}$ c. $K = \dfrac{[NO_2]^2[H_2]^7}{[NH_3]^2[H_2O]^4}$

    b. $K = \dfrac{[NH_3]^4[O_2]^7}{[NO_2]^4}$ d. $K = \dfrac{[PCl_3][Cl_2]}{[PCl_5]}$

4. a. $K = [Zn^{2+}]^2[Fe(CN)_6^{4-}]$ b. $K = \dfrac{1}{[H^+][OH^-]}$

5. $K_p = 1.6 \times 10^{12}\ \text{atm}^{-1}$

6. $K = 0.059$

7. $K = 1.4$

8. $K' = 71.9$

9. a. $K = \dfrac{[NH_3]^2}{[N_2][H_2]^3}$ b. $K = \dfrac{[ICl]^2}{[Cl_2]}$ c. $K = \dfrac{[BF_3]^2}{[F_2]^3}$

10. a. $K_p = 1.1 \times 10^3\ \text{atm}^{1/4}$ b. $K_p = 6.3 \times 10^{-13}\ \text{atm}$ c. $K_p = 7.9 \times 10^{-7}\ \text{atm}^{1/2}$

11. a. $K = 17$ b. $K = 4.1$ c. $K = 4.9 \times 10^3$

12. a. $K = [CO_2]$ b. $K = \dfrac{[H_2][CN^-]^2[Zn^{2+}]}{[HCN]^2}$ c. $K = [H_2O]^2[CO_2]^2$

13. $K = 10^{10}$ where $K = \dfrac{[\text{products}]}{[\text{reactants}]}$

14. $[H^+] = 1.0 \times 10^{-3}\ M$

15. $K' = 55.6$

16. $K = 1.45 \times 10^1$

17. $H^+ + CH_3COO^- \rightleftharpoons CH_3COOH$

18.

Condition	Speed of reaction	Position of equilibrium
Catalyst	Increases	No effect
Pressure	Depends on stoichiometry	Increase in pressure shifts reaction to side with fewer moles of gas—no effect on K_{eq}
Temperature	Generally increases	Generally increases K_{eq} as temperature increases
Concentration	Generally increases	Affects position of equilibrium, but not the value of K_{eq}

19. $K = K_p(RT)^{-\Delta n} = K_p(RT) \Rightarrow K = 3.9 \times 10^{13}$

20. $K_p = 1.79 \times 10^{-4}$; $K_c = 5.89 \times 10^{-1}$

21. $P_{CO_2} = 0.201$ atm; $P_{CO} = 0.549$ atm

22. $K_p = 1.5 \times 10^8$ atm^{-1}

23. $P_{PCl_5} = 0.023$ atm; $P_{PCl_3} = 0.177$ atm; $P_{Cl_2} = 0.277$ atm

24. $P_{N_2O_4} = 0.109$ atm; $P_{NO_2} = 0.111$ atm

25.   a.   to the left
       b.   to the right
       c.   no shift - the system is at equilibrium

26.   a. $Q < K$, system shifts to the right
       b. $Q = K$, system is at equilibrium
       c. $Q > K$, system shifts to the left

27. $Q = 0.065$

28.   a. $Q = 43$, system shifts to the left. (Remember water is a pure liquid.)
       b. $Q = 1.2 \times 10^{-3}$, system shifts to the right.
       c. $Q = 6.4$, system shifts to the left.

29. $[NO] = 6.9 \times 10^{-5}$ M; $[N_2] = [O_2] = 3.4 \times 10^{-3}$ M

30. $P_{PCl_3} = P_{Cl_2} = 0.0125$ atm; $P_{PCl_5} = 7.3 \times 10^{-5}$ atm

31. $[NO] = 1.4 \times 10^{-2}$ M; $[N_2] = 0.850$ M; $[O_2] = 0.560$ M

32. $P_{PCl_5} = 0.485$ atm; $P_{PCl_3} = 0.804$ atm; $P_{Cl_2} = 1.30$ atm

33. $[H^+] = 5.07 \times 10^{-5}$ M

34. $[H_2] = [F_2] = 0.167$ M; $[HF] = 4.2$ M

35. $[NH_3] = 0.198$ M; $[OH^-] = [NH_4^+] = 1.9 \times 10^{-3}$ M

36. $[OH^-] = 1.2 \times 10^{-5}$ M

37. 3.2×10^2

38. $4.5\ M$

39. $[SbCl_3] = 7.3 \times 10^{-3}\ M;\ [Cl_2] = 0.627\ M;\ [SbCl_5] = 0.173\ M$

40. $[SbCl_3] = [Cl_2] = 0.16\ M;\ [SbCl_5] = 1.09\ M$

41. The reaction would move left to right (fewer moles on the right side).

42.   a. The reaction shifts to the right.
    b. The reaction shifts to the right.
    c. The reaction shifts to the left.

43.   a. The reaction moves to the left.
    b. The reaction moves to the right.

44.   a. The reaction shifts to the left.
    b. The reaction shifts to the right.
    c. The reaction shifts to the right.

45. Adding $NH_3(g)$ drives the reaction to the left.

46.   a. Remove water as it is generated.
    b. Add more hydrogen and oxygen to the system. (This is the key to fuel cells on space flights.)

47. D	48. B	49. B	50. A	51. A	52. C
53. B	54. A	55. A	56. C	57. A	58. A
59. D	60. C	61. A	62. B	63. A	64. A
65. A	66. D	67. B	68. A	69. B	70. C
71. C	72. B	73. B			

Chapter 14

Acids and Bases

In this chapter you will learn about the essential properties of acids and bases. Your textbook also surveys common acids and bases.

14.1 The Nature of Acids and Bases

When you finish this section you will be able to:

- Define acids and bases using the Brønsted-Lowry model.
- Identify conjugate acid-base pairs.
- Write equilibrium expressions for acid dissociations.

Your textbook introduces two concepts of acids and bases. The **Arrhenius concept** says that an **acid supplies H^+ to an aqueous solution**. A **base supplies OH^- to an aqueous solution**. This concept is limiting because there are many bases that do not contain OH^-. A more global description of acids and bases is the **Brønsted-Lowry concept**. An **acid is a proton (H^+) donor**. A **base is a proton acceptor**.

The remainder of this section and the next several sections are devoted to the reactions of acids in water. The general reaction for an acid, HA, in water is

$$HA \; + \; H_2O \; \rightleftharpoons \; A^- \; + \; H_3O^+$$

$$\underset{\text{acid 1}}{\uparrow} \qquad \underset{\text{base 2}}{\uparrow} \qquad \underset{\substack{\text{conjugate} \\ \text{base 1}}}{\uparrow} \qquad \underset{\substack{\text{conjugate} \\ \text{acid 2}}}{\uparrow}$$

HA donates a proton to H_2O giving A^- and H_3O^+. HA and A^- are **conjugate pairs**. H_2O and H_3O^+ are also conjugate pairs.

Example 14.1 A   Conjugate Pairs

Write the dissociation reaction for each of the following acids in water, and identify the conjugate acid-base pairs:

- a.   formic acid (HCOOH)
- b.   perchloric acid ($HClO_4$)
- c.   the hydrated iron(III) ion $[Fe(H_2O)_6]^{3+}$

Solution

a. $HCOOH(aq)$ + $H_2O(l)$ $\rightleftharpoons$ $COOH^-(aq)$ + $H_3O^+(aq)$
        ↑   ↑   ↑   ↑
     acid 1   base 2   conjugate   conjugate
                                                 base 1   acid 2

Your textbook points out that it is common practice to eliminate the solvent, water, from the equation (it has a constant concentration). This leads to the more common form,

$$HCOOH(aq) \rightleftharpoons COOH^-(aq) + H^+(aq)$$

b.   long form: $HClO_4(aq)$ + $H_2O(l)$ $\rightleftharpoons$ $ClO_4^-(aq)$ + $H_3O^+(aq)$
              ↑   ↑   ↑   ↑
           acid 1   base 2   conjugate   conjugate
                                                       base 1   acid 2

common form: $$HClO_4(aq) \rightleftharpoons ClO_4^-(aq) + H^+(aq)$$

c.   long form: $[Fe(H_2O)_6]^{3+}(aq)$ + $H_2O(l)$ $\rightleftharpoons$ $[Fe(H_2O)_5OH]^{2+}(aq)$ + $H_3O^+(aq)$
                        ↑   ↑   ↑   ↑
                     acid 1   base 2   conjugate   conjugate
                                                                              base 1   acid 2

common form: $$[Fe(H_2O)_6]^{3+}(aq) \rightleftharpoons [Fe(H_2O)_5OH]^{2+}(aq) + H^+(aq)$$

Equilibrium expressions for acid dissociations are written using the same concepts as for any other chemical equations. However, we use the **short form** of the acid dissociation when writing these expressions. Recall from your textbook that the equilibrium constant, K, is known as K_a for acid dissociations.

Example 14.1 B   Equilibrium Expressions for Acid Dissociations

Write an equilibrium expression for each of the equations in Example 14.1 A.

Strategy

As always, $K = \dfrac{[\text{products}]}{[\text{reactants}]}$

Solution

a. $K_a = \dfrac{[COOH^-][H^+]}{[HCOOH]}$

b. $K_a = \dfrac{[ClO_4^-][H^+]}{[HClO_4]}$

c. $K_a = \dfrac{[Fe(H_2O)_5OH^{2+}][H^+]}{[Fe(H_2O)_6^{3+}]}$

14.2   Acid Strength

When you finish this study section you will be able to:

- Compare relative strengths of acids.
- Solve problems regarding the autoionization of water.

The key idea of this section is that **the strength of an acid is indicated by the equilibrium position of the dissociation reaction**. If the equilibrium lies **far to the left** (as indicated by the value of K_a), the acid **does not dissociate** very much and is called **weak**. If the equilibrium lies **far to the right**, the acid **strongly dissociates** and is called **strong**. This idea is summarized in Table 14.1 of your textbook. **You should memorize the strong acids discussed below Table 14.1**. Note that with the strongest acids, the position of equilibrium lies so far to the right that K_a cannot be measured.

Example 14.2 in your textbook illustrates that:

THE STRONGER THE ACID, THE WEAKER ITS CONJUGATE BASE.
THE STRONGER THE BASE, THE WEAKER ITS CONJUGATE ACID.

Example 14.2 A   Relative Acid and Base Strengths

Using Table 14.2 in your textbook and knowing that K_a for $H_2O = 1 \times 10^{-14}$ at 25°C, arrange the following acids in order of their strength. Then arrange their conjugate bases in order.

$$HOC_6H_5, \ H_2O, \ HSO_4^-, \ NH_4^+, \ HNO_3$$

Solution

Acid strength is reflected by K_a (except for very strong acids, where K_a is too large to measure accurately). Also, K_a for all but very strong acids is much less than one, indicating that the equilibrium lies far to the left. **HNO$_3$ is a very strong acid.** According to the values of K_a from Table 14.2, the list of acid strengths should read

$$HNO_3, \ HSO_4^-, \ NH_4^+, \ HOC_6H_5, \ H_2O$$
strongest $\longleftrightarrow$ weakest

The list of conjugate base strengths must be just the opposite:

$$OH^-, \ OC_6H_5^-, \ NH_3, \ SO_4^{-2}, \ NO_3^-$$
strongest $\longleftrightarrow$ weakest

Your textbook points out that water is an **amphoteric** substance (it can act as an acid OR a base). For the autoionization reaction

$$2H_2O \ \rightleftharpoons \ OH^- \ + \ H_3O^+$$
$$K_a = K_w = 1 \times 10^{-14} \text{ at } 25°C$$

Critical point: IN AQUEOUS SOLUTION, THE ION PRODUCT [H$^+$][OH$^-$] IS ALWAYS EQUAL TO 1×10^{-14} at 25°C. Therefore, if you know [H$^+$] in a solution, you always know [OH$^-$]. The reverse must also hold true.

Review the relationship given in your textbook between [H$^+$] and [OH$^-$] for acidic, basic, and neutral solutions; then try the next example.

Example 14.2 B   Conversion Between [H⁺] and [OH⁻]

At **10°C**, K_w for the autoionization of water equals 2.9×10^{-15}. Calculate $[H^+]$ or $[OH^-]$ as necessary under each of the following conditions:

  a.   Calculate $[H^+]$ if $[OH^-] = 9.3 \times 10^{-4}$ M. Is the solution acidic or basic?
  b.   Calculate $[H^+]$ and $[OH^-]$ for a neutral solution.
  c.   Calculate $[OH^-]$ if $[H^+] = 6.7 \times 10^{-11}$ M. Is the solution acidic or basic?

Solution

$$K_w = [H^+][OH^-]$$

(Remember, $[H_2O]$ is essentially constant, so it is incorporated into the equilibrium expression.)

  a. $[H^+] = K_w / [OH^-] = 2.9 \times 10^{-15}/9.3 \times 10^{-4} = \mathbf{3.1 \times 10^{-12}\ \mathit{M}}$
        $[OH^-] > [H^+]$, so this solution is **basic**.

  b.   For a neutral solution, $[H^+] = [OH^-]$. If we let "X" $= [H^+] = [OH^-]$, then
        $$K_w = (X)(X) = X^2 = 2.9 \times 10^{-15}$$
        $$X = [H^+] = [OH^-] = \mathbf{5.4 \times 10^{-8}\ \mathit{M}}$$

  c. $[OH^-] = K_w / [H^+] = 2.9 \times 10^{-15}/6.7 \times 10^{-11} = \mathbf{4.3 \times 10^{-5}\ \mathit{M}}$
        $[OH^-] > [H^+]$, so this solution is **basic**.

14.3   The pH Scale

When you finish this section you will be able to convert among pH, pOH, $[OH^-]$, and $[H^+]$.

This is the first time in our chemistry course that we need to make use of **logarithms**. (See Appendix 1, Part A1.2 in your textbook if you need a review.) As your textbook points out,

$$pH = -\log [H^+]$$
$$pOH = -\log [OH^-]$$

The operator "**p**" means "**−log of the concentration of**." In practice pH is not exactly equal to −log $[H^+]$, but it is pretty close—certainly close enough for our purposes.

Example 14.3 A   Understanding the Meaning of "p"

Translate the meaning of each of the following uses of "p," or convert to p, as necessary. (Forgive the puns to follow.)

  a.   What is "−log [soup]" in terms of "p?"
  b.   What is "pCabin" in terms of "−log?"
  c.   What is "−log [ter Pan]" in terms of "p?"
  d.   What is "pRolling" in terms of "−log?"

Solution

  a.   psoup (Aarrgh!)
  b.   −log [Cabin] (Oy!)
  c.   pter Pan (Enough!)
  d.   −log [Rolling] (Isn't chemistry wonderful?)

Note the discussion on the **number of significant figures** at the beginning of the section in your textbook. When you are comfortable with that and you understand the equations below, please try the following problem:

$$pX = -\log[X]$$
$$[X] = 10^{-pX}$$

Look over the discussion in <u>Example 14.6 in your textbook</u>. Then try the following problem.

Example 14.3 B   Converting Between p and Concentration

Calculate the "p" or "[ ]" as necessary for each of the following. (Remember to use the proper number of significant figures!)

    a.   Calculate $[Cl^-]$ if $pCl = 7.32$.
    b.   Calculate pAg if $[Ag^+] = 0.034\ M$.
    c.   Calculate pNO_3 if $[NO_3^-] = 15\ M$.
    d.   Calculate $[NH_4^+]$ if the $pNH_4 = 11.87$.

Solution

a. $[Cl^-] = 10^{-pCl} = 10^{-7.32}$ (Enter **7.32** into your calculator, press the +/– key, and press either **10^x** or **inv log**.)

$$[Cl^-] = 4.8 \times 10^{-8}\ M$$

b. $pAg = -\log(0.034)$ (Enter **0.034** into your calculator, press the **log** key and press the +/– key.)

$$pAg = 1.47$$

c. $pNO_3 = -\log(15) = -1.18$

d. $[NH_4^+] = 10^{-11.87} = 1.3 \times 10^{-12}\ M$

For the next problem, recall that $K_w = [H^+][OH^-]$, or $pK_w = pH + pOH$. At $25°C$, $K_w = 1.0 \times 10^{-14}$ and $pK_w = 14$. Therefore, at $25°C$, **the sum of pH and pOH must always equal 14.0.**

Example 14.3 C   Converting among pH, pOH, [H⁺], and [OH⁻]

Fill in the blanks in the following table.

	pH	pOH	$[H^+]$	$[OH^-]$	acid, base, or neutral?
Solution a	6.88	_____	_____	_____	_____
Solution b	_____	_____	_____	8.4×10^{-14}	_____
Solution c	_____	3.11	_____	_____	_____
Solution d	_____	_____	1.0×10^{-7}	_____	_____

Solution

a. $[H^+] = 10^{-pH} = 10^{-6.88} = 1.3 \times 10^{-7}\ M$
      $[OH^-] = K_w / [H^+] = 1.0 \times 10^{-14} / 1.3 \times 10^{-7} = 7.6 \times 10^{-8}\ M$
      $pOH = -\log[OH^-] = -\log(7.6 \times 10^{-8}) = 7.12$

b. $pOH = -\log(8.4 \times 10^{-14}) = 13.08$
      $pH = 14 - pOH = 14 - 13.08 = 0.92$
      $[H^+] = 10^{-pH} = 10^{-0.92} = 0.12\ M$

c. $pH + pOH = 14 \Rightarrow 14 - 3.11 = pH = \mathbf{10.89}$
$[OH^-] = 10^{-3.11} = \mathbf{7.8 \times 10^{-4}\ M}$
$[H^+] = K_w / 7.8 \times 10^{-4} = \mathbf{1.3 \times 10^{-11}\ M}$

d. $[OH^-] = 1.0 \times 10^{-14} / 1.0 \times 10^{-7} = \mathbf{1.0 \times 10^{-7}\ M}$
$pOH = -\log(1.0 \times 10^{-7}) = \mathbf{7.00}$
$pH = 14.00 - 7.00 = \mathbf{7.00}$

The completed table is:

	pH	pOH	$[H^+]$	$[OH^-]$	acid, base, or neutral?
Solution a	6.88	7.12	1.3×10^{-7}	7.6×10^{-8}	acid
Solution b	0.92	13.08	0.12	8.4×10^{-14}	acid
Solution c	10.89	3.11	1.3×10^{-11}	7.8×10^{-4}	base
Solution d	7.00	7.00	1.0×10^{-7}	1.0×10^{-7}	neutral

14.4   Calculating the pH of Strong Acid Solutions

When you finish this section you will be able to calculate the pH of strong acid solutions.

Your textbook introduces you to the idea of **multiple equilibria** in this section. In order to properly assess acid-base problems in aqueous solution, you must always:

a. **recognize that autoionization of water is ALWAYS occurring in an aqueous solution, and**
b. **be able to determine whether autoionization will contribute significantly to the acid-base character of a solution.**

Calculating the pH of strong acid solutions is in general fairly straightforward because the dissociation equilibrium lies so far to the right—that is, **the acid completely dissociates**. The autoionization is negligible as a contributor of H^+ to the solution. (See the discussion regarding Le Châtelier's principle in this section of your textbook.) The rare exception to this is when your strong acid is exceptionally dilute ($< 10^{-6}\ M$). In that case, water can contribute a **relatively** large proportion of H^+ to the solution.

The bottom line is that $[H^+]$ at equilibrium is $\approx$ [strong acid]$_0$, except in very dilute solutions.

Example 14.4 A   The pH of a Strong Acid

Calculate the **pH** and $[OH^-]$ of a $5 \times 10^{-3}\ M\ HClO_4$ solution.

Solution

$HClO_4$ is a very strong acid. It will completely dissociate. The autoionization of water will not contribute significantly to $[H^+]$.

$pH = -\log[HClO_4]_o = -\log[H^+] = -\log(5 \times 10^{-3}\ M)$
$pH = \mathbf{2.3}$

$[OH^-] = K_w/[H^+] = 1 \times 10^{-14} / 5 \times 10^{-3} = \mathbf{2 \times 10^{-12}\ M}$

Example 14.4 B   Practice with Strong Acids

A solution is prepared by adding 15.8 g of HCl to enough water to make a total volume of 400. mL. What is the pH of the solution? How much hydrogen ion is contributed by the autoionization of water?

Solution

Let's first find $[HCl]_0$.

$$\frac{\text{mol HCl}}{\text{L}} = \frac{15.8 \text{ g}}{0.400 \text{ L}} \times \frac{1 \text{ mol}}{36.5 \text{ g}} = \mathbf{1.08 \textit{ M}}$$

HCl is a strong acid, so it completely dissociates.

$$\mathbf{pH} = -\log[H^+] = -\log[HCl]_0 = -\log(1.08) = \mathbf{-0.033}$$

The next question had to do with $[H^+]$ due to the autoionization of water. We know that **the only source of OH^- is from the autoionization of water**. We can find $[OH^-]$ because

$$\mathbf{[OH^-]} = K_w / [H^+] = 1.0 \times 10^{-14} / 1.08 = \mathbf{9.3 \times 10^{-15} \textit{ M}}$$

In addition, $[H^+]$ due to H_2O autoionization must equal $[OH^-]$ due to H_2O autoionization (1:1 stoichiometry). Therefore,

$$[H^+]_{H_2O} = 9.3 \times 10^{-15} \textit{ M}$$

We can see Le Châtelier's principle at work here. If the autoionization of water were not suppressed, $[H^+]_{H_2O} = 1.0 \times 10^{-7} \textit{ M}$.

14.5   Calculating the pH of Weak Acid Solutions

When you finish this section you will be able to calculate the pH and percent dissociation of weak acid solutions.

A succinct strategy for solving weak acid problems is proposed in a Problem Solving Strategy Box before Example 14.8 in your textbook. The key points of the strategy are:

- Although there are often several reactions that can produce H^+, usually **only one predominates**. You can make the proper judgment based on the values of the equilibrium constants for the reactions.
- You must test any assumptions that you make regarding the extent of dissociation of a weak acid (i.e., $[HA] = [HA]_0$).

Let's work the following problem using the procedure given in your textbook:

Example 14.5 A   pH of a Weak Acid

Calculate the pH of a 0.500 M aqueous solution of formic acid, HCOOH ($K_a = 1.77 \times 10^{-4}$).

Solution

Step 1:   We recognize that HCOOH is a weak acid. The dissociation equilibrium lies far to the left. The same is true for the autoionization of H_2O ($K_w = 1.0 \times 10^{-14}$). The major species in solution are, therefore, **HCOOH and H_2O**.

Step 2:   Both HCOOH and H_2O can produce H^+.

$$HCOOH(aq) \rightleftharpoons COOH^-(aq) + H^+(aq) \qquad K_a = 1.77 \times 10^{-4}$$
$$H_2O(l) \rightleftharpoons H^+(aq) + OH^-(aq) \qquad K_w = 1.0 \times 10^{-14}$$

Step 3: The value of K_a for the dissociation of HCOOH is far greater than the value of K_w for the autoionization of water. Therefore, HCOOH dissociation will predominate as a source of H^+. (Technically, we have **assumed** that H_2O contributes a negligible amount of H^+. We will not test this assumption.)

Step 4: $K_a = \dfrac{[COOH^-][H^+]}{[HCOOH]}$

Steps 5, 6, and 7:

	initial (M)	change	final (M)
HCOOH	0.500	$-X$	$0.500 - X$
H^+	0	$+X$	X
$COOH^-$	0	$+X$	X

Step 8: $K_a = 1.77 \times 10^{-4} = \dfrac{X^2}{0.500 - X}$

Step 9: The equilibrium lies far to the left. Assume that the extent of dissociation, "X," **is negligible relative to 0.500 M.**

$$(0.500 - X = 0.500)$$
$$1.77 \times 10^{-4} = \frac{X^2}{0.500}$$
$$X = [H^+] = [COOH^-] = 9.41 \times 10^{-3}\ M$$

Step 10: Comparing X to $[HCOOH]_0$,

$$\frac{X}{[HCOOH]_0} \times 100\% = \frac{9.41 \times 10^{-3}}{0.500} \times 100\% = \textbf{1.9\%}$$

This value is less than 5%, so **our assumption of negligible dissociation is valid.**

Step 11: $[H^+] = 9.41 \times 10^{-3}\ M \Rightarrow \textbf{pH = 2.03}$

Example 14.5 B   Practice with Weak Acids

$K_a = 7.45 \times 10^{-4}$ for citric acid, $C_6H_{10}O_8$ (we'll call it "HCA"). Calculate the pH of a 0.200 M HCA solution.

Solution

The reactions of interest are:

$$HCA(aq) \rightleftharpoons CA^-(aq) + H^+(aq)$$
$$H_2O(l) \rightleftharpoons H^+(aq) + OH^-(aq)$$

K_a for HCA is much larger than K_w for water. Therefore the dissociation of HCA is the significant equilibrium. This equilibrium lies far to the left.

$$K_a = \frac{[CA^-][H^+]}{[HCA]}$$

	initial (M)	change	final (M)
HCA	0.200	$-X$	$0.200 - X$ (≈ 0.200)
H^+	0	$+X$	X
CA^-	0	$+X$	X

$$7.45 \times 10^{-4} = \frac{X^2}{0.200}$$

$$X = [H^+] = [CA^-] = 1.22 \times 10^{-2}\ M$$

$$\frac{X}{[HCA]} \times 100\% = \frac{1.22 \times 10^{-2}}{0.200} \times 100\% = 6.1\%$$

The percent of dissociation is greater than 5%; therefore, we **CANNOT "neglect X in comparison to 0.200."** (i.e., We are NOT making this assumption; therefore, we do not have to test it.) We must therefore solve the equation **using the quadratic equation**.

$$7.45 \times 10^{-4} = \frac{X^2}{0.200 - X}$$

Clearing the denominator,

$$1.49 \times 10^{-4} - 7.45 \times 10^{-4}(X) = X^2$$

Setting the equation equal to zero,

$$1X^2 + 7.45 \times 10^{-4}(X) - 1.49 \times 10^{-4} = 0$$
$$\quad \underset{a}{\uparrow} \qquad\qquad \underset{b}{\uparrow} \qquad\qquad \underset{c}{\uparrow}$$

$$X = \frac{-b \pm \sqrt{b^2 - 4ac}}{2a}$$

$$X = \frac{-7.45 \times 10^{-4} \pm \sqrt{(7.45 \times 10^{-4})^2 - (4)(1)(-1.49 \times 10^{-4})}}{(2)(1)}$$

$$X = \frac{-7.45 \times 10^{-4} + 0.02442}{2} = 1.18 \times 10^{-2}\ M$$

$$[H^+] = 1.18 \times 10^{-2}\ M \implies \textbf{pH = 1.93}$$

We may wish to check our results:

$$K_a = \frac{(1.18 \times 10^{-2})^2}{(0.200 - 1.18 \times 10^{-2})} = \frac{1.392 \times 10^{-4}}{0.1882} = \textbf{7.4} \times \textbf{10}^{\textbf{-4}}$$

This is O.K. within round-off error.

With regard to calculating the pH of a mixture of weak acids, the basic question remains: **Which is the dominant equilibrium among the several that are followed?** If you can resolve that, then the problem reduces to the pH of what is effectively one species in solution.

Example 14.5 C   The pH of a Mixture of Weak Acids

Calculate the pH of a mixture of 2.00 M formic acid (HCOOH, $K_a = 1.77 \times 10^{-4}$) and 1.50 M hypobromous acid (HOBr, $K_a = 2.06 \times 10^{-9}$). What are the concentrations of both the hypobromite ion (OBr^-) and hydroxide (OH^-) ion at equilibrium?

Solution

Like most other problems in equilibrium chemistry, this one is solved by **making proper assumptions**. Looking at the three sources of H^+,

$$HCOOH(aq) \rightleftharpoons H^+(aq) + COOH^-(aq) \qquad K_a = 1.77 \times 10^{-4}$$
$$HOBr(aq) \rightleftharpoons H^+(aq) + OBr^-(aq) \qquad K_a = 2.06 \times 10^{-9}$$
$$H_2O(aq) \rightleftharpoons H^+(aq) + OH^-(aq) \qquad K_w = 1.0 \times 10^{-14}$$

We see that **the dissociation of formic acid is by far the most important as a supplier of H^+.** Le Châtelier's principle will dictate that the dissociation of hypobromous acid and the autoionization of water will be suppressed. We have therefore reduced the problem to finding the pH of 2.00 M formic acid.

$$K_a = 1.77 \times 10^{-4} = \frac{[H^+][COOH^-]}{[HCOOH]}$$

	initial (M)	change	final (M)
HCOOH	2.00	$-X$	$2.00 - X$ (≈ 2.00)
H^+	0	$+X$	X
$COOH^-$	0	$+X$	X

$$1.77 \times 10^{-4} = \frac{X^2}{2.00}$$

$$X = 1.88 \times 10^{-2} M = [H^+] = [COOH^-]$$

To test the 5% rule,

$$\frac{0.0188}{2.00} \times 100\% = 0.94\%. \text{ Our assumption was O.K.}$$

$$[H^+] = 0.0188 \quad \Rightarrow \quad \textbf{pH = 1.73}$$

To find $[OBr^-]$, use its equilibrium expression:

$$K_a = \frac{[OBr^-][H^+]}{[HOBr]}$$

We know the K_a, $[H^+]$, and $[HOBr]$ (which $\approx [HOBr]_o$ because of negligible dissociation).

$$2.06 \times 10^{-9} = \frac{X(0.0188)}{1.50}$$

$$X = [OBr^-] = \textbf{1.6} \times \textbf{10}^{-7} M$$

To find $[OH^-]$, if pH = 1.73, pOH = 14 − 1.73 = 12.27

$$[OH^-] = 10^{-12.27} = \textbf{5.4} \times \textbf{10}^{-13} M$$

Example 14.5 D   Percent Dissociation

Determine the percent dissociation of the formic acid solution given in the previous problem.

Solution

$$\text{percent dissociated} = \frac{\text{amount dissociated } (M)}{\text{initial concentration } (M)} \times 100\%$$

The amount dissociated = "X" = $[COOH^-]$

$$\textbf{\% dissociated} = \frac{0.0188}{2.00} \times 100\% = \textbf{0.94\%}$$

You will note that we determined this as part of the previous problem!

Finally, your textbook deals with determination of K_a from the percent dissociated. This type of problem is straightforward if you recognize that **you can calculate equilibrium concentrations from the percent of dissociation** as is shown in our next example.

Example 14.5 E K_a from Percent Dissociation

In a 0.500 M solution, uric acid ($HC_5H_3N_4O_4$) is 1.6% dissociated. Calculate the value of K_a for uric acid.

Solution

$$\text{\% dissociated} = \frac{\text{amount dissociated}}{\text{initial concentration}} \times 100\%$$

The amount dissociated = $[C_5H_3N_4O_4^-]$ = $[H^+]$

$$1.6\% = \frac{[C_5H_3N_4O_4^-]}{0.500} \times 100\%$$

$$[C_5H_3N_4O_4^-] = [H^+] = 8.0 \times 10^{-3} \, M$$

We can now substitute into our equilibrium expression.

$$K_a = \frac{[C_5H_3N_4O_4^-][H^+]}{[HC_5H_3N_4O_4]} = \frac{(8.0 \times 10^{-3})^2}{0.500 - 0.008} = \textbf{1.3} \times \textbf{10}^{-4}$$

14.6   Bases

When you finish this section you will be able to calculate the pH of a variety of basic solutions.

The key to determining the pH of basic solutions is to recognize that, **in an equilibrium sense**, bases work in the same way that acids do. Just as there are both strong and weak acids, there are both strong and weak bases.

Strong bases completely dissociate. Using LiOH in water as an example,

$$LiOH(s) \rightarrow Li^+(aq) + OH^-(aq)$$

Therefore, one can consider that $[OH^-] \approx [LiOH]_o$. Once you know $[OH^-]$, you can use K_w to calculate $[H^+]$ and pH.

Your textbook points out that all alkali hydroxides are strongly basic. Alkaline earth hydroxides are also strongly basic, but are somewhat less soluble than alkali hydroxides.

Example 14.6 A   pH of a Strong Base

Calculate the pH of a solution made by putting 4.63 g of LiOH into water and diluting to a total volume of 400. mL.

Solution

As we discussed earlier, LiOH is a strong base. The equilibrium concentration of OH⁻ will be equal to the initial concentration of LiOH.

$$[LiOH]_0 = \frac{4.63 \text{ g LiOH}}{0.400 \text{ L}} \times \frac{1 \text{ mol LiOH}}{23.95 \text{ g}} = \mathbf{0.482 \ M}$$

$$[LiOH]_o = [OH^-] = \mathbf{0.482 \ M}$$

$$pOH = 0.316$$

Recall that $pK_w = 14.00 = pH + pOH$. Therefore,

$$\mathbf{pH} = 14 - pOH = 14.00 - 0.32 = \mathbf{13.68}$$

$[H^+] = 10^{-13.68} = 2.1 \times 10^{-14} \ M$. The only source of H^+ was the autoionization of water.

Weak bases react with water ("undergo hydrolysis") as described by the following equation.

$$B(aq) + H_2O(l) \rightleftharpoons BH^+(aq) + OH^-(aq)$$
$$\underset{\text{base 1}}{\uparrow} \quad \underset{\text{acid 2}}{\uparrow} \quad \underset{\text{acid 1}}{\uparrow} \quad \underset{\text{base 2}}{\uparrow}$$

For **weak bases**, as with weak acids, **the position of equilibrium lies far to the left**. The strategy for solving for the pH of weak bases (via pOH) is the same as for weak acids. Same steps. Same assumption. Same "5% test," as is shown in the following example.

Example 14.6 B   pH of a Weak Base

Calculate the pH of a 0.350 M solution of methylamine, CH_3NH_2 ($K_b = 4.38 \times 10^{-4}$).

Solution

We must proceed with the same problem-solving strategy as with weak acids. Given the value of K_b, the extent of equilibrium for the base hydrolysis will be far to the left.

$$CH_3NH_2(aq) + H_2O(l) \rightleftharpoons CH_3NH_3^+(aq) + OH^-(aq)$$

Although the autoionization of water can supply OH⁻ to the solution, the value of K_w is small relative to the K_b of methylamine, so $[OH^-]$ due to water can be neglected.

$$K_b = \frac{[CH_3NH_3^+][OH^-]}{[CH_3NH_2]}$$

	initial (M)	change	final (M)
CH_3NH_2	0.350	$-X$	$0.350 - X$ (≈ 0.350)
OH^-	0	$+X$	X
$CH_3NH_3^+$	0	$+X$	X

(The assumption, which must be tested, is that $[CH_3NH_2] = [CH_3NH_2]_0$. We will use the "5% rule" to verify this later.)

$$K_b = 4.38 \times 10^{-4} = \frac{X^2}{0.350}$$

$$X = [OH^-] = [CH_3NH_3^+] = \mathbf{0.0124\ M}$$

$$\frac{0.0124}{0.350} \times 100\% = 3.5\% \text{ hydrolysis, which passes the 5\% test.}$$

$$\mathbf{pOH} = -\log(0.0124) = \mathbf{1.91}$$
$$\mathbf{pH} = 14.00 - 1.91 = \mathbf{12.09}$$

Does the Answer Make Sense?

There are really two ways to answer the question. As a double-check of our math,

$$K_b = \frac{(0.0124)^2}{0.350} = 4.39 \times 10^{-4}$$

So far so good. The more significant question relates to the value for pH. We needed to have a pH > 7 for a base. We have that in this case. Therefore, the answer makes sense.

14.7   Polyprotic Acids

When you finish this section you will be able to solve for the pH and concentrations of species of polyprotic acids in aqueous solution.

A **polyprotic acid** can furnish **more than one proton** to a solution. Table 14.4 in your textbook lists a number of examples along with K_a values. Please note that in every case, $K_{a_1} \gg K_{a_2}$ for these acids. This means that for most of these polyprotic acids, the first proton comes off **relatively easily**. The second (and third, where applicable) does not. Another way of saying this is that the second and third dissociation reactions are generally so far "to the left" that we can neglect them. The beauty of this is that most polyprotic acid pH problems **reduce to finding the pH from a single, dominant equation**.

As you do the examples in your text and in this study guide, you will see that our problem-solving strategy is essentially the same as with our previous acid problems.

Example 14.7 A   pH of Oxalic Acid

Using the information in Table 14.4 in your textbook, **calculate the pH** of a 1.40 $M\ H_2C_2O_4$ (oxalic acid) solution and the **equilibrium concentrations of $H_2C_2O_4$, $HC_2O_4^-$, $C_2O_4^{2-}$, and OH^-.**

Solution

A.   pH, $[H_2C_2O_4]$, and $[HC_2O_4^-]$

The major species in solution are $\mathbf{H_2C_2O_4}$ **and** $\mathbf{H_2O}$. There are a number of equilibria that will occur. Based on the values of K_{a_1}, K_{a_2}, and K_w, by far the **most significant equilibrium** will be the dissociation of $H_2C_2O_4$.

$$H_2C_2O_4(aq) \rightleftharpoons HC_2O_4^-(aq) + H^+(aq)$$

$$K_{a_1} = \frac{[H^+][HC_2O_4^-]}{[H_2C_2O_4]}$$

	initial (M)	change	final (M)
$H_2C_2O_4$	1.40	$-X$	$1.40 - X$ (≈ 1.40)
H^+	0	$+X$	X
$HC_2O_4^-$	0	$+X$	X

As always, we will test our "negligible dissociation" assumption ($1.40 - X \approx 1.40$) later on.

$$6.5 \times 10^{-2} = \frac{[H^+][HC_2O_4^-]}{[H_2C_2O_4]} = \frac{X^2}{1.40}$$

$$X = [H^+] = [HC_2O_4^-] = \mathbf{0.302}$$

Testing the 5% rule,

$$\frac{0.302}{1.40} \times 100\% = 21.5\%!$$

Therefore, $[H_2C_2O_4] \neq [H_2C_2O_4]_0$ but rather equals "$1.40 - X$."

$$6.5 \times 10^{-2} = \frac{X^2}{1.40 - X}$$

Clearing the fraction and setting equal to zero so that we can use the quadratic formula,

$$X^2 + 0.065(X) - 0.091 = 0, \qquad a = 1, \ b = 0.065, \ c = -0.091$$

Solving,

$$X = \frac{-0.065 + \sqrt{(0.065)^2 - (4)(1)(-0.091)}}{(2)(1)}$$

$$X = \frac{-0.065 + 0.6068}{2} = \mathbf{0.271\ M}$$

$$[H^+] = [HC_2O_4^-] = \mathbf{0.27\ M}$$

$$[H_2C_2O_4] = 1.40 - 0.271 = \mathbf{1.13\ M}$$

$$pH = -\log(0.27) = \mathbf{0.57}$$

Checking our math,

$$K_{a_1} = \frac{(0.27)^2}{1.1} = \mathbf{0.066}, \ \text{O.K. within round-off error.}$$

B. $[C_2O_4^{2-}]$ and $[OH^-]$

We can find $[C_2O_4^{2-}]$ by using the second dissociation equilibrium of oxalic acid,

$$HC_2O_4^-(aq) \rightleftharpoons H^+(aq) + C_2O_4^{2-}(aq) \qquad K_{a_2} = \mathbf{6.1 \times 10^{-5}}$$

We know $[HC_2O_4^-]$, $[H^+]$ and K_{a_2}. We can therefore substitute into our equilibrium expression,

$$K_{a_2} = \frac{[H^+][C_2O_4^{2-}]}{[HC_2O_4^-]}$$

$$6.1 \times 10^{-5}\ M = \frac{(0.27)[C_2O_4^{2-}]}{0.27}$$

$$[C_2O_4^{2-}] = K_{a_2} = 6.1 \times 10^{-5} \, M$$

Note that $[C_2O_4^{2-}]$ is smaller than $[HC_2O_4^-]$ by a factor of 10^4. We can therefore conclude that the dissociation of $HC_2O_4^-$ was not a significant contributor of H^+ to the reaction.

We can find $[OH^-]$ by using $pK_w = pH + pOH$.

$$pOH = 14 - pH \qquad \Rightarrow \qquad pOH = 14 - 0.57 = 13.43$$
$$[OH^-] = 10^{-13.43} = 3.7 \times 10^{-14} \, M$$

The autoionization of water was clearly not important here (as evidenced by the low $[OH^-]$).

Keep in mind that even though the problem was long, it was all *based on the assumption* that only the first acid dissociation equilibrium was important.

Example 14.7 B   Phosphoric Acid

Using data from <u>Table 14.4 in your textbook</u>, calculate the pH, $[PO_4^{3-}]$, and $[OH^-]$ in a 6.0 M phosphoric acid (H_3PO_4) solution.

Solution

In order to calculate $[PO_4^{3-}]$, we must use K_{a_3}. We must, therefore, know $[HPO_4^{2-}]$ and $[H^+]$. To know $[HPO_4^{2-}]$, we must use K_{a_2}, which means knowing $[H_2PO_4^-]$. This can be determined using K_{a_1}. **The bottom line** is that we must proceed as in the previous problem, by determining pH and working our way down.[*] We can find $[OH^-]$ by determining the pH and using pK_w, as always. We will make the usual simplifying assumption that leads to

$$K_{a_1} = \frac{[H^+][H_2PO_4^-]}{[H_3PO_4]} \qquad \Rightarrow \qquad 7.5 \times 10^{-3} = \frac{X^2}{6.0}$$

$$X = [H^+] = [H_2PO_4^-] = 0.212 \, M$$

We must now run the 5% test,

$$\frac{0.212}{6.0} \times 100\% = 3.5\%,$$

which passes the test of "negligible dissociation."

To find $[HPO_4^{2-}]$,

$$K_{a_2} = \frac{[H^+][HPO_4^{2-}]}{[H_2PO_4^-]} \qquad \Rightarrow \qquad 6.2 \times 10^{-8} = \frac{(0.212)[HPO_4^{2-}]}{0.212}$$

$$[HPO_4^{2-}] = 6.2 \times 10^{-8} = K_{a_2}$$

---

[*] It can be shown that $[PO_4^{3-}] = \dfrac{K_{a_1} K_{a_2} K_{a_3} [H_3PO_4]_0}{[H^+]}$ but that is beyond the scope of this study guide.

To find $[PO_4^{3-}]$,

$$K_{a_3} = \frac{[H^+][PO_4^{3-}]}{[HPO_4^{2-}]} \quad\Rightarrow\quad 4.8\times10^{-13} = \frac{(0.212)[PO_4^{3-}]}{6.2\times10^{-8}}$$

$$[PO_4^{3-}] = 1.4\times10^{-19}\,M$$

We can find $[OH^-]$ in the usual way,

$$[OH^-] = K_w/[H^+] = 1.0\times10^{-14}/0.212$$

$$[OH^-] = 4.7\times10^{-14}\,M$$

Note the discussion regarding sulfuric acid along with Examples 14.16 and 14.17 in your textbook. Why does your textbook single out sulfuric acid for special consideration?

14.8   Acid-Base Properties of Salts

When you finish this section, you will be able to calculate the pH of a variety of salt solutions.

Salts are ionic compounds. They dissociate in water and may exhibit acid-base behavior. The **key question** in deciding whether a salt will act as an acidic, basic, or neutral species in solution is **"What are the acid-base properties, and strengths, of each component of the salt?"**

For example, sodium nitrite, $NaNO_2$, completely dissociates to give Na^+ and NO_2^- ions. These are the main species in solution in addition to H_2O. What are the acid-base properties of each of these species? Remember your conjugate acid-base relationships.

- Strong acids and bases have weak conjugates. **Na^+ and other alkali and alkaline earth metals exhibit no acid-base properties**.
- The nitrite ion, NO_2^-, is the conjugate base of the moderately weak acid HNO_2 ($K_a = 4.6\times10^{-4}$). Remember that "weak" is a relative term. HNO_2 is weak compared to HNO_3, which has a $K_a \gg 1$. But HNO_2 is far stronger than HCN ($K_a = 6.2\times10^{-10}$). **The NO_2^- ion will act as a very weak base**. (Your textbook points out that very weak bases, such as CN^- (and, here, NO_2^-), compete for protons against OH^-!)
- Water will have relatively little acid-base effect. Therefore **an aqueous solution of $NaNO_2$ should be slightly basic**.

Your textbook proves the relationship that, for an aqueous solution,

$$K_w = K_a \times K_b$$

You will generally find the K_a for a weak acid or K_b for a weak base in chemical data tables (such as Tables 14.2, 3, and 4 in your textbook). You will have to **calculate K_a or K_b for the ion of a salt** from given values for the conjugate neutral acid or base.

Example 14.8 A   Calculating K_a or K_b

Using the data given below, calculate K_a or K_b (as required), and write the reaction with water for each of the following aqueous ions:

a. NO_2^- (K_a for $HNO_2 = 4.0\times10^{-4}$)
b. F^- (K_a for HF $= 7.2\times10^{-4}$)
c. $C_6H_5NH_3^+$ (K_b for aniline, $C_6H_5NH_2$, equals 3.8×10^{-10})
d. $HC_3H_4N_2^+$ (K_b for imidazole, $C_3H_4N_2$, equals 1.11×10^{-7})

Solution

In each case, $K_w = K_a \times K_b$

a. $NO_2^-(aq) + H_2O(l) \rightleftharpoons HNO_2(aq) + OH^-(aq)$
   $K_b = K_w/K_a = 1.0 \times 10^{-14} / 4.0 \times 10^{-4} = \mathbf{2.5 \times 10^{-11}}$

b. $F^-(aq) + H_2O(l) \rightleftharpoons HF(aq) + OH^-(aq)$
   $K_b = K_w/K_a = 1.0 \times 10^{-14} / 7.2 \times 10^{-4} = \mathbf{1.4 \times 10^{-11}}$

c. $C_6H_5NH_3^+(aq) + H_2O(l) \rightleftharpoons C_6H_5NH_2(aq) + H_3O^+(aq)$

   OR

   $C_6H_5NH_3^+(aq) \rightleftharpoons C_6H_5NH_2(aq) + H^+(aq)$
   $K_a = K_w/K_b = 1.0 \times 10^{-14} / 3.8 \times 10^{-10} = \mathbf{2.6 \times 10^{-5}}$

d. $HC_3H_4N_2^+(aq) \rightleftharpoons C_3H_4N_2(aq) + H^+(aq)$ (short form)
   $K_a = K_w/K_b = 1.0 \times 10^{-14} / 9.0 \times 10^{-8} = \mathbf{1.1 \times 10^{-7}}$

Recalling the bottom line, you must decide how each part of the salt (cation and anion) will react with water. Your textbook points out a sticky case (such as $NH_4C_2H_3O_2$) where **both cation and anion exhibit acid-base behavior**. When this occurs, the overall pH of the solution is determined by the relative K_a and K_b values for the ions (see Table 14.5 in your textbook).

Example 14.8 B   *Predicting Acid-base Behavior*

Using data from Tables 14.2, 3, and 4 in your textbook, predict whether each of the following will create an acid, base, or neutral aqueous solution.

   a. Na_3PO_4 b. KI   c. HC_5H_5NCl (pyridinium chloride)   d. NH_4F

Solution

a. Na^+ exhibits no acid-base behavior. PO_4^{3-} is the most basic form of H_3PO_4 ($K_{b_1} = K_w/K_{a_3}$). This solution will be a **fairly strong base**.

b. K^+ exhibits no acid-base behavior. I^- also exhibits no acid-base behavior. This solution will be **neutral**.

c. Cl^- exhibits no acid-base behavior. Pyridinium ion, $HC_5H_5N^+$, is the conjugate acid of pyridine, C_5H_5N ($K_a = K_w/K_b$). This solution will be **acidic**.

d. NH_4^+ is the conjugate acid of NH_3 ($K_a = K_w/K_b = 5.6 \times 10^{-10}$). F^- is the conjugate base of HF ($K_b = K_w/K_a = 1.4 \times 10^{-11}$). $K_a > K_b$; therefore, the solution will be a **weak acid**.

Example 14.8 C   *pH of a Salt*

Calculate the pH of a 0.500 M $NaNO_2$ solution. (K_a for $HNO_2 = 4.0 \times 10^{-4}$)

Solution

As discussed in the previous example, $NaNO_2$ will act as a base in water due to the hydrolysis of NO_2^-.

$$NO_2^-(aq) + H_2O(l) \rightleftharpoons HNO_2(aq) + OH^-(aq)$$

$$K_b = \frac{[HNO_2][OH^-]}{[NO_2^-]}$$

$$K_b = K_w / K_a = 1.0 \times 10^{-14} / 4.0 \times 10^{-4} = \mathbf{2.5 \times 10^{-11}}$$

We may now proceed as with any other weak base problem.

	initial (M)	change	final (M)
NO_2^-	0.500	$-X$	$0.500 - X$ (≈ 0.500)
OH^-	0	$+X$	X
HNO_2	0	$+X$	X

$$2.5 \times 10^{-11} = \frac{X^2}{0.500}$$

$$X = [OH^-] = [HNO_2] = 3.5 \times 10^{-6}$$

This clearly passes the 5% rule (K_b is so small)!

$$pOH = -\log(3.5 \times 10^{-6}) = 5.45$$
$$\mathbf{pH} = 14 - 5.45 = \mathbf{8.55}$$

Reproduced here is a K_b table for use in this chapter.

Values of K_b for Some Common Weak Bases

Name	Formula	Conjugate Acid	K_b
Ammonia	NH_3	NH_4^+	1.8×10^{-5}
Methylamine	CH_3NH_2	$CH_3NH_3^+$	4.38×10^{-4}
Ethylamine	$C_2H_5NH_2$	$C_2H_5NH_3^+$	5.6×10^{-4}
Hydroxylamine	H_2ONH_2	$H_2ONH_3^+$	1.1×10^{-8}
Aniline	$C_6H_5NH_2$	$C_6H_5NH_3^+$	3.8×10^{-10}
Pyridine	C_5H_5N	$C_5H_5NH^+$	1.7×10^{-9}

14.9   The Effect of Structure on Acid-Base Properties

After you have read this section, answer the questions below. They will help you review the material.

1.   What are the two structural properties that determine if a molecule will act as an acid?
2.   Why is HF a weak acid compared to HCl?
3.   Explain why increasing the number of oxygens in an H—O—X grouping increases acid strength. (See Figure 14.9 in your textbook.)
4.   What is the relationship between the charge on a metal ion and the acidity of attached water molecules? Why?
5.   Why is HOCl a stronger acid than HOI?

Example 4.9   Structure and Properties

Given the two acids HOI and HIO_3, and two values for K_a, 0.17 and 2×10^{-11}, which value goes with which acid?

Solution

HIO_3, having more oxygens, will draw electrons away from the O—H (acidic) bond. Therefore, the O—H bond is weakened, and acid strength is substantially increased.

$$K_a \text{ for HOI} = 2 \times 10^{-11} \qquad\qquad K_a \text{ for } HIO_3 = 0.17$$

14.10 Acid-Base Properties of Oxides

After you have read this section, answer the questions below. They will help you review the material.

1. Why are some oxides covalent?
2. Under what circumstances will the O—X bond in H—O—X be ionic? What is the outcome in water of such an ionic bond?
3. Define **acidic oxide**.
4. Define **basic oxide**.
5. Which groups tend to produce acidic oxides? Basic oxides?

Example 14.10 Acidic and Basic Oxides

Determine whether each of the following oxides will give an acidic or basic solution when added to water.

a. K_2O
b. SO_3
c. MgO

Solution

a. The K—O bond is ionic. This bond can break so that an O—H bond can form. The solution is **basic**.
b. The S—O bond is covalent. O—H bonds will break to produce an **acidic solution**.
c. A **basic solution** is produced (see part "a").

14.11 The Lewis Acid-Base Model

When you finish this section, you will be able to identify the Lewis acids and bases in a reaction.

Your textbook points out that the **Lewis acid base** concept is **the most wide-ranging** of the three concepts introduced in this chapter.

Lewis acids are electron acceptors.
Lewis bases are electron donors.

Al^{3+}, H^+, and BF_3 are examples of Lewis acids.
NO_2^-, NH_3, and H_2O are examples of Lewis bases.

Note that many species that do not contain H^+ are Lewis acids. The all-encompassing nature of the Lewis concept is what makes it so useful. For example, for the reaction to form $Ag(NH_3)_2^+$:

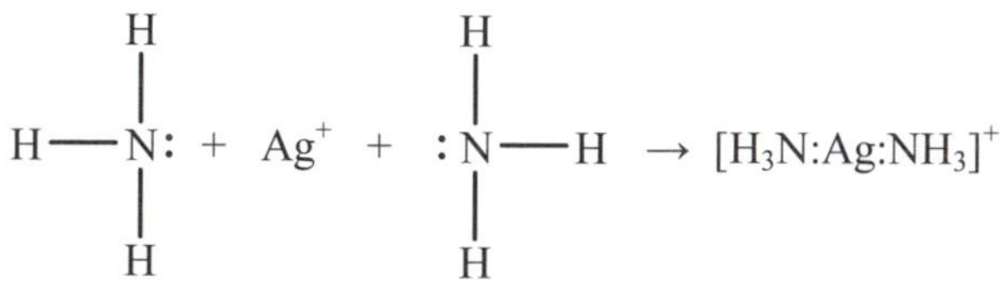

$$H_3N: + Ag^+ + :NH_3 \rightarrow [H_3N:Ag:NH_3]^+$$

Ag^+ is the Lewis acid because it **accepts electrons** from NH_3.
NH_3 is the Lewis base because it **donates electrons** to Ag^+.

Example 14.11   *Lewis Acids and Bases*

Identify the Lewis acid and base in each of the following reactions:

  a. $Cu^{2+}(aq) + 4NH_3(aq) \rightleftharpoons Cu(NH_3)_4^{2+}(aq)$
  b. $I^-(aq) + I_2(aq) \rightleftharpoons I_3^-(aq)$
  c. $Fe^{3+}(aq) + 6H_2O(l) \rightleftharpoons Fe(H_2O)_6^{3+}(aq)$

Solution

  a. $Cu^{2+} + 4{:}NH_3 \rightleftharpoons Cu(NH_3)_4^{2+}$

         ↑   ↑
       Lewis   Lewis
        acid   base

  b. $I^- \ + \ I_2 \ \rightleftharpoons \ I_3^-$

       ↑   ↑
     Lewis   Lewis
      base   acid

  c. $Fe^{3+} + \left[6 \ {:}O\begin{smallmatrix}H\\ \\H\end{smallmatrix} \right] \rightleftharpoons Fe(H_2O)_6^{3+}$

         ↑   ↑
       Lewis   Lewis
        acid   base

14.12 Strategy for Solving Acid-Base Problems: A Summary

Your textbook makes the key point here that we cannot merely memorize which formula to use to solve a given acid-base problem. There are too many variations and possible conditions. When doing such problems, we do not merely solve, we **PROBLEM SOLVE**.

Exercises

Section 14.1

1. List four strong acids and four strong bases.

2. Write the dissociation reaction for each of the following acids in water. Identify the conjugate acid-base pair in each case.

    a. C_6H_5COOH (benzoic acid)
    b. H_3BO_3 (boric acid)
    c. $H_2PO_4^-$ (dihydrogen phosphate)
    d. HNO_3 (nitric acid)

3. What is the conjugate base of the bicarbonate ion, HCO_3^-? Of formic acid, $HCHO_2$? Which is the stronger base? Why?

4. Write the dissociation reaction for each of the following acids in water, and identify the conjugate acid-base pairs:

    a. $HC_2H_2ClO_2$
    b. HCN
    c. NH_4Cl

5. Write the equilibrium expression for each of the reactions in Problem 2.

6. Write equilibrium expression for each of the equations in Problem 4.

Section 14.2

7. The values for K_a for the acids in Problem 2 are:

substance	K_a
C_6H_5COOH	6.14×10^{-5}
H_3BO_3	5.83×10^{-10}
$H_2PO_4^-$	6.3×10^{-8}
HNO_3	$\gg 1$

    Put the acids in order from strongest to weakest.

8. Predict which one of the bases in each pair is stronger:

    a. HCO_3^- or CO_3^{-2}
    b. NO_3^- or NO_2^-

9. Put the conjugate bases of the acids in Problem 7 in order from strongest to weakest.

Section 14.3

10. The pH of a solution is 11.93. What is $[H^+]$? $[OH^-]$? pOH?

11. For each of the following solutions at 25°C, calculate $[H^+]$ given $[OH^-]$ or $[OH^-]$ given $[H^+]$. Is the solution an acid or base?

    a. $[OH^-] = 1 \times 10^{-4}\ M$
    b. $[H^+] = 1 \times 10^{-6}\ M$
    c. $[H^+] = 1 \times 10^{-9}\ M$

12. A solution has $[OH^-] = 3.6 \times 10^{-1}$ M. Is this solution strongly or weakly acidic or basic?

13. What is the pH of a solution that has an H^+ concentration of 1.0×10^{-5} mol/L? Of 5.0×10^{-5} mol/L?

14. The gastric juice in our stomachs contains enough hydrochloric acid to make the hydrogen ion concentration about 0.01 mol/L. Calculate the approximate pH of gastric juice.

15. Calculate the pH for each of the following solutions at 25°C:

    a. $[H^+] = 3.8 \times 10^{-7}$ M
    b. $[H^+] = 7.2 \times 10^{-4}$ M
    c. $[H^+] = 4.1 \times 10^{-13}$ M

16. Calculate the pOH for each of the following solutions at 25°C:

    a. $[OH^-] = 2.0 \times 10^{-12}$ M
    b. $[OH^-] = 3.4 \times 10^{-11}$ M
    c. $[OH^-] = 9.2 \times 10^{-3}$ M

17. At 100°C, $K_w = 4.9 \times 10^{-13}$.

    a. What is the pH of a neutral solution at 100°C?
    b. Calculate the pH at 100°C if $[OH] = 6.3 \times 10^{-12}$ M.

18. Calculate the pH and pOH for each of the following solutions at 25°C:

    a. $[H^+] = 2.0 \times 10^{-12}$ M
    b. $[OH^-] = 3.9 \times 10^{-3}$ M
    c. $[H^+] = 7.7 \times 10^{-13}$ M
    d. $[OH^-] = 5.3 \times 10^{-9}$ M

19. If $[Cl^-] = 0.9$ M, what is pCl?

Section 14.4

20. Calculate the pH of a 7.0×10^{-2} M HCl solution.

21. Calculate the pH of a 2.8×10^{-5} M HNO$_3$ solution.

22. Fill in the missing information in the following table:

	pH	pOH	$[H^+]$	$[OH^-]$	acid, base, or neutral?
solution a	5.64				
solution b				3.9×10^{-6} M	
solution c			0.027 M		
solution d		1.7			

23. The pH of a solution of HClO$_4$ is 3.11. What is $[H^+]$?

24. The pOH of a 400-mL solution of HNO$_3$ is 12.44. How many grams of HNO$_3$ are in the solution?

25. If 0.10 mol of HCl is added to enough water to produce 1.0 L of solution, calculate the concentrations of H^+ and OH^- and the pH of the solution.

26. One liter of solution was prepared from water and 3.5×10^{-6} mol of HCl. Calculate the $[H^+]$, $[OH^-]$, and pH of the solution.

Section 14.5

27. A solution is made by dissolving 18.4 g of HNO_3 in enough water to make 662 mL of solution. Calculate the pH of the solution.

28. Calculate the $[H^+]$ and $[OH^-]$ in a 0.010 M solution of HCN. $K_a = 6.2 \times 10^{-10}$.

29. The K_a of chloroacetic acid ($ClCH_2COOH$) is 1.36×10^{-3}. Calculate the pH, the pOH, the $[H^+]$, and the $[OH^-]$ of a 1.00 M solution of chloroacetic acid.

30. What is the amount of hydrogen ion due to water in Problem 29? What is the percent dissociation of chloroacetic acid?

31. Calculate the pH of a 0.237 M solution of benzoic acid, C_6H_5COOH ($K_a = 6.14 \times 10^{-5}$).

32. A total of 0.0560 g of acetic acid is added to enough water to make 50 mL of solution. Calculate $[H^+]$, $[CH_3COO^-]$, $[CH_3COOH]$, and the pH at equilibrium. (K_a for acetic acid is 1.8×10^{-5}.)

33. Calculate K_a of a weak acid "HW" if a solution with an initial concentration of 0.200 M has a pH of 3.15.

34. Calculate the K_a of a 0.060 M weak monoprotic acid with a pH of 3.44.

35. Calculate the pH of a solution the contains 0.250 M H_2SO_4 and 1.00 M CH_3COOH ($K_a = 1.8 \times 10^{-5}$).

36. Calculate the original molarity of a solution of formic acid, HCOOH ($K_a = 1.9 \times 10^{-4}$) whose pH is 3.26 at equilibrium.

37. What is the percent dissociation of the benzoic acid in Problem 31?

38. Calculate the percent dissociation of 0.20 M benzoic acid whose $K_a = 6.4 \times 10^{-5}$.

39. Calculate the pH of a 0.20 M NH_4Cl solution ($K_a = 5.6 \times 10^{-10}$).

Section 14.6

40. The pH of a 300-g solution of NaOH is 12.97. The density of the solution is 1.10 g/mL. How many grams of NaOH are in the solution?

41. Add 0.0150 mol of LiOH to sufficient water to make 1 L of solution. Calculate the $[LiOH]_0$, $[H^+]$, $[OH^-]$, and pH of the solution.

42. Calculate the hydrogen ion and hydroxide ion concentration in:

    a. 0.005 M nitric acid,   b. 0.005 M KOH

43. How many grams of KOH are necessary to prepare 800 mL of a solution of pH = 11.56?

44. How many grams of $Ba(OH)_2$ are necessary to prepare 400 mL of a solution of pH = 12.46?

45. Calculate the pH of a 2.8×10^{-4} M $Ba(OH)_2$ solution, assuming complete dissociation of the $Ba(OH)_2$.

46. How many grams of NaOH are needed to prepare a 546-mL solution with pH of 10.00?

47. The pH of a 0.30 M solution of a weak base is 10.66. Calculate the K_b of the base.

48. A solution of ammonia, $K_b = 1.8 \times 10^{-5}$, has a pH of 11.22. Calculate the molarity of the solution.

49. Calculate the pH of a solution made by putting 11.17 g of KOH into water and diluting it to a volume of 600 mL.

50. Calculate the pH of a 0.500 M solution of dimethylamine, $(CH_3)_2NH$ ($K_b = 5.9 \times 10^{-4}$).

51. What is the percent of hydrolysis of the base in Problem 50? What is the concentration of hydrogen as a result of the autoionization of water?

52. Calculate the pH of a 0.76 M KOH solution.

53. Calculate how many grams of $HONH_2$ are needed to dissolve in enough water to make 250 mL of solution with a pH of 10.00. ($K_b = 1.1 \times 10^{-8}$)

54. Calculate the pH of a 0.15 M solution of CH_3COONa ($K_b = 5.6 \times 10^{-10}$).

Section 14.7

55. What is the pH of a 0.100 M solution of arsenic acid?

$$H_3AsO_4 \rightleftharpoons H_2AsO_4^- + H^+ \qquad K_a = 6.0 \times 10^{-3}$$
$$H_2AsO_4^- \rightleftharpoons HAsO_4^{2-} + H^+ \qquad K_a = 1.05 \times 10^{-7}$$
$$HAsO_4^{2-} \rightleftharpoons AsO_4^{3-} + H^+ \qquad K_a = 3.0 \times 10^{-12}$$

What is the concentration of AsO_4^{3-} ion? (Hint: Use the quadratic equation.)

56. Calculate the concentrations of CO_3^{2-}, HCO_3^-, and H^+ in a 0.025 M H_2CO_3 solution. (See Table 14.4 in your textbook for K_a values of carbonic acid.)

57. Calculate the equilibrium concentration of PO_4^{3-} in a solution prepared by adding 716.2 g of H_3PO_4 to water and diluting to a total volume of 750 mL. (See Table 14.4 in your textbook for K_a values of phosphoric acid.)

58. Calculate the concentrations of H_2CO_3, HCO_3^-, CO_3^{2-}, H^+, and OH^- and the pH at equilibrium in a solution that is initially 0.010 M H_2CO_3.

59. Calculate the concentrations of $C_2H_2O_4$, $C_2HO_4^-$, $C_2O_4^{2-}$, and H^+ in a 0.10 M of oxalic acid solution.

Section 14.8

60. Would solutions of the following salts act as acids or as bases when dissolved in liquid HCN?
    a. $NaNO_3$
    b. $NaOH$
    c. $NaCN$
    d. $NaCl$
    e. $NaNH_2$

61. Using data from Table 14.2 in your textbook, determine the values for K_b of
    a. CN^-
    b. NH_3
    c. $C_6H_5O^-$

62. Arrange the bases in Problem 61 in order from strongest to weakest.

63. Calculate the pH of a 0.450 M solution of sodium propionate, $NaOCH_2CH_3$ (K_a for propionic acid, $HOCH_2CH_3 = 1.34 \times 10^{-5}$).

64. Calculate the pH of a 0.20 M K_2S solution (K_{a_2} for $H_2S = 1 \times 10^{-13}$).

65. Determine the equilibrium concentration of HCN in a 0.0500 M NaCN solution (K_a for HCN = 6.2×10^{-10}).

66. Predict whether an aqueous solution of ammonium acetate (CH_3COONH_4) will be acidic, basic, or neutral.

Multiple Choice Questions

67. Which of the following substances does not fit the definition of an Arrhenius base?

    A. NH_3 B. NaOH   C. KOH   D. H_2O

68. The conjugate acid of NH_3 is:

    A. NH_2^- B. NHOH   C. NH_4^+ D. NH_2^+

69. Which of the following is the correct equilibrium expression for this reaction:

$$HCN(aq) + H_2O(l) \rightleftharpoons H_3O^+(aq) + CN^-(aq)$$

    A. $\dfrac{[CN^-][H_3O^+]}{[H_2O][HCN]}$ B. $\dfrac{[H_3O^+][CN^-]}{[HCN]}$ C. $\dfrac{[HCN]}{[H_3O^+][CN^-]}$ D. $\dfrac{[H_2O][HCN]}{[CN^-][H_3O^+]}$

70. Which one of the following acids would produce the weakest conjugate base?

    A. Sulfuric acid   B. Ammonium ion   C. Phenol   D. Acetic acid

71. What is K_b if K_a of an acid is 2.0×10^{-3}?

    A. 5.0×10^{11} B. 1×10^{-3} C. 2.0×10^{-11} D. 5.0×10^{-12}

72. What is the value of pK_w at 25°C?

    A. 7.0   B. 14   C. 1×10^{-14} D. 1×10^{-10}

73. What is the pK_a of the acid in Problem #71?

    A. 2.7   B. −2.6   C. 6.2   D. −6.21

74. What is the $[OH^-]$ of a solution with pH = 3?

    A. 3×10^{11} B. 1×10^{-11} C. 1.8×10^5 D. 11

75. The pOH of a 2.0 M HCl solution is:

    A. 12   B. 14   C. 13.7   D. 0.3

76. The pOH of a 0.300 M HI solution is:

    A. 13.5   B. 0.522   C. 0.150   D. 3.5

77. What is the final pH of a 0.100 M hydrochloric acid solution that is diluted by a factor of 10?

    A. 2   B. 13   C. 1   B. 4

78. What is the final pH of a sulfuric acid solution prepared by mixing together 392.0 g of H_2SO_4 with enough water to reach 2.0 liters? After it is diluted by a factor of 15?

    A. −0.30, 0.88   B. −0.6, 0.57   C. 1, 6.3   D. 0.30, 0.88

79. Calculate the pH of a 0.25 M HCN solution. $K_a = 6.2 \times 10^{-10}$.

    A. 4.7   B. 9.3   C. 2.5   D. 5.0

80. 0.276 g of HCN are dissolved in 50.0 mL of solution. What is the pH of this acidic solution?

    A. 2.0   B. 4.7   C. 5.0   D. 6.5

81. A 0.50 M HX solution is 0.30% ionized. What is its pH?

    A. 0.30   B. 0.0016   C. 0.65   D. 2.8

82. What is the pH of a 0.44 M solution of an acid with $K_a = 7.5 \times 10^{-5}$?

    A. 2.24   B. −0.54   C. 11.8   D. 4.48

83. What is the percent ionization of a 0.55 M HW solution with a $K_a = 1.5 \times 10^{-2}$?

    A. 15.0%   B. 30.0%   C. 0.30%   D. 65.0%

84. What is the pOH of a 0.50 M ethylamine, $C_2H_5NH_2$, solution? $pK_b = 5.3$.

    A. 11.2   B. 2.8   C. 0.28   D. 0.0016

85. Calculate the pH of a 0.0075 M solution of a weak base, $K_b = 1.6 \times 10^{-6}$.

    A. 10.04   B. 3.96   C. 0.00124   D. 13.98

86. Calculate the pK_a of the conjugate acid of methylamine (CH_3NH_2), if the pH of a 0.750 M methylamine solution is 11.8.

    A. 9.72   B. −9.59   C. 11.8   D. 2.32

87. Calculate the pH of a 0.10 M H_2S solution. $K_{a_1} = 1.3 \times 10^{-7}$, $K_{a_2} = 1.0 \times 10^{-13}$.

    A. 4.2   B. 6.1   C. 3.9   D. 1.1

88. What is the pH of a 0.1 M phosphoric acid solution? $K_{a_1} = 7.5 \times 10^{-3}$, $K_{a_2} = 6.2 \times 10^{-8}$, $K_{a_3} = 4.8 \times 10^{-13}$.

    A. 4.0   B. 1.5   C. 2.0   D. 1.0

89. Which of the following salts would produce an acidic solution?

    A. NaCl   B. CH_3NH_3Cl C. CH_3NHNa D. $NaNO_3$

90. Which one of the following substances is most acidic?

    A. CH_3Cl B. CH_2ClO C. H_3CCOOH D. HI

91. Which one of the following substances has the strongest conjugate base?

    A. HI   B. HBr   C. HF   D. HCl

92. Which one of the following substances is the strongest base?

    A. H_2O B. NH_3 C. NH_2^- D. CH_4

93. Which one of the following oxides is acidic?

    A. CaO   B. K_2O C. SO_2 D. MgO

94. A solution prepared by mixing water with the oxide of which one of the following groups of elements would yield an acid?

    A. Group IA   B. Group IB   C. Group VIA   D. Group VIIA

95. A Lewis acid:

    A. accepts a proton   B. releases hydrogens   C. accepts electrons   D. donates electrons

Answers to Exercises

1. Acids: HCl, HNO$_3$, HClO$_4$, H$_2$SO$_4$. Bases: NaOH, KOH, Ba(OH)$_2$, LiOH.

2.

	acid 1	base 2		base 1	acid 2
a.	C$_6$H$_5$COOH	+ H$_2$O	$\rightleftharpoons$	C$_6$H$_5$COO$^-$	+ H$_3$O$^+$
b.	H$_3$BO$_3$	+ H$_2$O	$\rightleftharpoons$	H$_2$BO$_3^-$	+ H$_3$O$^+$
c.	H$_2$PO$_4^-$	+ H$_2$O	$\rightleftharpoons$	HPO$_4^{2-}$	+ H$_3$O$^+$
d.	HNO$_3$	+ H$_2$O	$\rightleftharpoons$	NO$_3^-$	+ H$_3$O$^+$

3. Bicarbonate ion   Conjugate base: CO$_3^{2-}$
   Formic acid   Conjugate base: HCOO$^-$ Bicarbonate ion is the stronger base.

   By looking at the pK_a's of each, this can be determined. The higher the pK_a, the weaker the acid; the weaker the acid, the stronger the base. pK_a formic acid = 3.74, pK_a bicarbonate ion = 10.32; therefore, bicarbonate ion is a weaker acid and a stronger base than formic acid.

4.

	acid 1	base 2		base 1	acid 2
a.	HC$_2$H$_2$ClO$_2$(aq)	+ H$_2$O(l)	$\rightleftharpoons$	C$_2$H$_2$ClO$_2^-$ (aq)	+ H$_3$O$^+$(aq)
b.	HCN(aq)	+ H$_2$O(l)	$\rightleftharpoons$	CN$^-$(aq)	+ H$_3$O$^+$(aq)
c.	NH$_4$Cl(aq)	+ H$_2$O(l)	$\rightleftharpoons$	NH$_3$Cl$^-$(aq)	+ H$_3$O$^+$(aq)

5. a. $K_a = \dfrac{[C_6H_5COO^-][H^+]}{[C_6H_5COOH]}$ c. $K_a = \dfrac{[HPO_4^{2-}][H^+]}{[H_2PO_4^-]}$

   b. $K_a = \dfrac{[H_2BO_3^-][H^+]}{[H_3BO_3]}$ d. $K_a = \dfrac{[NO_3^-][H^+]}{[HNO_3]}$ $(K \gg 1)$

6. a. $K_a = \dfrac{[C_2H_2ClO_2^-][H^+]}{[HC_2H_2ClO_2]}$ c. $K_a = \dfrac{[NH_3Cl^-][H^+]}{[NH_4Cl]}$

   b. $K_a = \dfrac{[CN^-][H^+]}{[HCN]}$

7. HNO$_3$, C$_6$H$_5$COOH, H$_2$PO$_4^-$, H$_3$BO$_3$
   strongest $\longleftarrow$ $\longrightarrow$ weakest

8. a. CO$_3^{2-}$ b. NO$_2^-$

9. H$_2$BO$_3^-$, HPO$_4^{2-}$, C$_6$H$_5$COO$^-$, NO$_3^-$
   strongest $\longleftarrow$ $\longrightarrow$ weakest

10. [H$^+$] = 1.2×10^{-12} M; [OH$^-$] = 8.5×10^{-3} M; pOH = 2.07

11. a. [H$^+$] = 1×10^{-10}; the solution is basic.
    b. [OH$^-$] = 1×10^{-8}; the solution is acidic.
    c. [OH$^-$] = 1×10^{-5}; the solution is basic.

12. strongly basic

13. a. pH = 5   b. pH = 4.3

14. pH = 2.0

15. a. pH = 6.42   b. pH = 3.14   c. pH = 12.39

16. a. pOH = 11.70   b. pOH = 10.47   c. pOH = 2.04

17. a. pH = 6.15   b. pH = 1.11

18. a. pH = 11.70; pOH = 2.30   c. pH = 12.11; pOH = 1.89
    b. pH = 11.59; pOH = 2.41   d. pH = 5.72; pOH = 8.28

19. pCl = 0.05

20. pH = 1.15

21. pH = 4.55

22.

	pH	pOH	$[H^+]$	$[OH^-]$	acid, base, or neutral?
solution a	5.64	8.36	$2.3 \times 10^{-6}\ M$	$4.4 \times 10^{-9}\ M$	acid
solution b	8.59	5.41	$2.6 \times 10^{-9}\ M$	$3.9 \times 10^{-6}\ M$	base
solution c	1.57	12.43	$0.027\ M$	$3.7 \times 10^{-13}\ M$	acid
solution d	12.3	1.7	$5.\ \times 10^{-13}\ M$	$2.\ \times 10^{-2}\ M$	base

23. $[HClO_4] = 7.8 \times 10^{-4}\ M$

24. 0.69 g HNO_3

25. $[H^+] = 0.10\ M$; $[OH^-] = 1.0 \times 10^{-13}\ M$; pH = 1.00

26. $[H^+] = 3.5 \times 10^{-6}\ M$; $[OH^-] = 2.9 \times 10^{-9}\ M$; pH = 5.45

27. pH = 0.355

28. $[OH^-] = 4.55 \times 10^{-9}\ M$; $[H^+] = 2.2 \times 10^{-6}\ M$

29. pH = 1.43; pOH = 12.57; $[OH^-] = 2.7 \times 10^{-13}\ M$; $[H^+] = 3.7 \times 10^{-2}\ M$

30. $[H^+]_{H_2O} = [OH^-]_{total} = 2.7 \times 10^{-13}\ M$; 3.7%

31. pH = 2.42

32. $[H^+] = [CH_3COO^-] = 5.8 \times 10^{-4}\ M$; $[CH_3COOH] = 0.018\ M$; pH = 3.24

33. $K_a = 2.5 \times 10^{-6}$

34. $K_a = 2.2 \times 10^{-6}$

35. pH = 0.53

36. $1.6 \times 10^{-3}\ M$

37. percent dissociation = 0.72%

38. 1.8%

39. pH = 4.98

40. 1.02 g NaOH

41. $[LiOH]_o = [OH^-] = 0.015$; $[H^+] = 6.7 \times 10^{-13}\ M$; pH = 12.18

42. a. $[OH-] = 2.0 \times 10^{-12}\ M$; $[H+] = 0.0050\ M$ b. $[OH^-] = 0.0050\ M$; $[H+] = 2.0 \times 10^{-12}\ M$

43. 0.16 g KOH

44. 1.0 g $Ba(OH)_2$

45. pH = 10.75

46. 2.2×10^{-3} g

47. 7.0×10^{-7}

48. 0.15 M

49. pH = 13.52

50. pH = 12.23

51. percent hydrolysis = 3.4%; $[H^+]_{H_2O} = [H^+]_{total} = 5.9 \times 10^{-13}\ M$

52. pH = 13.88

53. 7.5 g

54. pH = 8.96

55. pH = 1.66; $[AsO_4^{3-}] = 1.5 \times 10^{-17}\ M$

56. $[CO_3^{2-}] = 5.6 \times 10^{-11}\ M$; $[HCO_3^-] = [H^+] = 1.0 \times 10^{-4}\ M$

57. Calculate $[H^+]$, and work your way down: $[H^+] = 0.27$, (pH = 0.57); $[PO_4^{3-}] = 1.1 \times 10^{-19}\ M$

58. $[H_2CO_3] = 0.010\ M$; $[HCO_3^-] = [H^+] = 6.6 \times 10^{-5}\ M$; $[CO_3^{2-}] = 5.6 \times 10^{-11}\ M$;
     $[OH^-] = 1.5 \times 10^{-10}\ M$; pH = 4.18

59. $[C_2H_2O_4] = 0.046\ M$; $[C_2HO_4^-] = [H^+] = 0.054\ M$; $[C_2O_4^{2-}] = 6.1 \times 10^{-5}\ M$

60. a. Weak base   c. Base   e. Base
     b. Strong base   d. Base

61. a. $K_b = 1.6 \times 10^{-5}$ b. $K_b = 1.8 \times 10^{-5}$ c. $K_b = 6.3 \times 10^{-5}$

62. $C_6H_5O^- > NH_3 > CN^-$

63. pH = 9.26

64. pH = 13.0

65. $[\text{HCN}] = 9.0 \times 10^{-4}\ M$

66.   Neutral (pH = 7.0)

67. A	68. C	69. B	70. A	71. D	72. B
73. A	74. B	75. C	76. A	77. A	78. A
79. D	80. C	81. D	82. A	83. A	84. B
85. B	86. A	87. C	88. B	89. B	90. D
91. C	92. C	93. C	94. C	95. C	

Chapter 15

Acid-Base Equilibria

Section	Contents	Page in this Study Guide
15.1	Solutions of Acids or Bases Containing a Common Ion	335
15.2	Buffered Solutions	336
15.3	Buffer Capacity	341
15.4	Titrations and pH Curves	343
15.5	Acid-Base Indicators	349
	Exercises	351
	Multiple Choice Questions	353
	Answers to Exercises	356

This chapter covers a great deal of material. The beauty is that **you already have the knowledge to handle the material**. Very little is "new." Most of the material represents **applications of equilibrium** theory as applied to acid-base chemistry. If you **think through** the various problems, you should do just fine.

15.1 Solutions of Acids or Bases Containing a Common Ion

When you finish this section you will be able to perform calculations on acidic solutions that involve a common ion.

The important concept in this section is how the addition of a "common ion" to a solution can depress the dissociation of an acid. This is based on Le Châtelier's principle. To illustrate the point, acetic acid ($HC_2H_3O_2$) dissociates in water as follows:

$$HC_2H_3O_2(aq) \xrightleftharpoons{K_a} H^+(aq) + C_2H_3O_2^-(aq)$$

Let us add sodium acetate ($NaC_2H_3O_2$), which completely dissociates to form $Na^+(aq)$ and $C_2H_3O_2^-(aq)$. The acetate ion, $C_2H_3O_2^-$, is *common* to the solution (see the above equation). It is called the *common ion. The common ion imposes a stress on the system.* The system responds by shifting **to the left**. *Therefore, there is much less dissociation than before.* The general effect of a common ion is to depress the **amount of dissociation.**

Because the dissociation is depressed, we can assume that the equilibrium concentration of acetic acid equals the initial concentration. The same is true of the acetate ion.

Let's see how this applies to the pH of a weak acid.

Example 15.1 The Common Ion Effect

Calculate the pH and the percent dissociation of the acid in each of the following solutions:

a. 0.200 M $HC_2H_3O_2$ ($K_a = 1.8 \times 10^{-5}$)
b. 0.200 M $HC_2H_3O_2$ in the presence of 0.500 M $NaC_2H_3O_2$

Solution

a. The major species are acetic acid, $HC_2H_3O_3$, and H_2O. As we saw in the previous chapter, K_w for the autoionization of water is insignificant compared to the K_a for acetic acid dissociation. We need only calculate based on the acid dissociation.

$$HC_2H_3O_2(aq) \rightleftharpoons H^+(aq) + C_2H_3O_2^-(aq)$$

$$K_a = \frac{[H^+][C_2H_3O_2^-]}{[HC_2H_3O_2]}$$

Using techniques from Chapter 14,

$$1.8 \times 10^{-5} = \frac{X^2}{0.200}$$

$$X = [H^+] = [C_2H_3O_2^-] = 1.9 \times 10^{-3}\ M, \qquad pH = 2.7$$

$$\% \text{ dissociation} = \frac{1.9 \times 10^{-3}}{0.200} \times 100\% = 0.95\%$$

b. **The difference here is that the presence of the acetate ion, $C_2H_3O_2^-$,** ($NaC_2H_3O_2$ is a strong electrolyte, as discussed previously) will *depress the dissociation of $HC_2H_3O_2$* as predicted by Le Châtelier's principle.

We can assume (and test) that because there is so little dissociation,

$$[HC_2H_3O_2]_0 \approx [HC_2H_3O_2] = 0.200\ M$$
$$\text{and}$$
$$[C_2H_3O_2^-]_0 \approx [C_2H_3O_2^-] = 0.500\ M$$

This makes our problem remarkably easy to solve. Only $[H^+]$ is unknown,

$$K_a = \frac{[H^+][C_2H_3O_2^-]}{[HC_2H_3O_2]} \qquad \Rightarrow \qquad 1.8 \times 10^{-5} = \frac{[H^+](0.500)}{0.200}$$

$$[H^+] = 7.2 \times 10^{-6}\ M, \qquad pH = 5.14$$

Note that the **pH increased markedly** due to the presence of the **basic salt**. The extent of dissociation is now reflected by $[H^+]$ because the only significant source of H^+ ions is the dissociation of $HC_2H_3O_2$.

$$\% \text{ dissociation} = \frac{7.2 \times 10^{-6}}{0.200} \times 100\% = 3.6 \times 10^{-3}\ \%$$

Therefore, the dissociation was substantially depressed, and our assumption was very good.

15.2   *Buffered Solutions*

When you finish this section you will be able to:

- Calculate the pH of buffer solutions.
- Determine the effect of the addition of acids and bases on the pH of buffer solutions.

A buffered solution resists change in its pH upon the addition of a strong acid or strong base. The last section set us up for doing buffer problems by introducing us to the common ion effect on equilibrium. We saw that the

extent of equilibrium is depressed by the addition of a common ion. A buffered solution generally contains a weak acid and its salt or a weak base and its salt. Try the following buffer calculation, and think about

1. the common ion effect and
2. the dominant equilibrium.

Example 15.2 A   pH of a Buffered Solution

Calculate the pH of a solution that contains 0.250 M formic acid, HCOOH ($K_a = 1.8 \times 10^{-4}$), and 0.100 M sodium formate, HCOONa.

Strategy

The major species in solution are **HCOOH, Na$^+$, HCOO$^-$, and H$_2$O.**

Na$^+$	has no acid-base properties.
H$_2$O	has $K_w = 1.0 \times 10^{-14}$, and is thus weak relative to HCOOH.
HCOO$^-$	has a $K_b = K_w/K_a = 5.6 \times 10^{-11}$. There will be very little hydrolysis of HCOO$^-$. It will be depressed further by the common ion effect.
HCOOH	has $K_a = 1.8 \times 10^{-4}$. Although this *acid dissociation is depressed* by the presence of the common ion HCOO$^-$, *it will be the dominant equilibrium.*

$$HCOOH(aq) \rightleftharpoons H^+(aq) + HCOO^-(aq)$$

	initial (M)	change	final (M)	
HCOOH	0.250	$-X$	$0.250 - X$	($\approx 0.250\ M$)
H$^+$	0	$+X$	X	
HCOO$^-$	0.100	$+X$	$0.100 + X$	($\approx 0.100\ M$)

As shown in our previous chapter, it is a virtual certainty that we can ignore "X" relative to the initial concentrations of HCOOH and HCOO$^-$. We will test this with the 5% rule.

Keep in mind that buffer problems are among the easiest to solve because the **initial and final concentrations** of the acid and conjugate bases **do not change significantly**, and they are usually known.

Solution

$$K_a = \frac{[H^+][HCOO^-]}{[HCOOH]} \quad \Rightarrow \quad 1.8 \times 10^{-4} = \frac{[H^+](0.100)}{0.250}$$

$$[H^+] = 4.5 \times 10^{-4}\ M, \quad \text{pH} = 3.35$$

The actual percent dissociation of HCOOH, which is equal to [H$^+$], is small enough to easily pass the 5% rule.

The real utility of buffers comes when we add a strong acid or base. A buffered solution should *resist pH change* when a substantial amount of H$^+$ or OH$^-$ ions are added. ("Substantial amount" will vary from solution to solution, depending upon the "buffer capacity." We will consider that in Section 15.3.)

If we add a strong base to a solution that contains a weak acid, the reaction is complete.

$$\textbf{HA}(aq) + \textbf{OH}^-(aq) \rightarrow \textbf{A}^-(aq) + \textbf{H}_2\textbf{O}(l)$$

Similarly, adding a strong acid to a weak base gives a complete reaction,

$$\textbf{A}^-(aq) + \textbf{H}^+(aq) \rightarrow \textbf{HA}(aq)$$

Notice that in these equations, unless **excess** strong acid or base is added, **NO STRONG ACID OR BASE BUILDS UP.** Only weak species remain. This is why a buffer works so well. Let's use this idea in our next problem.

Example 15.2 B Addition of a Strong Base to a Buffer

Calculate the pH of the 0.250 *M* HCOOH/0.100 *M* HCOONa buffer used in the last problem after the addition of 10.0 mL of 6.00 *M* NaOH to the **original buffered solution volume of 500.0 mL.**

Strategy

As your textbook points out, there are really two parts to this problem. The first is to *determine the moles* of weak acid and base existing *after the addition of the strong base ("stoichiometry part").* The second part is to *calculate the pH of the buffered solution* after you have your new equilibrium concentrations *("equilibrium part").*

Solution

a. **Stoichiometry Part**

We need to work in moles (or mmoles).

$$\text{mmoles HCOOH}_{\text{initial}} = 0.250 \text{ mmol/mL} \times 500.0 \text{ mL} = \textbf{125 mmol}$$
$$\text{mmoles HCOO}^-_{\text{initial}} = 0.100 \text{ mmol/mL} \times 500.0 \text{ mL} = \textbf{50.0 mmol}$$
$$\text{mmol OH}^- \text{ added} = 6.00 \text{ mmol/mL} \times 10.0 \text{ mL} = \textbf{60.0 mmol}$$

	$\text{HCOOH}(aq)$ +	$\text{OH}^-(aq)$ →	$\text{HCOO}^-(aq)$
initial (mmol)	125	60.0	50.0
final (mmol)	65.0	≈0	110

After addition of NaOH.

$$[\text{HCOOH}] = 65.0 \text{ mmol/510 mL} = 0.128 \text{ } M$$
$$(500 + 10) \qquad \text{equilibrium concentration}$$
$$[\text{HCOO}^-] = 110 \text{ mmol/510 mL} = 0.216 \text{ } M$$

b. **Equilibrium Part**

As in the previous problem, the acid dissociation of HCOOH is the significant factor here. As before,

$$K_a = \frac{[\text{H}^+][\text{HCOO}^-]}{[\text{HCOOH}]} \qquad \Rightarrow \qquad 1.8 \times 10^{-4} = \frac{[\text{H}^+](0.216)}{(0.128)}$$

$$[\text{H}^+] = \textbf{1.1 x } 10^{-4}, \qquad \textbf{pH = 4.0} \quad \text{(The 5\% rule easily holds.)}$$

Comparing this result to that in the previous problem, we note that the *pH has gone from 3.4 to 4.0* after addition of a substantial quantity of OH⁻. The *pH change of 0.6 units* indicates that our buffer has neutralized the base well. If we added just *60 mmoles of OH⁻ to 500 mL of water,* [OH⁻] would equal 0.92, and *pH would equal 13.* The presence of the buffered solution allowed us to maintain an acid pH in spite of the strong base that was added.

Example 15.2 C   Practice with Buffers

A solution is prepared by adding 31.56 g NaCN and 22.30 g HCN to 600.0 mL of water (K_a for HCN = 6.2×10^{-10}).

    a.   What is the pH of this solution?
    b.   What is the pH after the addition of 50.0 mL of 3.00 *M* HCl?
    c.   What is the pH after a **further** addition of 80.0 mL of 4.00 *M* NaOH?

Strategy

We have the following major species in solution: **Na^+, H_2O, HCN,** and **CN^-:**

    HCN   is a weak acid, $K_a = 6.2 \times 10^{-10}$.
    CN^-   its conjugate base, has $K_b = K_w/K_a = 1.0 \times 10^{-14}/6.2 \times 10^{-10} = 1.6 \times 10^{-5}$.
    H_2O   with its low K_w, does not affect the acid-base characteristics of the system.

Because **CN^-** is the **strongest acid-base substance in our solution**, its behavior will dominate. The base hydrolysis reaction is

$$CN^-(aq) + H_2O(l) \rightleftharpoons HCN(aq) + OH^-(aq)$$

and

$$K_b = \frac{[HCN][OH^-]}{[CN^-]}$$

As always in problems involving buffered solutions, we can make our usual assumption regarding the lack of reaction of CN^- due to the presence of the common species CN^- (in the form of HCN).

After calculating the pH of this solution (part a), *we need to assess the pH after addition of HCl*, a strong acid, for *part b*. In general terms, *the pH should go down, but only a small amount*, because we have a buffer. The exception to this is if we *exceed* the buffer capacity. We must use our *two-part approach*, taking into account the initial complete reaction of CN^- with H^+ (from HCl).

$$CN^-(aq) + H^+(aq) \rightarrow HCN(aq)$$

As long as we have some CN^- ion left, we will maintain our buffered solution. We may then use our *equilibrium expression* to assess pH (via pOH).

In *part c, we must consider the addition of a strong base, OH^- ion, to the system*. It will react completely with our weak acid, HCN, to form CN^-.

$$HCN(aq) + OH^-(aq) \rightarrow CN^-(aq) + H_2O(l)$$

As long as we have some HCN left, we will maintain a buffered solution. *The pH of the solution should rise, but not much,* if we maintain our buffer. We may use our *equilibrium expression* to determine the pOH, and then the pH.

Solution

    a.   Let's calculate [HCN] and [CN^-].

$$[HCN] = \frac{mol\ HCN}{L} = \frac{1\ mol}{27.0\ g} \times \frac{22.30\ g}{0.600\ L} = 1.38\ M$$

$$[CN^-] = \frac{mol\ NaCN}{L} = \frac{1\ mol}{49.0\ g} \times \frac{31.56\ g}{0.600\ L} = 1.07\ M$$

Using the equilibrium expression for the buffered solution, as discussed above,

$$K_b = \frac{[HCN][OH^-]}{[CN^-]} \qquad \Rightarrow \qquad 1.6 \times 10^{-5} = \frac{(1.38)[OH^-]}{1.07}$$

$$[OH^-] = 1.25 \times 10^{-5}, \quad pOH = 4.90, \quad pH = 9.10$$

b. **Stoichiometry Part**

mmoles HCN$_{initial}$ $= 1.38 \text{ mmol/mL} \times 600 \text{ mL} = $ **828 mmol**
mmoles CN$^-_{initial}$ $= 1.07 \text{ mmol/mL} \times 600 \text{ mL} = $ **642 mmol**
mmoles H$^+$ added $= 3.00 \text{ mmol/mL} \times 50.0 \text{ mL} = $ **150 mmol**

$$CN^-(aq) + H^+(aq) \rightarrow HCN(aq)$$

initial (mmol)	642	150	828
final (mmol)	492	≈ 0	978

$$[CN^-] = 492 \text{ mmol}/(600 + 50) \text{ mL} = \mathbf{0.757}\ M$$
$$[HCN] = 978 \text{ mmol}/650 \text{ mL} = \mathbf{1.50}\ M$$

Equilibrium Part (Use the **equilibrium** expression.)

$$K_b = \frac{[HCN][OH^-]}{[CN^-]} \qquad \Rightarrow \qquad 1.6 \times 10^{-5} = \frac{(1.50)[OH^-]}{0.757}$$

$$[OH^-] = 8.1 \times 10^{-6}\ M, \quad pOH = 5.09, \quad pH = 8.91$$

c. **Stoichiometry Part** (from part b)

$$\text{mmol HCN}_{initial} = 978 \text{ mmol}$$
$$\text{mmol CN}^-_{initial} = 492 \text{ mmol}$$

We are adding $80.0 \text{ mL} \times 4.00 \text{ mmol H}^+/\text{mL} = $ **320 mmol OH$^-$: This will react with HCN, as discussed in our strategy section.**

$$HCN(aq) + OH^-(aq) \rightarrow CN^- + H_2O(l)$$

initial (mmol)	978	320	492
final (mmol)	658	≈ 0	812

$$[HCN] = 658 \text{ mmol}/(650 + 80) \text{ mL} = \mathbf{0.901}\ M$$
$$[CN^-] = 812 \text{ mmol}/730 \text{ mL} = \mathbf{1.11}\ M$$

Equilibrium Part (Use the **equilibrium** expression.)

$$K_b = \frac{[HCN][OH^-]}{[CN^-]} \qquad \Rightarrow \qquad 1.6 \times 10^{-5} = \frac{(0.901)[OH^-]}{1.11}$$

$$[OH^-] = 2.0 \times 10^{-5} M, \quad pOH = 4.71, \quad pH = 9.29$$

Does the Answer Make Sense?

The initial pH was **basic** (9.10). This makes sense because CN$^-$ is a stronger base than HCN is an acid. The pH **dropped** (8.91) after addition of a strong acid. This makes sense. The pH **rose** (9.29) after the subsequent addition of a strong base. This makes sense. In neither part b nor part c did we exceed our buffer capacity, so the pH changes were not drastic.

Example 15.2 D   pH and pK_a

The K_a of propionic acid, $HC_3H_5O_2$, is 1.34×10^{-5} (p$K_a = 4.87$). What is the pH when $[HC_3H_5O_2] = [C_3H_5O_2^-]$?

Solution

Let's look at the problem from a strictly mathematical point of view. If we use the equation for the acid dissociation of propionic acid,

$$HC_3H_5O_2(aq) \rightleftharpoons C_3H_5O_2^-(aq) + H^+(aq)$$

$$K_a = \frac{[C_3H_5O_2^-][H^+]}{[HC_3H_5O_2]}$$

If $[HC_3H_5O_2] = [C_3H_5O_2^-]$, $K_a = [H^+] = 1.34 \times 10^{-5}\ M$ and **pK_a = pH**; therefore, **pH = 4.87.**

If we use the equation for the base hydrolysis of the propionate ion, $C_3H_5O_2^-$,

$$C_3H_5O_2^-(aq) + H_2O(l) \rightleftharpoons HC_3H_5O_2(aq) + OH^-(aq)$$

$$K_b = \frac{[HC_3H_5O_2][OH^-]}{[C_3H_5O_2^-]}$$

If $[C_3H_5O_2^-] = [HC_3H_5O_2]$,

$$[OH^-] = K_b = K_w/K_a = 1.0 \times 10^{-14}/1.34 \times 10^{-5} = 7.5 \times 10^{-10}$$

$$pOH = 9.13, \quad pH = 4.87$$

Although we would normally use the K_a expression to solve the pH in this situation, two important points need to be made from this problem:

1. **If [HA] = [A$^-$], then pH = pK_a and pOH = pK_b.**
2. In buffer problems, **uniquely, either the K_a or K_b expressions can be used,** and the same pH will result. This is because **BOTH [HA] and [A$^-$] are explicitly known.**

15.3   Buffer Capacity

When you finish this section you will be able to:

- Calculate the pH of a solution where the buffer capacity has been exceeded.
- Choose among alternatives the best buffer system for a given pH.

Your textbook emphasizes the two important measures involved in buffered solutions.

1. The pH is determined by the **ratio** of [HA]/[A$^-$].
2. The **buffer capacity** is determined by the **magnitudes** of [HA] and [A$^-$]. The more you have, the more strong acid or base that can be neutralized.

Example 15.7 in your textbook compares the pH change brought about by adding a strong acid to solutions with different buffering capacities. Let's try a problem where we **exceed** the buffering capacity of a solution.

Example 15.3 A   Buffering Capacity

Calculate the pH of a 0.500-L solution that contains 0.15 M HCOOH ($K_a = 1.8 \times 10^{-4}$) and 0.20 M HCOONa. Then, calculate the pH of the solution after the addition of 10.0 mL of 12.0 M NaOH.

Solution

We can solve the problems using the same strategy as in the previous sections. The major species are **HCOOH, Na$^+$, HCOO$^-$**, and **H$_2$O**. The dominant equilibrium will involve the acid dissociation of HCOOH.

$$HCOOH(aq) \rightleftharpoons H^+(aq) + HCOO^-(aq)$$

$$K_a = \frac{[H^+][HCOO^-]}{[HCOOH]} \quad \Rightarrow \quad 1.8 \times 10^{-4} = \frac{[H^+](0.20)}{0.15}$$

$$[H^+] = 1.4 \times 10^{-4}\, M, \quad pH = 3.87$$

When we add the strong base, we have the reaction between OH$^-$ ions and HCOOH,

$$HCOOH(aq) + OH^-(aq) \rightarrow HCOO^-(aq) + H_2O(l)$$

We must proceed with our two-step approach.

1. *Stoichiometry Part*

   mmoles HCOOH$_{initial}$ = 0.15 mmol/mL × 500 mL = **75 mmol**
   mmoles HCOO$^-$$_{initial}$ = 0.20 mmol/mL × 500 mL = **100 mmol**
   mmoles OH$^-$ added = 12.00 mmol/mL × 10.0 mL = **120 mmol**

   We have more OH$^-$ than is needed to neutralize the HCOOH. HCOOH is the **limiting reactant** in this solution.

	HCOOH(aq)	+ OH$^-$(aq)	→ HCOO$^-$(aq)	+ H$_2$O(l)
initial (mmol)	75	120	100	
final (mmol)	≈0	45	175	

   We have 45 mmoles of excess OH$^-$! There is no **"equilibrium part" in this problem!** OH$^-$ ion is a far stronger base than HCOO$^-$ and will thus determine the pH.

   $$[OH^-] = 45\ mmol/(500 + 10)\ mL = \mathbf{0.088\ M}$$
   $$pOH = 1.05 \qquad \mathbf{pH\ 12.95}$$

When we *exceeded the buffer capacity* of the solution, *the pH changed drastically*. If we would have doubled the amounts of formic acid and sodium formate, we would not have exceeded the buffer capacity of the solution.

All other things being equal, the closer the [HA]/[A$^-$] ratio is to one, the better will be the buffer capacity of the solution. This will happen when [H$^+$] = K_a, or pH = pK_a. This means that, when selecting a buffer that will be used to maintain the pH of a solution in a particular small range,

**The pK_a of the buffer should be as close as
possible to the desired pH of the solution.**

Example 15.3 B   Selecting a Buffer

We wish to buffer a solution at pH = 10.07. Which one of the following bases (and conjugate acid salts) would be most useful?

   a.   NH$_3$ ($K_b = 1.8 \times 10^{-5}$)
   b.   C$_6$H$_5$NH$_2$ ($K_b = 4.2 \times 10^{-10}$)
   c.   N$_2$H$_4$ ($K_b = 9.6 \times 10^{-7}$)

Solution

$$K_a \text{ for } NH_4^+ = K_w/K_b = 1.0 \times 10^{-14}/1.8 \times 10^{-5} = 5.6 \times 10^{-10}$$
$$\mathbf{pK_a \ (NH_4) \ = \ 9.26}$$

$$K_a \text{ for } C_6H_5NH_3^+ = K_w/K_b = 1.0 \times 10^{-14}/4.2 \times 10^{-10} = 2.4 \times 10^{-5}$$
$$\mathbf{pK_a \ (C_6H_5NH_3^+) \ = \ 4.62}$$

$$K_a \text{ for } N_2H_5^+ = K_w/K_b = 1.0 \times 10^{-14}/9.6 \times 10^{-7} = 1.0 \times 10^{-8}$$
$$\mathbf{pK_a \ (N_2H_5^+) \ = \ 7.98}$$

NH_3 would be the best choice for a buffer system. (Please keep in mind that a good deal of thought must not only go into the mathematics, but the **chemistry as well.** That is, how will NH_3 affect your experiment?)

15.4   Titrations and pH Curves

When you finish this section you will be able to calculate the pH at any point along a curve for the following titrations:

- strong acid - strong base
- weak acid - strong base
- weak base - strong acid

As a precursor to the calculations in this chapter, you should be able to define the following terms: titrant, buret, indicator, titration curve, and endpoint.

In order for a titration to be feasible, it must be **complete** and **fast.** In order to be complete, the reaction should have a value of $K \geq 10^7$ **or so.**

Strong Acid - Strong Base

Let's consider the titration of a **HCl** solution with **NaOH.** We say that the process is complete; however, we can calculate an equilibrium constant for the reaction.

$$H^+(aq) + OH^-(aq) \rightleftharpoons H_2O(l)$$

The reaction is the reverse of the autoionization of water therefore

$$K = 1/K_w = 1 \times 10^{14}$$

This reaction is certainly complete! So much so, in fact, that this type of titration is less an equilibrium problem than a stoichiometry one.

Example 15.4 A   Strong Acid - Strong Base Titration

Calculate the pH after the following **total** volumes of **0.2500 M HCl** have been **added** to **50.00 mL of 0.1500 M NaOH.**

a. 0.00 mL	c. 29.50 mL	e. 30.50 mL
b. 4.00 mL	d. 30.00 mL	f. 40.00 mL

Solution

a.   **0.00 mL:** We have a 0.1500 M NaOH solution. Because it completely dissociates, and OH^- ion is a strong base,

$$[OH^-] = 0.1500 \ M, \qquad pOH = 0.82, \qquad \mathbf{pH = 13.18}$$

b. **4.00 mL total:** As discussed previously, the reaction between H^+ and OH^- is complete. Calculating moles of each reactant is always the preferred way to begin.

$$\text{mmol } OH^-_{\text{initial}} = 0.1500 \text{ mmol/mL} \times 50.00 \text{ mL} = \textbf{7.500 mmol}$$
$$\text{mmol } H^+ \text{ added} = 0.2500 \text{ mmol/mL} \times 4.00 \text{ mL} = \textbf{1.000 mmol}$$

$$H^+(aq) \ + \ OH^-(aq) \ \rightarrow \ H_2O(l)$$

	H^+	OH^-	
initial (mmol)	1.000	7.500	excess
final (mmol)	≈0	6.500	excess

The relatively small amount of H^+ ions added were neutralized by OH^- ions. We have 6.500 mmol of OH^- ion excess, which determines the pH.

$$[OH^-] = 6.500 \text{ mmol}/(50.00 + 4.00) \text{ mL} = \textbf{0.1203 } \textbf{\textit{M}}$$
$$pOH = 0.92, \qquad \textbf{pH = 13.08}$$

The pH has not declined much because we still have an excess of OH^- ion.

c. **29.50 mL total:** Proceeding as above,

$$\text{mmol } OH^-_{\text{initial}} = \textbf{7.500 mmol}$$
$$\text{mmol } H^+ \text{ added} = 0.2500 \text{ mmol/mL} \times 29.50 \text{ mL} = \textbf{7.375 mmol}$$

$$H^+(aq) \ + \ OH^-(aq) \ \rightarrow \ H_2O(l)$$

	H^+	OH^-	
initial (mmol)	7.375	7.500	excess
final (mmol)	≈0	0.125	excess

$$[OH^-] = 0.125 \text{ mmol}/(50.00 + 29.50) \text{ mL} = \textbf{1.57} \times \textbf{10}^{-3} \textbf{\textit{M}}$$
$$pOH = 2.80, \qquad \textbf{pH = 11.20}$$

The pH has begun to come down somewhat, but the solution is still quite basic.

d. **30.00 mL total:** As above,

$$\text{mmol } OH^-_{\text{initial}} = \textbf{7.500 mmol}$$
$$\text{mmol } H^+ \text{ added} = 0.2500 \text{ mmol/mL} \times 30.00 \text{ mL} = \textbf{7.500 mmol}$$

$$H^+(aq) \ + \ OH^-(aq) \ \rightarrow \ H_2O(l)$$

	H^+	OH^-	
initial (mmol)	7.500	7.500	excess
final (mmol)	≈0	≈0	excess

We have added **exactly** enough H^+ ions to neutralize all the OH^- ions. We have only water, Na^+, and Cl^- in the solution. Of these, only **water** has any acid-base properties. This neutral water solution, as always, has $K_w = 1.0 \times 10^{-14}$. The **pH = 7.00**.

CRITICAL POINT: THE ONLY CASE IN WHICH THE pH AT THE EQUIVALENCE POINT WILL BE EQUAL TO 7.00 WILL BE THAT OF A STRONG ACID - STRONG BASE TITRATION!

Also notice the rather sharp pH change (4.20 units) within 0.500 mL of the equivalence point. This is typical of a strong acid-strong base titration.

e. **30.50 mL total:** Proceeding as always,

$$H^+(aq) \ + \ OH^-(aq) \ \rightarrow \ H_2O(l)$$

	H^+	OH^-	
initial (mmol)	7.625	7.500	excess
final (mmol)	0.125	≈0	excess

We have **passed** the equivalence point. (This is called the "post-equivalence point region.") We have excess H^+ ion. The solution is now acidic.

$$[H^+] = 0.125 \text{ mmol}/80.50 \text{ mL} = \mathbf{1.55 \times 10^{-3}} \; \boldsymbol{M}$$
$$\mathbf{pH = 2.81}$$

Being just 0.500 mL beyond the equivalence point caused a sharp drop in the pH. Generally, the stronger the species being titrated, the sharper the pH drop.

f.   **40.00 mL total:** We are now well past the equivalence point.

$$H^+(aq) \; + \; OH^-(aq) \; \rightarrow \; H_2O(l)$$

	H^+	OH^-	H_2O
initial (mmol)	10.00	7.500	excess
final (mmol)	2.50	≈ 0	excess

$$[H^+] = 2.50 \text{ mmol}/90.00 \text{ mL} = \mathbf{0.0278} \; \boldsymbol{M}$$
$$\mathbf{pH = 1.56}$$

If you plot the mL of HCl added vs. pH for these data, your plot should look like that in <u>Figure 15.2 in your textbook.</u>

Weak Acid - Strong Base

When doing calculations for this type of titration it is useful to relate our thinking back to our sections on **buffered solutions (15.2 and 15.3).**

$$HA(aq) \; + \; OH^-(aq) \; \rightarrow \; A^-(aq) \; + \; H_2O(l)$$

You will be generating a buffer (weak acid and conjugate base) as a result of the addition of OH^- ions to the solution. You will have a buffered solution until you exceed the buffer capacity of the solution, and then the pH will rise dramatically.

Let's use the titration of formic acid, HCOOH ($K_a = 1.8 \times 10^{-4}$) with NaOH as an illustrative example. **The equilibrium constant for the titration can be calculated.** The reaction

$$\mathbf{HCOOH}(aq) \; + \; \mathbf{OH^-}(aq) \; \rightleftharpoons \; \mathbf{HCOO^-}(aq) \; + \; \mathbf{H_2O}(l)$$

can be expressed as a **combination** of

1. $HCOOH(aq) \; \rightleftharpoons \; H^+(aq) \; + \; HCOO^-(aq)$ $\qquad$ $K_a = 1.8 \times 10^{-4}$

and 2. $H^+(aq) \; + \; OH^+(aq) \; \rightleftharpoons \; H_2O(l)$ $\qquad$ $1 \backslash K_w = 1.0 \times 10^{14}$

K for the overall reaction $= K_1 \times K_2 = K_a \times 1/K_w = K_a/K_w$

$$\boldsymbol{K} = 1.8 \times 10^{-4}/1.0 \times 10^{-14} = \mathbf{1.8 \times 10^{10}}$$

This is why we say that the titration goes completely.

Example 15.4 B   Weak Acid - Strong Base Titration

Calculate the pH after the following **total** volumes of 0.4000 M NaOH are added to 50.00 mL of 0.200 M HCOOH.

a. 0.00 mL	d. 24.50 mL	f. 25.50 mL
b. 5.00 mL	e. 25.00 mL	g. 40.00 mL
c. 12.50 mL		

Solution

a.   **0.00 mL:** We have a weak acid solution, 0.2000 M HCOOH. We have done these kinds of problems many times and will present a **shortened version here.** The equilibrium of interest is

$$HCOOH(aq) \; \rightleftharpoons \; H^+(aq) \; + \; HCOO^-(aq)$$

$$K_a = \frac{[H^+][HCOO^-]}{[HCOOH]} \quad \Rightarrow \quad 1.8 \times 10^{-4} = \frac{X^2}{0.2000}$$

$$[H^+] = 6.0 \times 10^{-3}, \quad \textbf{pH} = \textbf{2.22}$$

In titration problems, if the titration is feasible (large K), our "5% test" will virtually always hold true.

b. **5.00 mL:** As pointed out previously, the titration reaction of interest is

$$\textbf{HCOOH}(aq) + \textbf{OH}^-(aq) \rightleftharpoons \textbf{HCOO}^-(aq) + \textbf{H}_2\textbf{O}(l)$$

The reaction is essentially complete ($K = 1.8 \times 10^{10}$). After this reaction is completed, we will have a **buffered solution** containing HCOOH and $HCOO^-$ ion. You must then calculate the pH of the buffered solution.

Stoichiometry Part:

$$\textbf{mmol HCOOH}_{\text{initial}} = 0.2000 \text{ mmol/mL} \times 50.00 \text{ mL} = \textbf{10.00 mmol}$$
$$\textbf{mmol OH}^- \textbf{ added} = 0.4000 \text{ mmol/mL} \times 5.00 \text{ mL} = \textbf{2.00 mmol}$$

	$HCOOH(aq)$	+ $OH^-(aq)$	$\rightarrow$ $HCOO^-(aq)$	+ $H_2O(l)$
initial (mmol)	10.00	2.00	≈ 0	excess
final (mmol)	8.00	≈ 0	2.00	excess

Equilibrium Part:

Formic acid is a stronger acid than formate ion is a base. Therefore we can use the acid dissociation equilibrium to solve the problem (but in this so-called **"buffer region"** the base hydrolysis would work as well). (See Section 15.2 if you need a review of buffered solutions.)

$$[\textbf{HCOOH}] = 8.00 \text{ mmol/(50.00 + 5.00) mL} = \textbf{0.145 } M$$
$$[\textbf{HCOO}^-] = 2.00 \text{ mmol/55.00 mL} = \textbf{0.0364 } M$$
$$\textbf{HCOOH}(aq) \rightleftharpoons \textbf{H}^+(aq) + \textbf{HCOO}^-(aq)$$

$$K_a = \frac{[H^+][HCOO^-]}{[HCOOH]} \quad \Rightarrow \quad 1.8 \times 10^{-4} = \frac{[H^+](0.0364)}{0.145}$$

$$[H^+] = 7.2 \times 10^{-4}, \quad \textbf{pH} = \textbf{3.14}$$

The pH does not rise much in the buffer region of a titration.

c. **12.50 mL:** We shall proceed as in part b.

Stoichiometry Part:

$$\textbf{mmol HCOOH}_{\text{initial}} = \textbf{10.00 mmol}$$
$$\textbf{mmol OH}^- \textbf{ added} = 0.4000 \text{ mmol/mL} \times 12.50 \text{ mL} = \textbf{5.000 mmol}$$

	$HCOOH(aq)$	+ $OH^-(aq)$	$\rightarrow$ $HCOO^-(aq)$	+ $H_2O(l)$
initial (mmol)	10.00	5.000	≈ 0	excess
final (mmol)	5.00	≈ 0	5.00	excess

Equilibrium Part:

$$[\textbf{HCOOH}] = 5.00 \text{ mmol/(50.00 + 12.50) mL} = \textbf{0.0800 } M$$
$$[\textbf{HCOO}^-] = 5.00 \text{ mmol/62.50 mL} = \textbf{0.800 } M$$

$$K_a = \frac{[H^+][HCOO^-]}{[HCOOH]} \quad \Rightarrow \quad 1.8 \times 10^{-4} = \frac{[H^+](0.0800)}{0.0800}$$

$$[\textbf{H}^+] = K_a = 1.8 \times 10^{-4}, \quad \textbf{pH} = \textbf{3.74}$$

Recall from Section 15.3 that when $[HA] = [A^-]$, $pK_a = pH$. This is the situation at this point in our titration. This is called the **titration midpoint**. We are **halfway** to the equivalence point. The titration midpoint is an especially important experimental point because if we know pH, we can find pK_a for the weak acid.

d. **24.50 mL:** Proceeding as always,

> **mmol HCOOH$_{initial}$ = 10.00 mmol**
> **mmol OH$^-$ added** = 0.4000 mmol/mL $\times$ 24.50 mL = **9.800 mmol**

	HCOOH(aq)	+ OH$^-$(aq)	$\rightarrow$ HCOO$^-$(aq)	+ H$_2$O(l)
initial (mmol)	10.00	9.800	≈ 0	excess
final (mmol)	0.20	≈ 0	9.800	excess

We are near the limit of our buffer capacity.

> **[HCOOH]** = 0.20 mmol/(50.00 + 24.50) mL = **2.67×10^{-3} M**
> **[HCOO$^-$]** = 9.80 mmol/74.50 mL = **0.132 M**

$$K_a = \frac{[H^+][HCOO^-]}{[HCOOH]} \qquad \Rightarrow \qquad 1.8 \times 10^{-4} = \frac{[H^+](0.132)}{2.67 \times 10^{-3}}$$

$$[H^+] = 3.6 \times 10^{-6}, \qquad \textbf{pH = 5.44}$$

e. **25.00 mL:**

> **mmol HCOOH$_{initial}$ = 10.00 mmol**
> **mmol OH$^-$ added = 10.00 mmol**

	HCOOH(aq)	+ OH$^-$(aq)	$\rightarrow$ HCOO$^-$(aq)	+ H$_2$O(l)
initial (mmol)	10.00	10.00	≈ 0	excess
final (mmol)	≈ 0	≈ 0	10.00	excess

We have reached the equivalence point. We now have a solution that contains a **base** that can undergo hydrolysis,

$$HCOO^-(aq) + H_2O(l) \rightleftharpoons HCOOH(aq) + OH^-(aq)$$

$$K_b = K_w/K_a = \frac{1.0 \times 10^{-14}}{1.8 \times 10^{-4}} = 5.6 \times 10^{-11} = \frac{[HCOOH][OH^-]}{[HCOO^-]}$$

> **[HCOO$^-$]** = 10.00 mmol/(50.00 + 25.00) mL = **0.133 M**

Solving this weak base problem in the usual way,

$$5.6 \times 10^{-11} = \frac{X^2}{0.133} \qquad \Rightarrow \qquad X = [OH^-] = 2.73 \times 10^{-6} \text{ M}$$

$$pOH = 5.56, \qquad \textbf{pH = 8.44}$$

We have a mildly basic solution, and that is reflected in the pH value.

f. **25.50 mL:**

> **mmoles HCOOH$_{initial}$ = 10.00 mmol**
> **mmoles OH$^-$ added = 10.20 mmol**

We have neutralized all of the weak acid and have **0.20 mmol of strong base remaining**. This is the **post-equivalence point region**. The pH will be determined by the excess strong base.

$$[OH^-] = 0.20 \text{ mmol}/(50.00 + 25.50) \text{ mL} = \textbf{2.65} \times \textbf{10}^{-3} \textbf{ M}$$
$$pOH = 2.58, \qquad \textbf{pH = 11.42}$$

Notice that the pH has risen sharply now that we have **surpassed the buffer capacity** of our formic acid/formate buffered solution.

g. **40.00 mL:** You should be able to show that we have **6.00 mmol of excess OH^- ions** and

$$[OH^-] = 6.00 \text{ mmol}/90.00 \text{ mL} = 0.0667 \text{ } M$$
$$pOH = 1.18, \qquad \textbf{pH} = \textbf{12.82}$$

Summing up, note that our "pH break" around the equivalence point region was somewhat less sharp than with the strong acid - strong base titration. The smaller the titration equilibrium constant, the less sharp the break at the equivalence point. If you plot a titration curve, it should look similar to <u>Figure 15.3 in your textbook.</u>

Weak Base - Strong Acid

These kinds of titrations are handled using the same strategies as those involving weak acids and strong bases. We have the **buffer region, equivalence point, and post-equivalence point**, as in the previous titrations. Think about the dominant reactions in each part of the problem, be careful regarding math, and **check to make sure that your answer makes sense.**

Example 15.4 C   Weak Base - Strong Acid Titration

Calculate the pH at each of the following points in the titration of **50.00 mL of a 0.01000 M** sodium phenolate ($NaOC_6H_5$) solution with **1.000 M HCl solution** (K_a for $HOC_6H_5 = 1.05 \times 10^{-10}$):

   a.   initially   b.   midpoint   c.   equivalence point

Solution

Sodium phenolate is a strong electrolyte. It will completely dissociate to give Na^+ and $\textbf{OC}_6\textbf{H}_5^-$.

$$\textbf{K}_{\textbf{b}} \text{ for } OC_6H_5^- = K_w/K_a = 1.0 \times 10^{-14} / 1.05 \times 10^{-10} = \textbf{9.5} \times \textbf{10}^{-5}$$

a. This weak base undergoes hydrolysis in the usual way,

$$OC_6H_5^-(aq) + H_2O(l) \rightleftharpoons HOC_6H_5(aq) + OH^-(aq)$$

$$K_b = \frac{[HOC_6H_5][OH^-]}{[OC_6H_5^-]} \qquad \Rightarrow \qquad 9.5 \times 10^{-5} = \frac{X^2}{0.01000}$$

$$X = [OH^-] = 9.8 \times 10^{-4} \text{ } M$$
$$pOH = 3.0, \qquad \textbf{pH} = \textbf{11.0}$$

b. By definition, the pH at the titration midpoint equals pK_a for the acid. Therefore, **pOH = pK_b for the base.** For our solution,

$$pK_b = \log K_b = \log(9.5 \times 10^{-5}) = \textbf{4.0}$$
$$pOH = 4.0, \qquad \textbf{pH} = \textbf{10.0}$$

c. All of our base has been converted to acid at the equivalence point. Although we know how to find the pH of a very weak acid, *we need to know its concentration* at the equivalence point. After all, a small *dilution* occurred as a result of adding HCl. A key question then is, *"How many mL of HCl were required to reach the equivalence point?"*

$$\textbf{mmoles } OC_6H_5^- \text{ }_{\text{initial}} = 0.0100 \text{ mmol/mL} \times 50.00 \text{ mL} = \textbf{0.500 mmol}$$
$$\textbf{mmoles HCl required to just neutralize} = \textbf{0.5000 mmol}$$

$$\text{mL HCl required to just neutralize} = 0.5000 \text{ mmol} \times \frac{1 \text{ mL}}{1.000 \text{ mmol}} = \mathbf{0.5000 \text{ mL}}$$

Our **total volume** is now $50.00 + 0.5000 = \mathbf{50.50 \text{ mL}}$

$$\mathbf{[HOC_6H_5]} = 0.500 \text{ mmol}/50.50 \text{ mL} = \mathbf{9.90 \times 10^{-3} \, M}$$

To solve the pH of this weak acid,

$$HOC_6H_5(aq) \rightleftharpoons H^+(aq) \, OC_6H_5^-(aq)$$

$$K_a = \frac{[H^+][OC_6H_5^-]}{[HOC_6H_5]} \quad \Rightarrow \quad 1.05 \times 10^{-10} = \frac{X^2}{9.90 \times 10^{-3}}$$

$$[H^+] = 1.02 \times 10^{-6}, \qquad \mathbf{pH = 5.99}$$

Notice that the pH indicates that we have a dilute acidic solution. In the sense that our pH started high and ended low, our answers make sense. However, the initial solution was so dilute that our pH did not change a great deal. As such, this system would be considered marginal for a titration.

15.5   Acid-Base Indicators

When you finish this section you will be able to select the proper indicator for an acid-base titration.

The purpose of an **indicator** is to allow you to know when you have reached the equivalence point in a titration. Acid-base indicators are normally organic acids (see Figure 15.6 in your textbook). Thus, when you add an indicator to a solution and perform a titration, **you are titrating the indicator along with your substance of interest.** This requires some small extra amount of titrant.

The volume of titrant at which you visually detect the equivalence point is called the **endpoint. The volume at the equivalence point is never exactly equal to the endpoint, but it is usually quite close.** There are two criteria that are used to choose an indicator for a given titration:

- The pK_a of the indicator should be within ± 1 unit of the pH of the solution at the equivalence point.
- The color change at the endpoint should be clearly distinguishable.

Example 15.5 A   Proper Indicator Choice

A $0.100 \, M \, NH_3$ solution is being titrated with $0.200 \, M \, HCl$. Using data from Figure 15.8 in your textbook, select a suitable indicator for the titration.

Solution

The reaction of interest is

$$NH_3(aq) + H^+(aq) \rightarrow NH_4^+(aq)$$

The important question is, "What is the pH at the equivalence point?" We can estimate it by noting that since the HCl solution is **twice as concentrated** as the NH_3 solution, it will take **1/2 the volume** of HCl solution to neutralize the NH_3 solution. That is,

$$\textbf{Volume at the equivalence point} = V_0 + {}^1/_2V_0 = {}^3/_2V_0$$

where V_0 is the original volume of NH_3 solution.

Because the volume is 3/2 as much, the NH_4^+ solution will be 2/3 as concentrated as the original NH_3 solution.

$$[NH_4^+] = 0.100\ M \times 2/3 = \mathbf{0.067\ M}$$

$$K_{a_{NH_4^+}} = K_w / K_{b_{NH_3}} = 5.6 \times 10^{-10} = \frac{[H^+][NH_3]}{[NH_4^+]} = \frac{X^2}{0.0667}$$

$$X = [H^+] = 6.1 \times 10^{-6}\ M, \qquad \mathbf{pH = 5.21}$$

The indicator should have a pK_a close to 5.2. **Bromcresol green is a suitable choice.** The color change, blue to yellow (base to acid), is easy to distinguish.

Example 15.5 B   Indicator Colors

A solution has a pH of 7.0. What will be the color of the solution with each of the following indicators? (See Figure 15.8 in your textbook.)

a.   phenolphthalein
b.   methyl orange
c.   bromthymol blue
d.   thymol blue

Solution

a.   clear
b.   yellow
c.   green-blue
d.   yellow

Exercises

Section 15.1

1. Qualitatively, what is the effect on pH of adding sodium benzoate to an aqueous solution of benzoic acid? Why does this change occur?

2. Calculate the pH and the percent dissociation of the acid in each of the following solutions:

    a.   0.150 M C_6H_5COOH (benzoic acid, $K_a = 6.14 \times 10^{-5}$)
    b.   0.150 M C_6H_5COOH in the presence of 0.350 M C_6H_5COONa

3. Calculate the pH and the percent hydrolysis of the base in each of the following solutions:

    a.   0.438 M NH_3 ($K_b = 1.8 \times 10^{-5}$)
    b.   0.438 M NH_3 to which has been added 0.300 M NH_4Cl

4. A solution is prepared from 0.150 mol of CH_3COOH, 0.0100 mol of CH_3COO^- (from CH_3COONa), and enough water to make a total volume of 1.00 L. Calculate the value of [H^+] at equilibrium (acetic acid, $K_a = 1.8 \times 10^{-5}$).

Section 15.2

K_a or K_b for the following problems can be found in Appendix 5 in your textbook.

5. A solution contains 0.350 mol of $(CH_3)_3N$, 0.050 mol of $(CH_3)_3NH^+$, and enough water to make a total volume of 1.00 L. Calculate the value of [OH^-] (trimethylamine, $K_b = 6.25 \times 10^{-5}$).

6. Determine the pH in a solution containing 0.085 M NH_3 and 0.247 M NH_4Cl.

7. Calculate the pH of the buffer system with 0.10 M Na_2HPO_4/0.15 M KH_2PO_4.

8. Is a 6 M HCl solution a good buffer? Is it biologically useful?

9. What is the pH of a 200 mL solution containing 21.46 g of benzoic acid and 37.68 g of sodium benzoate,

    a.   initially?
    b.   after 30.0 mL of 5.00 M HCl have been added?
    c.   Has this buffer exceeded its buffer capacity after addition of the acid?

10. A buffer is prepared by adding 20.5 g of CH_3COOH and 17.8 g of CH_3COONa to enough water to make 5.00×10^2 mL of solution. Calculate the pH of this buffer solution.

11. Calculate the pH of 1.00 L of a buffer solution containing 0.10 M HCN and 0.12 M CN^-.

12. What must be the ratio of acetic acid to sodium acetate to prepare a buffer whose pH = 4.81?

13. What is the pH of the buffer containing 0.35 M NH_4Cl and 0.15 M NH_3?

Section 15.3

14. Calculate the pH if 0.01 mol of HCl is added to the buffer in Problem 11.

15. A solution contains 0.300 moles of acetic acid and 0.200 moles of sodium acetate in a total volume of 500. mL. How much 6.00 M NaOH must be added so that the pH of the solution equals the pK_a of acetic acid?

16. Calculate the pH if 0.0200 mol of NaOH is added to the original buffer in Problem 11.

17. A solution contains 0.216 moles of a base and 0.614 moles of the conjugate acid in a total volume of 800 mL. The pH of the solution is 9.65. What is the value for K_b of the base?

18. How many grams of sodium acetate must be dissolved in a 0.200 M acetic acid solution to make a 400. mL buffer solution with a pH of 4.56? (Assume that the volume of the solution remains constant.)

19. The pH of a bicarbonate-carbonic acid buffer is known to be 8.00. Calculate the ratio of the concentration of carbonic acid to that of the bicarbonate ion (K_a for $H_2CO_3 = 4.3 \times 10^{-7}$).

20. What would the pH of the solution in Problem 18 be if we added 40.0 mL of 0.30 M HCl?

21. We wish to buffer a solution at pH = 4.58. Which of the following acids (and conjugate base salts) would be most useful?

    a. CH_3COOH $(K_a = 1.8 \times 10^{-5})$
    b. C_6H_5COOH $(K_a = 6.14 \times 10^{-5})$
    c. $ClCH_2COOH$ $(K_a = 1.36 \times 10^{-3})$
    d. C_6H_5OH $(K_a = 1.6 \times 10^{-10})$

Section 15.4

22. Calculate K for the titration of HCl with NaOH.

23. A volume of 17.8 mL of a 0.344 M H_2SO_4 solution is required to completely neutralize 20.0 mL of a KOH solution. Calculate the concentration (in molarity) of the KOH solution.

24. If 37.42 mL of 0.1078 M NaOH solution is required to completely neutralize 25.00 mL of an HCl solution, what is the concentration (in molarity) of the acid solution?

25. Calculate the pH after the following **total** volumes of 0.3000 M NaOH have been added to 40.00 mL of 0.6000 M HCl.

    a. 0.00 mL   d. 79.40 mL   f. 80.50 mL
    b. 5.00 mL   e. 80.00 mL   g. 90.00 mL
    c. 40.00 mL

26. A sample of 50.0 mL of a commercial vinegar solution (which contains acetic acid) is titrated with a 1.00 M NaOH solution. What is the concentration of acetic acid in vinegar if 5.75 mL of the base was required for the titration?

27. Calculate the pH after the following **total** volumes of 0.3200 M HCl have been added to 25.00 mL of 0.1600 M NaOH.

    a. 0.00 mL   c. 12.40 mL   e. 12.60 mL
    b. 1.00 mL   d. 12.50 mL   f. 15.00 mL

28. If a 25.00-mL sample of base "X" requires 18.34 mL of 0.100 M HCl to reach the equivalence point, what is the concentration of base "X"? (Assume that the acid-base stoichiometry is 1:1.)

29. If 12.5 mL of 0.500 M H_2SO_4 exactly neutralizes 50.0 mL of NaOH, what is the concentration of the NaOH solution?

30. Calculate the pH at the equivalence point for the titration of 0.10 M NH_3 with 0.10 M HCl.

31. Calculate the pH at the equivalence point for the titration of 0.01 M CH$_3$COOH with 0.10 M NaOH.

32. A solution containing 100.0 mL of 0.1350 M CH$_3$COOH ($K_a = 1.8 \times 10^{-5}$) is being titrated with 0.5400 M NaOH. Calculate the pH:

    a. initially.
    b. halfway to the equivalence point.
    c. at the equivalence point.
    d. 5 mL past the equivalence point.

33. Calculate K for the titration in Problem 32.

34. Calculate the number of milliliters of a 0.153 M sodium hydroxide solution that must be added to a 20.0-mL sample of a solution of hydrochloric acid whose pH is 0.747 to reach the equivalence point.

35. The equivalence point for a solution of acid "A" occurs after 46.7 mL of base has been added. After 34.2 mL of base had been added (earlier in the titration) the pH was 4.64. What is the K_a for acid "A"?

36. A solution containing 50.00 mL of 0.1800 M NH$_3$ ($K_b = 1.8 \times 10^{-5}$) is being titrated with 0.3600 M HCl. Calculate the pH:

    a. initially.
    b. after the addition of 5.00 mL of HCl.
    c. after the addition of a total volume of 12.50 mL of HCl.
    d. after the addition of a total volume of 25.00 mL of HCl.
    e. after the addition of 26.00 mL of HCl.

37. Calculate the equilibrium constant for the titration of benzoic acid (C$_6$H$_5$COOH) with NH$_3$. Is this titration feasible? Why or why not?

38. A solution containing 2.049 g of a weak acid required 43.88 mL of 0.1207 M NaOH to reach the equivalence point. What is the molar mass of the acid?

Section 15.5

39. Using information found in <u>Figure 15.8 in your textbook</u>, select a suitable indicator **other than phenolphthalein** for the titrations in Problems 25 and 27.

40. Pick the better indicator for the titration in Problem 32.

41. Pick a suitable indicator for the titration in Problem 36.

42. Pick a suitable indicator for the titration in Problem 35.

43. What indicator would you choose for the titration of a water solution of HCN (concentration unknown) with a 0.100 M solution of NaOH (K_a for HCN = 6.2×10^{-10})?

Multiple Choice Questions

44. Calculate the pH of a solution prepared by mixing 40.0 mL of a 0.02 M HCl solution with 200.0 mL of 0.20 M HCN solution. Assume volumes to be additive. K_a for HCN = 1.0×10^{-10}.

    A. 2.4   B. 10   C. 2.0   D. 5.6

45.   Calculate the pH of a solution prepared by mixing 60.0 mL of a 0.200 M NaOH solution with 60.0 mL of a 0.200 M CH_3NH_2 solution. Assume the volumes to be additive. K_b for $CH_3NH_2 = 4.3 \times 10^{-5}$.

A. 11.8   B. 13.0   C. 1.0   D. 0.7

46.   What is the pOH of the solution prepared in the previous problem?

A. 13.3   B. 2.2   C. 1.0   D. 13.0

47.   What is the pOH of a solution prepared by mixing 0.30 mol of HCN and 0.50 mol of NaCN? $pK_a = 9.40$.

A. 7.00   B. 10.9   C. 3.36   D. 4.38

48.   40.0 g of NaF and 40.0 g of HF are mixed in 900 mL of solution. What is the pH of this buffer pair? $K_a = 7.2 \times 10^{-4}$.

A. 2.8   B. 4.9   C. 1.6   D. 3.1

49.   What is the pH of a buffer solution prepared by adding 20.0 g of KClO to 1050 mL of a 1.10 M HClO solution? Assume no volume change. $K_a = 3.5 \times 10^{-8}$.

A. 6.5   B. 7.8   C. 3.9   D. 5.0

50.   A student is asked to prepare a buffer solution composed of KH_2PO_4 and K_2HPO_4 with a pH of 6.00. What ratio of base to acid is required to prepare this buffer solution? $K_a = 6.2 \times 10^{-8}$.

A. 10 : 1   B. 0.55 : 1   C. 1.1 : 1   D. 0.062 : 1

51.   By what factor is the $[H^+]$ of a solution lowered if the pH changes from 10.40 to 7.40?

A. 100   B. 30.0   C. 3.00   D. 1000

52.   How many grams of NaF must be added to 120 mL of 1.00 M HF solution in order to adjust the pH to 3.40? Assume no change in volume upon addition of solid NaF. $K_a = 7.2 \times 10^{-4}$.

A. 0.69   B. 4.3   C. 1.6   D. 6.9

53.   A 100.0-mL sample of solution that is 0.20 M in both NaF and HF has 4.0 mL of 1.00 M HCl added to it. Calculate the pH change of this solution. $K_a = 7.2 \times 10^{-4}$.

A. 1.8   B. −0.18   C. 0.16   D. 0.0020

54.   4.00 mL of a 0.80 M NaOH solution is added to 130.0 mL of 0.20 M HCN/0.20 M NaCN buffer solution. Calculate the change in pH of this buffer solution.

A. 0.11   B. 1.2   C. −1.2   D. 0.46

55.   A particular amount of 0.20 M NaOH is accidentally spilled into 32.5 mL of a 0.050 M HCN/0.050 M NaCN buffer solution. The pH changed by +0.11 units. How many mL of NaOH were spilled into the buffer solution?

A. 4.0   B. 2.4   C. 3.9   D. 1.3

56.   40.0 mL of 0.320 M benzoic acid is titrated with 60.0 mL of 0.20 M NaOH. Calculate the pH of the resulting solution at the equivalence point. $K_a = 6.3 \times 10^{-5}$.

A. 5.5   B. 8.5   C. 7.7   D. 3.5

57.   50.0 mL of 0.020 M HCN is titrated with 100.0 mL of 0.010 M LiOH. Calculate the pH of the resulting solution at the equivalence point. $K_a = 4.9 \times 10^{-9}$.

A. 9.9   B. 4.1   C. 7.7   D. 5.2

58. What is the pH of the resulting solution when 50.0 mL of 1.10 M LiOH is titrated with 52.0 mL of 1.10 M HCl?

   A. 7.00   B. 13.0   C. 2.00   D. −0.260

59. 1.024 g of an unknown monoprotic strong acid is dissolved in 70.0 mL of solution. This solution is then titrated to the equivalence point with 80.0 mL of 0.200 M KOH. Calculate the molar mass of this acid.

   A. 128 g/mol   B. 81.0 g/mol   C. 64.0 g/mol   D. 211 g/mol

60. How many mL of a 1.06 M calcium hydroxide solution are required to titrate 70.0 mL of 0.88 M phosphoric acid solution to the third equivalence point?

   A. 87.2   B. 70.0   C. 22.6   D. 43.9

Answers to Exercises

1. The pH is raised due to the common-ion effect.

2. a. pH = 2.52; 2% dissociated.
   b. pH = 4.58; less than 2.0×10^{-2} % dissociated.

3. a. pH = 11.45; 0.6% hydrolyzed.
   b. pH = 9.42; less than 6×10^{-3} % hydrolyzed.

4. $[H^+] = 2.7 \times 10^{-4} \, M$

5. $[OH^-] = 4.4 \times 10^{-4} \, M$

6. pH = 8.79

7. pH = 7.03

8. Yes, a 6 *M* HCl solution is a superior buffer in that it resists changes in pH when an acid or base is added. However, it does not meet the traditional definition of a buffer, which contains a conjugate acid/ base pair or pairs. It is not biologically useful as it is far too acidic to be biologically safe.

9. a. pH = 4.37
   b. pH = 3.73
   c. No, the buffer has not exceeded its capacity.

10. pH = 4.55

11. pH = 9.29

12. The ratio must be 0.86 to 1.00.

13. pH = 8.88

14. pH = 9.21

15. A total of 8.33 mL of 6.0 *M* NaOH must be added.

16. pH = 9.45

17. $K_b = 1.27 \times 10^{-4}$

18. 4.3 g

19. 1:0.023

20. pH = 4.39

21. CH_3COOH would be the most useful ($pK_a = 4.75$).

22. $K = 1/K_w = 1 \times 10^{14}$

23. 0.612 *M* KOH

24.   0.1614 M HCl

25.   a.   pH = 0.22   d.   pH = 2.82   f.   pH = 11.10
       b.   pH = 0.30   e.   pH = 7.00   g.   pH = 12.36
       c.   pH = 0.82

26.   0.115 M

27.   a.   pH = 13.20   c.   pH = 10.93   e.   pH = 3.07
       b.   pH = 13.15   d.   pH = 7.00   f.   pH = 1.70

28. 7.34×10^{-2} M

29.   0.250 M

30.   pH = 5.28

31.   pH = 8.35

32.   a.   pH = 2.81   c.   pH = 8.89
       b.   pH = 4.74   d.   pH = 12.32

33. $K = 1.8 \times 10^{9}$

34.   23.4 mL

35. $K_a = 6.3 \times 10^{-5}$

36.   a.   pH = 11.26   c.   pH = 9.26   e.   pH = 2.32
       b.   pH = 9.86   d.   pH = 5.09

37. $K = K_a K_b / K_w = 1.1 \times 10^5$. This titration is not feasible. The value for K is too low (below 10^7 or so).

38.   molar mass = 386.9 g/mol

39.   Bromthymol blue or thymol blue are suitable choices.

40.   Thymol blue or phenolphthalein are suitable choices.

41.   Bromcresol green is a suitable choice.

42.   methyl orange

43.   alizarin yellow R

44. A   45. B   46. C   47. D   48. A   49. A
50. D   51. D   52. B   53. B   54. A   55. D
56. B   57. A   58. C   59. C   60. A

Chapter 16

Solubility and Complex Ion Equilibria

Solubility and complex ion equilibria are extensions of the same equilibrium concepts we have been considering for the past 3 chapters. The applications are different, but the questions we ask and our approach to finding answers to those questions (and making sure our answers make sense!) are the same.

16.1   Solubility Equilibria and the Solubility Product

When you finish this section you will be able to:

* Interconvert between solubility and K_{sp}.
* Solve problems relating to the common ion effect.

As with many of the previous sections, solubility equilibria use principles you have used before. This section deals with **solubility**, the amount of a salt that can be dissolved in water. Your textbook points out that the **solubility of a salt is variable. The solubility product ("K_{sp}") is constant** (at a given temperature).

The solubility equilibrium expression is set up as any other. For example, the equilibrium expression for the dissolution of Ag_2S in water is

$$Ag_2S(s) \rightleftharpoons 2Ag^+(aq) + S^{2-}(aq) \qquad K_{sp} = 1.6 \times 10^{-49}$$
$$K_{sp} = [Ag^+]^2[S^{2-}]$$

Remember that the pure solid, Ag_2S, is not included in the equilibrium expression.

Example 16.1 A   Solubility Equilibrium Expressions

Write products and equilibrium expressions for the following dissolution reactions:

a. $Ba(OH)_2(s)$ __

b. $PbCO_3(s)$ __

c. $Ag_2CrO_4(s)$ __

d. $Ca_3(PO_4)_2(s)$

__

Solution

a. $Ba(OH)_2(s) \rightleftharpoons Ba^{2+}(aq) + 2OH^-(aq)$
   $K_{sp} = [Ba^{2+}][OH^-]^2$

b. $PbCO_3(s) \rightleftharpoons Pb^{2+}(aq) + CO_3^{2-}(aq)$
   $K_{sp} = [Pb^{2+}][CO_3^{2-}]$

c. $Ag_2CrO_4(s) \rightleftharpoons 2Ag^+(aq) + CrO_4^{2-}(aq)$
   $K_{sp} = [Ag^+]^2[CrO_4^{2-}]$

d. $Ca_3(PO_4)_2(s) \rightleftharpoons 3Ca^{2+}(aq) + 2PO_4^{3-}(aq)$
   $K_{sp} = [Ca^{2+}]^3[PO_4^{3-}]^2$

Example 16.1 B K_{sp} from Solubility Data

Silver sulfide (Ag_2S) has a **solubility of 3.4×10^{-17} M** at 25°C. Calculate K_{sp} for Ag_2S.

Solution

The key here is *how we define solubility*. For the reaction

$$Ag_2S(s) \rightleftharpoons 2Ag^+(aq) + S^{2-}(aq),$$

one mole of S^{2-} is produced for every mole of Ag_2S that dissolves. Therefore the solubility equals the concentration of S^{2-} in solution.

$$\text{solubility} = s = [S^{2-}] = 3.4 \times 10^{-17} \, M$$

The stoichiometry of the reaction indicates that *2 moles of Ag^+ ion are produced for each mole of S^{2-} ion.* Therefore,

$$[Ag^+] = 2[S^{2-}] = 6.8 \times 10^{-17} \, M$$
$$K_{sp} = [Ag^+]^2 [S^{2-}] = (6.8 \times 10^{-17})^2(3.4 \times 10^{-17}) = 1.6 \times 10^{-49}$$

The keys to solving solubility problems are to

* properly define solubility, and
* properly use the reaction stoichiometry.

Example 16.1 C   Solubility from K_{sp} Data

Calculate the solubility of each of the following in moles per liter and grams per liter.

   a. $NiCO_3$ $K_{sp} = 1.4 \times 10^{-7}$
   b. $Ba_3(PO_4)_2$ $K_{sp} = 6 \times 10^{-39}$
   c. $PhBr_2$ $K_{sp} = 4.6 \times 10^{-6}$

Solution

We need to properly write dissolution reactions and equilibrium expressions for each. We must then carefully define solubility for each salt.

   a. $NiCO_3(s) \rightleftharpoons Ni^{2+}(aq) + CO_3^{2-}(aq)$

   $$s = [Ni^{2+}] = [CO_3^{2-}]$$
   $$K_{sp} = [Ni^{2+}][CO_3^{2-}]$$

$$1.4 \times 10^{-7} = (s)(s) = s^2 \quad \Rightarrow \quad s = 3.7 \times 10^{-4} \, M$$

This means that **3.7×10^{-4} mol/L of NiCO₃** will go into solution.

$$g/L = 3.7 \times 10^{-4} \, \text{mol/L} \times 118.7 \, \text{g/mol} = s = \textbf{0.44 g/L NiCO}_3$$

b.   **$Ba_3(PO_4)_2(s) \rightleftharpoons 3Ba^{2+}(aq) + 2PO_4^{3-}(aq)$**

Solubility = s = **moles of $Ba_3(PO_4)_2$** that go into solution. The stoichiometry of the reaction dictates that

$$[Ba^{2+}] = 3s \qquad [PO_4^{3-}] = 2s$$
$$K_{sp} = [Ba^{2+}]^3[PO_4^{3-}]^2$$
$$6 \times 10^{-39} = (3s)^3(2s)^2 = 108s^5$$
$$s^5 = 5.55 \times 10^{-41} \quad \Rightarrow \quad s = 9 \times 10^{-9} \, M$$
$$g/L = 9 \times 10^{-9} \, \text{mol/L} \times 601.8 \, \text{g/mol} = s = \textbf{5} \times \textbf{10}^{-6} \, \textbf{g/L}$$

c.   **$PbBr_2(s) \rightleftharpoons Pb^{2+}(aq) + 2Br^-(aq)$**

$$[Pb^{2+}] = s \qquad [Br^-] = 2s$$
$$K_{sp} = [Pb^{2+}][Br^-]^2$$
$$4.6 \times 10^{-6} = (s)(2s)^2 \quad \Rightarrow \quad s = 1.0 \times 10^{-2} \, M$$
$$g/L = 1.0 \times 10^{-2} \, \text{mol/L} \times 367 \, \text{g/mol} = s = \textbf{3.7 g/L}$$

Your textbook points out that **relative solubilities** among different compounds **cannot** be measured simply by comparing K_{sp} values. You must take the **composition of the salt** into account as illustrated by the next example.

Example 16.1 D   Relative Solubilities

Which of the following compounds is the most soluble?

AgCl	$K_{sp} = 1.5 \times 10^{-10}$
Ag₂CrO₄	$K_{sp} = 9.0 \times 10^{-12}$
Ag₃PO₄	$K_{sp} = 1.8 \times 10^{-18}$

Solution

$$\textbf{AgCl}(s) \rightleftharpoons \textbf{Ag}^+(aq) + \textbf{Cl}^-(aq)$$

$$s = [Ag^+] = [Cl^-]$$
$$K_{sp} = [Ag^+][Cl^-] = s^2 = 1.5 \times 10^{-10}$$
$$s = \textbf{1.2} \times \textbf{10}^{-5} \, \textbf{M}$$

$$\textbf{Ag}_2\textbf{CrO}_4(s) \rightleftharpoons 2\textbf{Ag}^+(aq) + \textbf{CrO}_4^{2-}(aq)$$

$$s = [CrO_4^{2-}] \qquad 2s = [Ag^+]$$
$$K_{sp} = [Ag^+]^2[CrO_4^{2-}] = (2s)^2(s) = 4s^3 = 9.0 \times 10^{-12}$$
$$s = \textbf{1.3} \times \textbf{10}^{-4} \, \textbf{M}$$

$$\textbf{Ag}_3\textbf{PO}_4(s) \rightleftharpoons 3\textbf{Ag}^+(aq) + \textbf{PO}_4^{3-}(aq)$$

$$s = [PO_4^{3-}] \qquad 3s = [Ag^+]$$
$$K_{sp} = [Ag^+]^3[PO_4^{3-}] = (3s)^3(s) = 27s^4 = 1.8 \times 10^{-18}$$
$$s = \textbf{1.6} \times \textbf{10}^{-5} \, \textbf{M}$$

Ag₂CrO₄ is the most soluble.

Common Ion Effect

We have encountered this before. Recall that **Le Châtelier's principle** predicts that **adding a common ion to the solution shifts the equilibrium to the left** in solubility equations. In other words **adding a common ion** (that doesn't react with other species in the solution) **reduces the solubility.**

Example 16.1 D   Common Ion Effect

Calculate the solubility of SrF_2 ($K_{sp} = 7.9 \times 10^{-10}$) in

    a.   pure water
    b.   0.100 M $Sr(NO_3)_2$
    c.   0.400 M NaF

Solution

The reaction of interest is

$$SrF_2(s) \rightleftharpoons Sr^{2+}(aq) + 2F^-(aq)$$
$$K_{sp} = [Sr^{2+}][F^-]^2$$

a.   In pure water, solubility of SrF_2 equals

$$s = [Sr^{2+}] \qquad 2s = [F^-]$$
$$7.9 \times 10^{-10} = (s)(2s)^2 = 4s^3 \quad \Rightarrow \quad s = 5.8 \times 10^{-4}\ M$$

b. $Sr(NO_3)_2$ is a strong electrolyte. This means that a 0.100 M solution will supply **0.100 M Sr^{2+} ion** to the solution.

$$[Sr^{2+}] = 0.100 + s \approx 0.100\ M$$

We are assuming that s is negligible relative to 0.100 M. We can test this with the 5% test.

$$[F^-] = 2s$$
$$K_{sp} = 7.9 \times 10^{-10} = (0.100)(2s)^2 = 0.4s^2$$
$$s = 4.4 \times 10^{-5}\ M$$

A solubility of 4.4×10^{-5} M is <5% of 0.100. The solubility was thus *decreased by a factor of 13* due to the presence of the common ion.

c.   NaF is a strong electrolyte. The dissociation of 0.400 M NaF will give

$$[F^-] = 0.400\ M + 2s \approx 0.400$$
$$[Sr^{2+}] = s$$
$$K_{sp} = 7.9 \times 10^{-10} = s(0.400)^2 = 0.16s$$
$$s = 4.9 \times 10^{-9}\ M$$

The solubility was markedly *decreased* by the presence of the common F^- ion.

Although the rigorous solutions are beyond the scope of your textbook, generally substances such as acids or bases that can combine with one of the ions of the salt will increase the solubility of the salt by removing product and thus forcing the reaction to the right.

16.2 Precipitation and Qualitative Analysis

When you finish this study section you will be able to:

- Determine whether the mixing of two solutions will produce a precipitate.
- Calculate the concentration of each ion in a solution in which a precipitate is produced.

This section asks the musical question, "If two solutions are mixed, will a precipitate form?" Then, "If it forms, what will be the concentration of each ion in solution?"

Your textbook introduces "Q," **the ion product** of the initial ion concentrations. **If $Q > K_{sp}$, precipitation will occur.**

Example 16.2 A   The Ion Product

A 200.0-mL solution of 1.3×10^{-3} M $AgNO_3$ is mixed with 100.0 mL of a 4.5×10^{-5} M Na_2S solution. Will precipitation occur?

Solution

Na^+ and NO_3^- ions are completely soluble, so we would presume that the precipitate would be **Ag_2S.** The reaction of interest is

$$2Ag^+(aq) + S^{2-}(aq) \rightleftharpoons Ag_2S(s)$$

The ion product is

$$Q = [Ag^+]_0^2 [S^{2-}]_0$$

$$[Ag^+]_0 = 1.3 \times 10^{-3} \, M \times \frac{200.0 \text{ mL (original volume)}}{(200.0 + 100.0) \text{ mL (total soln volume)}} = \mathbf{8.7 \times 10^{-4} \, M}$$

$$[S^{2-}]_0 = 4.5 \times 10^{-5} \, M \times \frac{100.0 \text{ mL}}{300.0 \text{ mL}} = \mathbf{1.5 \times 10^{-5} \, M}$$

$$Q = (8.7 \times 10^{-4})^2 (1.5 \times 10^{-5}) = \mathbf{1.1 \times 10^{-11}}$$

$K_{sp} = 1.6 \times 10^{-49}$. $Q > K_{sp}$, so **precipitation will occur.**

Once we determine that precipitation will occur, we are faced with the problem of determining the **equilibrium concentrations** of each of our ions of interest.

The general strategy involves **assuming** that because K_{sp} is so low, **if a precipitate forms, it will do so quantitatively.** We can then use the equilibrium expression involving the solubility of the salt (with a common ion) to solve for the equilibrium concentration of each ion. Let's illustrate this with the next example.

Example 16.2 B   Solubility from Mixing Solutions

Calculate the equilibrium concentration of each ion in a solution obtained by mixing 50.0 mL of 6.0×10^{-3} M $CaCl_2$ with 30.0 mL of 0.040 M NaF. (K_{sp} for $CaF_2 = 4.0 \times 10^{-11}$)

Solution

First, let's verify that precipitation in fact occurs,

$$[Ca^{2+}]_0 = 6.0 \times 10^{-3} \, M \times \frac{50.0 \text{ mL}}{(50.0 + 30.0) \text{ mL}} = 3.8 \times 10^{-3} \, M$$

$$[F^-]_0 = 0.040 \, M \times \frac{30.0 \text{ mL}}{80.0 \text{ mL}} = 0.015 \, M$$

$$Ca^{2+}(aq) + 2F^-(aq) \rightleftharpoons CaF_2(s)$$

$$Q = [Ca^{2+}][F^-]^2 = (3.8 \times 10^{-3})(0.015)^2 = 8.4 \times 10^{-7}$$

$$Q > K_{sp}, \text{ so precipitation occurs.}$$

We can assume a **quantitative ("stoichiometric") reaction between** Ca^{2+} **and** F^- to form CaF_2 solid. We can then see how much of which ion remains as excess in solution.

mmol $Ca^{2+}_{initial}$ $= 6.0 \times 10^{-3}$ mmol/mL $\times 50.0$ mL $=$ **0.30 mmol**
mmol $F^-_{initial}$ $= 0.040$ mmol/mL $\times 30.0$ mL $=$ **1.2 mmol**

	$Ca^{2+}(aq)$	$+$	$2F^-(aq)$	$\rightleftharpoons$	$CaF_2(s)$
initial (mmol)	0.30		1.2		0
final (mmol)	≈ 0		0.60		0.30

$$[Ca^{2+}] = s$$
$$[F^-] = 0.60 \text{ mmol}/80.0 \text{ mL} = 7.5 \times 10^{-3} \, M + 2s \approx 7.5 \times 10^{-3} \, M$$

$$K_{sp} = [Ca^{2+}][F^-]^2$$
$$4.0 \times 10^{-11} = s(7.5 \times 10^{-3})^2$$
$$s = 7.1 \times 10^{-7} \, M = [Ca^{2+}]$$
$$7.5 \times 10^{-3} \, M = [F^-]$$

Remember again that although the solubility of different ions can change depending upon what is in solution, **the solubility product (K_{sp}) remains the same** at a given temperature.

16.3   Equilibria Involving Complex Ions

When you finish this section you will be able to:

- Calculate the concentrations of species in a solution involving complex ions.
- Determine the increase in solubility of an insoluble salt by adding a complex ion.

Your textbook introduces several new terms here. You should be able to define: **complex ion, ligand,** and **formation constant.** The ultimate goal of this section is to demonstrate that **introducing a Lewis base** into a solution **enhances the solubility** of an otherwise insoluble salt.

The key idea here is that to avoid being drowned in a sea of equations, **we must make (and test) simplifying assumptions where possible!**

Let's do the following example to demonstrate the idea of **complex ion formation.**

Example 16.3 A   Complex Ion Formation

Calculate the concentrations of Ag^+ and $Ag(CN)_2^-$ in a solution prepared by mixing **100.0 mL of** $5.0 \times 10^{-3} \, M \, AgNO_3$ with **100.0 mL of 2.00 M KCN.**

$$Ag^+(aq) + 2CN^-(aq) \rightleftharpoons Ag(CN)_2^-(aq) \qquad K_1 = 1.3 \times 10^{21}$$

Solution

Note that K_1 equals the equilibrium constant for the reaction, and it is overwhelmingly large. This means that **virtually all of the Ag^+ ion will end up as $Ag(CN)_2^-$.** The other important point to note is that **our ligand, CN^-, is present in huge excess.** Therefore,

$$[CN^-] \approx [CN^-]_0$$

The equilibrium expression is

$$K_1 = \frac{[Ag(CN)_2^-]}{[Ag^+][CN^-]^2}$$

To get equilibrium concentrations,

$$[CN^-] \approx [CN^-]_0 = 1.00 \ M$$

(after mixing, the volume doubles, so the concentration is halved). Virtually all of the silver is present as $Ag(CN)_2^-$.

Therefore,

$$[Ag(CN)_2^-] \approx [Ag^+]_0 = 2.5 \times 10^{-3} \ M$$

(Remember the dilution!) Rearranging our overall equilibrium expression,

$$[Ag^+] = \frac{[Ag(CN)_2^-]}{K_1[CN^-]^2} = \frac{2.5 \times 10^{-3}}{1.3 \times 10^{21} \ (1.00)^2} = 1.9 \times 10^{-24} \ M$$

The important question in all of this is, "How does complex ion formation affect solubility?" Using our first example (16.3 A), we note that most of the silver was complexed as $Ag(CN)_2^-$. That means that as free Ag^+ ion is being produced by an insoluble salt (such as AgI, $K_{sp} = 1.5 \times 10^{-16}$), it is being **removed by complexing with CN^- ions.** This is pulling the dissolution of AgI(s) **to the right.**

**FORMATION OF A COMPLEX ION
INCREASES THE SOLUBILITY OF AN "INSOLUBLE" SALT.**

Example 16.3 B   Solubility And Complex Ion Formation

Calculate the solubility of AgI(s) in $1.00 \ M \ CN^-$ ion (K_{sp} for AgI $= 1.5 \times 10^{-16}$).

Solution

We have, up until now, said that the solubility of AgI equals

$$s = [Ag^+] = [I^-]$$

However, Ag^+ ion forms complexes that enhance the solubility. A more precise statement is that

$$s = [I^-] = [\text{all Ag species}] = [Ag^+] + [AgCN] + [Ag(CN)_2^-]$$

We know, based on the formation constant, that virtually all of the Ag^+ ion is present as $Ag(CN)_2^-$. Therefore,

$$s = [I^-] \approx [Ag(CN)_2^-]$$

The overall equilibrium of AgI(s) and $CN^-(aq)$ can be represented as

$$AgI(s) + 2CN^-(aq) \rightleftharpoons Ag(CN)_2^- + I^-(aq)$$

$$K = K_{sp}K_1 = \frac{[Ag(CN)_2^-][I^-]}{[CN^-]^2} = 1.5 \times 10^{-16} (1.3 \times 10^{21})$$

$$K = 1.9_5 \times 10^5$$

$[CN^-] = 2.00 - 2X$ (original concentration − amount complexed)

$[Ag(CN)_2^-] = X$ (amount formed, X − solubility, s)

$[I^-] = X$ (X − solubility, s)

$$K = \frac{[Ag(CN)_2^-][I^-]}{[CN^-]^2}$$

$$1.95 \times 10^5 = \frac{X^2}{(2.00 - 2X)^2}$$

$$\sqrt{1.95 \times 10^5} = \frac{X}{2.00 - 2X}$$

Getting rid of the denominator,

$$883 - 883X = X$$
$$X = \textbf{solubility} = \textbf{0.999 } \textit{\textbf{M}}$$

Note how our solubility has increased from 1.2×10^{-8} M to 0.999 M due to the presence of the ligand.

Exercises

Section 16.1

1. Write products and equilibrium expressions for the following dissolution reactions.

    a. $PbI_2(s) \rightleftharpoons$

    b. $Sr_3(PO_4)_2(s) \rightleftharpoons$

    c. $MnS(s) \rightleftharpoons$

2. Calculate $[I^-]$ in an AgI solution with $[Ag^+] = 1.2 \times 10^{-8}$ M. (AgI has $K_{sp} = 1.5 \times 10^{-16}$.)

3. Manganese sulfide (MnS) has a $K_{sp} = 2.3 \times 10^{-13}$. Calculate the solubility of MnS.

4. Calculate the solubility product (K_{sp}) of silver sulfate if its molar solubility is 1.5×10^{-2} mol/L.

5. The K_{sp} for manganese(II) carbonate is 1.8×10^{-11}. Calculate the solubility of $MnCO_3$.

6. Strontium phosphate ($Sr_3(PO_4)_2$) has a $K_{sp} = 1 \times 10^{-31}$. Calculate the solubility of $Sr_3(PO_4)_2$.

7. Calculate the K_{sp} for calcium sulfate if its solubility is 0.67 g/L.

8. The solubility of AgBr is 7.1×10^{-7} M. Calculate K_{sp} for AgBr.

9. Calculate the solubility (in g/L) of $Fe(OH)_3$. The K_{sp} for iron(III) hydroxide $[Fe(OH)_3]$ is 1.8×10^{-15}.

10. The solubility of SrF_2 is 5.8×10^{-4} M. Calculate K_{sp} for SrF_2.

11. The solubility of the ionic compound M_2X_3, having a molar mass of 288 g, is 3.6×10^{-17} g/L. Calculate the K_{sp} of the compound.

12. Calculate the concentration of Ag^+ in a saturated solution of Ag_2CrO_4 ($K_{sp} = 9.0 \times 10^{-12}$).

13. What is the solubility of silver chloride (in g/L) ($K_{sp} = 1.6 \times 10^{-10}$) in a 6.5×10^{-3} M silver nitrate solution?

14. Calculate the number of grams of ZnS ($K_{sp} = 2.5 \times 10^{-22}$) that will dissolve in 3.0×10^2 mL of 0.050 M $Zn(NO_3)_2$.

15. Calculate the solubility of $Co(OH)_2$ ($K_{sp} = 2.5 \times 10^{-16}$) at pH 11.50.

16. Calculate the solubility of $PbCO_3$ ($K_{sp} = 1.5 \times 10^{-15}$) in

    a. pure water

    b. 0.0400 M $Pb(NO_3)_2$

17. If 50.0 mL solution of 2.0×10^{-3} M $CaCl_2$ is mixed with 100.0 mL of 5.0×10^{-2} M NaF, will precipitation occur?

Section 16.2

18. Determine if a precipitate will form when exactly 200 mL of 0.0040 M $BaCl_2$ is added to 600 mL of 0.0080 M K_2SO_4.

19. Calculate the number of moles of Ag_2CrO_4($K_{sp} = 9.0 \times 10^{-12}$) that will dissolve in 1.00 L of 0.010 M K_2CrO_4 solution. What will be the ion concentrations at equilibrium?

20. Calculate the equilibrium concentration of each ion in a solution obtained by mixing 75.0 mL of 2.5×10^{-2} M $AgNO_3$ with 25.0 mL of 3.2×10^{-4} M KI.

21. Will $BaCO_3$ ($K_{sp} = 1.6 \times 10^{-9}$) precipitate if a sample of 20.0 mL of 0.10 M $Ba(NO_3)_2$ is added to 50.0 mL of 0.10 M Na_2CO_3?

22. Write equations for the stepwise and overall formation of $Cd(NH_3)_4{}^{2+}$.

23. What is the molar solubility of silver chloride in a 1.0 M NH_3 solution?

Section 16.3

24. Calculate the concentrations of Ag^+ and $Ag(NH_3)_2{}^+$ in a solution with an initial Ag^+ concentration of 4.0×10^{-3} M and an initial NH_3 concentration of 0.500 M ($K_1 = 2.1 \times 10^3$, $K_2 = 8.2 \times 10^3$).

Multiple Choice Questions

25. Calculate the molar solubility of $BaCO_3$ in water. $K_{sp} = 8.1 \times 10^{-9}$.

   A. 9.0×10^{-5} B. 4.9×10^{-9} C. 8.5×10^{-9} D. 4.5×10^{-5}

26. A 500.0 mL sample of solution saturated with AgCl is allowed to evaporate to dryness. 0.966 mg of AgCl is recovered. Calculate K_{sp} for AgCl.

   A. 1.80×10^{-10} B. 3.74×10^{-8} C. 2.22×10^{-7} D. 3.74×10^{-6}

27. Calculate the molar solubility of magnesium hydroxide in pure water. $K_{sp} = 1.5 \times 10^{-11}$.

   A. 3.74×10^{-6} B. 1.6×10^{-4} C. 2.5×10^{-6} D. 1.9×10^{-6}

28. Calculate the number of grams of magnesium hydroxide present in 1300 mL of a saturated magnesium hydroxide solution. $K_{sp} = 1.5 \times 10^{-11}$.

   A. 0.019   B. 0.14   C. 0.012   D. 0.0093

29. 200.0 mL PbI_2 saturated solution is evaporated to dryness. Calculate K_{sp} for this substance if 120 mg of it were recovered after evaporation.

   A. 8.80×10^{-9} B. 4.35×10^{-9} C. 6.00×10^{-5} D. 9.43×10^{-5}

30. Calculate the molar solubility of AgCl in 0.11 M NaCl solution.

   A. 6.6×10^{-11} B. 1.5×10^{-9} C. 7.2×10^{-10} D. 2.0×10^{-11}

31. Calculate the molar solubility of Ag_2CrO_4 in a 0.080 M Na_2CrO_4 solution.

   A. 4.8×10^{-7} B. 7.00×10^{-12} C. 1.1×10^{-5} D. 2.4×10^{-7}

32. How many milligrams of FeS per liter will dissolve in a 0.20 M Na_2S solution? $K_{sp} = 4.9 \times 10^{-18}$.

   A. 1.6×10^{-13} B. 4.6×10^{-9} C. 8.6×10^{-12} D. 2.4×10^{-18}

33. Calculate the formation constant, K_f, for the following complex ion:

$$Ag^+ + 2Br^- \rightleftharpoons AgBr_2{}^-$$

   where the concentrations of free silver, free bromide, and complex ion are 1.56×10^{-6}, 0.20, and 0.20 M, respectively.

   A. 3.2×10^6 B. 7.8×10^6 C. 1.3×10^7 D. 7.8×10^{-7}

Answers to Exercises

1. a. $PbI_2(s) \rightleftharpoons Pb^{2+}(aq) + 2I^-(aq)$ $K_{sp} = [Pb^{2+}][I^-]^2$
   b. $Sr_3(PO_4)_2(s) \rightleftharpoons 3Sr^{2+}(aq) + 2PO_4^{3-}(aq)$ $K_{sp} = [Sr^{2+}]^3[PO_4^{3-}]^2$
   c. $MnS(s) \rightleftharpoons Mn^{2+}(aq) + S^{2-}(aq)$ $K_{sp} = [Mn^{2+}][S^{2-}]$

2. $1.3 \times 10^{-8}\ M$

3. solubility $= 4.8 \times 10^{-7}\ M$

4. $K_{sp} = 1.4 \times 10^{-5}$

5. solubility $= 4.2 \times 10^{-6}\ M$

6. solubility $= 2 \times 10^{-7}\ M$

7. $K_{sp} = 2.4 \times 10^{-5}$

8. $K_{sp} = 5.0 \times 10^{-13}$

9. solubility $= 9.7 \times 10^{-3}$

10. $K_{sp} = 7.8 \times 10^{-10}$ (actual $= 7.9 \times 10^{-10}$)

11. $K_{sp} = 3.3 \times 10^{-93}$

12. $[Ag^+] = 2.6 \times 10^{-4}\ M$

13. solubility $= 3.6 \times 10^{-6}$ g/L

14. 1.5×10^{-19}g ZnS

15. solubility $= 2.5 \times 10^{-11}\ M$

16. a.   solubility $= 3.9 \times 10^{-8}\ M$ b.   solubility $= 3.8 \times 10^{-14}\ M$

17. Yes, $Q = 7.4 \times 10^{-7}$ ($> 4.0 \times 10^{-11}$, the K_{sp} of CaF_2)

18. Yes, $Q = 6.0 \times 10^{-6}$. $BaSO_4$ will precipitate out of solution until $[Ba^{2+}][SO_4^{2-}] > 1.5 \times 10^{-9}$, the K_{sp} of $BaSO_4$.

19. 1.5×10^{-5} moles; solubility $= 1.5 \times 10^{-5}$ mol/L; $[Ag^+] = 3.0 \times 10^{-5}\ M$; $[CrO_4^{2-}] = 0.010\ M$

20. $[Ag^+] = 1.9 \times 10^{-2}\ M$; $[I^-] = 8.0 \times 10^{-15}\ M$.

21. Yes, $BaCO_3$ will precipitate out of solution.

22. stepwise: $Cd^{2+} + NH_3 \rightleftharpoons Cd(NH_3)^{2+}$ K_1
    $Cd(NH_3)^{2+} + NH_3 \rightleftharpoons Cd(NH_3)_2^{2+}$ K_2
    $Cd(NH_3)_2^{2+} + NH_3 \rightleftharpoons Cd(NH_3)_3^{2+}$ K_3
    $Cd(NH_3)_3^{2+} + NH_3 \rightleftharpoons Cd(NH_3)_4^{2+}$ K_4
    overall $Cd^{2+} + 4NH_3 \rightleftharpoons Cd(NH_3)_4^{2+}$ $K_1K_2K_3K_4$

23. solubility $= 0.50$ M

24. $[Ag^+] = 9.6 \times 10^{-10}$; $[Ag(NH_3)_2^+] = 4.0 \times 10^{-3}$

25. A   26. A   27. B   28. A   29. A   30. B

31. C   32. A   33. A

Chapter 17

Spontaneity, Entropy, and Free Energy

Recall the first law of thermodynamics from Chapter 6. It says that energy can neither be created nor destroyed - just changed from one form to another. The law describes energy changes in chemical reactions but does not answer the more fundamental question, "Can we use thermodynamics to predict **IF** reactions will occur?" This chapter shows how such predictions can be made.

17.1   Spontaneous Processes and Entropy

When you finish this section you will be able to:

- Define entropy.
- Choose among alternatives that which has the greatest positional entropy.

Your textbook makes several points regarding **spontaneous** reactions.

- **Spontaneous** means occurring **without outside intervention**.
- Rate of reaction is **irrelevant** to spontaneity. Spontaneous means it will happen, not necessarily quickly, or even in Earth's lifetime.
- Spontaneous processes **increase the entropy of the universe**. (The entropy of a **system can decrease** if that of the **surroundings increases**.)

⇒ Entropy is a complex mathematical function that describes the number of possible arrangements (**positional probability**) of the states of a substance.

⇒ Each arrangement available to a substance is called a **microstate**.

⇒ Gases (in general) have a much higher positional entropy than liquids or solids.

Let's have some practice with microstates and entropy.

Example 17.1 A   Microstates and Entropy

You have three identical atoms, "A," "B," and "C." They can go back and forth from one "underline" to the other via the "wall" as shown.

$$\underline{AB} \quad \| \quad \underline{C} \qquad \text{(for example)}$$
$$\text{side 1} \quad \rightleftarrows \quad \text{side 2}$$

a. List all the possible microstates that these three atoms can have on our two dashes. In other words, list all the ways they can arrange themselves.

b. How much more probable is arrangement II than arrangement I?

       arrangement I   •••   ‖   __
       arrangement II   ••   ‖   •

c. Does this make sense in terms of our understanding of entropy?

Solution

a. Microstates

    $\underline{ABC}$ ‖ $\underline{}$ (I) $\underline{A}$ ‖ $\underline{BC}$

    $\underline{AB}$ ‖ $\underline{C}$ (II) $\underline{B}$ ‖ $\underline{AC}$

    $\underline{AC}$ ‖ $\underline{B}$ (II) $\underline{C}$ ‖ $\underline{AB}$

    $\underline{BC}$ ‖ $\underline{A}$ (II) $\underline{}$ ‖ $\underline{ABC}$

b. Arrangement I can happen **one way**. Arrangement II can happen three ways. Arrangement II is thus three times as likely as arrangement I.

c. The "atoms" are **more likely** to spread than to be confined to one side of the wall. That agrees with our concept of entropy as positional probability.

Example 17.1 B   Positional Entropy

Which of the following pairs is likely to have the higher positional entropy per mole at a given temperature?

    a. Solid or gaseous phosphorus
    b. $CH_4(g)$ or $C_3H_8(g)$
    c. $KOH(s)$ or $KOH(aq)$

Solution

a. Gaseous phosphorus will have the higher positional entropy. It is not as constrained by intramolecular bonds as solid phosphorus is.

b. All other things being equal, larger molecules containing many single bonds have more positional possibilities than smaller ones. $C_3H_8(g)$ has the higher entropy.

c. In general (all other things being equal), liquids have **slightly** higher entropies than solids do. **However**, in this case, the positional probabilities of $KOH(aq)$ are constrained due to hydrogen bonding interactions. $KOH(s)$ has the higher entropy.

17.2   Entropy and the Second Law of Thermodynamics

The following questions will test your understanding of the "second law."

1.   State the second law of thermodynamics.
2.   State the second law in terms of the system and surroundings.
3.   Under what circumstance can the **entropy of the system decrease** for a spontaneous process?
4.   How do the first and second laws fundamentally differ?

17.3   The Effect of Temperature on Spontaneity

When you finish this section you will be able to:

•   State the importance of exothermic reactions to entropy.
•   Calculate the change in entropy of the surroundings from the heat of reaction of the solution.

Exothermic reactions **give off energy to the surroundings.** Therefore, **random motions** of particles in the surroundings **increase.** When random motions increase, **positional probabilities increase.** The <u>key point</u> from all this is that **exothermic reactions increase the entropy of the surroundings (ΔS_{surr}).**

The **magnitude** of the increase in ΔS_{surr} **depends on the temperature.** (See the money-related discussion in your textbook.) Recall from Chapter 6 that we think of **heat flow** in terms of **the system.**

> **Exothermic reaction (at constant pressure), $\Delta H = -$**
> **Endothermic reaction (at constant pressure), $\Delta H = +$**

$$\Delta S_{surr} = \frac{-\Delta H}{T} \text{ (in Kelvin)}$$

Example 17.3 A ΔS_{surr} and the Heat of Reaction

Calculate ΔS_{surr} for each of the following reactions at 25°C and 1 atm.

a. $C_3H_8(g) + 5O_2(g) \rightarrow 3CO_2(g) + 4H_2O(g)$ $\Delta H = -2045$ kJ
b. $(NH_4)_2Cr_2O_7(s) \rightarrow N_2(g) + Cr_2O_3(s) + 4H_2O(g)$ $\Delta H = -315$ kJ
c. $H_2O(l) \rightarrow H_2O(g)$ $\Delta H = +44$ kJ

Solution

a. $\Delta S_{surr} = \dfrac{-\Delta H}{T}$

$T = (25 + 273) = 298$ K

$\Delta S_{surr} = \dfrac{-(-2045 \text{ kJ})}{298 \text{ K}} = 6.86 \text{ kJ/K} = \mathbf{6860 \text{ J/K}}$

There is an increase because the reaction is exothermic.

b. $\Delta S_{surr} = \dfrac{-(-315 \text{ kJ})}{298 \text{ K}} = 1.06 \text{ kJ/K} = \mathbf{1060 \text{ J/K}}$

There is an increase here as well because the reaction is exothermic.

c. $\Delta S_{surr} = \dfrac{-(44 \text{ kJ})}{298 \text{ K}} = -0.15 \text{ kJ/K} = \mathbf{-150 \text{ J/K}}$

Energy must be taken from the surroundings to the system in this reaction. Therefore, ΔS_{surr} will be negative.

The relationship between **entropy change** and reaction spontaneity is summarized in <u>Table 17.3 in your textbook</u>. Consider the information in that table, and try the following example.

Example 17.3 B   Reaction Spontaneity

Determine if the values for entropy in each of the following will produce a spontaneous process. Also, which of the following processes is endothermic (from the perspective of the system)?

a. $\Delta S_{sys} = 30 \text{ J/K}$ $\Delta S_{surr} = 50 \text{ J/K}$
b. $\Delta S_{sys} = -27 \text{ J/K}$ $\Delta S_{surr} = 40 \text{ J/K}$
c. $\Delta S_{sys} = 140 \text{ J/K}$ $\Delta S_{surr} = -85 \text{ J/K}$
d. $\Delta S_{sys} = 60 \text{ J/K}$ $\Delta S_{surr} = -85 \text{ J/K}$

Solution

$$\Delta S_{univ} = \Delta S_{sys} + \Delta S_{surr}$$

A reaction will be spontaneous if $\Delta S_{univ} > 0$.
A reaction will not be spontaneous if $\Delta S_{surr} < 0$.

a. $\Delta S_{univ} = \quad 30 + 50 \quad = 80 \text{ J/K}$, **spontaneous, exothermic.**
b. $\Delta S_{univ} = \quad -27 + 40 \quad = 13 \text{ J/K}$, **spontaneous, exothermic.**
c. $\Delta S_{univ} = 140 + (-85) = 55 \text{ J/K}$, **spontaneous, endothermic.**
d. $\Delta S_{univ} = \quad 60 + (-85) = -25 \text{ J/K}$, **not spontaneous, endothermic.**

Notice also from Table 17.3 that there are circumstances under which **temperature plays the deciding factor in whether a reaction will be spontaneous.**

17.4   Free Energy

When you finish this section you will be able to relate free energy and spontaneity.

Free energy is a mathematical term that describes **unequivocally** whether a reaction will be spontaneous. It is experimentally useful because **it reflects ΔS_{univ}.** Your textbook states two important relationships.

$$\Delta G = \Delta H - T\Delta S$$

(When no subscript appears, it is assumed that we are referring to the system.)

$$\Delta S_{univ} = \dfrac{-\Delta G}{T}$$

The first equation gives us an explicit way of calculating free energy. It further says that there are circumstances under which **temperature will determine whether a reaction is spontaneous.** (See <u>Table 17.5 in your textbook</u>.) Note also that since the temperature in Kelvin will always be greater than zero, $-\Delta G = +\Delta S_{univ}$ **always.**

The case of the temperature dependency of ice melting, as described in your textbook, is a perfect example of this. You have two opposing entropy factors. On the one hand, the reaction is **endothermic**, which **opposes the process** ($\Delta S_{surr} = -$). On the other hand, melting **increases the positional probability of the system** ($\Delta S_{sys} = +$). The temperature will determine which process will dominate (whether ice melts).

The second equation says that the ΔG must be **negative** (< 0) in order for a reaction to proceed.

Example 17.4 A   Free Energy and Spontaneity

Given the values for ΔH, ΔS, and T, determine whether each of the following sets of data represent spontaneous or nonspontaneous processes.

	ΔH (kJ)	ΔS (J/K)	T (K)
a.	40	300	130
b.	40	300	150
c.	40	−300	150
d.	−40	−300	130
e.	−40	300	150

Solution

$$\Delta G = \Delta H - T\Delta S$$

For a reaction to be spontaneous, ΔG must be < 0. When you do the calculations, make sure that you either **change ΔH to joules or ΔS to kilojoules!**

a. $\Delta G =$ 40 kJ − 0.300 kJ/K (130 K) = +1 kJ, **nonspontaneous**
b. $\Delta G =$ 40 kJ − 0.300 kJ/K (150 K) = −5 kJ, **spontaneous**
c. $\Delta G =$ 40 kJ − (−0.300 kJ/K)(150 K) = +85 kJ, **nonspontaneous**
d. $\Delta G =$ −40 kJ − (−0.300 kJ/K)(130 K) = −1 kJ, **spontaneous**
e. $\Delta G =$ −40 kJ − 0.300 kJ/K (150 K) = −85 kJ, **spontaneous**

Note that temperature is NOT important to having a spontaneous reaction when the reaction is exothermic and there is an increase in entropy (it will **always** be spontaneous), or when the reaction is endothermic with a decrease in entropy (it will **always** be nonspontaneous).

Example 17.4 B   Free Energy and Temperature

You know that the boiling point of water is 373 K. See how this compares to the minimum temperature for reaction that you determine thermodynamically for the phase change:

$$H_2O(l) \rightarrow H_2O(g)$$

where $\Delta H = 44$ **kJ** and $\Delta S = 119$ **J/K**.

Solution

The criterion for spontaneity is $\Delta G < 0$. This means that $\Delta H - T\Delta S < 0$.

Adding $T\Delta S$ to both sides,

$$\Delta H < T\Delta S.$$

Dividing both sides by ΔS,

$$\frac{\Delta H}{\Delta S} < T.$$

Using the data from our problem, $T > \dfrac{4.4 \times 10^4 \text{ J}}{119 \text{ J/K}}$ or $T > 370$ **K**, which is close to the actual value.

17.5   Entropy Changes in Chemical Reactions

When you finish this section you will be able to:

- Predict the sign of entropy changes for a given reaction.
- Calculate ΔS from thermodynamic data tables.

This section in your textbook begins by reminding us that ΔS_{surr} is related to **heat flow** from the system. BUT ΔS_{sys} is related to the positional **probabilities** for each of the reactants. For example,

$$2H_2(g) + O_2(g) \rightarrow 2H_2O(g)$$

$\Delta S = -89$ kJ (at 25°C). We lose entropy because **3 total moles** of the gases on the left have **more** positional possibilities than **2 moles** of vapor on the right.

- **For a chemical reaction involving only the gas phase, entropy is related to the total number of moles on either side of the equation. A decrease means lower entropy, an increase means higher entropy.**
- **For a chemical reaction involving different phases, the production of a gas will (in general) increase the entropy much more than an increase in the number of moles of a liquid or solid.**

For example,

$$2HNO_3(aq) + Na_2CO_3(s) \rightarrow 2NaNO_3(aq) + H_2O(l) + CO_2(g)$$
$$\Delta S = +88 \text{ kJ (at 25°C)}.$$

Example 17.5 A   The Sign of Entropy Changes

Predict the sign of $\Delta S°$ for each of the following reactions:

   a. $(NH_4)_2Cr_2O_7(s) \rightarrow Cr_2O_3(s) + 4H_2O(l) + N_2(g)$
   b. $Mg(OH)_2(s) \rightarrow MgO(s) + H_2O(g)$
   c. $PCl_5(g) \rightarrow PCl_3(g) + Cl_2(g)$

Solution

   a.   There are many more moles on the right-hand side. In addition, a gas is formed on the right-hand side. **The change in entropy will be positive.**

   b.   A gas is formed on the right-hand side. **The change in entropy will be positive**.

   c.   More moles of gas are present on the right-hand side. **The change in entropy will be positive**.

The **third law of thermodynamics** says that **the entropy of a perfect crystal at 0 K is zero**. This means that the absolute entropy of substances can be explicitly measured. (See Appendix 4 in your textbook.)

As was true with $\Delta H°$ (a state function), $\Delta S°$ (also a state function) can be determined as the difference between the sum of the entropy of products minus the sum of the entropy of the reactants.

$$\Delta S°_{reaction} = \Sigma n_p S°_{products} - \Sigma n_r S°_{reactants}$$

Example 17.5 B   Entropy of Reaction

Calculate $\Delta S°$ for each of the following reactions using data from Appendix 4 in your textbook.

   a. $N_2O_4(g) \rightarrow 2NO_2(g)$
   b. $Fe_2O_3(s) + 2Al(s) \rightarrow 2Fe(s) + Al_2O_3(s)$
   c. $4NH_3(g) + 5O_2(g) \rightarrow 4NO(g) + 6H_2O(g)$

Solution

$$\Delta S^{\circ}_{reaction} = \sum n_p S^{\circ}_{products} - \sum n_r S^{\circ}_{reactants}$$

a. $\Delta S^{\circ}_{rxn} = 2S^{\circ}_{NO_2(g)} - S^{\circ}_{N_2O_4(g)}$

         $= 2$ mol $(240$ J/K mol$) - 1$ mol $(304$ J/K mol$)$

     $\Delta S^{\circ}_{rxn} = \mathbf{176}$ **J/K** (The entropy should increase due to an increase in the number of moles of gas.)

b. $\Delta S^{\circ}_{rxn} = [2S^{\circ}_{Fe(s)} + S^{\circ}_{Al_2O_3(g)}] - [S^{\circ}_{Fe_2O_3(s)} + 2S^{\circ}_{Al(s)}]$

         $= [2$ mol $(27$ J/K mol$) + 1$ mol $(51$ J/K mol$)] - [1$ mol $(90$ J/K mol$) + 2$ mol $(28$ J/K mol$)]$

     $\Delta S^{\circ}_{rxn} = \mathbf{-41}$ **J/K** (This is fairly small because there were no net phase changes.)

c. $\Delta S^{\circ}_{rxn} = [4S^{\circ}_{NO(g)} + 6S^{\circ}_{H_2O(g)}] - [4S^{\circ}_{NH_3(g)} + 5S^{\circ}_{O_2(g)}]$

         $= [4$ mol $(211$ J/K mol$) + 6$ mol $(189$ J/K mol$)] - [4$ mol $(193$ J/K mol$)$

         $+ 5$ mol $(205$ J/K mol$)]$

     $\Delta S^{\circ}_{rxn} = \mathbf{181}$ **J/K** (We increased the number of moles of gas, which is reflected by the increase in entropy.)

17.6   Free Energy and Chemical Reactions

When you finish this section you will be able to calculate the standard free energy of formation and use it to predict spontaneity of chemical reactions.

The **standard free energy change (ΔG°)** is the free energy change that occurs if reactants **in their standard states (1 atm, 25°C)** are converted to products in their standard states.

Your textbook points out that ΔG° *cannot be measured directly*, but is an important value because it represents a *standard set of conditions* at which to compare properties of reactions (as we will see later).

Three methods of calculating ΔG° are introduced in this section.

1. $\Delta G^{\circ} = \Delta H^{\circ} - T\Delta S^{\circ}$
2. **by manipulating known equations, as in Hess's law problems for ΔH°**
3. $\Delta G^{\circ} = \sum n_p \Delta G^{\circ}_{f(products)} - \sum n_r \Delta G^{\circ}_{f(reactants)}$

The following examples illustrate each of these methods:

Example 17.6 A   Standard Free Energy from Entropy and Enthalpy

Using data for **ΔH° and ΔS°**, calculate ΔG° for the following reactions at 25°C and 1 atm.

    a. $Cr_2O_3(s) + 2Al(s) \rightarrow Al_2O_3(s) + 2Cr(s)$
    b. $C_3H_8(g) + 5O_2(g) \rightarrow 3CO_2(g) + 4H_2O(g)$

Solution

$$\Delta G^{\circ} = \Delta H^{\circ} - T\Delta S^{\circ}$$

We need to calculate ΔH° and ΔS° for each reaction.

$$\Delta H^{\circ} = \sum n_p \Delta H^{\circ}_{f(products)} - \sum n_r \Delta H^{\circ}_{f(reactants)}$$

$$\Delta S^{\circ} = \sum n_p \Delta S^{\circ}_{products} - \sum n_r \Delta S^{\circ}_{reactants}$$

a. $\Delta H° = [1 \text{ mol}(-1676 \text{ kJ/mol}) + 2 \text{ mol } (0 \text{ kJ/mol})] - [1 \text{ mol } (-1128 \text{ kJ/mol}) + 2 \text{ mol } (0 \text{ kJ/mol})]$

$\hspace{3.5cm} \uparrow \hspace{3.5cm} \uparrow \hspace{3.5cm} \uparrow \hspace{3.5cm} \uparrow$

$\hspace{3.2cm} Al_2O_3 \hspace{3.2cm} Cr \hspace{3.2cm} Cr_2O_3 \hspace{3.2cm} Al$

$\Delta H° = -548 \text{ kJ}$

$\Delta S° = [1 \text{ mol } (51 \text{ J/K mol}) + 2 \text{ mol } (24 \text{ J/K mol})] - [1 \text{ mol } (81 \text{ J/K mol}) + 2 \text{ mol } (28 \text{ J/K mol})]$

$\hspace{3.5cm} \uparrow \hspace{3.5cm} \uparrow \hspace{3.5cm} \uparrow \hspace{3.5cm} \uparrow$

$\hspace{3.2cm} Al_2O_3 \hspace{3.2cm} Cr \hspace{3.2cm} Cr_2O_3 \hspace{3.2cm} Al$

$\Delta S° = -38 \text{ J/K}$

$\Delta G° = \Delta H° - T\Delta S° = -548 \text{ kJ} - 298 \text{ K } (-0.038 \text{ kJ/K})$

$\Delta G° = -537 \text{ kJ}$; the reaction is spontaneous.

b. $\Delta H° = [3 \text{ mol } (-393.5 \text{ kJ/mol}) + 4 \text{ mol } (-242 \text{ kJ/mol})] - [5 \text{ mol } (0 \text{ kJ/mol}) + 1 \text{ mol } (-104 \text{ kJ/mol})]$

$\hspace{3.5cm} \uparrow \hspace{3.5cm} \uparrow \hspace{3.5cm} \uparrow \hspace{3.5cm} \uparrow$

$\hspace{3.2cm} CO_2 \hspace{3.2cm} H_2O \hspace{3.2cm} O_2 \hspace{3.2cm} C_3H_8$

$\Delta H° = -2044 \text{ kJ}$

$\Delta S° = [3 \text{ mol}(214 \text{ J/K mol}) + 4 \text{ mol } (189 \text{ J/K mol})] - [5 \text{ mol } (205 \text{ J/K mol}) + 1 \text{ mol } (270 \text{ J/K mol})]$

$\hspace{3.5cm} \uparrow \hspace{3.5cm} \uparrow \hspace{3.5cm} \uparrow \hspace{3.5cm} \uparrow$

$\hspace{3.2cm} CO_2 \hspace{3.2cm} H_2O \hspace{3.2cm} O_2 \hspace{3.2cm} C_3H_8$

$\Delta S° = 103 \text{ J/K}$

$\Delta G° = \Delta H° - T\Delta S° = -2044 \text{ kJ} - 298 \text{ K } (0.103 \text{ kJ/K})$

$\Delta G° = -2075 \text{ kJ}$; this combustion is highly spontaneous.

Example 17.6 B   Standard Free Energy by Combining Equations

Given the following data:

(Equation 1) $\quad S(s) + {}^3/_2O_2(g) \rightarrow SO_3(g) \hspace{1.5cm} \Delta G° = -371 \text{ kJ}$

(Equation 2) $\quad 2SO_2(g) + O_2(g) \rightarrow 2SO_3(g) \hspace{1cm} \Delta G° = -142 \text{ kJ}$

Calculate $\Delta G°$

(Goal Equation) $\quad S(s) + O_2(g) \rightarrow SO_2(g)$

Is the reaction spontaneous?

Solution

We solve this problem just as we would a Hess's law problem (see Chapter 6). We must manipulate Equations 1 and 2 so that we can combine them to get the reaction of interest. (Remember what you do to the equations must also be done to the $\Delta G°$ values!)

One mole of S(s) appears only once on the left-hand side of the **goal equation**, and **once** on the left-hand side of Equation 1.

Equation 1 must remain as it is.

One mole of SO₂(g) appears on the **right-hand side** of the goal equation. **Two moles** of $SO_2(g)$ are present on the **left-hand side** of Equation 2. We must therefore

multiply Equation 2 by −½

to get one mole of $SO_2(g)$ on the right-hand side. (Remember to multiply $\Delta G°$ by −½ as well!)

$$\begin{array}{ll} S(s) + {}^3/_2O_2(g) \rightarrow SO_3(g) & \Delta G° = -371 \text{ kJ} \\ \underline{SO_3(g) + SO_2(g) \rightarrow {}^1/_2O_2(g)} & \underline{\Delta G° = +71 \text{ kJ}} \\ S(s) + O_2(g) \rightarrow SO_2(g) & \Delta G° = -300 \text{ kJ} \end{array}$$

Example 17.6 C   Standard Free Energy from "Products - Reactants"

Calculate ΔG° for the reaction

$$C_3H_8(g) \; + \; 5O_2(g) \; \rightleftharpoons \; 3CO_2(g) \; + \; 4H_2O(g)$$

using ΔG° data from Appendix 4. Compare the answer with that from our Example 17.6 A, part b.

Solution

$$\Delta G^\circ \; = \; \sum n_p \Delta G^\circ_{f(products)} \; - \; \sum n_r \Delta G^\circ_{f(reactants)}$$

$$= [\,3\Delta G^\circ_{f(CO_2(g))} \; + \; 4\Delta G^\circ_{f(H_2O(g))}\,] - [\,\Delta G^\circ_{f(C_3H_8(g))} \; + \; 5\Delta G^\circ_{f(O_2(g))}\,]$$

$$= [3 \; \text{mol} \; (-394 \; \text{kJ/mol}) + 4 \; \text{mol} \; (-229 \; \text{kJ/mol})] - [1 \; \text{mol} \; (-24 \; \text{kJ/mol}) + 5 \; \text{mol} \; (0 \; \text{kJ/mol})]$$

$$\Delta G^\circ \; = -2074 \; \text{kJ}$$

The ΔG° values agree within about 0.1%. Also note that as with ΔH°, ΔG° for elements in their standard states equals zero.

17.7   The Dependence of Free Energy on Pressure

When you finish this section you will be able to calculate ΔG at pressures other than 1 atmosphere.

We have until now assumed standard conditions. This section deals with **free energy at nonstandard pressures.** For an ideal gas, enthalpy is not pressure-dependent. **Entropy, however, is affected by pressure.** More positions are possible at lower pressure than higher pressure, therefore

$$S_{\text{low pressure}} > S_{\text{high pressure}}$$

Your textbook derives the relationship

$$\Delta G \; = \; \Delta G^\circ \; + \; RT \ln(Q)$$

where $R = 8.3145$ J/K mol
T = temperature in Kelvins
Q = reaction quotient (the mass action expression relating to **initial** quantities)

It is important to learn to use **and interpret** the results from this equation. Through a similar equation (to be introduced in the next section) we can relate equilibrium constants to ΔG.

Example 17.7   Relating Free Energy and Pressure

Calculate ΔG at 700 K for the following reaction:

$$C(s, \text{graphite}) \; + \; H_2O(g) \; \rightleftharpoons \; CO(g) + H_2(g)$$

Initial pressures are $P_{H_2O} = 0.85$ atm, $P_{CO} = 1.0 \times 10^{-4}$ atm, $P_{H_2} = 2.0 \times 10^{-4}$ atm.

Solution

Our first task is to determine ΔG°.

$$\Delta G^\circ = [\,\Delta G^\circ_{f(CO(g))} \; + \; \Delta G^\circ_{f(H_2(g))}\,] - [\,\Delta G^\circ_{f(C\,(s, \text{ graphite}))} \; + \; \Delta G^\circ_{f(H_2O(g))}\,]$$

$$\Delta G^\circ = [1 \; \text{mol} \; [(-137 \; \text{kJ/mol}) + 1 \; \text{mol} \; (0 \; \text{kJ/mol})] - [1 \; \text{mol} \; (0 \; \text{kJ/mol}) + 1 \; \text{mol} \; (-229 \; \text{kJ/mol})]$$

$$\Delta G^\circ = +92 \; \text{kJ}$$

This tells us that the reaction is not spontaneous **under standard conditions**. To evaluate at 700 K and the given pressures,

$$Q = \frac{P_{H_2} P_{CO}}{P_{H_2O}} = \frac{(2.0 \times 10^{-4} \text{ atm})(1.0 \times 10^{-4} \text{ atm})}{0.85 \text{ atm}} = \mathbf{2.35 \times 10^{-8} \text{ atm}}$$

$$\Delta G = \Delta G^\circ + RT \ln(Q) = 9.2 \times 10^4 \text{ J} + 8.3145 \text{ J/K mol } (700 \text{ K}) \ln(2.35 \times 10^{-8})$$
$$= 9.2 \times 10^4 \text{ J} + (-1.022 \times 10^5 \text{J})$$
$$= -1.02 \times 10^4 \text{ J}$$
$$\Delta G = -10 \text{ kJ}^*$$

Under conditions of high temperature and low product pressure, the reaction becomes spontaneous.

The final subheading in <u>Section 17.7 of your textbook</u> addresses "The Meaning of ΔG for a Chemical Reaction." It stresses that a spontaneous reaction will not necessarily go to completion. Rather, there may be some intermediate point that reflects the lowest possible ΔG value for the reaction.

17.8   Free Energy and Equilibrium

When you finish this section you will be able to:

- Interpret the direction of a reaction given appropriate data.
- Interconvert between K and ΔG°.

Equilibrium occurs at the lowest free energy available to the system. In Chapter 13, we defined equilibrium as occurring when the forward rate of reaction is equal to the reverse rate of reaction. That is still valid. Thermodynamically, your textbook defines equilibrium as occurring when

$$G_{\text{forward reaction}} = G_{\text{reverse reaction}}$$

In other words, **$\Delta G = 0$ at equilibrium.**

If $\Delta G < 0$, it means that $G_{\text{reactants}} > G_{\text{products}}$, and the reaction will go **to the right until $G_{\text{reactants}} = G_{\text{products}}$.**
If $\Delta G > 0$, it means that $G_{\text{reactants}} < G_{\text{products}}$, and the reaction will **go to the left until $G_{\text{reactants}} = G_{\text{products}}$.**

<u>Table 17.6 in your textbook</u> mathematically relates **ΔG° to K**. Remember that if $K = 1$ (the equilibrium condition), $\ln K = 0$ and $\Delta G = \Delta G^\circ$. At equilibrium, therefore,

$$\Delta G^\circ = -RT \ln(K)$$

Example 17.8 A   Free Energy and the Direction of Reaction

Using the same reaction as in our Example 17.7,

$$C(s, \text{graphite}) + H_2O(g) \rightleftharpoons CO(g) + H_2(g)$$

where **$T = 700$ K and $\Delta G^\circ = 92$ kJ**, determine the direction of reaction if the following initial pressure of each gas is

$$P_{H_2O} = \mathbf{0.67 \text{ atm}}, \quad P_{CO} = \mathbf{0.23 \text{ atm}}, \quad P_{H_2} = \mathbf{0.51 \text{ atm}}$$

---

*Note the discussion on units in the margin next to <u>Example 17.13 in your textbook</u>.

Solution

The goal is to calculate ΔG.

$$\Delta G = \Delta G° + RT \ln K$$

$$K = \frac{P_{H_2} P_{CO}}{P_{H_2O}} = \frac{(0.51 \text{ atm})(0.23 \text{ atm})}{0.67 \text{ atm}} = \textbf{0.175 atm}$$

$$\Delta G = 9.2 \times 10^4 \text{ J} + 8.3145 \text{ J/K mol } (700 \text{ K}) \ln(0.175)$$

$$\boldsymbol{\Delta G} = 82000 \text{ J} = \textbf{82 kJ}$$

This reaction will go very far to the left.

Example 17.8 B   Free Energy and Equilibrium Constant

Given the values for $\Delta G°$ that you calculated in Example 17.6 A, calculate K for the following reaction (at 25°C):

$$Cr_2O_3(s) + 2Al(s) \rightleftharpoons Al_2O_3(s) + 2Cr(s)$$

Solution

$\Delta G° = -537 \text{ kJ}$

$$\Delta G° = -RT \ln K$$

$$\ln K = \frac{\Delta G°}{-RT}, \quad \text{or} \quad K = e^{-\Delta G°/RT}$$

$$\ln K = \frac{-5.37 \times 10^5 \text{ J}}{(-8.3145 \text{ J/K mol})(298 \text{ K})} = 216.73$$

$$\boldsymbol{K} = e^{216.73} = \textbf{1.3} \times \textbf{10}^{\textbf{94}}, \text{ so this reaction is quite spontaneous.}$$

Example 17.8 C   Summing It All Up

We have previously (Chapter 15) studied the weak base-strong acid titration

$$\textbf{NH}_3\textbf{(}\textit{aq}\textbf{)} + \textbf{H}^+\textbf{(}\textit{aq}\textbf{)} \rightleftharpoons \textbf{NH}_4^+\textbf{(}\textit{aq}\textbf{)}$$

We said that you can look at it as the sum of two reactions (for purposes of calculating K).

1. $NH_3(aq) + H_2O(l) \rightleftharpoons NH_4^+(aq) + OH^-(aq)$ $(K_b = 1.8 \times 10^{-5})$
2. $H^+(aq) + OH^-(aq) \rightleftharpoons H_2O(l)$ $(1/K_w = 1.0 \times 10^{+14})$

Please do the following:

a. Calculate K for the titration reaction using the two equilibrium expressions.
b. Calculate $\Delta H°$, $\Delta S°$, $\Delta G°$, and K for the titration reaction using data from <u>Appendix 4 of your textbook</u>.
c. How do the calculated K values compare?
d. Is the titration spontaneous?

Solution

a. Summing the reactions requires multiplying the equilibrium constants

$$K_{titration} = K_b \times 1/K_w = 1.8 \times 10^9$$

b. $\Delta H°$ and $\Delta S°$ = the thermodynamic values for the products − those of the reactants

$$\Delta H° = [1 \text{ mol } (-132 \text{ kJ/mol})] - [1 \text{ mol } (-80 \text{ kJ/mol}) + 1 \text{ mol } (0 \text{ kJ/mol})]$$
$$\qquad\qquad\qquad \uparrow \atop NH_4{}^+ \qquad\qquad\qquad\qquad \uparrow \atop NH_3 \qquad\qquad\qquad \uparrow \atop H^+$$
$$\Delta H° = -52 \text{ kJ}$$

$$\Delta S° = [1 \text{ mol } (113 \text{ J/K mol})] - [1 \text{ mol}(111 \text{ J/K mol}) + 1 \text{ mol } (0 \text{ J/K mol})]$$
$$\Delta S° = 2 \text{ J/K}$$

$$\Delta G° = \Delta H° - T\Delta S° = -52000 \text{ J} - 298 \text{ K } (2 \text{ J/K}) = -52596 \text{ J}$$
$$\Delta G° = -53 \text{ kJ}$$

$$\ln K = \frac{\Delta G°}{-RT} = \frac{+52596 \text{ J}}{(8.3145 \text{ J/K mol})(298 \text{ K})} = 21.23$$

$$K = e^{21.23} = 1.7 \times 10^9$$

c. The K values are comparable (within about 5%).

d. The titration is spontaneous (as we know from experience).

17.9   *Free Energy and Work*

Your textbook makes the critical point that a thermodynamically favorable reaction might be made faster via a catalyst, but a catalyst would not be effective in one in which $\Delta G = +$. What are the two solutions to q_p, and why are they different?

Exercises

Section 17.1

1. Given 8 molecules in the two-bulb set up described in <u>Table 17.1 in your textbook</u>, calculate the relative probability of finding all 8 molecules in the left-hand bulb. What does this tell you regarding entropy and probability?

2. Which of the following pairs of substances is likely to have the higher positional entropy?

    a. $HCl(aq)$ or $HCl(g)$
    b. $P_4(s)$ or $P_4O_{10}(g)$
    c. $O_2(g)$ or $P_4O_{10}(g)$
    d. $H_2O(s)$ or $H_2O(l)$
    e. $NO_2(g)$ or $N_2O_4(g)$
    f. $Ar(g)$ at 5 atm or $Ar(g)$ at 0.30 atm.

3. Predict the sign of the entropy change for each of the following processes.

    a. Potassium hydroxide pellets are dissolved in water.
    b. Solid ammonium dichromate is burned to give solid chromium oxide, water vapor, and nitrogen gas.
    c. Saturated calcium acetate is mixed with ethanol to form a gel.

4. Predict the sign of the entropy change for each of the following reactions:

    a. $Ag^+(aq) + Cl^-(aq) \rightarrow AgCl(s)$
    b. $NH_4Cl(s) \rightarrow NH_3(g) + HCl(g)$
    c. $H_2(g) + Br_2(g) \rightarrow 2HBr(g)$

5. When water freezes, is there an increase in entropy? Explain.

Section 17.3

6. Calculate ΔS_{surr} for each of the following reactions at 25°C and 1 atm:

    a. $Br(l) \rightarrow Br(g)$ $\Delta H = +31$ kJ
    b. $2C_2H_6(g) + 7O_2(g) \rightarrow 4CO_2(g) + 6H_2O(g)$ $\Delta H = -2857$ kJ

7. A chemical reaction gives a change in entropy of the universe of −48 J/K. Is the process spontaneous? Why or why not?

8. Which of the following values represent spontaneous processes? Which ones are exothermic (from the point of view of the system)?

    a. $\Delta S_{sys} = +358$ J/K, $\Delta S_{surr} = -358$ J/K
    b. $\Delta S_{sys} = -358$ J/K, $\Delta S_{surr} = -52$ J/K
    c. $\Delta S_{sys} = -358$ J/K, $\Delta S_{surr} = +463$ J/K
    d. $\Delta S_{sys} = +358$ J/K, $\Delta S_{surr} = -463$ J/K

Section 17.4

9. Given the following values for ΔH, ΔS, and T, determine whether each of the following sets of data represent spontaneous or nonspontaneous processes.

	ΔH(kJ)	ΔS(J/K)	T(K)
a.	−16	50	300
b.	12	40	300
c.	− 5	−20	200
d.	− 5	20	200
e.	− 5	−20	500

10. Given the following ΔH and ΔS values, determine the temperature at which the reactions would be spontaneous:

   a. $\Delta H = 10.5$ kJ $\Delta S = 30$ J/K
   b. $\Delta H = 1.8$ kJ $\Delta S = 113$ J/K
   c. $\Delta H = -11.7$ kJ $\Delta S = -105$ J/K

11. Predict the sign of the entropy change for each of the following processes:

   a. evaporating a beaker of ethanol at room temperature
   b. cooling nitrogen gas from 80°C to 20°C
   c. freezing liquid bromine below its melting point (−7.2°C)

12. The heat of fusion for actinium is 10.50 kJ/mol. The entropy of fusion is 9.6 J/K mol. Calculate the melting point of actinium.

13. At a constant temperature of 298 K, calculate ΔS_{sys} and ΔS_{univ} for the free expansion of 3.0 L of an ideal gas at 1.0 atm to 11.0 L. (101.3 J = 1 L atm)

14. The heat of vaporization for protactinium is 481 kJ/mol. The entropy of vaporization is 109 J/K mol. Calculate the boiling point of protactinium. Compare with the actual value of approximately 4500 K.

15. If the molar heat of vaporization of ethanol is 39.3 kJ/mol and its boiling point is 78.3°C, calculate ΔS for the vaporization of 0.50 mol ethanol.

16. The normal boiling point of diethyl ether is 308 K. The enthalpy of vaporization is 27.2 kJ/mol. Calculate ΔS for the vaporization of 1.0 mol of diethyl ether under these conditions.

17. The melting point of silicon is 1683 K. The heat of fusion is 46.4 kJ/mol. Calculate the entropy of fusion of silicon.

18. Determine whether the following chemical change is spontaneous:

$$SO_2(g) + NO_2(g) \rightarrow SO_3(g) + NO(g)$$

Section 17.5

19. Predict the sign of $\Delta S°$ for each of the following reactions:

   a. $Sr(g) + \frac{1}{2}O_2(g) \rightarrow SrO(c)$ (c = crystalline)
   b. $2Al(s) + 3F_2(g) \rightarrow 2AlF_3(s)$

20. Using Appendix 4 in your textbook, calculate the standard enthalpy changes for these reactions at 25°C:

   a. $CaCO_3(s) \rightarrow CaO(s) + CO_2(g)$
   b. $N_2(g) + 3H_2(g) \rightarrow 2NH_3(g)$
   c. $H_2(g) + Cl_2(g) \rightarrow 2HCl(g)$

21. Using data from Appendix 4 in your textbook, calculate $\Delta S°$ for each of the following reactions:

   a. $CH_4(g) + N_2(g) \rightarrow HCN(g) + NH_3(g)$
   b. $2Ag_2O(s) \rightarrow 4Ag(s) + O_2(g)$
   c. $Cd(s) + \frac{1}{2}O_2(g) \rightarrow CdO(s)$

22. Using data for $\Delta H°$ and $\Delta S°$ in Appendix 4 in your textbook, calculate $\Delta G°$ (at 25°C) for each of the reactions in Problem 21.

23. Calculate $\Delta G°$ for each of the reactions in Problem 21 using $\Delta G°$ data from Appendix 4 in your textbook. How do these compare with your answers from Problem 22?

Section 17.6

24. Using <u>Appendix 4 in your textbook</u>, calculate $\Delta G°$ for the combustion of ethane (C_2H_6):

$$2C_2H_6(g) + 7O_2(g) \rightarrow 4CO_2(g) + 6H_2O(l)$$

25. Calculate $\Delta G°$ for the following reaction:

$$C_2H_4(g) + 3O_2(g) \rightarrow 2CO_2(g) + 2H_2O(l)$$

26. Calculate $\Delta G°$ for the following reactions at 25°C:

    a. $2MgO(s) \rightarrow 2Mg(s) + O_2(g)$
    b. $CH_4(g) + 2O_2(g) \rightarrow CO_2(g) + 2H_2O(l)$

27. Calculate $\Delta G°$ for the following reactions:

    a. $2Mg(s) + O_2(g) \rightarrow 2MgO(s)$
    b. $2C_2H_2(g) + 5O_2(g) \rightarrow 4CO_2(g) + 2H_2O(l)$
    c. $N_2(g) + O_2(g) \rightarrow 2NO(g)$

28. Given the following data:

    a. $N_2(g) + 2O_2(g) \rightarrow 2NO_2(g)$ $\Delta G° = 104$ kJ
    b. $2NO(g) + O_2(g) \rightarrow 2NO_2(g)$ $\Delta G° = -70$ kJ

    Calculate $\Delta G°$ for $N_2(g) + O_2(g) \rightarrow 2NO(g)$.

29. Using <u>Appendix 4 in your textbook</u>, calculate $\Delta G°$ for each of the following reactions at 298 K:

    a. $2Cu_2O(s) + O_2(g) \rightarrow 4CuO(s)$
    b. $C_2H_5OH(l) \rightarrow C_2H_4(g) + H_2O(g)$

30. State the enthalpy and entropy conditions under which $\Delta G°$ for a reaction must be "–".

31. Calculate $\Delta G°$ for the following reaction at 25°C (use <u>Appendix 4 in your textbook</u>):

$$2H_2(g) + O_2(g) \rightleftharpoons 2H_2O(l)$$

32. Given the following data:

    a. $2H_2(g) + C(s) \rightarrow CH_4(g)$ $\Delta G° = -51$ kJ
    b. $2H_2(g) + O_2(g) \rightarrow 2H_2O(l)$ $\Delta G° = -474$ kJ
    c. $C(s) + O_2(g) \rightarrow CO_2(g)$ $\Delta G° = -394$ kJ

    Calculate $\Delta G°$ for $CH_4(g) + 2O_2(g) \rightarrow CO_2(g) + 2H_2O(l)$.

Section 17.7

33. Calculate ΔG at 600 K for the following reaction:

$$P_4(g) + 5O_2(g) \rightleftharpoons P_4O_{10}(s),$$

    where the initial pressures are $P_{P_4} = 0.52$ atm and $P_{O_2} = 2.1 \times 10^{-3}$ atm.

34. Calculate the equilibrium constant, K, at 25°C for each of the reactions in Problem 21.

Section 17.8

35. Calculate ΔG_f° (in kcal/mol), and determine whether the reaction will occur spontaneously.

$$I_2(s) + Cl_2(g) \rightleftharpoons 2ICl(g)$$

36. If $R = 1.99$ cal/mol K, calculate K_p for the dissociation of HCl given that:

$$\tfrac{1}{2}H_2(g) + \tfrac{1}{2}Cl_2(g) \rightleftharpoons HCl(g) \qquad \Delta G_f^\circ = -22.7 \text{ kcal}$$

What does this say about the tendency of this dissociation?

37. The value of the equilibrium constant for a given reaction is $K = 6 \times 10^{-23}$. What does that indicate about the spontaneity of the reaction?

38. The value of the equilibrium constant for a given reaction is $K = 8 \times 10^{58}$. What does this tell us regarding the speed of the reaction?

39. We said in Chapter 14 that at 25°C, $K_w = 1.0 \times 10^{-14}$. Calculate K_w thermodynamically, and compare it to our Chapter 14 value for the reaction:

$$H_2O(l) \rightleftharpoons H^+(aq) + OH^-(aq)$$

40. Calculate K_p for the following reaction at 25°C: $2H_2O(g) \rightleftharpoons 2H_2(g) + O_2(g)$.

Multiple Choice Questions

41. A state of higher entropy means:

   A. A lower number of possible arrangements C. Lower probabilities to reach a possible state
   B. A higher number of possible arrangements D. An exothermic process

42. Which of the following processes would result in a decrease of entropy?

   A. Freezing   B. Melting   C. Evaporating   D. An expanding gas

43. Four distinct vessels (each with a different capacity) hold 2.3 moles of a particular gas each. Which system has the highest total entropy?

   A. 1.0 L   B. 0.2 L   C. 0.15 L   D. 0.5 L

44. Which of the following processes has the lowest probability of being achieved?

   A. A feather flying away from the ground   C. Water freezing into ice at 273 K
   B. A rock rolling down the hill   D. A piece of paper flying away from the ground

45. Which of the following processes must be spontaneous?

   A. $\Delta S_{surr} > 0, \Delta S_{sys} > 0$ B. $\Delta S_{surr} > 0, \Delta S_{sys} < 0$ C. $\Delta S_{univ} < 0, \Delta S_{sys} > 0$ D. $\Delta S_{surr} < 0$

46. Which of the following conditions would ensure a spontaneous process?

   A. $\Delta S_{surr} > 0$ B. $\Delta S_{sys} < 0$ C. $\Delta S_{sys} < 0$ D. $\Delta S_{univ} > 0$

47. Heat is released during a particular process. This means that:

   A. The process is spontaneous under all conditions   C. The process tends to be spontaneous
   B. $\Delta S_{surr} > 0$ D. $\Delta S_{sys} > 0$

48. Which of the following processes would you expect to be spontaneous?

A. $\Delta S_{surr} = 25$ J/K, $\Delta S_{sys} = -27$ J/K
B. $\Delta S_{surr} = 25$ J/K, $\Delta S_{sys} = 27$ J/K
C. $\Delta S_{univ} = -20$ J/K, $\Delta S_{sys} = -20$ J/K
D. $\Delta S_{surr} = -80$ J/K, $\Delta S_{sys} = 20$ J/K

49. Which of the following processes do you expect to be the most spontaneous at the respective temperature, if $\Delta S_{sys} = 0$ J/K?

A. $\Delta H = 25$ kJ, $T = 28°C$
B. $\Delta H = -475$ kJ, $T = 28°C$
C. $\Delta H = -260$ kJ, $T = 273$ K
D. $\Delta H = -300$ kJ, $T = 208$ K

50. Which of the following processes do you expect to be the most spontaneous at the respective temperature, if $\Delta S_{sys} = 0$ J/K?

A. $\Delta H = -75$ kJ, $T = 38°C$
B. $\Delta H = -650$ kJ, $T = -7°C$
C. $\Delta H = -260$ kJ, $T = 273$ K
D. $\Delta H = -1500$ kJ, $T = 358$ K

51. Which of the following conditions must be met for a process to be spontaneous?

A. $\Delta G < 0$
B. $\Delta H < 0$
C. $\Delta S_{surr} > 0$
D. $\Delta S_{sys} > 0$

52. Which of the following system conditions would allow a process to be spontaneous at all temperatures?

A. $\Delta S > 0, \Delta H < 0$
B. $\Delta S > 0, \Delta H > 0$
C. $\Delta S < 0, \Delta H > 0$
D. $\Delta S < 0, \Delta H < 0$

53. Calculate one temperature that would allow the following process to be spontaneous:
$\Delta S = 30$ J/K, $\Delta H = 120$ kJ

A. 400°C
B. 4002 K
C. 360°C
D. 3600 K

54. Calculate the standard absolute entropy, in J/mol K, of Mg:

$$2NO_2(g) + 2MgO(s) \rightarrow Mg(NO_3)_2(s) + Mg(s) \qquad \Delta S° = -462.1 \text{ J/K}$$

$S°_{NO_2(g)} = 239.9$ J/mol K $\qquad S°_{MgO(s)} = 27.0$ J/mol K $\qquad S°_{Mg(NO_3)_2(s)} = 39.2$ J/mol K

A. 32.5
B. 234.3
C. 27.8
D. 0.00

55. Calculate $\Delta S°$ for the following reaction:

$$H_3AsO_4(aq) \rightarrow 3H^+(aq) + AsO_4{}^{3-}(aq)$$

$S°_{H^+(aq)} = 0.00$ J/mol K $\qquad S°_{H_3AsO_4(aq)} = 44.0$ J/mol K $\qquad S°_{AsO_4{}^{3-}(aq)} = -38.9$ J/mol K

A. 5.10 J/K
B. −82.9 J/K
C. −5.1 J/K
D. −72.7 J/K

56. Based on the following data, calculate the entropy change for the given reaction:

$$\begin{aligned}
Fe(s) + 5CO(g) &\rightarrow Fe(CO)_5(g) & \Delta S° &= ? \\
Fe(CO)_5(l) &\rightarrow Fe(CO)_5(g) & \Delta S° &= 107.2 \text{ J/K} \\
Fe(s) + 5CO(g) &\rightarrow Fe(CO)_5(l) & \Delta S° &= -677.3 \text{ J/mol K}
\end{aligned}$$

A. 57.00 J/K
B. 784.2 J/mol K
C. 507.3 J/mol K
D. −570.1 J/mol K

57. Under standard conditions (all gases at $P = 1$ atm), will the following reaction take place under sunlight?

$$3Cl_2(g) + 2CH_4(g) \rightarrow CH_3Cl(g) + CH_2Cl_2(g) + 3HCl(g)$$

$G°_{f(CH_4(g))} = -50.72$ kJ/mol $\qquad G°_{f(CH_3Cl(g))} = -57.37$ kJ/mol

$G°_{f(CH_2Cl_2(g))} = -68.85$ kJ/mol $\qquad G°_{f(HCl(g))} = -95.30$ kJ/mol

A. Yes
B. No
C. Under sunlight only
D. Not in any conditions

58. For the following reaction,

$$3Cl_2(g) + 2CH_4(g) \rightarrow CH_3Cl(g) + CH_2Cl_2(g) + 3HCl(g) \quad \Delta G° = -307.6 \text{ kJ}$$

calculate $\Delta H°$, in kJ/mol, for HCl, given that

Molecule	$\Delta H°$ (kJ/mol)	$S°$ (J/mol K)
Cl_2	0.00	223.1
CH_4	−74.81	186.3
CH_3Cl	−80.83	234.6
CH_2Cl_2	−92.47	270.2
HCl	?	186.9

A. −270.2   B. −92.31   C. 93.31   D. 270.2

59. Calculate the free energy change ΔG, in kJ, for the following reaction at 298 K:

$$N_2(g) + 2O_2(g) \rightarrow 2NO_2(g)$$
$$\text{0.2 atm} \quad \text{0.2 atm} \quad \text{0.39 atm}$$

A. 109   B. 37.32   C. 13.98   D. −37.32

60. The value of K_{eq} goes down by a factor of 100.0. Compute the value, in kJ, of the change in $\Delta G°$ for the reaction at 298 K.

A. 1.37   B. 11.41   C. 0.382   D. −11.41

61. If the free energy is equal to 10.0 kJ, and it is four times as big as the standard free energy, what must the value of Q be under standard conditions?

A. 20.6   B. 0.003   C. 3.25   D. 1.00

62. Calculate the standard free energy of a specific reaction if the formation constant is 1.47×10^9.

A. −54.0 kJ   B. 23.5 kJ   C. 54.0 kJ   D. 2.51 kJ

63. Which one of the following processes is the fastest?

A. $\Delta G = 23.5$ kJ
B. $\Delta G = 58.5$ kJ
C. $\Delta G = -5800$ kJ
D. Speed of reaction is unrelated to its free energy change.

Answers to Exercises

1. Probability $= 1/2^8 = 1/256$. This says that the probability dictates that molecules will be distributed more evenly throughout the bulbs—in other words, a higher degree of entropy.

2. a. $HCl(g)$
    b. $P_4O_{10}(g)$
    c. $P_4O_{10}(g)$ (though the difference is quite small)
    d. $H_2O(l)$
    e. $N_2O_4(g)$
    f. $Ar(g)$ at 0.30 atm.

3. a. positive   b. sharply positive   c. negative

4. a. ΔS = negative
    b. ΔS = positive
    c. We cannot accurately predict the sign of ΔS but we know that the change is very small.

5. No, entropy decreases as molecules are less randomly distributed, as in freezing.

6. a. $\Delta S_{surr} = -100$ J/K (2 sig figs)   b. $\Delta S_{surr} = 9590$ J/K

7. The entropy of the universe decreased. The process is therefore **nonspontaneous.**

8. Only process "c" is spontaneous. Process "c" is the only exothermic process.

9. a. spontaneous
    b. nonspontaneous
    c. spontaneous
    d. spontaneous
    e. nonspontaneous

10. a. spontaneous at 350 K or above
    b. spontaneous at 16 K or above
    c. spontaneous at 111 K or below

11. a. $\Delta S > 0$ (positive)   b. $\Delta S < 0$ (negative)   c. $\Delta S < 0$ (negative)

12. melting point $= 1100$ K

13. $\Delta S_{sys} = \Delta S_{univ} = 0.027$ L atm/K $= 2.7$ J/K

14. boiling point $= 4410$ K

15. $\Delta S = 56$ J/K

16. $\Delta S = 88$ J/K

17. entropy of fusion $= 27.6$ J/K mol

18. Yes, spontaneous ($\Delta G° = -36$ kJ)

19. a. negative   b. negative

20. a. 178 kJ
    b. −92 kJ
    c. −184 kJ

21. a. +17 J/K   b. +133 J/K   c. −100 J/K

22. a. +159 kJ   b. +22 kJ   c. −228 kJ

23. a. +159 kJ   b. +22 kJ   c. −228 kJ

24. $\Delta G° = -2932$ kJ

25. $\Delta G° = -1331$ kJ

26. a. $\Delta G° = 1138$ kJ   b. $\Delta G° = -817$ kJ

27. a. −1138 kJ   b. −2468 kJ   c. 174 kJ

28. $\Delta G° = 174$ kJ

29. a. $\Delta G° = -216$ kJ   b. $\Delta G° = 14$ kJ

30. $\Delta H = -$, $\Delta S = +$

31. $\Delta G° = -474$ kJ

32. $\Delta G° = -817$ kJ

33. $\Delta G° = -2541$ kJ

34. a. $K = 1.4 \times 10^{-28}$ b. $K = 1.4 \times 10^{-4}$ c. $K = 9.2 \times 10^{39}$

35. $\Delta G_f^° = -1.32$ kcal/mol ICl; spontaneous

36. $K_p = 2.1 \times 10^{-17}$. This very small value of K_p shows that hardly any of the reactants are present at equilibrium compared to the amount of HCl present.

37. The reaction is not spontaneous.

38. The equilibrium constant says <u>nothing</u> about the rate of reaction.

39. $K = 9.5 \times 10^{-15}$. This is very close to our value of K_w of 1.0×10^{-14}.

40. $K_p = 5.3 \times 10^{-81}$

41. B   42. A   43. A   44. D   45. A   46. D

47. B   48. B   49. B   50. D   51. A   52. A

53. B   54. A   55. B   56. D   57. A   58. B

59. A   60. B   61. A   62. A   63. D

Chapter 18

Electrochemistry

Your textbook defines **electrochemistry** as **the study of the interchange of chemical and electrical energy**. In this chapter, we are concerned with the **use of chemical reactions to generate an electric current** and **the use of electric current to produce chemical reactions**.

18.1   Balancing Oxidation-Reduction Equations

When you finish this section you will be able to balance redox equations using the half-reaction method in acidic or basic solutions.

In Chapter 4 you learned how balance redox equations using the oxidation states method, now you will to learn how to **balance redox equations using the half-reaction method**. When thinking about electrochemistry you will find that splitting redox reactions into oxidation and reduction halves is very convenient for describing what happens at different electrodes. The **half-reaction method** is therefore especially important to master.

The steps involved in balancing redox reactions by this method are given in your textbook. When you work redox problems, always remember that **you can only add substances to your equation that are already in solution** (such as H_2O, H^+ in acid solution, and OH^- in basic solution). Let's try to balance the following equation together.

Example 18.1 A   Balancing Redox Equations by the Half-reaction Method

Balance the following equation **in acid solution** using the half-reaction method.

$$Cu(s) + HNO_3(aq) \rightarrow Cu^{2+}(aq) + NO(g)$$

Solution

Step 1: **Identify and write equations for the half-reactions**.

$$\underset{0}{\textbf{Cu}}(s) \;+\; \underset{+5}{\textbf{HNO}_3}(aq) \;\rightarrow\; \underset{+2}{\textbf{Cu}^{2+}}(aq) \;+\; \underset{+2}{\textbf{NO}}(g)$$

Copper is being oxidized: $Cu \rightarrow Cu^{2+}$
Nitrogen is being reduced: $HNO_3 \rightarrow NO \quad (N^{5+} \rightarrow N^{2+})$

Step 2: **Balance each half-reaction**.

The oxidation is balanced atomically. We need to add two electrons to the right-hand side to balance it electronically.

    i.   (oxidation) $Cu \rightarrow Cu^{2+} + \textbf{2e}^-$
    ii. (reduction) $HNO_3 \rightarrow NO$

    a. **Balance all atoms that are neither oxygen nor hydrogen**. (Nitrogens are already balanced).

    b. **Balance oxygens** by adding water to the side that needs oxygen. (The left-hand side has 3 oxygens, the right-hand side has 1, so **2** waters must be added to the right-hand side.)

$$HNO_3 \rightarrow NO + \textbf{2H}_2\textbf{O}$$

    c. **Balance hydrogens** by adding H^+ to the side that needs hydrogen. (The left-hand side has 1 hydrogen, the right-hand side has 4, so **3** hydrogen ions must be added to the left-hand side.)

$$HNO_3 + \textbf{3H}^+ \rightarrow NO + 2H_2O$$

The half-reaction is now balanced atomically, but not electronically.

    d. **Balance charges** by adding **electrons** to the side that is more positive. (The left-hand side has 3 positives; the right-hand side is neutral. Therefore, we need to add **3** electrons to the left-hand side.)

$$HNO_3 + 3H^+ + \textbf{3e}^- \rightarrow NO + 2H_2O$$

Both half-reactions are now balanced.

Step 3: **Equalize electron transfer**.

The same number of electrons must be gained as are lost in the reaction. Therefore, we must multiply each reaction by numbers that will allow both reactions to have **the same** number of electrons exchanged.

With our reactions, the lowest common denominator of electrons is **6**. Therefore, we would multiply the oxidation by **3**, and multiply the reduction by **2**.

$$3Cu \rightarrow 3Cu^{2+} + 6e^-$$
$$2HNO_3 + 6H^+ + 6e^- \rightarrow 2NO + 4H_2O$$

Step 4: **Add the half-reactions, and cancel appropriately to get a complete redox reaction**.

$$3Cu \rightarrow 3Cu^{2+} + 6e^-$$
$$\underline{2HNO_3 + 6H^+ + 6e^- \rightarrow 2NO + 4H_2O}$$
$$3Cu + 2HNO_3 + 6H^+ + 6e^- \rightarrow 3Cu^{2+} + 2NO + 4H_2O + 6e^-$$

Canceling electrons, we get our final result:

$$\textbf{3Cu} + \textbf{2HNO}_3 + \textbf{6H}^+ \rightarrow \textbf{3Cu}^{2+} + \textbf{2NO} + \textbf{4H}_2\textbf{O}$$

Double Check

Do we have the same number of each kind of atom on both sides? Yes, there are 3 coppers, 2 nitrogens, 6 oxygens, and 8 hydrogens on each side. Are the charges the same on both sides? Yes, +6. The equation is balanced.

Example 18.1 B   Practice with the Half-Reaction Method

Balance the following equation in acidic solution:

$$Cr_2O_7^{2-}(aq) + NO(g) \rightarrow Cr^{3+}(aq) + NO_3^-(aq)$$

Solution

Step 1: $Cr_2O_7^{2-} + NO \rightarrow Cr^{3+} + NO_3^-$

Nitrogen is being oxidized: N^{2+} to N^{5+}
Chromium is being reduced: Cr^{6+} to Cr^{3+}

Step 2: $Cr_2O_7^{2-} \rightarrow Cr^{3+}$

$$Cr_2O_7^{2-} \rightarrow 2Cr^{3+}$$
$$Cr_2O_7^{2-} \rightarrow 2Cr^{3+} + 7H_2O$$
$$Cr_2O_7^{2-} + 14H^+ \rightarrow 2Cr^{3+} + 7H_2O$$
$$Cr_2O_7^{2-} + 14H^+ + 6e^- \rightarrow 2Cr^{3+} + 7H_2O \quad \textbf{(balanced)}$$

$$NO \rightarrow NO_3^-$$
$$NO + 2H_2O \rightarrow NO_3^-$$
$$NO + 2H_2O \rightarrow NO_3^- + 4H^+$$
$$NO + 2H_2O \rightarrow NO_3^- + 4H^+ + 3e^- \quad \textbf{(balanced)}$$

Step 3: $2\,[NO + 2H_2O \rightarrow NO_3^- + 4H^+ + 3e^-]$

$$2NO + 4H_2O \rightarrow 2NO_3^- + 8H^+ + 6e^- \qquad \text{(oxidation)}$$
$$\underline{Cr_2O_7^{2-} + 14H^+ + 6e^- \rightarrow 2Cr^{3+} + 7H_2O} \qquad \text{(reduction)}$$
$$\mathbf{Cr_2O_7^{2-}(aq) + 2NO(g) + 6H^+(aq) \rightarrow 2Cr^{3+}(aq) + 2NO_3^-(aq) + 3H_2O(l)}$$

Double Check

There are 2 chromiums, 2 nitrogens, 9 oxygens, and 6 hydrogens on each side. Also, each side has a total charge of +4. The equation is balanced.

One of the conditions of balancing equations is that you can only add to equations what is actually in the solution. In **acid** solutions, you have a lot of H^+ ions. In **basic** solutions, however, you have a lot of OH^- ions. **We need to add OH^- ions to balance for hydrogen when balancing a reaction that takes place in basic solution.** However, adding OH^- to balance hydrogens puts oxygens out of balance. To get around that,

1. Balance basic solutions as if they were acidic.
2. Add a number of OH^- ions to both sides so that you just balance excess H^+ ions.
3. $H^+ + OH^-$ will form H_2O on the side with excess H^+. Free OH^- will appear on one side of the equation.
4. Double check atoms and charges, as always.

Try this technique on the equation in the previous example.

Example 18.1 C   Balancing Redox Equations in Basic Solution

Balance the following equation (<u>IT IS ALREADY</u> balanced in acid) assuming it is now in basic solution.

$$Cr_2O_7^{2-}(aq) + 2NO(g) + 6H^+(aq) \rightarrow 2Cr^{3+}(aq) + 2NO_3^-(aq) + 3H_2O(l)$$

Solution

We need to get rid of the excess H^+, because OH^- is the dominant acid-base related species. Therefore, **add $6OH^-$ to both sides**. (Whatever is done to the left side must be done to the right, if the equation is already balanced.)

$$Cr_2O_7^{2-} + 2NO + 6H^+ + \mathbf{6OH^-} \rightarrow 2Cr^{3+} + 2NO_3^- + 3H_2O + \mathbf{6OH^-}$$

The H^+ will combine with the OH^-, giving H_2O.

$$Cr_2O_7^{2-} + 2NO + \mathbf{6H_2O} \rightarrow 2Cr^{3+} + 2NO_3^- + 3H_2O + 6OH^-$$

The 3 H_2O's on the right-hand side can be canceled (with 3 of the waters on the left-hand side), which gives the final balanced equation.

$$\mathbf{Cr_2O_7^{2-}(aq) + 2NO(g) + 3H_2O(l) \rightarrow 2Cr^{3+}(aq) + 2NO_3^-(aq) + 6OH^-(aq)}$$

18.2   Galvanic Cells

When you finish this section you will be able to define some of the terms that are commonly used with electrochemistry.

We have discussed reduction-oxidation (redox) reactions in Section 4.9. The following terms should be reviewed: **reduction, oxidation, reducing agent, and oxidizing agent. A galvanic cell** is a device in which **chemical energy is converted to electrical energy**. There is a need to physically separate the oxidizing and reducing agents in galvanic cells so that energy of reaction can be used. <u>Figures 18.2 and 18.3 in your textbook</u> show the important features of a galvanic cell. You should know the function of the following:

- **cathode** - reduction occurs here. Species undergoing reduction ("oxidizing agent") receive electrons from the cathode.
- **anode** - oxidation occurs here. Species undergoing oxidation ("reducing agent") lose electrons here.
- **salt bridge** (or porous disk) - allows exchange of ions to keep electric neutrality while electroactive solutions remain separated.

Also, know the following terms:

- **electromotive force (emf)** - the driving force with which electrons are pulled through a wire.
- **volt (V)** - the unit of electrical potential. It equals 1 joule/coulomb.

In constructing galvanic cells, keep in mind the **direction of electron flow**. Species undergoing *reduction receive electrons from the cathode*. Species undergoing *oxidation donate electrons to the anode*. The direction of electron flow is therefore **from the anode to the cathode**.

Example 18.2   A Bit of Review

Identify the species in each of the following equations that would receive electrons from the cathode and that would lose electrons at the anode in each of the following galvanic cells.

a. $Mg(s) + 2H^+(aq) \rightarrow Mg^{2+}(aq) + H_2(g)$
b. $MnO_4^-(aq) + 5Fe^{2+}(aq) + 8H^+(aq) \rightarrow Mn^{2+}(aq) + 5Fe^{3+}(aq) + H_2O(l)$

Solution

It is quite useful to separate balanced reactions into half-reactions. The reduction half-reaction receives electrons at the **cathode**. The **oxidation** half-reaction loses electrons at the **anode**.

    a.   oxidation (anode): $Mg(s) \rightarrow Mg^{2+}(aq) + 2e^-$
            reduction (cathode): $2H^+(aq) + 2e^- \rightarrow H_2(g)$

    b.   oxidation (anode): $Fe^{2+}(aq) \rightarrow Fe^{3+}(aq) + e^-$
            reduction (cathode): $MnO_4^-(aq) + 8H^+(aq) + 5e^- \rightarrow Mn^{2+}(aq) + 4H_2O(l)$

As in all galvanic cells, the direction of electron flow is from the anode to the cathode.

18.3   *Standard Reduction Potentials*

When you finish this section you will be able to:

- Determine the proper cell reaction and $E°$ value, given two half-reactions.
- Describe the make up of a galvanic cell involving a given chemical reaction.

The electromotive force (emf) of a galvanic cell is a **combination** of the potentials of two half-reactions. Because the cathodic potentials are described relative to anodic reactions, we need one absolute standard against which all other half-reactions can be compared. The standard is the **standard hydrogen** electrode,

$$2H^+ + 2e^- \rightarrow H_2 \qquad E° = 0.00 \text{ V (exactly, by definition)}$$

This is an arbitrary, but necessary, assignment. Table 18.1 in your textbook has a list of standard reduction potentials for many half-reactions. **Standard means 298 K and 1 atmosphere**.

- Galvanic cells require $E°_{cell} > 0$ V.
- One of the tabulated reduction potentials will have to be reversed (to form an oxidation half-reaction) in every $E°$ calculation.
- To determine which reaction is to be reversed, the **sum** of the **oxidation and reduction half-reactions must be > 0 V in a galvanic cell**.
- When you **reverse** a reaction, $E°$ **gets the opposite sign**.
- When you **multiply** a reaction by a coefficient (for purposes of balancing), **the $E°$ is NOT changed!**

Let's have some practice with calculating $E°_{cell}$ and determining the reduction and oxidation reactions in a galvanic cell.

Example 18.3 A   Cell Emf

Using data from Table 18.1 in your textbook, calculate the emf values ($E°_{cell}$) for each of the following reactions. State which are galvanic.

    a. $Mg(s) + 2H^+(aq) \rightarrow Mg^{2+}(aq) + H_2(g)$
    b. $Cu^{2+}(aq) + 2Ag(s) \rightarrow Cu(s) + 2Ag^+(aq)$
    c. $2Zn^{2+}(aq) + 4OH^-(aq) \rightarrow 2Zn(s) + O_2(g) + 2H_2O(l)$

Solution

We need to split each reaction into two half-reactions and look up their reduction potentials. **(Remember that when a half-reaction is an oxidation, $E°$ must change sign!)**

    a.   reduction: $2H^+ + 2e^- \rightarrow H_2$ $E°$ = 0.00 V
           oxidation: $Mg \rightarrow Mg^{2+} + 2e^-$ $E°$ = <u>+2.37 V</u>
                                         $E°_{cell}$ = **+2.37 V, galvanic**

b.　reduction: $Cu^{2+} + 2e^- \rightarrow Cu$　　　　$E^\circ = +0.34$ V
　　oxidation: $2Ag \rightarrow 2Ag^+ + 2e^-$　　　$E^\circ = \underline{-0.80\ \text{V}}$
　　　　　　　　　　　　　　　　　　　$E^\circ_{cell} = -0.46$ V, not galvanic - will not
　　　　　　　　　　　　　　　　　　　　　　　　　　run in this direction

c.　reduction: $2Zn^{2+} + 4e^- \rightarrow 2Zn$　　　$E^\circ = -0.76$ V
　　oxidation: $4OH^- \rightarrow O_2 + 2H_2O + 4e^-$　$E^\circ = \underline{-0.40\ \text{V}}$
　　　　　　　　　　　　　　　　　　　$E^\circ_{cell} = -1.16$ V, not galvanic

Another key point regarding the magnitude and sign of E° for half-reactions is that **THE MORE POSITIVE THE E° VALUE, THE MORE LIKELY THE SPECIES IS TO BE REDUCED.** (It is a stronger oxidizing agent.) (The contrary with regard to oxidation is also true). For example, Br_2 is more likely to be reduced than I_2 because Br_2 has the more positive E°.

Example 18.3 B Reduction Strength

Place the following in order of increasing strength as oxidizing agents.

$$Fe^{2+},\ ClO_2,\ F_2,\ AgCl$$

Solution

From looking at the E° values in Table 18.1 of your textbook, the order is **$Fe^{2+} < AgCl < ClO_2 < F_2$.**

Example 18.3 C Composing Galvanic Cells

Given the following half-cells, decide which is the anode and the cathode, balance, write the overall cell reaction, and calculate E°_{cell} if the cells are galvanic.

a.　$Ni^{2+} + 2e^- \rightarrow Ni$　　　　　$E^\circ = -0.23$ V
　　$O_2 + 4H^+ + 4e^- \rightarrow 2H_2O$　　$E^\circ = +1.23$ V

b.　$Ce^{4+} + e^- \rightarrow Ce^{3+}$　　　　$E^\circ = +1.70$ V
　　$Sn^{2+} + 2e^- \rightarrow Sn$　　　　　$E^\circ = -0.14$ V

Solution

The goal is to see which arrangement of half-reactions will make $E^\circ_{cell} > 0$.

a.　Reversing the Ni^{2+} reduction will create a galvanic cell. Note that to balance the overall reaction, it must also be multiplied by 2.

cathode: $O_2 + 4H^+ + 4e^- \rightarrow 2H_2O$	$E^\circ = 1.23$ V
anode: 　$2Ni \rightarrow 2Ni^{2+} + 4e^-$	$E^\circ = 0.23$ V
overall: $O_2(g) + 4H^+(aq) + 2Ni(s) \rightarrow 2H_2O(l) + 2Ni^{2+}(aq)$	$E^\circ_{cell} = 1.46$ V

b.　Reversing the Sn^{2+} half-reaction will create a galvanic cell. The cerium half-reaction must be multiplied by 2 to balance the overall reaction.

cathode: $2Ce^{4+} + 2e^- \rightarrow 2Ce^{3+}$	$E^\circ = 1.70$ V
anode: 　$Sn \rightarrow Sn^{2+} + 2e^-$	$E^\circ = 0.14$ V
overall: $2Ce^{4+}(aq) + Sn(s) \rightarrow Sn^{2+}(aq) + 2Ce^{3+}(aq)$	$E^\circ_{cell} = 1.84$ V

Immediately after <u>Example 18.3 in your textbook</u>, a method of representing electrochemical cells is introduced. This **line notation** is a shorthand that chemists use to describe cells. The general line notation formation is:

$$\underset{\substack{\text{porous disk or}\\\text{salt bridge}}}{\underset{\text{side}}{\text{anode}} \ \middle\| \ \underset{\text{side}}{\text{cathode}}}$$

Each side will have a "phase boundary" separating, for example, a solid electrode from ions in solution, such as $Cu(s)$ and $Cu^{2+}(aq)$. Such a boundary is denoted by a single vertical line (" $|$ ") with the *solid phase on the outside*. Therefore, for the copper/zinc reactions shown in <u>Figure 18.6 in your textbook</u>:

$$\underset{\text{(oxidation)}}{\text{anode:}} \quad Zn \ \rightarrow \ Zn^{2+} + 2e^-$$

$$\underset{\text{(reduction)}}{\text{cathode:}} \quad Cu^{2+} + 2e^- \rightarrow Cu$$

The line notation is

$$Zn(s) \ \middle| \ Zn^{2+}(aq) \ \middle\| \ Cu^{2+}(aq) \ \middle| \ Cu(s)$$

Your textbook points out that, on occasion, only ions (rather than solids) will be involved in the redox process. In that case, "inert" electrodes are used. The line notation is given for such an example in <u>Example 18.3(b) in your textbook</u>.

Example 18.3 D   Line Notations

Write line notations for each of the following galvanic cells:

a. $Hg^{2+} + Cd \rightarrow Hg + Cd^{2+}$
b. $Pb + 2Cr^{3+} \rightarrow Pb^{2+} + 2Cr^{2+}$ (a platinum electrode is used at the cathode)
c. $Cu^{2+} + 2Pu^{4+} + 4H_2O \rightarrow Cu + 2PuO_2^+ + 8H^+$ (a platinum electrode is used at the anode)

Solution

A good strategy is to separate the redox equation into its half reactions, so what happens at the anode (oxidation) and cathode (reduction) become apparent.

a. anode: $Cd \rightarrow Cd^{2+} + 2e^-$
    cathode: $Hg^{2+} + 2e^- \rightarrow Hg$

        line notation: $Cd(s) \ \middle| \ Cd^{2+}(aq) \ \middle\| \ Hg^{2+}(aq) \ \middle| \ Hg$

b. anode: $Pb \rightarrow Pb^{2+} + 2e^-$
    cathode: $2Cr^{3+} + 2e^- \rightarrow 2Cr^{2+}$

    The cathodic reaction involves only ions, so we need an inert electrode (Pt) to conduct electricity from one side of the cell to the other. The line notation is

        line notation: $Pb(s) \ \middle| \ Pb^{2+} \ \middle\| \ Cr^{3+}, Cr^{2+} \ \middle| \ Pt$
                           ↖ ↗
                    order here is not
                      important

c. This seems complex, but becomes clearer when you separate the equation into its half-reactions:

    anode: $2Pu^{4+} + 4H_2O \rightarrow 2PuO_2^+ + 8H^+ + 2e^-$
    cathode: $Cu^{2+} + 2e^- \rightarrow Cu$

    Remember that a Pt electrode is used at the anode!

        line notation: $Pt \ \middle| \ Pu^{4+}, PuO_2^+, H^+ \ \middle\| \ Cu^{2+} \ \middle| \ Cu$

We are now most of the way toward completely characterizing a galvanic cell. We have:

- calculated cell potentials
- described the direction of electron flow
- designated the anode and cathode

We have not yet "*designated the electrode and ions present in each compartment*." Normally, the solid metal in a half-reaction will serve as the electrode. If a half-reaction contains only ions, a nonreacting conductor (usually platinum) will be the electrode.

Example 18.3 E   Describing Galvanic Cells

Describe a galvanic cell based on the two half-reactions below.

$$Cu^{2+} + 2e^- \rightarrow Cu \qquad\qquad E° = 0.34 \text{ V}$$
$$Cr_2O_7^{2-} + 14H^+ + 6e^- \rightarrow 2Cr^{3+} + 7H_2O \qquad E° = 1.33 \text{ V}$$

Solution

In order to have a galvanic cell, $E°$ must be > 0. We must reverse the copper reduction. We also need to balance the entire cell reaction electrically by multiplying the copper equation by 3.

anode: $3Cu \rightarrow 3Cu^{2+} + 6e^-$ $\qquad\qquad E° = -0.34$ V
cathode: $Cr_2O_7^{2-} + 14H^+ 6e^- \rightarrow 2Cr^{3+} + 7H_2O$ $\qquad E° = +1.33$ V

overall: $3Cu(s) + Cr_2O_7^{2-}(aq) + 14H^+(aq) \rightarrow 3Cu^{2+}(aq) + 2Cr^{3+}(aq) + 7H_2O(l)$ $\quad E°_{cell} = 0.99$ V

A **copper metal electrode** will be in the Cu/Cu^{2+} compartment on the **anode** side. A **platinum electrode** will be in the $Cr_2O_7^{2-}/Cr^{3+}$ compartment on the **cathode** side. The line notation for this cell is

line notation: $Cu \mid Cu^{2+} \parallel Cr_2O_7^{2-}, Cr^{3+}, H^+ \mid Cu$

The electron flow (from anode to cathode) will be from **copper to platinum**.

18.4   Cell Potential, Electrical Work, and Free Energy

When you finish this section you will be able to interconvert between free energy and cell voltage.

The first part of this section emphasizes that the **actual work** that can be achieved is **always less than** the **theoretical work** available. Nonetheless, your textbook presents the relationship between **free energy** (maximum is assumed though not attainable) and **cell potential**.

$$\Delta G = -nFE$$

at standard conditions,

$$\Delta G° = -nFE°$$

where ΔG = free energy (in Joules)
n = moles of electrons exchanged in the redox reaction
F = the Faraday, a constant (96,486 Coulombs per mole of electrons)
E = cell voltage (Joules/Coulomb).

Note that ΔG and E have **opposite signs**. For a spontaneous process, ΔG is "−" and E is "+."

Example 18.4 A   Cell Voltage and Free Energy

Calculate the $\Delta G°$ for the reaction

$$Zn^{2+}(aq) + Cu(s) \rightleftharpoons Zn(s) + Cu^{2+}(aq)$$

Will zinc ions plate out on a copper strip?

Solution

The two pertinent half-reactions are

$$Zn^{2+} + 2e^- \rightarrow Zn \qquad E° = -0.76 \text{ V}$$
$$Cu \rightarrow Cu^{2+} + 2e^- \qquad E° = \underline{-0.34 \text{ V}}$$
$$E°_{cell} = -1.10 \text{ V}$$

$\Delta G° = -nF\,E° = -2 \text{ mol } e^- \, (96486 \text{ C/mol } e^-)(-1.10 \text{ J/C})$
$\Delta G° = 2.12 \times 10^5 \text{ J} = \textbf{212 kJ}$

Zinc ions will not plate out on a copper strip at standard conditions. (Note that this result is also indicated by the fact that $E°_{cell} < 0$.)

Example 18.4 B   Practice with Cell Voltage and Free Energy

Vanadium(V) can be reduced to vanadium(IV) by reaction with a "Jones reductor," a Zn-Hg amalgam. The reactions of interest are:

$$VO_2^+ + 2H^+ + e^- \rightarrow VO^{2+} + H_2O \qquad E° = +1.00 \text{ V}$$
$$Zn^{2+} + 2e^- \rightarrow Zn \qquad E° = -0.76 \text{ V}$$

Calculate $J°$ and $\Delta G°$ for the reaction.

Solution

For a spontaneous reduction of vanadium, Zn is oxidized. The reactions are (watch the electron balance!):

$$2VO_2^+ + 4H^+ + 2e^- \rightarrow 2VO^{2+} + 2H_2O \qquad E° = +1.00 \text{ V}$$
$$\underline{Zn \rightarrow Zn^{2+} + 2 e^-} \qquad\qquad\qquad\qquad E° = +0.76 \text{ V}$$
$$2VO_2^+(aq) + 4H^+(aq) + Zn(s) \rightarrow 2VO^{2+}(aq) + 2H_2O(l) + Zn^{2+}(aq) \qquad E°_{cell} = 1.76 \text{ V}$$

$\Delta G° = -2 \text{ mol } e^- \, (96{,}486 \text{ C/mol } e^-)(1.76 \text{ J/C})$
$\Delta G° = -3.40 \times 10^5 \text{ J} = \textbf{-340 kJ}$

18.5   The Dependence of Cell Potential on Concentration

When you finish this section you will be able to:

- Predict the redox processes in concentration cells.
- Use the Nernst equation to solve for cell voltage at nonstandard conditions.
- Calculate equilibrium constants from cell voltages.

This section considers the calculation of cell voltages at nonstandard concentrations (i.e., not 1 M). Your textbook introduces the **concentration cell**, a cell in which current flows due only to a **difference in concentration** of an ion in two different compartments of a cell. **Le Châtelier's principle** is applicable here. In a cell where there is an equal concentration of metal ion on both sides, $E^{\circ}_{cell} = 0$.

However, if the concentrations are different, a "stress" is put on the system that will be equalized by electron flow to allow reduction and oxidation to occur. (See <u>Figures 18.9 and 18.10 in your textbook</u>.) When the concentrations in the half-cells become equal, $E^{\circ}_{cell} = 0$ and the system is at equilibrium.

Example 18.5 A   Concentration Cells

A cell has on its left side a 0.20 $M\,Cu^{2+}$ solution. The right side has a 0.050 $M\,Cu^{2+}$ solution. The compartments are connected by Cu electrodes and a salt bridge. Designate the cathode, anode, and direction of current.

Solution

Current will flow in this cell until the concentration of Cu^{2+} is equal in both compartments. This means that the concentration of Cu^{2+} in the **left-hand side (0.20 M) must be reduced** by

$$Cu^{2+} + 2e^- \rightarrow Cu$$

The left hand-side will be the **cathode**. The right-hand side (0.050 $M\,Cu^{2+}$) will be the **anode**. Current will flow from right to left (anode to cathode).

The point of introducing concentration cells (which, as your textbook points out, produce a very small voltage) is really to illustrate the fact that **nonstandard concentrations produce a cell voltage that is different from that at standard concentrations**.

$$E_{cell} \text{ (nonstandard)} \neq E^{\circ}_{cell}$$

The Nernst equation is derived in your textbook. We use the equation to **calculate the cell voltage at nonstandard concentrations**.

$$E_{cell} = E^{\circ}_{cell} - \frac{0.0591}{n} \log(Q) \qquad \text{(at 25°C)}$$

where Q has its usual significance as $[products]_0 / [reactants]_0$

The mathematics of the Nernst equation are that if $[reactants]_0$ **are higher than** $[products]_0$, $\log(Q)$ will be < 0 and E_{cell} **will be** $> E^{\circ}_{cell}$. This is consistent with Le Châtelier's principle and concentration cells. Also, when a battery is fully discharged, it is at equilibrium, and $E_{cell} = 0$.

Example 18.5 B   The Nernst Equation

Calculate E_{cell} for a galvanic cell based on the following half-reactions at 25°C.

$$Cd^{2+} + 2e^- \rightarrow Cd \qquad\qquad E^{\circ} = -0.40 \text{ V}$$
$$Pb^{2+} + 2e^- \rightarrow Pb \qquad\qquad E^{\circ} = -0.13 V$$

where $[Cd^{2+}] = 0.010$ M and $[Pb^{2+}] = 0.100\ M$.

Solution

In order for the cell to be galvanic, E° must be greater than 0 V. This means that the cadmium must undergo oxidation, giving a net cell reaction

$$Cd(s) + Pb^{2+}(aq) \rightarrow Cd^{2+}(aq) + Pb(s) \qquad\qquad E^{\circ}_{cell} = +0.27 \text{ V}$$

The Nernst equation for this cell is

$$E = E° - \frac{0.0591}{2} \log\left(\frac{[Cd^{2+}]}{[Pb^{2+}]}\right) = 0.27 - 0.02955 \log\left(\frac{0.010}{0.100}\right)$$

$$E = 0.27 - 0.02955 \log(0.10) = 0.27 + 0.02955$$

$$E_{cell} = \mathbf{0.30\ V}$$

Example 18.5 C   Nernst Equation

Calculate E_{cell} for a galvanic cell based on the following half-reactions at 25°C.

(Equation 1)	$FeO_4^{2-} + 8H^+ + 3e^- \rightarrow Fe^{3+} + 4H_2O$	$E° = +2.20\ V$
(Equation 2)	$O_2 + 4H^+ + 4e^- \rightarrow 2H_2O$	$E° = +1.23\ V$

where $[FeO_4^{2-}] = 2.0 \times 10^{-3}\ M$ $[O_2] = 1.0 \times 10^{-5}\ M$
$[Fe^{3+}] = 1.0 \times 10^{-3}\ M$ pH $= 5.2$

Solution

The ferrate ion, FeO_4^{2-}, has a higher reduction potential so it will be reduced while water is oxidized to oxygen. To balance electrically, Equation 1 must be multiplied by 4 and Equation 2 must be reversed and multiplied by 3.

$$4FeO_4^{2-} + 32H^+ + 12e^- \rightarrow 4Fe^{3+} + 16H_2O \qquad\qquad E° \;= +2.20\ V$$
$$6H_2O \rightarrow 3O_2 + 12H^+ + 12e^- \qquad\qquad\qquad\qquad E° \;= -1.23\ V$$
$$\overline{4FeO_4^{2-}(aq) + 20H^+(aq) \rightarrow 4Fe^{3+}(aq) + 3O_2(g) + 10H_2O(l) \qquad E_{cell}° = \mathbf{+0.97\ V}}$$

$$E = E_{cell}° - \frac{0.0591}{n} \log(Q) = 0.97 - \frac{0.0591}{12} \log\left(\frac{[Fe^{3+}]^4[O_2]^3}{[FeO_4^{2-}]^4[H^+]^{20}}\right)$$

$$E = 0.97 - \frac{0.0591}{12} \log\left(\frac{(1.0\times10^{-3})^4\,(1.0\times10^{-5})^3}{(2.0\times10^{-3})^4\,(6.31\times10^{-6})^{20}}\right)$$
$$\overset{\uparrow}{\text{(pH = 5.2)}}$$

Notice the extreme H^+ ion dependence! Many calculators can take 6.31×10^{-6} to the 20th power. If your calculator cannot, you may split the value to solve:

$$(6.31 \times 10^{-6})^{20} = (6.31)^{20} \times (10^{-6})^{20}$$
$$= 1.0 \times 10^{16} \times 10^{-120}$$
$$= \mathbf{1.0 \times 10^{-104}}$$

$$E = 0.97 - 4.9\times10^{-3} \log\left(\frac{1.0\times10^{-27}}{1.6\times10^{-115}}\right)$$
$$= 0.97 - 4.9 \times 10^{-3} \log(6.3 \times 10^{87})$$
$$= 0.97 - 0.43$$
$$E = \mathbf{0.54\ V}$$

Recall that **at equilibrium both E and $\Delta G = 0$.** Your textbook uses this information along with the relationship between ΔG and K to derive a formula that **relates E to K.**

$$\log(K) = \frac{nE^\circ}{0.0591} \qquad \text{(at 25°C)}$$

Remember, we are dealing with **equilibrium conditions** in this case.

Example 18.5 D   Equilibrium Constants and Cell Potential

Calculate the equilibrium constant for the reaction in Example 18.3 E,

$$3Cu(s) + Cr_2O_7^{2-}(aq) + 14H^+(aq) \rightarrow 3Cu^{2+}(aq) + 2Cr^{3+}(aq) + 7H_2O(l) \qquad E^\circ_{cell} = \textbf{0.99 V}$$

Solution

There are 6 moles of electrons transferred in this redox reaction.

$$\log(K) = \frac{nE^\circ}{0.0591} = \frac{6\,(0.99)}{0.0591} = 100.51$$

$$K = 10^{100.51} = 10^{0.51} \times 10^{100} = 3 \times 10^{100}$$

Example 18.5 E   Summing It All Up

Consider the reaction

$$Ni^{2+}(aq) + Sn(s) \rightarrow Ni(s) + Sn^{2+}(aq)$$

Calculate the following: E°_{cell}, ΔG°, and K at 25°C. In addition, determine the minimum **ratio of $[Sn^{2+}]/[Ni^{2+}]$** necessary in order to make the reaction spontaneous as written.

Solution

a.   The half-reactions are:

$$Ni^2 + 2e^- \rightarrow Ni \qquad E^\circ = -0.23 \text{ V}$$
$$Sn \rightarrow Sn^{2+} + 2e^- \qquad E^\circ = +0.14 \text{ V}$$
$$\overline{\qquad\qquad\qquad\qquad E^\circ_{cell} = -0.09 \text{ V}}$$

The reaction is not spontaneous under standard conditions.

b. $\Delta G^\circ = -nFE^\circ = -2 \text{ mol } e^- \,(96{,}486 \text{ C/mol } e^-)(-0.09 \text{ J/C}) = 1.74 \times 10^4 \text{ J}$
   $\Delta G^\circ = \textbf{20 kJ (rounded to 1 sig fig)}$

c.   We can calculate K using either **$\log K = nE^\circ/0.059$** or **$\ln K = -\Delta G^\circ/RT$.** Let's do it both ways.

$$\log K = \frac{2\,(-0.09)}{0.0591} = -3.05 \quad \Rightarrow \quad K = 9 \times 10^{-4}$$

$$\ln K = \frac{-17400 \text{ J}}{(8.3148 \text{ J/K mol})\,(298 \text{ K})} = -7.022 \quad \Rightarrow \quad K = 9 \times 10^{-4}$$

d.   In order for the reaction to be spontaneous, $E > 0$. Using the Nernst equation,

$$E > 0 > E^\circ_{cell} - \frac{0.0591}{2} \log\left(\frac{[Sn^{2+}]}{[Ni^{2+}]}\right)$$

reducing, $$0 > -0.09 - 0.0296 \log\left(\frac{[Sn^{2+}]}{[Ni^{2+}]}\right)$$

$$\frac{0.09}{-0.0296} \ (= -3.05) > \log\left(\frac{[Sn^{2+}]}{[Ni^{2+}]}\right)$$

$$\frac{[Sn^{2+}]}{[Ni^{2+}]} < 9 \times 10^{-4}$$

The ratio of Sn^{2+} to Ni^{2+} must be less than 9×10^{-4} for this reaction to be spontaneous.

18.6   Batteries

The following review questions will help you test your knowledge of the material in this section.

1.   Define battery.
2.   Why is a lead-storage battery especially useful in automobiles?
3.   Give the anode, cathode, and overall cell reactions in the lead-storage battery.
4.   Calculate E° for the lead storage battery.
5.   Describe the idea behind how the extent of battery discharge can be measured.
6.   Why can "jump starting" a car be dangerous?
7.   Describe the parts of a typical dry-cell battery.
8.   Why does an alkaline dry cell last longer than an acid dry cell?
9.   Give the anode, cathode, and overall cell reactions for the acid and alkaline dry cells.
10. Why is the nickel-cadmium battery rechargeable?
11. Define "fuel cell."
12. Give the anode, cathode, and overall reactions for the hydrogen-oxygen fuel cell.

18.7   Corrosion

The following review questions will help you test your knowledge of the material in this section.

1.   Define corrosion.
2.   Why is corrosion so ubiquitous? (Consider the reducing potentials in Table 18.1.)
3.   If most native metals **theoretically** oxidize in air, why don't these metals **actually** oxidize?
4.   List the reactions, and calculate the E° for the corrosion of iron.
5.   What is the role of water and salt in the rusting of iron? Why do cars rust more in wet, cold areas of the United States than in dry, warm areas?
6.   What is the role of chromium in corrosion protection?
7.   How does galvanizing iron help prevent rust?
8.   How does magnesium act as **cathodic protection** for iron pipes?

18.8   Electrolysis

When you finish this section you will be able to use Faraday's constant and current to relate time to extent of electrolysis.

This section begins with the definition of **electrolysis** as **the process of forcing a current through a cell to produce a chemical change** for which the cell potential is negative. In order for electrolysis to occur, you must apply an external voltage that is **greater than** the potential of the galvanic cell if you want to force the reaction in the opposite (electrolytic) direction.

Example 18.8 A   Electrolytic Cells

What voltage is necessary to force the following electrolysis reaction to occur?

$$2I^-(aq) + Cu^{2+}(aq) \rightarrow I_2(s) + Cu(s)$$

Which process would occur at the anode? cathode? Assuming the iodine oxidation takes place at a platinum electrode, what is the direction of electron flow in this cell?

Solution

$$
\begin{array}{lll}
\text{anode (oxidation):} & 2I^- \rightarrow I_2 + 2e^- & E^\circ = -0.54 \text{ V} \\
\text{cathode (reduction):} & Cu^{2+} + 2e^- \rightarrow Cu & E^\circ = +0.34 \text{ V} \\
& & E^\circ_{\text{cell}} = -0.20 \text{ V}
\end{array}
$$

More than 0.20 V must be externally applied to make this reaction proceed. The direction of electron flow is always <u>F</u>rom <u>A</u>node <u>T</u>o <u>CAT</u>hode ("FAT CAT"), so it is from the platinum electrode to the copper electrode.

If you put a potential on a system that contains one metal ion, and the **potential is above that at which the metal ion will reduce**, you will plate out that metal. The amount of metal reduced is directly related to the current, in **amps (Coulombs/sec)**, that flows in the system. In practice, this is not the case because (as pointed out several times in your textbook) **electrochemistry is not a perfectly efficient process**. However, for purposes of problem solving, we will assume it is.

Dimensional analysis works wonders with electrolysis problems. Your goal is often to find **moles of electrons** of the metal.

$$\text{moles of } e^- = \underbrace{\frac{1 \text{ mol } e^-}{96,486 \text{ C}}}_{\text{Faraday}} \times \underbrace{\frac{C}{s}}_{\text{amps}} \times \underbrace{s}_{\text{time}}$$

Example 18.8 B   Electrolysis

How many grams of copper can be reduced by applying a 3.00-A current for 16.2 min to a solution containing Cu^{2+} ions?

Strategy

Our goal is to find grams. As always, we want to work through **moles**, and in this case, **moles of electrons**.

$$\textbf{current} \xrightarrow{\text{time}} \textbf{Coulombs} \Rightarrow \textbf{moles of } e^- \Rightarrow \textbf{moles of Cu} \Rightarrow \textbf{g of Cu}$$

Solution

$$\text{g Cu} = \frac{3.00\ \text{C}}{\text{s}} \times \frac{60\ \text{s}}{\text{min}} \times 16.2\ \text{min} \times \frac{1\ \text{mol e}^-}{96{,}486\ \text{C}} \times \frac{1\ \text{mol Cu}}{2\ \text{mol e}^-} \times \frac{63.54\ \text{g Cu}}{1\ \text{mol Cu}} = \textbf{0.96 g Cu}$$

Electrolysis can be used to separate a mixture of ions if the reduction potentials are fairly far apart. Remember that **the metal ion with the highest reduction potential is the easiest to reduce**.

Example 18.8 C Order of Reduction

Using Table 18.1 in your textbook, predict the order of reduction and which of the following ions will reduce first at the cathode of an electrolytic cell:

$$Ag^+, \quad Zn^{2+}, \quad IO_3^-$$

Solution

$$
\begin{array}{ll}
Ag^+ + e^- \rightarrow Ag & E^\circ = +0.80\ \text{V} \\
Zn^{2+} + 2e^- \rightarrow Zn & E^\circ = -0.76\ \text{V} \\
IO_3^- + 6H^+ + 5e^- \rightarrow \tfrac{1}{2}I_2 + 3H_2O & E^\circ = +2.10\ \text{V}
\end{array}
$$

The IO_3^- will reduce first, followed by Ag^+ and Zn^{2+} as the voltage is increased.

Example 18.8 D Electrolysis of Water

What volume of $H_2(g)$ and $O_2(g)$ is produced by electrolyzing water at a current of 4.00 A for 12.0 minutes (assuming ideal conditions)?

Strategy

The overall reaction is

$$2H_2O(l) \rightarrow 2H_2(g) + O_2(g)$$

In practice, the actual ratio is not exactly 2:1, for a variety of reasons including oxygen solubility. We can, however, run through the calculation for practice.

We are asked to find the **volume** of hydrogen and oxygen. The **ideal gas law** relates moles of a gas to volume. Our strategy must therefore be to **find moles of each gas**; then use the ideal gas relationship that **1 mole of a gas occupies 22.4 L** (under ideal conditions) to find volume. Also, for every mole of water, **2 moles of electrons** are exchanged.

$$\text{mol H}_2 = \frac{4.00\ \text{C}}{\text{s}} \times \frac{60\ \text{s}}{\text{min}} \times 12.0\ \text{min} \times \frac{1\ \text{mol e}^-}{96{,}486\ \text{C}} \times \frac{1\ \text{mol H}_2}{2\ \text{mol e}^-} = \textbf{0.0149 mol H}_2$$

$$\text{mol O}_2 = 0.0149\ \text{mol H}_2 \times \frac{1\ \text{mol O}_2}{2\ \text{mol H}_2} = \textbf{0.00746 mol O}_2$$

$$\text{L H}_2 = \frac{22.4\ \text{L}}{\text{mol}} \times 0.0149\ \text{mol} = \textbf{0.334 L H}_2 \quad \Rightarrow \quad \textbf{0.167 L O}_2$$

18.9   Commercial Electrolytic Processes

Your textbook begins this section by noting that metals are "such good reducing agents. . . ." What is the consequence of the ease with which they are oxidized? Listed below are other questions to guide your study of this section:

1. Why did the Hall-Héroult process allow the price of aluminum to drop precipitously?
2. In the Hall-Héroult process, the carbon rods must on occasion be replaced. Why?
3. Why is it necessary to go to such great lengths to electrolytically produce aluminum metal? That is, why can't it be electrolyzed from aqueous solution?
4. In the electrolysis of aqueous sodium chloride, why isn't sodium produced at the cathode?

Exercises

Section 18.1

1. Balance the following redox reactions that take place in an acid solution:

   a. $H_3AsO_4 + Zn \rightarrow AsH_3 + Zn^{2+}$
   b. $HS_2O_3^- \rightarrow S + HSO_4^-$
   c. $Cr_2O_7^{2-} + Cl^- \rightarrow Cr^{3+} + Cl_2$
   d. $MnO_2 + Hg + Cl^- \rightarrow Mn^{2+} + Hg_2Cl_2$

2. Balance the following in basic solution:

   a. $P_4 \rightarrow PH_3 + HPO_3^{2-}$
   b. $Cl_2 + OH^- \rightarrow Cl^- + ClO_3^-$
   c. $Zn + NO_3^- \rightarrow Zn^{2+} + NH_3$

3. Balance the following redox reactions that take place in a basic solution:

   a. $HXeO_4^- + Pb \rightarrow Xe + HPbO_2^-$
   b. $ClO_4^- + I^- \rightarrow ClO_3^- + IO_3^-$
   c. $Co(OH)_3 + Sn \rightarrow Co(OH)_2 + HSnO_2^-$

Section 18.2

4. In each of the following half-reactions, give the species being reduced and the number of electrons needed to balance the half-reaction:

   a. $AgBrO_3 + ?e^- \rightarrow Ag + BrO_3^-$
   b. $HCrO_4^- + 7H^+ + ?e^- \rightarrow Cr^{3+} + 4H_2O$
   c. $WO_3 + 6H^+ + ?e^- \rightarrow W + 3H_2O$

5. Identify the species in each of the following reactions that would receive electrons from the cathode and that would lose electrons at the anode in each of the following galvanic cells:

   a. $Au^{3+}(aq) + Zn(s) \rightleftharpoons Au^+(aq) + Zn^{2+}(aq)$
   b. $3Pu^{6+}(aq) + 2Cr^{3+}(aq) \rightleftharpoons 2Cr^{6+}(aq) + 3Pu^{4+}(aq)$

Section 18.3

Use the following reactions and potentials taken from the Chemical Rubber Company *Handbook of Physics and Chemistry* for Problems 6-11.

Reaction	Potential (volts)
$2SO_4^{2-} + 4H^+ + 2e^- \rightleftharpoons S_2O_6^{2-} + 2H_2O$	−0.2
$AuBr_4^- + 3e^- \rightleftharpoons Au + 4Br^-$	+0.858
$O_3 + 2H^+ + 2e^- \rightleftharpoons O_2 + H_2O$	+2.07
$Ba^{2+} + 2e^- \rightleftharpoons Ba$	−2.09
$In^{3+} + e^- \rightleftharpoons In^{2+}$	−0.49

6. Select the strongest oxidizing agent from the reactions above.

7. Select the strongest reducing agent from the reactions above.

8. Using Table 18.1 in your textbook, arrange the following species in order of increasing strength as oxidizing agents (assume all species are in their standard states):

$$Sn^{2+}, \quad Au^{3+}, \quad Fe^{3+}, \quad Fe^{2+}, \quad Ca^{2+}$$

9. Write a balanced equation, and calculate the value of $E°$ for the reaction of $AuBr_4^-$ with In^{2+}.

10. Write a balanced equation, and calculate the value of $E°$ for the reaction of Ba^{2+} with O_2 to form O_3. Would this reaction be spontaneous? Why or why not?

11. Write the balanced reaction for the galvanic cell that would yield the highest voltage given the reactions listed above.

12. A galvanic cell consists of an Ag electrode in a 1.0 M AgNO$_3$ solution and a Ni electrode in a 1.0 M Ni(NO$_3$)$_2$ solution. Calculate the standard emf of this electrochemical cell at 25°C. (Use Table 18.1 in your textbook.)

13. Regarding the following reaction:

$$F_2(g) + 2I^-(aq) \rightarrow 2F^-(aq) + I_2(s)$$

   a. List the species being oxidized.
   b. List the species being reduced.
   c. Calculate $E°$ for this cell. (See Table 18.1 in your textbook.)
   d. Which species receives electrons from the cathode?
   e. Which species donates electrons to the anode?

14. Answer the same questions that were posed in Problem 13 for the following reaction:

$$Hg_2^{2+}(aq) + Zn(s) \rightleftharpoons 2Hg(s) + Zn^{2+}(aq)$$

15. Answer the same questions that were posed in Problem 13 for the following reaction:

$$Fe^{2+}(aq) + Ce^{4+}(aq) \rightleftharpoons Fe^{3+}(aq) + Ce^{3+}(aq)$$

16. Given the following half-reactions:

$$Co^{2+} + 2e^- \rightleftharpoons Co \qquad E° = -0.277 \text{ V}$$
$$Ce^{4+} + e^- \rightleftharpoons Ce^{3+} \qquad E° = 1.61 \text{ V}$$

   a. Write the overall equation for the galvanic cell.
   b. Calculate $E°$ for the cell.

17. Regarding the galvanic cell you composed for the reaction in Problem 16:

   a. Designate the anode and cathode.
   b. Describe the direction of electron flow.

18. Given the following half-reactions:

$$[PtCl_4]^{2-} + 2e^- \rightleftharpoons Pt + 4Cl^- \qquad E° = 0.755 \text{ V}$$
$$Fe^{3+} + 3e^- \rightleftharpoons Fe \qquad E° = -0.037 \text{ V}$$

   a. Write the overall equation for the galvanic cell.
   b. Calculate $E°$ for the cell.

19. Regarding the galvanic cell you composed for the equation in Problem 18, designate the anode and cathode.

20. Given the following half-reactions:

$$Ag + 2CN^- \rightleftharpoons Ag(CN)_2^- + e^- \qquad E° = +0.31 \text{ V}$$
$$Ag \rightleftharpoons Ag^+ + e^- \qquad E° = -0.80 \text{ V}$$

Calculate $E°$ for the reaction:

$$Ag^+(aq) + 2CN^-(aq) \rightleftharpoons Ag(CN)_2^-(aq)$$

21. Given the following half-reactions:

$$La^{3+} + 3e^- \rightleftharpoons La \qquad E° = -2.52 \text{ V}$$
$$Fe^{3+} + e^- \rightleftharpoons Fe^{2+} \qquad E° = 0.77 \text{ V}$$

a. Write the overall equation for the galvanic cell.
b. Calculate $E°$ for the cell.

22. Regarding the galvanic cell you composed in Problem 21, designate the anode and the cathode.

23. Given the following half-reactions:

$$PuO_2^+ + 4H^+ + e^- \rightleftharpoons Pu^{4+} + 2H_2O \qquad E° = +1.15 \text{ V}$$
$$Cu^{2+} + 2e^- \rightleftharpoons Cu \qquad E° = +0.34 \text{ V}$$

a. Write the overall equation for the galvanic cell.
b. Calculate $E°$ for the cell.

24. Regarding the galvanic cell you composed for the reaction in Problem 23:

a. Designate the anode and cathode
b. Describe the direction of electron flow
c. Describe the electrodes in each compartment

Section 18.4

25. Calculate the standard free-energy change for the following reaction (use Table 18.1 in your textbook):

$$2Hg^{2+} + Mn \rightleftharpoons Hg_2^{2+} + Mn^{2+}$$

26. Calculate the standard free-energy change for the following reaction:

$$Na^+ + Cu \rightleftharpoons Na + Cu^+$$

27. Calculate the standard free-energy change for the following reaction:

$$Cr^{3+} + ClO_2^- \rightleftharpoons Cr^{2+} + ClO_2$$

28. Using data from Table 18.1 in your textbook, calculate $\Delta G°$ for the reaction:

$$2Al(s) + 6H^+(aq) \rightleftharpoons 2Al^{3+}(aq) + 3H_2(g)$$

Will the reaction be spontaneous (i.e., will hydrogen gas bubble from solution)?

29. Calculate $\Delta G°$ for the following reaction (use Table 18.1 in your textbook):

$$2K^+ + Cu \rightleftharpoons 2K + Cu^{2+}$$

30. Calculate ΔG for the following reaction:

$$3Hg_2^{2+} + 2Cr \rightleftharpoons 6Hg + 2Cr^{3+}$$

31. Calculate $\Delta G°$ for the following reaction:

$$Br_2 + Sn \rightleftharpoons 2Br^- + Sn^{2+}$$

32. Calculate $E°$ and $\Delta G°$ for the reaction:

$$UO_2^{2+}(aq) + 4H^+ + 2Ag(s) + 2Cl^-(aq) \rightleftharpoons U^{4+}(aq) + 2AgCl(s) + 2H_2O(l)$$

Given the following half-reactions:

$$UO_2^{2+} + 4H^+ + 2e^- \rightleftharpoons U^{4+} + 2H_2O \qquad E° = +0.334 \text{ V}$$
$$AgCl + e^- \rightleftharpoons Ag + Cl^- \qquad\qquad E° = +0.222 \text{ V}$$

Section 18.5

33. A cell has on its left side a $1.0 \times 10^{-3}\ M\ Zn^{2+}$ solution. The right side has a $0.030\ M\ Zn^{2+}$ solution. The compartments are connected by Zn electrodes and a salt bridge. Designate the cathode, anode, and direction of current.

34. Calculate the emf for each of the following half-reactions:

a. $Cr_2O_7^{2-}(0.350\ M) + 14H^+(0.0100\ M) + 6e^- \rightleftharpoons 2Cr^{3+}(1 \times 10^{-3}\ M) + 7H_2O \qquad E° = +1.33 \text{ V}$
b. $AuBr_2^-(0.084\ M) + e^- \rightleftharpoons Au + 2Br^-(0.1443\ M) \qquad\qquad\qquad\qquad\qquad E° = +0.959\text{V}$

35. For a galvanic cell based on the following half-reactions at 25°C:

$$Cl_2 + 2e^- \rightarrow 2Cl^- \qquad E° = 1.36 \text{ V}$$
$$Ni^{2+} + 2e^- \rightarrow Ni \qquad E° = -0.23 \text{ V}$$

Calculate E_{cell} where $[Cl_2] = 0.5$ atm, $[Cl^-] = 1.0\ M$, and $[Ni^{2+}] = 1.0\ M$.

36. For a galvanic cell based on the following half-reactions at 25°C:

$$Pb^{2+} + 2e^- \rightarrow Pb \qquad E° = -0.13$$
$$Zn^{2+} + 2e^- \rightarrow Zn \qquad E° = -0.76$$

Calculate E_{cell} where $[Pb^{2+}] = 1.0 \times 10^{-2}\ M$ and $[Zn^{2+}] = 3.0 \times 10^{-2}\ M$.

37. Calculate the cell voltage at 25°C for the equation in Problem 28 given the following equilibrium concentrations:

$$[Al^{3+}] = 0.025\ M,\ P_{H_2} = 1.0 \text{ atm, pH} = 3.50.$$

38. Calculate the cell voltage for the reaction:

$$Hg_2^{2+}(aq) + Cd(s) \rightleftharpoons 2Hg(s) + Cd^{2+}(aq)$$

where $[Hg_2^{2+}] = 1.0 \times 10^{-3}\ M$ and $[Cd^{2+}] = 5.0 \times 10^{-3}\ M$

$$Cd^{2+} + 2e^- \rightleftharpoons Cd \qquad E° = -0.40 \text{ V}$$
$$Hg_2^{2+} + 2e^- \rightleftharpoons 2Hg \qquad E° = +0.80 \text{ V}$$

39. Calculate the equilibrium constant, K, for the following reaction at 25°C:

$$PbO_2 + 4H^+ + 2Hg + 2Cl^- \rightleftharpoons Pb^{2+} + 2H_2O + Hg_2Cl_2 \qquad\qquad E°_{cell} = 1.12 \text{ V}$$

40.   Given the data in Problem 38, calculate the equilibrium constant, K, for the reaction:

$$2Hg(s) + Cd^{2+}(aq) \rightleftharpoons Hg_2^{2+} + Cd(s)$$

How does the value of K in this problem relate to the value for the reverse reaction (as presented in Problem 38)?

41.   Calculate K for the reaction of aluminum with hydrogen given in Problem 28.

42.   Calculate the electromotive force at 25°C for a zinc-copper cell in which the zinc sulfate concentration is 0.010 molar, and the copper sulfate concentration is 0.10 molar.

43.   Calculate the value of the equilibrium constant at 25°C for the reaction:

$$Zn + 2H^+(aq) \rightleftharpoons Zn^{2+}(aq) + H_2$$

Section 18.8

44.   Bismuth can be electrolytically reduced according to the following reaction:

$$BiO^+ + 2H^+ + 3e^- \rightleftharpoons Bi + H_2O$$

How many grams of bismuth can be reduced by applying a 5.60-A current for 28.3 min. to a solution containing BiO^+ ions (assuming 100% efficiency)?

45.   A sample containing nitric acid was titrated by electrolytically reducing water to form OH^- ion. How many moles of H^+ ion were originally in the sample if the hydroxide required 356.1 sec of generation at 9.07×10^{-3} A?

Multiple Choice Questions

46.   What is the initial oxidation half-reaction for the following reaction?

$$\_CuS + \_H^+ + \_SO_4^{2-} \rightarrow \_Cu^{2+} + \_SO_2 + \_H_2O + \_S$$

A. $CuS \rightarrow Cu^{2+} + S$ B. $SO_4^{2-} \rightarrow SO_2$ C. $CuS \rightarrow Cu^{2+}$ D. $SO_4^{2-} \rightarrow S$

47.   What is the proper set of coefficients for the following reaction?

$$\_CuS + \_H^+ + \_SO_4^{2-} \rightarrow \_Cu^{2+} + \_SO_2 + \_H_2O + \_S$$

A.   1, 2, 1, 3, 3, 2, 1   B.   1, 4, 1, 1, 1, 2, 1   C.   3, 2, 1, 1, 1, 3, 3   D.   4, 1, 1, 1, 2, 4, 4

48.   In the following reaction, which element or substance is the oxidizing agent?

$$PbSO_4(s) + H_2O(l) \rightarrow Pb(s) + PbO_2(s) + SO_4^{2-}(aq) + H^+(aq)$$

A. PbO_2 B. $PbSO_4$ C. $Pb(s)$ D. H_2O

49.   What chemical process takes place in the cathode?

A.   oxidation   B.   reduction   C.   ionic flow   D.   anion buildup

50.   The purpose of the salt bridge is to:

A.   allow electron flow   C.   allow chemical oxidation
B.   allow chemical reduction   D.   allow ion flow

51.   Which of the following half-reactions has a standard reduction potential equal to zero?

   A. $Zn^{2+}(aq) + 2e^- \rightarrow Zn(s)$
   B. $2H^+(aq)$ [1.0 M] $+ 2e^- \rightarrow H_2(g)$ [1.0 atm]
   C. $Cu^{2+}(aq) + 2e^- \rightarrow Cu(s)$
   D. $2H^+(aq)$ [0.10 M] $+ 2e^- \rightarrow H_2(g)$

52.   Calculate $E°$ for the following reaction:

$$Sn(s) + Sn^{4+}(aq) \rightarrow 2Sn^{2+}(aq)$$

   $Sn^{2+}(aq) + 2e^- \rightarrow Sn(s)$ $E° = -0.14$ V
   $Sn^{4+}(aq) + 2e^- \rightarrow Sn^{2+}(s)$ $E° = 0.15$ V

   A. 0.01V   B. 0.29 V   C. 0.16 V   D. −0.13 V

53.   Calculate $E°$ for the following reaction:

$$Fe(s) + 2Fe^{3+}(aq) \rightarrow 3Fe^{2+}(aq)$$

   $Fe^{3+}(aq) + e^- \rightarrow Fe^{2+}(aq)$ $E° = 0.77$ V
   $Fe^{2+}(aq) + 2e^- \rightarrow Fe(s)$ $E° = -0.41$ V

   A. 1.18 V   B. −0.36 V   C. 1.59 V   D. −0.05 V

54.   Calculate the potential for the reduction of iron(III) to iron(II) based on the following information:

   $Fe^{3+}(aq) + 3e^- \rightarrow Fe$ $E° = -0.04$ V
   $Fe^{2+}(aq) + 2e^- \rightarrow Fe$ $E° = -0.44$ V

   A. −0.48 V   B. −0.40 V   C. 0.76 V   D. 0.40 V

55.   Calculate $\Delta G°$ for the following reaction:

$$Sn(s) + Sn^{4+}(aq) \rightarrow 2Sn^{2+}(aq)$$

   $Sn^{2+}(aq) + 2e^- \rightarrow Sn(s)$ $E° = -0.14$ V
   $Sn^{4+}(aq) + 2e^- \rightarrow Sn^{2+}(s)$ $E° = 0.15$ V

   A. 55.0 kJ   B. −56.0 kJ   C. −227 kJ   D. −25.2 kJ

56.   Calculate $\Delta G°$ for the following reaction:

$$Fe(s) + 2Fe^{3+}(aq) \rightarrow 3Fe^{2+}(aq)$$

   $Fe^{3+}(aq) + e^- \rightarrow Fe^{2+}(aq)$ $E° = 0.77$ V
   $Fe^{2+}(aq) + 2e^- \rightarrow Fe(s)$ $E° = -0.41$ V

   A. −227 kJ   B. 112 kJ   C. −112 kJ   D. 30.9 kJ

57.   Calculate $E°$ for a reaction that has a $\Delta G° = -770$ kJ and transfers 4 electrons.

   A. 22.4 V   B. 2.99 V   C. 1.99 V   D. 0.33 V

58.   At 298 K, a cell reaction exhibits a standard emf of 0.04 V. The equilibrium constant for the cell reaction is 22.4. What is the value of n for the cell reaction?

   A. 5   B. 1   C. 4   D. 2

59. Calculate the emf of the cell that utilizes the following reaction:

$$2Co^{3+}(aq) + 2H_2O(l) \rightarrow H_2O_2(aq) + 2H^+(aq) + 2Co^{2+}(aq)$$

at 298 K, when $[Co^{2+}] = 1.0\ M$, $[H^+] = 0.20\ M$, $[H_2O_2] = 1.3\ M$, and $[Co^{3+}] = 0.50\ M$

  A. −0.087 V   B. 0.064 V   C. 0.06 V   D. −0.007 V

60. Calculate K_{sp} for iron(III) sulfide given the following data:

$$FeS(s) + 2e^- \rightarrow Fe(s) + S^{2-}(aq) \qquad E° = -1.01\ V$$
$$Fe^{2+}(aq) + 2e^- \rightarrow Fe(s) \qquad E° = -0.44\ V$$

  A. 5.54×10^{-20} B. 7.64×10^{-10} C. 8.90×10^{-26} D. 2.82×10^{-17}

61. A cell is based on the following reaction

$$Fe(s) + 2Fe^{3+}(aq) \rightarrow 3Fe^{2+}(aq) \qquad E° = 1.18\ V$$

Calculate the concentration of iron(II), in M, if the cell emf is 1.28 V when $[Fe^{3+}] = 0.50\ M$

  A. 0.47 M B. 0.24 M C. 0.047 M D. 0.71 M

62. Calculate the pH of the cathode compartment for the following reaction if $E = 3.01$ V when $[Cr^{3+}] = 0.15\ M$, $[Al^{3+}] = 0.30\ M$, and $[(Cr_2O_7)^{2-}] = 0.55\ M$.

$$2Al(s) + (Cr_2O_7)^{2-}(aq) + 14H^+ \rightarrow 2Al^{3+}(aq) + 2Cr^{3+}(aq) + 7H_2O(l)$$

  A. 0   B. 1   C. 1.5   D. 3.6

63. In a lead-storage battery, what element or compound serves as the anode?

  A. lead   B. hydrogen sulfate   C. lead oxide   D. hydrogen

64. If a battery consists of 10 cells, each producing 1.5 V, what is the total output of the battery?

  A. 0.15 V   B. 15 V   C. 7.5 V   D. 12.24 V

65. In a fuel cell that uses the reaction of hydrogen and oxygen to form water, the reduced species is:

  A. oxygen   B. hydrogen   C. hydroxide ion   D. hydronium ion

66. Which of the following elements would you expect to corrode most easily?

  A. Ag   B. Au   C. Al   D. Fe

67. You are asked to protect an iron surface from corrosion using cathodic protection. Which one of the following elements would you consider to be the best protector?

  A. Al   B. Mn   C. Na   D. Zn

68. An aqueous copper(II) chloride solution is electrolyzed for a period of 156 minutes using a current of 9.00 A. If inert electrodes are used in the process, how many grams of copper are removed from the solution?

  A. 27.8 g   B. 55.6 g   C. 31.8 g   D. 15.4 g

69. How long, in hours, does it take to remove all of the polonium from an aqueous $PoCl_4$ solution that contains 1958 g of $PoCl_4$ using a current of 6.80 A?

  A. 32.4   B. 16.2   C. 34.0   D. 88

70.   An aqueous lead(II) chloride solution contains 927.0 g of lead chloride. What current, in A, will be necessary to remove all the lead from the solution in 48.0 hours?

   A. 3.73   B. 5.00   C. 10.0   D. 7.47

71.   What is the name of the electrolytic process for producing aluminum?

   A. Hall-Héroult   B. Davy   C. Galium   D. Deville

72.   Which of the following statements about the electrolysis of brine is not true?

   A. Hydrogen is produced at the cathode.
   B. Electrolysis of NaCl will produce NaOH.
   C. Contamination of NaOH by NaCl is eliminated by using mercury as the conductor at the cathode.
   D. Contamination of NaCl by NaOH is eliminated by using mercury as the conductor at the anode.

Answers to Exercises

1. a. $P_4 + 2H_2O + 4OH^- \rightarrow 2PH_3 + 2HPO_3^{2-}$
    b. $3Cl_2 + 6OH^- \rightarrow 5Cl^- + ClO_3^- + 3H_2O$
    c. $4Zn + NO_3^- + 6H_2O \rightarrow 4Zn^{2+} + NH_3 + 9OH^-$

2. a. $H_3AsO_4 + 4Zn + 8H^+ \rightarrow AsH_3 + 4Zn^{2+} + 4H_2O$
    b. $3HS_2O_3^- + H^+ \rightarrow 4S + 2HSO_4^- + H_2O$
    c. $Cr_2O_7^{2-} + 6Cl^- + 14H^+ \rightarrow 2Cr^{3+} + 3Cl_2 + 7H_2O$
    d. $MnO_2 + 2Hg + 2Cl^- + 4H^+ \rightarrow Mn^{2+} + Hg_2Cl_2 + 2H_2O$

3. a. $HXeO_4^- + 3Pb + 2OH^- \rightarrow Xe + 3HPbO_2^-$
    b. $3ClO_4^- + I^- \rightarrow 3ClO_3^- + IO_3^-$
    c. $2Co(OH)_3 + Sn + OH^- \rightarrow 2Co(OH)_2 + HSnO_2^- + H_2O$

4. a. $Ag^+ + e^- \rightarrow Ag$
    b. $HCrO_4^- + 7H^+ + 3e^- \rightarrow Cr^{3+} + 4H_2O \ (Cr^{6+} \rightarrow Cr^{3+})$
    c. $WO_3 + 6H^+ + 6e^- \rightarrow W + 3H_2O \ (W^{6+} \rightarrow W)$

5.
Receives electrons	Loses electrons
a. Au^{3+}	Zn
b. Pu^{6+}	Cr^{3+}

6. O_3 is the strongest oxidizing agent.

7. Ba is the strongest reducing agent.

8. $Ca^{2+} < Fe^{2+} < Sn^{2+} < Fe^{3+} < Au^{3+}$

9. $AuBr_4^- + 3In^{2+} \rightleftharpoons Au + 4Br^- + 3In^{3+}$ $E° = +1.35$ V

10. $Ba^{2+} + O_2 + H_2O \rightleftharpoons Ba + O_3 + 2H^+$ $E° = -4.16$ V $E° < 0$; this reaction is nonspontaneous

11. $Ba + O_3 + 2H^+ \rightleftharpoons Ba^{2+} + O_2 + H_2O$

12. 1.03 V

13. a. I^- is being oxidized
     b. F_2 is being reduced
     c. $E° = +2.33$ V
     d. F_2 receives electrons
     e. I^- donates electrons to the anode

14. a. Zn is being oxidized
     b. Hg_2^{2+} is being reduced
     c. $E° = +1.56$ V
     d. Hg_2^{2+} receives electrons
     e. Zn donates electrons

15. a. Fe^{2+} is being oxidized
     b. Ce^{4+} is being reduced
     c. $E° = +0.93$ V
     d. Ce^{4+} receives electrons
     e. Fe^{2+} donates electrons

16. a. $2Ce^{4+} + Co \rightleftharpoons Co^{2+} + 2Ce^{3+}$ b. $E° = 1.89$ V

17. a. Oxidation of Co takes place at the anode. Reduction of Ce^{4+} takes place at the cathode.
     b. Electrons flow from the anode to the cathode.

18. a. $3[PtCl_4]^{2-} + 2Fe \rightleftharpoons 3Pt + 12 Cl^- + 2Fe^{3+}$ b. $E° = 0.792$ V

19. Oxidation of Fe takes place at the anode. Reduction of $[PtCl_4]^{2-}$ takes place at the cathode.

20. $E° = 1.11$ V

21. a. $3Fe^{3+} + La \rightleftharpoons 3Fe^{2+} + La^{3+}$ b. $E° = 3.29$ V

22. Oxidation of La takes place at the anode. Reduction of Fe^{3+} takes place at the cathode.

23. a. $2PuO_2^+ + Cu + 8H^+ \rightleftharpoons 2Pu^{4+} + Cu^{2+} + 4H_2O$ b. $E° = +0.81$ V

24. a. Oxidation of copper takes place at the anode. Reduction of PuO_2^+ takes place at the cathode.
     b. Electrons flow from the anode to the cathode.
     c. Copper can act as the anode. Platinum can act as the cathode.

25. $\Delta G° = -403$ kJ

26. $\Delta G° = 312$ kJ

27. $\Delta G° = 140$ kJ

28. $\Delta G° = -961$ kJ; Yes, the reaction will be spontaneous.

29. $\Delta G° = 629$ kJ

30. $\Delta G° = -886$ kJ

31. $\Delta G° = -237$ kJ

32. $E° = +0.112$ V; $\Delta G° = -21.6$ kJ

33. The right side will be the cathode. The left side will be the anode. The current will flow from left to right (anode to cathode).

34. a. $E = +1.11$ V   b. $E = +1.00$ V

35. $E_{cell} = 1.58$ V

36. $E_{cell} = 0.62$ V

37. $E = +1.48$ V

38. $E = +1.18$ V

39. $K = 8 \times 10^{37}$ (using 0.0591)

40. $K = 3 \times 10^{-41}$. K for this problem $= 1/K$ for Problem 38.

41. $K = 1 \times 10^{168}$

42. 1.13 volt

43. 5.0×10^{25}

44. 6.87 g of Bi

45. 3.35×10^{-5} moles of H^+

46. A	47. B	48. B	49. B	50. D	51. B
52. B	53. A	54. D	55. B	56. A	57. C
58. D	59. C	60. A	61. C	62. A	63. A
64. B	65. A	66. C	67. D	68. A	69. D
70. A	71. A	72. D			

Chapter 19

The Nucleus: A Chemist's View

In our study of chemistry, we have been dealing so far with the movement of electrons within and between atoms. We have been relatively unconcerned about nuclear structure—until now. This chapter deals with **nuclear transformations**. An important goal is to relate **radioactive decay** to nuclear energy and to the age of terrestrial objects.

19.1   Nuclear Stability and Radioactive Decay

When you finish this section you will be able to:

- Define each mechanism of nuclear decay.
- List the products, given the mechanism of nuclear decay.
- Determine each member of a decay series, given appropriate information.

Radioactive decay is the process by which a nucleus decomposes to form a different nucleus along with additional particles.

The key to determining nuclear decay products (from a general chemistry point of view) is to remember that *the sum of the protons after the decay equals that before the decay. The same is true with the number of neutrons.* For example, neptunium undergoes decay as follows:

$$^{237}_{93}\text{Np} \ \rightarrow \ ^{4}_{2}\text{He} \ + \ ^{233}_{91}\text{Pa}$$

Notice that the **sum** of protons and neutrons is the same on both sides of the equation as is the number of protons.

Your textbook discusses **six types of radioactive processes**. Each involves isotopes of various atoms. Note that for *individual atoms*, we prefer the term **nuclide.** These are, briefly,

- **α-particle production** results in the release of $^{4}_{2}\text{He}$.

    example: $^{240}_{94}\text{Pu} \ \rightarrow \ ^{4}_{2}\text{He} \ + \ ^{236}_{92}\text{U}$

- **β-particle production** results in the release of $_{-1}^{0}e$.

$$\text{example: } {}_{76}^{194}\text{Os} \rightarrow {}_{77}^{194}\text{Ir} + {}_{-1}^{0}e$$

- **γ-ray production** (High energy radiation released as other nuclear transformations occur.)

- **positron production** results in the release of $_{1}^{0}e$.

$$\text{example: } {}_{9}^{17}\text{F} \rightarrow {}_{8}^{17}\text{O} + {}_{1}^{0}e$$

- **electron capture** is accomplished when the nucleus captures an inner orbital electron.

$$\text{example: } {}_{92}^{229}\text{U} + {}_{-1}^{0}e \rightarrow {}_{91}^{229}\text{Pa}$$

- **spontaneous fission** is the decomposition of the nucleus into two relatively large fractions.

The following example tests your knowledge of the various nuclear transformation processes.

Example 19.1 A   Nuclear Decay

Write equations for each of the following processes:

    a. ${}_{96}^{241}\text{Cf}$ undergoes electron capture.

    b. ${}_{95}^{241}\text{Am}$ produces an α particle.

    c. ${}_{54}^{121}\text{Xe}$ produces a positron.

    d. ${}_{53}^{138}\text{I}$ produces a β particle.

Solution

Remember that the **sum** of protons and neutrons, as well as the **number** of protons is **consistent** on both sides of the equation. The key then is to calculate these values, then determine the element that has the values associated with it.

    a. ${}_{96}^{241}\text{Cf} + {}_{-1}^{0}e \rightarrow {}_{95}^{241}\text{Am}$

    b. ${}_{95}^{241}\text{Am} \rightarrow {}_{93}^{237}\text{Np} + {}_{2}^{4}\text{He}$

    c. ${}_{54}^{121}\text{Xe} \rightarrow {}_{53}^{121}\text{I} + {}_{1}^{0}e$

    d. ${}_{53}^{138}\text{I} \rightarrow {}_{54}^{138}\text{Xe} + {}_{-1}^{0}e$

Example 19.1 B   Practice with Nuclear Decay

Fill in the missing particle in each of the following equations:

    a. ${}_{67}^{164}\text{Ho} + {}_{-1}^{0}e \rightarrow ?$

    b. ${}_{67}^{158}? \rightarrow {}_{66}^{158}\text{Dy} + {}_{1}^{0}e$

    c. ${}_{94}^{242}\text{Pu} \rightarrow ? + {}_{92}^{238}\text{U}$

Solution

    a. ${}_{67}^{164}\text{Ho} + {}_{-1}^{0}e \rightarrow {}_{66}^{164}\textbf{Dy}$

    b. ${}_{67}^{158}\textbf{Ho} \rightarrow {}_{66}^{158}\text{Dy} + {}_{1}^{0}e$

    c. ${}_{94}^{242}\text{Pu} \rightarrow {}_{2}^{4}\textbf{He} + {}_{92}^{238}\text{U}$

There are a limited number of ways that a nucleus can decay. It may release many particles before finally becoming stable.

A **decay series is a number of sequential decays** by an unstable nuclide. The series continues until a stable nuclide is formed. For example,

$$^{241}_{95}\text{Am} \rightarrow \, ^{237}_{93}\text{Np} + \, ^{4}_{2}\text{He} \rightarrow \, ^{233}_{91}\text{Pa} + \, ^{4}_{2}\text{He}$$

Each step is consistent with the previous one in terms of proton and neutron totals. Keep in mind that when a particle (such as α) is released, it is lost to the system, so it should not enter into the calculation.

Example 19.1 C   Decay Series

$^{247}_{97}\text{Bk}$ undergoes decay to $^{208}_{82}\text{Pb}$ in the following order:

$$\alpha, \, \alpha, \, \beta, \, \alpha, \, \alpha, \, \beta, \, \alpha, \, \beta, \, \alpha, \, \alpha, \, \alpha, \, \alpha, \, \beta, \, \beta, \, \alpha$$

Write equations for the first **six** steps.

Solution

1. $^{247}_{97}\text{Bk} \rightarrow \, ^{243}_{95}\text{Am} + \, ^{4}_{2}\text{He}$

2. $^{243}_{95}\text{Am} \rightarrow \, ^{239}_{93}\text{Np} + \, ^{4}_{2}\text{He}$

3. $^{239}_{93}\text{Np} \rightarrow \, ^{239}_{94}\text{Pu} + \, ^{0}_{-1}\text{e}$

4. $^{239}_{94}\text{Pu} \rightarrow \, ^{235}_{92}\text{U} + \, ^{4}_{2}\text{He}$

5. $^{235}_{92}\text{U} \rightarrow \, ^{231}_{90}\text{Th} + \, ^{4}_{2}\text{He}$

6. $^{231}_{90}\text{Th} \rightarrow \, ^{231}_{91}\text{Pa} + \, ^{0}_{-1}\text{e}$

19.2   The Kinetics of Radioactive Decay

When you finish this section you will be able to interconvert between the half-life of a nuclide and the amount of that nuclide remaining at time "*t*."

The decay of nuclides follows a **first-order rate law.** That means that the kinetics are governed by **the same equations** that were introduced for first-order kinetics in Section 12.3. These are

$$\ln\left(\frac{N}{N_0}\right) = -kt$$

where N_0 = the original number (mass) of nuclides
N = the number remaining at time t
k = the first-order rate constant
t = the time

Also,

$$t_{1/2} = \frac{0.693}{k}$$

where $t_{1/2}$ is the half-life of the nuclide.

The following examples are intended to show how you can apply these equations directly to radioactive decay data.

Example 19.2 A   Half-life and Concentration

The half-life of $^{239}_{94}Pu$ is 2.411×10^4 years. How many years will elapse before 99.9% of a given sample decomposes?

Strategy

We have no specific amounts. However, we do know that (from a **fractional** point of view) **0.999 of our original 1.000** decomposes, leaving **0.001** remaining. We can thus establish the **ratio N/N_0 as 0.001/1.000**. We can find k from $t_{1/2}$.

Solution

$$k = \frac{0.693}{t_{1/2}} = \frac{0.693}{2.411 \times 10^4 \ yr} = 2.87 \times 10^{-5}/yr$$

$$\ln\left(\frac{N}{N_0}\right) = -kt$$

$$\ln(0.001) = 2.87 \times 10^{-5}/yr \ (t)$$

$$t = \mathbf{2.4 \times 10^5 \ yr}$$

Does the Answer Make Sense?

Our total time, 2.4×10^5 yr, is about **10 half-lives**. This means that we should have about $1/2^{10}$, (or 0.001) of our original material remaining. This is in fact the case, so the answer makes sense.

Example 19.2 B   Practice with Half-Life and Concentration

The half-life of protactinium-217 is 4.9×10^{-3}s. How much of a 3.50-mg sample of $^{217}_{91}Pa$ will remain after 1.000 sec?

Solution

This can either be solved using $t_{1/2}$ or k. Let's use $t_{1/2}$ in this problem (we used k in the previous one). The key question is "How many half-lives have passed in 1.000 sec?"

$$\# \text{ half-lives} = \frac{1 \text{ half-life}}{4.9 \times 10^{-3} \ s} \times 1.000 \ s = 204$$

The fraction of the original sample remaining will be $1/2^{204} = \mathbf{3.9 \times 10^{-62}}$. The final amount will be $3.50 \ (3.9 \times 10^{-62}) = \mathbf{1.4 \ H \ 10^{-61} \ mg}$. Because of the short half-life, essentially none of the original nuclide remains after one second.

19.3   Nuclear Transformations

The following problem is intended to give you practice with the particles involved in nuclear transformations. Remember that in this case nuclides are not spontaneously decaying, but are instead being bombarded with particles (either neutrons or other nuclides) that **cause a heavier element to form**.

Example 19.3   *Nuclear Transformations*

Fill in the missing particles in each of the following nuclear transformations:

a. $? + {}^{4}_{2}He \rightarrow {}^{243}_{97}Bk + 2{}^{1}_{0}n$

b. ${}^{253}_{99}Es + {}^{4}_{2}He \rightarrow ? + {}^{1}_{0}n$

c. ${}^{250}_{98}Cf + {}^{11}_{5}B \rightarrow {}^{257}_{103}Lr + ?$

Solution

a. ${}^{241}_{95}Am + {}^{4}_{2}He \rightarrow {}^{243}_{97}Bk + 2{}^{1}_{0}n$

b. ${}^{253}_{99}Es + {}^{4}_{2}He \rightarrow {}^{256}_{101}Md + {}^{1}_{0}n$

c. ${}^{250}_{98}Cf + {}^{11}_{5}B \rightarrow {}^{257}_{103}Lr + 4{}^{1}_{0}n$

19.4 Detection and Uses of Radioactivity

When you finish this section you will be able to solve problems involving radioactive dating of materials.

The **key assumptions** when using radioactivity to date objects using **radiocarbon dating** are:

- the ratio of ${}^{14}C/{}^{12}C$ has been constant over the years, and
- living systems have that **same constant ratio** until they die.

The mathematical basis of the technique is that once the living thing dies, the **ratio of ${}^{14}C/{}^{12}C$ diminishes** with a **half-life of 5730 years**. This means that after 17,190 years (3 half-lives), the ${}^{14}C/{}^{12}C$ ratio (be it amount, or "disintegrations per second" or whatever) will be $1/2^{3} = 1/8$ what it was before.

Example 19.4 A   *Radiocarbon Dating*

A sample of bone taken from an archeological dig was determined by radiocarbon dating to be 12,000 years old. If we assume that a constant atmospheric ${}^{14}C/{}^{12}C$ ratio has **13.6 disintegrations per minute per gram of carbon, how many disintegrations per minute per gram** does our 12,000 year old sample give off ($t_{1/2}$ for ${}^{14}C = 5730$ years)?

Solution

We know the age of our sample, and we know the half-life. We can therefore determine, by using our first-order decay formula, the ratio "N/N_0," which in this case represents the ratios of disintegrations per minute per gram.

$$k = 0.693/t_{1/2}$$
$$= 0.693/5730 \text{ yr}$$
$$= 1.21 \times 10^{-4}/\text{yr}$$

$$\ln(N/N_0) = -kt$$
$$= -(1.21 \times 10^{-4}/\text{yr})(12,000 \text{ yr})$$
$$= -1.45$$

$$N/N_0 = e^{-1.45} = \mathbf{0.234}$$

$N_0 = 13.6$ disintegrations; therefore, $N = 0.234 (13.6)$

$N = 3.2$ disintegrations per minute per gram

In determining the age of exceptionally old objects (billions of years), your textbook discusses the nuclear transformation

$$^{238}_{92}U \ \rightarrow \ ^{206}_{82}Pb \qquad\qquad t_{1/2} = 4.5 \times 10^9 \ yr$$

Example 19.6 in your textbook deals with the age of a rock based on comparing **atoms** of $^{238}_{92}U$ and $^{206}_{82}Pb$. There is a 1:1 stoichiometry in this relationship. Let's take this one step further and compare **grams** of U and Pb. The relationship in this case is **not** 1:1, but rather **238 g U/206 g Pb.**

Example 19.4 B   Dating via Uranium

A rock contains 0.141 g of $^{206}_{82}Pb$ for every 1.000 g $^{238}_{92}U$. How old is the rock ($t_{1/2} \ ^{238}_{92}U = 4.5 \times 10^9$ yr, and you are to assume that the intermediate nuclides decay instantaneously)?

Solution

The goal is really to find the ratio N/N_0 for uranium decay. The amount of lead found is a **direct reflection** of the amount of uranium that has decayed.

$$^{238}_{92}U \textbf{ decayed } = 0.141 \text{ g Pb} \ \times \ \frac{238 \text{ g U}}{206 \text{ g Pb}} = \textbf{0.163 g } ^{238}_{92}U$$

The original amount of $^{238}_{92}U$ = g found + g decayed = $1.000 + 0.163$ = **1.163 g.**

$$k = \frac{0.693}{t_{1/2}} = \frac{0.693}{4.5 \times 10^9 \ yr} = \textbf{1.54} \times \textbf{10}^{-10}\textbf{/yr}$$

$$\ln(N/N_0) = -kt$$
$$\ln(1.000/1.163) = 1.54 \times 10^{-10}/yr \ (t)$$
$$t = \textbf{9.8} \times \textbf{10}^8 \textbf{ years old}$$

19.5   Thermodynamic Stability of the Nucleus

When you finish this section you will be able to use the **mass defect** of an atom **to calculate its binding energy**.

A nucleus is a bundle of protons and neutrons bound together. The nucleus is more energetically stable than the individual array of protons and neutrons. This is known because the **sum of the masses** of protons and neutrons in a nucleus is **less than** the sum of the masses of individual nucleons. The difference in the mass is called the **mass defect ("Δm")**. The mass has been converted to energy, as predicted by

$$E = mc^2, \text{ or } \Delta E = \Delta mc^2$$

"ΔE" is called the binding energy. For example,

$$^{187}_{77}Ir = 187.958830 \text{ g/mol}$$

1 mole of 77 (protons + electrons) + 110 neutrons =

$$\begin{array}{ll} 77 \times 1.007825 \text{ g per proton or electron} & = \quad 77.602525 \text{ g} \\ \underline{110 \times 1.008665 \text{ g per neutron}} & \underline{= 110.95315 \ \text{ g}} \\ & \quad 188.555675 \text{ g} \end{array}$$

The **mass defect, Δm** = 188.555675 g − 187.958830 g = **0.596845 g/mol**. That is the mass that has been converted to energy. To find out how much, remember that

$$1 \text{ Joule} = 1 \text{ kg m}^2/\text{s}^2$$
$$\text{speed of light} = c = 3.00 \times 10^8 \text{ m/s}$$

$$\Delta E = \Delta m c^2 = \frac{-0.596845 \text{ g}}{\text{mol}} \times \frac{1 \text{ kg}}{1000 \text{ g}} \times \left(\frac{3.00 \times 10^8 \text{ m}}{\text{s}}\right)^2 = 5.37 \times 10^{13} \text{ kg m}^2/\text{s}^2 \text{ mol}$$

$$\Delta E = -5.37 \times 10^{13} \text{ J/mol} \qquad \text{(The "−" indicated energy is released [exothermic process] when the nucleus is formed.)}$$

In order to calculate ΔE per nucleus, we must divide by Avogadro's number.

$$\Delta E = \frac{-5.37 \times 10^{13} \text{ J}}{\text{mol}} \times \frac{1 \text{ mol}}{6.02 \times 10^{23} \text{ nuclei}} = -8.92 \times 10^{-11} \text{ J/nucleus}$$

In order to convert to J/nucleon, divide by 187 nucleons in the $^{187}_{77}\text{Ir}$ nucleus.

$$\Delta E = \frac{-8.92 \times 10^{-11} \text{ J}}{\text{nucleus}} \times \frac{1 \text{ nucleus}}{187 \text{ nucleons}} = -4.77 \times 10^{-13} \text{ J/nucleon}$$

Finally the binding energy per nucleon in "million electron volts" (MeV),

$$\Delta E = \frac{-4.77 \times 10^{-13} \text{ J}}{\text{nucleon}} \times \frac{1 \text{ MeV}}{1.60 \times 10^{-13} \text{ J}} = -2.98 \text{ MeV/nucleon}$$

This means that 2.98 MeV **per nucleon** would be released if the $^{187}_{77}\text{Ir}$ nucleus were formed from individual protons and neutrons (with electrons surrounding the nucleus).

Example 19.5 *Mass Defect and Binding Energy*

Determine the binding energy in **J/mol** and **MeV/nucleon** for $^{101}_{46}\text{Pd}$ (atomic mass = 100.908287 g/mol).

Solution

Mass of individual nucleons.

$$
\begin{array}{ll}
46 \times 1.007825 \text{ g/proton or electron} = & 46.35995 \text{ g} \\
+ \quad 55 \times 1.008665 \text{ g/neutron} \qquad = & 55.476575 \text{ g} \\
\hline
& 101.836525 \text{ g/mol}
\end{array}
$$

$$\text{mass defect } (\Delta m) = 101.836525 \text{ g} - 100.908287 \text{ g} = \mathbf{0.928238 \text{ g}}$$

$$\Delta E = \Delta m c^2 = (-9.28238 \times 10^{-4} \text{ kg})(3.00 \times 10^8 \text{ m/s})^2$$
$$\text{(minus sign because mass is } \underline{\text{lost}} \text{ in forming the nuclide)}$$

$$\Delta E = -8.35 \times 10^{13} \text{ J/mol}$$

$$\frac{\text{MeV}}{\text{nucleon}} = \frac{-8.35 \times 10^{13} \text{ J}}{\text{mol}} \times \frac{1 \text{ MeV}}{1.60 \times 10^{-13} \text{ J}} \times \frac{1 \text{ mol}}{6.02 \times 10^{23} \text{ nuclei}} \times \frac{1 \text{ nuclide}}{101 \text{ nucleons}}$$

$$\Delta E = -8.59 \text{ MeV/nucleon}$$

19.6   Nuclear Fission and Nuclear Fusion

The following questions will help you review the material in this section of your textbook.

1. Define **fusion.**
2. Define **fission.**
3. Is $^{235}_{92}U + ^{1}_{0}n \rightarrow ^{137}_{52}Te + ^{97}_{40}Zr + 2 ^{1}_{0}n$ an example of fission or fusion?
4. How does a **chain reaction** work?
5. Describe the process of fission in a nuclear warhead (also, define **subcritical, critical,** and **supercritical).**
6. Outline how nuclear reactors work. Focus on the functions of the **reactor core, moderator, control rods, and cooling systems.**
7. What are the consequences for failure of each of the systems in a nuclear reactor?
8. What is the principle behind a **breeder reactor?**
9. Why is fusion a process worth exploring?
10. What is the difficulty with doing fusion reactions on Earth?

19.7   Effects of Radiation

The following questions will help you review the material in this section of your textbook.

1. Why is any energy source potentially dangerous?
2. Why does exposure to radioactivity seem less hazardous than it really can be? (Discuss the exposure "per event.")
3. Define **somatic damage** and **genetic damage.**
4. What variables affect the degree of damage that radiation can cause? Discuss each one.
5. Your textbook says that our exposure to natural radiation is much greater than to man-made sources. Why are we so worried about a relatively small exposure level?
6. Describe the **linear model** and **threshold model** of radiation damage.

Exercises

Section 19.1

1.   Write equations for each of the following processes:

   a. $^{73}_{31}Ga$ produces a β particle.

   b. $^{68}_{31}Ga$ undergoes electron capture.

   c. $^{192}_{78}Pt$ produces an α particle.

2.   What must be the end product of the natural radioactive series that begins with ^{235}U and involves 7 alpha and 4 beta emissions?

3.   Write equations for each of the following processes:

   a. $^{207}_{87}Fr$ produces an α particle.

   b. $^{234}_{90}Th$ produces a β particle.

   c. $^{62}_{29}Cu$ produces a positron.

4.   Write the nuclear equation for each of the following processes:

   a.   α-decay of $^{222}_{88}Ra$

   b.   Positron decay of $^{66}_{31}Ga$

   c.   β-decay of $^{39}_{17}Cl$

5.   Fill in the missing particle in each of the following equations:

   a. $^{129}_{51}Sb \rightarrow$ _____ $+ \ ^{0}_{-1}e$

   b.   _____ $+ \ ^{0}_{-1}e \rightarrow \ ^{7}_{3}Li$

   c. $^{205}_{83}Bi \rightarrow \ ^{205}_{82}Pb \ +$ _____

   d. $^{206}_{87}Fr \rightarrow$ _____ $+ \ ^{4}_{2}He$

6.   Fill in the missing particle in each of the following equations:

   a.   _____ $+ \ ^{0}_{-1}e \rightarrow \ ^{212}_{86}Rn$

   b.   _____ $\rightarrow \ ^{208}_{85}At \ + \ ^{4}_{2}He$

   c. $^{226}_{90}Th \rightarrow$ _____ $+ \ ^{4}_{2}He$

   d. $^{186}_{77}Ir \rightarrow$ _____ $+ \ ^{0}_{1}e$

7.   Fill in the proper mass and charge numbers in the nuclear equations:

$$^{238}U + n \rightarrow U \rightarrow Np + β \rightarrow Pu + β$$

8.   The radioactive isotope $^{242}_{96}Cm$ undergoes decay to $^{206}_{32}Pb$ in the following order:

$$α, \ α, \ α, \ α, \ α, \ α, \ α, \ β, \ β, \ α, \ β, \ β, \ α.$$

   Write equations for the first <u>six</u> steps.

9.   The sixth step in the previous problem should leave you with $^{218}_{84}\text{Po}$. Continue writing equations in the series until you reach $^{206}_{82}\text{Pb}$.

10. Which is likely to have a greater number of stable isotopes, indium (In), tin (Sn), or antimony (Sb)?

Section 19.2

11. Calculate the rate constant for each of the following values of half-life:

    a. $^{182}_{72}\text{Hf}$, $t_{1/2} = 9 \times 10^6$ yr

    b. $^{228}_{91}\text{Pa}$, $t_{1/2} = 26$ hr

    c. $^{225}_{88}\text{Ra}$, $t_{1/2} = 14.8$ dy

    d. $^{181}_{78}\text{Pt}$, $t_{1/2} = 51$ s

12. The half-life of $^{161}_{65}\text{Tb}$ is 6.9 days. How many grams of an original 3.000-g sample will remain after two weeks?

13. The half-life of $^{129}_{53}\text{I}$ is 1.7×10^7 yr. How many grams of an original 50.00-g sample will remain after 5.0×10^8 yr?

14. How long will it take for 98.6% of a sample of $^{188}_{79}\text{Au}$ to decompose ($t_{1/2} = 8.8$ min)?

15. How long will it take for 99.99% of a sample of $^{199}_{84}\text{Po}$ to decompose ($t_{1/2} = 5.2$ min)?

16. How many half-lives have passed if 87.5% of a substance has decomposed? How many if 99.999% has decomposed?

17. One gram of $^{198}_{79}\text{Au}$ decays by β-emission to produce stable mercury.

    a.   Write a nuclear equation for the process

    b.   If the half-life of $^{198}_{79}\text{Au}$ is 65 hours, how much gold will be left after 130 hours?

    c.   How much mercury will be present after 260 hours?

Section 19.4

18. Upon what principle does carbon-14 dating rest?

19. Let us assume a constant $^{14}\text{C}/^{12}\text{C}$ ratio of 13.6 disintegrations per minute per gram of living matter. A sample of a petrified tree was found to give 1.2 disintegrations per minute per gram. How old is the tree? ($t_{1/2} = {}^{14}\text{C} = 5730$ years)

20. The radioactive isotope $^{237}_{90}\text{Th}$ has a rate constant, $k = 4.91 \times 10^{-11}$ yr^{-1}. Is this nuclide useful for determining the age of bone samples? Why or why not?

21. A rock contains 0.688 g of $^{206}_{82}\text{Pb}$ for every 1.00 g of $^{238}_{92}\text{U}$. How old is the rock? ($t_{1/2}$ $^{238}_{92}\text{U}$ = 4.5×10^9 yr. Assume intermediate nuclides decay instantaneously to the stable Pb nuclide.)

Section 19.5

22. Calculate the mass defect in grams for one mole of each of the following:

    a. $^{235}_{92}U$, atomic mass = 235.0439 g/mol

    b. $^{127}_{53}I$, atomic mass = 126.9004 g/mol

    c. $^{75}_{33}As$, atomic mass = 74.9216 g/mol

23. Determine the binding energy in J/mol and MeV/nucleon for $^{66}_{30}Zn$ (atomic mass = 65.9260 g/mol).

24. Determine the binding energy in J/mol and MeV/nucleon for $^{150}_{60}Nd$ (atomic mass = 149.9207 g/mol).

Section 19.6

25. Write a nuclear reaction for each of the following processes:

    a. Neutron-initiated fission of ^{235}U. Assume that two neutrons are produced and that one fission product is $^{144}_{54}Xe$.

    b. The fusion of a tritium nucleus and a deuterium to produce a helium nucleus and a neutron.

Multiple Choice Questions

26. Which one of the following types of radioactive decay does not change the mass number or atomic number of an element?

    A. Gamma rays   B. Electron capture   C. Alpha decay   D. Positron emission

27. Which one of the following decay modes has not been observed?

    A. Neutron emission   B. Positron emission   C. Alpha emission   D. Electron capture

28. A ^{238}U nucleus decays by alpha emission. What is the product of this reaction?

    A. ^{234}U B. ^{234}Th C. ^{242}Pu D. ^{242}Ra

29. What is the final product of the following decay series of ^{230}Th?

    $$\alpha, \alpha, \alpha, \beta, \alpha, \beta$$

    A. ^{218}Po B. ^{214}Pb C. ^{210}Po D. ^{214}Po

30. Nuclei situated above the belt of stability usually decay through which of the following types of radioactive decay?

    A. Alpha emission   B. Gamma rays   C. Beta emission   D. Neutron emission

31. Positron production results in:

    A. Higher proton/neutron ratio   C. Smaller proton/neutron ratio
    B. Same proton/neutron ratio   D. Smaller neutron/proton ratio

32. Which one of the following decay series would change ^{210}Pb to ^{206}Pb?

    A. $\alpha, \alpha, \alpha, \beta, \alpha, \gamma$ B. β, β, α C. $\beta, \beta, \alpha, \alpha$ D. $\alpha, \beta, \alpha, \gamma$

33. How long, in years, will it take for ^{208}Po activity to be reduced by 90.0%? The half-life of ^{208}Po is 2.83 years.

     A. 9.40   B. 26.0   C. 28.3   D. 3.14

34. How much of a 2.00-g ^{239}Pu sample would have decayed after 1.6×10^4 years? The half-life for ^{239}Pu is 2.4×10^4 years.

     A. 1.98 g   B. 0.74 g   C. 0.2 g   D. 1.28 g

35. It took 109.8 years for a 300.0 mg sample of an unknown radioactive material to completely disintegrate. Calculate the mass of the element (in grams/mole) assuming that the disintegrations per second are constant over the life-time of the sample, and the sample is labeled 100.0 Ci (Curie = 3.7×10^{10} disintegrations per second).

     A. 14.0   B. 90.0   C. 209   D. 238

36. How long, in years, will it take for ^{208}Po activity to be reduced down to 0.01% of the original activity? The half-life of ^{208}Po is 2.83 years.

     A. 283   B. 18.8   C. 2.82   D. 37.6

37. How many half-lives have passed if a sample's activity is reduced to 0.0488% of the sample's initial activity?

     A. 11   B. 9   C. 10   D. 5

38. The number of disintegrations per second of ^{14}C in a living organism is 31 for every two grams of carbon. How long, in years, will it take for the level of activity to reach 29 disintegrations per second per gram of carbon? The half-life of ^{14}C is 5730 years.

     A. 11.0   B. 551   C. 110   D. 5.36×10^3

39. If the number of disintegrations of ^{14}C in a living organism is 930 disintegrations per second per gram of carbon, how old is the content of a clay amphora that displays 9.22 disintegrations per minute per gram of carbon?

     A. 2100 years   B. 1200 years   C. 2150 years   D. 4300 years

40. Calculate the total binding energy, in joules, in one atom of ^{32}S (atomic mass = 31.97207 amu). m_p = 1.00728 amu and m_n = 1.00866 amu.

     A. 6.82 H 10^{-11} B. 4.10 H 10^{-5} C. 4.10 H 10^{-11} D. 1.36 H 10^{-12}

41. Calculate the mass defect in 2 moles of ^{32}S (atomic mass = 31.97207 amu). m_p = 1.00728 amu and m_n = 1.00866 amu.

     A. 0.58351 g   B. 0.84891 g   C. 1.2203 g   D. 0.62355 g

42. Calculate the binding energy per nucleon, in MeV per mole, for ^{33}S nucleus (atomic mass = 32.97146 amu). m_p = 1.00728 amu and m_n = 1.00866 amu.

     A. 8.52   B. 8.25   C. 9.01   D. 30.2

43. In nuclear fission, a critical mass means that:

     A. Less than one neutron per fission is available.
     B. The process is capable of sustaining itself at a constant rate fission.
     C. The process is capable of expanding its rate of fission.
     D. More than one neutron per fission is available.

44. The role of the moderator in a nuclear reactor is to:

   A. Absorb neutrons and thereby regulate the rate of reaction.
   B. Speed up neutrons released by the reactor core so the control rods can capture them more efficiently.
   C. Slow down neutrons so the uranium fuel can capture them more efficiently.
   D. Cool down the reactor core and prevent possible explosions from heat.

45. One obstacle to be overcome in fusion is:

   A. Finding nuclei that can fuse together.
   B. Bringing nuclei together to fuse.
   C. Finding nuclei that give substantial energy when fused.
   D. Designing systems that will produce the required temperatures to allow fusion.

46. Which one of the following forms of radiation is most penetrating?

   A. alpha particles   B. protons   C. electrons   D. gamma rays

47. According to the linear model of exposure:

   A. The higher the dose of radiation, the higher the danger.
   B. Radiation is dangerous only above a certain threshold dose.
   C. The higher the dose and the shorter the time of exposure, the smaller the danger.
   D. Radiation is dangerous only at one certain dose.

48. Which one of the following types of radiation cause the greatest damage due to ionization ability?

   A. alpha particles   B. beta particles   C. gamma rays   D. positrons

Answers to Exercises

1. a. $^{73}_{31}\text{Ga} \rightarrow ^{73}_{32}\text{Ge} + ^{0}_{-1}\text{e}$

    b. $^{68}_{31}\text{Ga} + ^{0}_{-1}\text{e} \rightarrow ^{68}_{30}\text{Zn}$

    c. $^{192}_{78}\text{Pt} \rightarrow ^{188}_{76}\text{Os} + ^{4}_{2}\text{He}$

2. $^{207}_{82}\text{Pb}$

3. a. $^{207}_{87}\text{Fr} \rightarrow ^{203}_{85}\text{At} + ^{4}_{2}\text{He}$

    b. $^{234}_{90}\text{Th} \rightarrow ^{234}_{91}\text{Pa} + ^{0}_{-1}\text{e}$

    c. $^{62}_{29}\text{Cu} \rightarrow ^{62}_{28}\text{Ni} + ^{0}_{1}\text{e}$

4. a. $^{222}_{88}\text{Ra} \rightarrow ^{4}_{2}\text{He} + ^{218}_{86}\text{Rn}$

    b. $^{66}_{31}\text{Ga} \rightarrow ^{0}_{1}\text{e} + ^{66}_{30}\text{Zn}$

    c. $^{39}_{17}\text{Cl} \rightarrow ^{0}_{-1}\text{e} + ^{39}_{18}\text{Ar}$

5. a. $^{129}_{51}\text{Sb} \rightarrow ^{129}_{52}\text{Te} + ^{0}_{-1}\text{e}$

    b. $^{7}_{4}\text{Be} + ^{0}_{-1}\text{e} \rightarrow ^{7}_{3}\text{Li}$

    c. $^{205}_{83}\text{Bi} \rightarrow ^{205}_{82}\text{Pb} + ^{0}_{1}\text{e}$

    d. $^{206}_{87}\text{Fr} \rightarrow ^{202}_{85}\text{At} + ^{4}_{2}\text{He}$

6. a. $^{212}_{87}\text{Fr} + ^{0}_{-1}\text{e} \rightarrow ^{212}_{86}\text{Rn}$

    b. $^{212}_{87}\text{Fr} \rightarrow ^{208}_{85}\text{At} + ^{4}_{2}\text{He}$

    c. $^{226}_{90}\text{Th} \rightarrow ^{222}_{88}\text{Ra} + ^{4}_{2}\text{He}$

    d. $^{186}_{77}\text{Ir} \rightarrow ^{186}_{76}\text{Os} + ^{0}_{1}\text{e}$

7. $^{238}_{92}\text{U} + ^{1}_{0}\text{n} \rightarrow ^{239}_{92}\text{U} \rightarrow ^{239}_{93}\text{Np} + ^{0}_{-1}\text{e} \rightarrow ^{239}_{94}\text{Pu} + ^{0}_{-1}\text{e}$

8. a. $^{242}_{96}\text{Cm} \rightarrow ^{238}_{94}\text{Pu} + ^{4}_{2}\text{He}$

    b. $^{238}_{94}\text{Pu} \rightarrow ^{234}_{92}\text{U} + ^{4}_{2}\text{He}$

    c. $^{234}_{92}\text{U} \rightarrow ^{230}_{90}\text{Th} + ^{4}_{2}\text{He}$

    d. $^{230}_{90}\text{Th} \rightarrow ^{226}_{88}\text{Ra} + ^{4}_{2}\text{He}$

    e. $^{226}_{88}\text{Ra} \rightarrow ^{222}_{86}\text{Rn} + ^{4}_{2}\text{He}$

    f. $^{222}_{86}\text{Rn} \rightarrow ^{218}_{84}\text{Po} + ^{4}_{2}\text{He}$

9. a. $^{218}_{84}\text{Po} \rightarrow ^{214}_{82}\text{Pb} + ^{4}_{2}\text{He}$

    b. $^{214}_{82}\text{Pb} \rightarrow ^{214}_{83}\text{Bi} + ^{0}_{-1}\text{e}$

    c. $^{214}_{83}\text{Bi} \rightarrow ^{214}_{84}\text{Po} + ^{0}_{-1}\text{e}$

    d. $^{214}_{84}\text{Po} \rightarrow ^{210}_{82}\text{Pb} + ^{4}_{2}\text{He}$

    e. $^{210}_{82}\text{Pb} \rightarrow ^{210}_{83}\text{Bi} + ^{0}_{-1}\text{e}$

    f. $^{210}_{83}\text{Bi} \rightarrow ^{210}_{84}\text{Po} + ^{0}_{-1}\text{e}$

    g. $^{210}_{84}\text{Po} \rightarrow ^{206}_{82}\text{Pb} + ^{4}_{2}\text{He}$

10. Antimony (Sb)

11. a. $k = 7.7 \times 10^{-8}/\text{yr}$

    b. $k = 0.0267/\text{hr}$

    c. $k = 0.04683/\text{dy}$

    d. $k = 0.0136/\text{s}$

12. 0.74 g will remain.

13. 7.0×10^{-8} g will remain.

14. 54 min.

15. 69 min

16. a. 3 half-lives   b. 19.93 half-lives

17. a. $^{198}_{79}Au \rightarrow ^{0}_{-1}e + ^{198}_{80}Hg$
    b. 0.25 g
    c. 0.94 g

18. C-14 is present in the atmosphere in a constant amount ($^{14}CO_2$), because it is continually being produced by cosmic ray activity that results in neutron capture by a nitrogen atom and the subsequent expulsion of a proton. Therefore, at all times, a constant small quantity of $^{14}CO_2$ is available to growing plants and other organisms. Once the growing process stops and C-14 is no longer taken up by the organism, the amount of it in organismal tissues begins to diminish through radioactive decay. By measuring the radioactivity due to C-14 in living organisms (e.g. wood) and comparing this to the radioactivity in preserved organisms (e.g. wood or charcoal), one can calculate from the known half-life of C-14 (5600 years) the time that must have elapsed to reduce the radioactivity to that of the preserved object.

19. The tree is 20,000 years old.

20. No, it is not. If we assume that bones may be as old as 2×10^6 yr, this is far less than one half-life, and therefore probably undetectable.

21. The rock is 3.8×10^9 years old.

22. a. mass defect = 1.915095 g   c. mass defect = 0.700555 g
    b. mass defect = 1.55535 g

23. ΔE = binding energy = -5.59×10^{13} J/mol and -8.79 MeV/nucleon

24. $\Delta E = -1.20 \times 10^{14}$ J/mol and -8.31 MeV/nucleon

25. a. $^{235}_{92}U + ^{1}_{0}n \rightarrow ^{90}_{38}Sr + ^{144}_{54}Xe + 2^{1}_{0}n$
    b. $^{3}_{1}H + ^{2}_{1}H \rightarrow ^{4}_{2}He + ^{1}_{0}n$

26.	A	27.	A	28.	B	29.	D	30.	C	31.	C
32.	B	33.	A	34.	B	35.	A	36.	D	37.	A
38.	B	39.	D	40.	C	41.	A	42.	A	43.	B
44.	C	45.	D	46.	D	47.	A	48.	A		

Chapter 20

The Representative Elements

The chemistry of the representative elements can be explained by their electronic structures. As you go through the chapter, note how elements within a group show similar reactive properties.

20.1   A Survey of the Representative Elements

When you finish this section you will be able to:
- Describe the effect of the size of an atom on reactivity.
- List some methods of preparing elements.

This section begins by reviewing the positions of various types of elements in the periodic table.

- **representative elements** are those in which *s* **and** *p* **orbitals are being filled.**
- **transition elements** are those in which the *d* **orbitals are being filled**.
- lanthanides and actinides (sometimes called "inner-transition metals") are those in which the **4f or 5f orbitals are being filled**.

You should be able to determine the type of element by its position on the periodic table. (Review Section 7.11 if you have forgotten!) You may also wish to review electron configurations to confirm your assignments. Try the following review examples.

Example 20.1 A   Electron Configuration Review

Write the electron configurations (either shorthand or longhand) for each of the following elements:

    a. yttrium   b. silver   c. bismuth   d. iodine

Solution

We can either use our configuration triangle (Chapter 7) or use the position in the periodic table to establish the electron configuration.

a. Yttrium has atomic number 39. It is in **Period 5** and **Group 3B**. It must, therefore, be **[Kr] $5s^2 4d^1$**. Checking with the longhand form,

$$\text{Y: } 1s^2 2s^2 2p^6 3s^2 3p^6 4s^2 3d^{10} 4p^6 5s^2 4d^1$$

b. Silver has atomic number 47. It is in **Period 5** and **Group 1B**. It is **[Kr] $5s^1 4d^{10}$**.

$$\text{Ag: } 1s^2 2s^2 2p^6 3s^2 3p^6 4s^2 3d^{10} 4p^6 5s^1 4d^{10}$$

c. Bismuth has atomic number 83. Elements with high atomic numbers get rather difficult to handle by using the configuration triangle. It is easier to use your knowledge of the periodic table. Bismuth is in **Period 6** and **Group 5A**. The shorthand form is **[Xe] $6s^2 4f^{14} 5d^{10} 6p^3$**. The longhand form is

$$\text{Bi: } 1s^2 2s^2 2p^6 3s^2 3p^6 4s^2 3d^{10} 4p^6 5s^2 4d^{10} 5p^6 6s^2 4f^{14} 5d^{10} 6p^3$$

d. Iodine has atomic number 53. It is in **Period 5** and **Group 7A**. It is **[Kr] $5s^2 4d^{10} 5p^5$**.

$$\text{I: } 1s^2 2s^2 2p^6 3s^2 3p^6 4s^2 3d^{10} 4p^6 5s^2 4d^{10} 5p^5$$

Example 20.1 B Types of Elements

Classify each of the following elements as **representative** or **transition**:

a. iron   b. sodium   c. argon   d. nickel

Solution

a. transition   b. representative   c. representative   d. transition

The section focuses attention on chemical differences within groups caused by size differences in atoms. In particular **the first member of a group tends to have different properties than the rest of the group because it is so much smaller than the rest of the atoms in the group.**

-   Hydrogen is a nonmetal and forms covalent bonds. Lithium is a metal and forms ionic bonds.
-   Beryllium oxide is amphoteric. Other Group 2A oxides are basic.
-   Boron behaves as a nonmetal or semimetal. Other Group 3A elements are metal.
-   Carbon can form π bonds due to effective overlap between $2p$ orbitals of carbon and carbon, carbon and nitrogen, and carbon and oxygen. Other Group 4A elements do not show this behavior.
-   Nitrogen can form π bonds because of its size, just as carbon can. Other Group 5A elements normally do not. (A few phosphorus double-bonded compounds have been known to exist.)
-   A similar situation exists with oxygen in Group 6A.
-   Fluorine has a very weak F−F bond because the atoms are so small that the nuclei approach each other very closely. This causes large electron-electron repulsions.

The section ends with a discussion of some methods of preparation of elements. Read over that section, and try the next example.

Example 20.1 C Preparation of Elements

Write equations for the following preparations:

a.   The preparation of gaseous potassium from the reaction of liquid sodium and liquid potassium chloride.
b.   The preparation of xenon gas from the reaction of aqueous xenon difluoride and water to form gaseous hydrogen fluoride, oxygen, and xenon.
c.   The preparation of solid chromium by reducing solid chromium(III) oxide with aluminum metal.

Solution

a. $Na(l) + KCl(l) \rightarrow NaCl(l) + \mathbf{K(g)}$
b. $XeF_2(aq) + 2H_2O(l) \rightarrow \mathbf{Xe(g)} + 2HF(g) + O_2(g)$
c. $Cr_2O_3(s) + 2Al(s) \rightarrow \mathbf{2Cr(s)} + Al_2O_3(s)$

20.2   The Group 1A Elements

The following questions and examples will help you review the material in this section.

1.   What electron configuration is common to each of the elements in this group?
2.   Your textbook says that although lithium is a **stronger reducing agent**, it reacts **more slowly** with water than potassium or sodium. What does this illustrate?
3.   Why do alkali metals react so well with nonmetals?

Example 20.2 A   Preparation of Alkali Metals

Write the reaction for the preparation of each of the following Group 1A metals.

a.   Liquid sodium by electrolysis of molten sodium chloride.
b.   Solid rubidium by the reaction of solid calcium and liquid rubidium chloride.

Solution

a.   Chapter 18 deals with electrolysis. The reaction of interest (done at 600°C) is

$$2NaCl(l) \rightarrow 2Na(l) + Cl_2(g)$$

b.   Potassium, rubidium, and cesium have relatively low melting points and cannot be made by electrolysis.

$$Ca(s) + 2RbCl(l) \rightarrow 2Rb(s) + CaCl_2(l)$$

As a wrap-up to this section, study the material in Table 20.5 in your textbook, and try the following example.

Example 20.2 B   Practice with Reactions

Predict the products of the following reactions:

a. $2Rb(s) + Cl_2(g) \rightarrow$?
b. $Rb(s) + O_2(g) \rightarrow$?
c. $12K(s) + P_4(s) \rightarrow$?
d. $2Na(s) + 2H_2O(l) \rightarrow$?

Solution

a. $2Rb(s) + Cl_2(g) \rightarrow 2RbCl(s)$
b. $Rb(s) + O_2(g) \rightarrow RbO_2(s)$
c. $12K(s) + P_4(s) \rightarrow 4K_3P(s)$
d. $2Na(s) + 2H_2O(l) \rightarrow 2NaOH(aq) + H_2(g)$

4. Lithium reacts more slowly with water than potassium because

    A. Lithium is a less powerful reducing agent than potassium.
    B. Lithium is a less powerful oxidizing agent than potassium.
    C. $Li(s)$ has a lower melting point than $K(s)$.
    D. $Li(s)$ has a higher melting point than $K(s)$.

5. What is the product of the following reaction?

    $$Na(s) + O_2(g) \text{ (limiting reagent)} \rightarrow ?$$

    A. NaO_2 B. Na_2O_2 C. Na_2O D. Na^+, O

Answers to review questions are at the end of this study guide chapter.

20.3 Hydrogen

The following questions will help you review the material in this section.

1. Why does hydrogen act so differently from the metals in this group?
2. Why does hydrogen have such low boiling and melting points?
3. Write the combustion reaction between hydrogen and oxygen to form water.
4. Write the reaction between methane and water to produce hydrogen.
5. Why isn't electrolysis a very practical method of producing hydrogen?
6. List two commercial uses of hydrogen.
7. Define **hydride, ionic hydride, covalent hydride,** and **interstitial hydride.**
8. How is palladium used to help purify hydrogen?
9. Why might interstitial hydrides be used for storage of hydrogen gas?

10. Which one of the following statements about hydrogen is not true?

    A. Its major industrial use is in the Haber process.
    B. The major industrial source of hydrogen is water electrolysis.
    C. It forms covalent hydrides when it reacts with other nonmetals.
    D. In interstitial hydrides, hydrogen atoms occupy small holes within the metallic structure.

11. Which one of the following hydrides is an ionic hydride?

    A. NaH B. H_2O C. HCl D. NH_3

12. How many moles of hydrogen would be produced if 17.0 g of potassium reacted with an excess of water?

    A. 0.868   B. 0.448   C. 0.217   D. 0.218

13. With which of the following elements would hydrogen act as a reducing agent?

    A. $Li(s)$ B. $Ca(s)$ C. $Na(s)$ D. $F_2(g)$

Answers to review questions are at the end of this study guide chapter.

20.4 The Group 2A Elements

The following questions and examples will help you review the material in this section.

1. Why are the elements in this group called **alkaline earth** metals?
2. Why does beryllium display very different properties from the rest of the Group 2A elements?

Example 20.4 A   Reactions of Group 2A Elements

Based on the information in Table 20.7 in your textbook, predict the products of the following reactions:

    a. $Sr(s) + 2H_2O(l) \rightarrow ?$
    b. $CaO(s) + H_2O(l) \rightarrow ?$
    c. $Be(s) + 2H_2O(l) \rightarrow ?$

Solution

    a. $Sr(s) + 2H_2O(l) \rightarrow Sr^{2+}(aq) + 2OH^{1}(aq) + H_2(g)$
    b. $CaO(s) + H_2O(g) \rightarrow Ca^{2+}(aq) + 2OH^-(aq)$
    c. $Be(s) + 2H_2O(l) \rightarrow$ No observable reaction!

3.   List some practical uses of alkaline earth metals.

Example 20.4 B   Practice with Group 2A Reactions

Based on the information in Table 20.7 in your textbook, predict the **reactants** given the following **products**.

    a. $? + ? \rightarrow Ca_3N_2(s)$
    b. $? + ? \rightarrow 2RaO(s)$
    c. $? + ? + ? \rightarrow Be(OH)_4{}^{2-}(aq) + H_2(g)$
    d. $? + ? \rightarrow SrH_2(s)$

Solution

    a. $\mathbf{3Ca(s) + N_2(g) \rightarrow Ca_3N_2(s)}$
    b. $\mathbf{2Ra(s) + O_2(g) \rightarrow 2RaO(s)}$
    c. $\mathbf{Be(s) + 2OH^-(aq) + 2H_2O(l) \rightarrow Be(OH)_4{}^{2-}(aq) + H_2(g)}$
    d. $\mathbf{Sr(s) + H_2(g) \rightarrow SrH_2(s)}$

4.   What is "hard water"?

5.   Describe the **ion exchange method** for softening hard water.

6.   Which one of the following elements does not react vigorously with water?

    A. Be   B. Ca   C. Sr   D. Ba

7. Sr_3N_2 can be produced by reacting which one of the following pairs of reagents?

    A. $Sr(s)$ with HNO_3 B. $Sr(s)$ with N_2 C. $Sr(OH)_2$ with N_2 D. $SrCO_3$ with HNO_3

8.   Which of the following is one use of magnesium?

    A. It is necessary for muscle function.   C. It is used in producing fluorescent light.
    B. It is necessary for renal function.   D. It is used in softening hard water.

Answers to review questions are at the end of this study guide chapter.

20.5   The Group 3A Elements

The following questions and examples will help you review the material in this section.

1. Why does boron exhibit different chemical behavior than the other elements in this group?
2. What is the simplest stable borane?
3. Describe the bonding in boranes.
4. Why are boranes highly reactive?
5. Why is aluminum amphoteric?
6. Why is gallium so useful in thermometers that measure a wide range of temperatures?

Example 20.5   Reactions

Predict the products of each of the following reactions:

   a. $2Al(s) + N_2(g) \rightarrow$?
   b. $2Ga(s) + 2OH^-(aq) + 6H_2O(l) \rightarrow$?
   c. $2In(s) + 6H^+(aq) \rightarrow$?
   d. $4Tl(s) + 3O_2(g) \rightarrow$? (low temperatures)

Solution

   a. $2Al(s) + N_2(g) \rightarrow 2AlN(s)$
   b. $2Ga(s) + 2OH^-(aq) + 6H_2O(l) \rightarrow 2Ga(OH)_4^-(aq) + 3H_2(g)$
   c. $2In(s) + 6H^+(aq) \rightarrow 2In^{3+}(aq) + 3H_2(g)$
   d. $4Tl(s) + 3O_2(g) \rightarrow 2Tl_2O_3(s)$ (low temperatures)

7. Which one of the following statements about aluminum is incorrect?

   A.   It bonds covalently to nonmetals.
   B.   Its +3 oxide is amphoteric.
   C.   It reacts with chlorine to form $AlCl_2$.
   D.   It has high electrical conductivity, like metals.

8. Which one of the following oxides is amphoteric?

   A. Tl_2O_3 B. Ga_2O_3 C. AlO_2 D. GaO_2

Answers to review questions are at the end of this study guide chapter.

20.6   The Group 4A Elements

The following questions and examples will help you review the material in this section.

Example 20.6 A   Electron Configuration

Write shorthand electron configurations for each of the following Group 4A elements:

   a.   carbon   b.   germanium   c.   tin

Solution

a. carbon: [He] $2s^2 2p^2$
b. germanium: [Ar] $4s^2 3d^{10} 4p^2$
c. tin: [Kr] $5s^2 4d^{10} 5p^2$

Notice that Group 4A elements have $\boldsymbol{ns^2 np^2}$ electron configurations.

1. What is the hybridization on the central atom in molecules such as $PbCl_4$ and $GeBr_4$?
2. Why can't carbon form a compound such as CCl_6^{2-} while tin can form $SnCl_6^{2-}$?
3. Why can carbon form π bonds with other **Period 2** nonmetals while other Group 4A elements cannot?
4. Why, with regard to bond energies, does the silicon to oxygen bond dominate silicon chemistry rather than a silicon to silicon bond?
5. Define **allotrope**.
6. What are the three allotropic forms of carbon?
7. What is the major industrial use of silicon? Germanium?
8. Under what conditions is each of the allotropes of tin stable?
9. What is **tin disease**?
10. What is the common source of lead?
11. Your textbook says that lead may have contributed to the demise of the Roman civilization. Why?
12. What has been a significant cause of lead poisoning in the twentieth century?
13. Why are there very few known lead(IV) compounds?

Example 20.6 B   Reactions of Group 4A Elements

Predict the products of each of the following reactions:

a. $Sn(s) + 2H^+(aq) \rightarrow$?
b. $Ge(s) + O_2(g) \rightarrow$?
c. $Pb(s) + Cl_2(g) \rightarrow$?

Solution

a. $Sn(s) + 2H^+(aq) \rightarrow Sn^{2+}(aq) + H_2(g)$
b. $Ge(s) + O_2(g) \rightarrow GeO_2(s)$
c. $Pb(s) + Cl_2(g) \rightarrow PbCl_2(s)$ (note that Pb is the only Group 4A element that forms the dihalide. Others form the tetrahalide. MX_4.)

14. Which one of the following statements about the elements of Group IVA is incorrect?

A. All of these elements can form covalent compounds with nonmetals.
B. All these elements except carbon can behave as Lewis acids.
C. Only carbon and silicon can form stable compounds with sp^2 hybridization.
D. Tin and lead are the two metallic elements of the group.

Answers to review questions are at the end of this study guide chapter.

20.7   The Group 5A Elements

The following questions and examples will help you review the chemistry of Group 5A elements.

Example 20.7 A   Electron Configurations

Give the shorthand configuration for each of the following elements:

    a.   nitrogen   b.   arsenic   c.   bismuth

Solution

    a.   nitrogen: **[He]** $2s^2 2p^3$
    b.   arsenic: **[Ar]** $4s^2 3d^{10} 4p^3$
    c.   bismuth: **[Xe]** $6s^2 4f^{14} 5d^{10} 6p^3$

Notice the $ns^2 np^3$ configuration on each of these Group 5A elements.

1.   Why are there no known ionic compounds containing bismuth(V) or antimony(V)?
2.   Justify the fact that NH_3 and PH_3 act as Lewis bases.
3.   What is the hybridization of As in $AsCl_3$?
4.   Why can't nitrogen form compounds with five ligands?
5.   Discuss the structure of compounds such as PCl_5.

6.   Which one of the following compounds can act as a Lewis base?

    A. BiF_5 B. $SbCl_5$ C. $AsCl_3$ D. PF_5

Example 20.7 B   Reactions of Group 5A Elements

Predict the products of the following reactions:

    a. $P_4O_{10}(s) + 10C(s) \rightarrow ?$
    b. $2Ca_3(PO_4)_2(s) + 6SiO_2(s) \rightarrow ?$ (decomposition)

Solution

    a. $P_4O_{10}(s) + 10C(s) \rightarrow 4P(s) + 10CO(g)$
    b. $2Ca_3(PO_4)_2(s) + 6SiO_2(s) \rightarrow 6CaSiO_3(s) + P_4O_{10}(s)$

Answers to review questions are at the end of this study guide chapter.

20.8   The Chemistry of Nitrogen

The following questions and examples will help you review the material in this section.

1.   Why is N_2 such a stable molecule?
2.   Why is nitrogen used as an atmosphere for reactants that normally react with water or oxygen?

Example 20.8 A   Thermodynamics of Nitrogen Compounds

Using data from <u>Appendix 4 in your textbook</u>, calculate $\Delta S°$ and $\Delta G°$ for each of the following reactions.

a. $N_2O(g) \rightarrow N_2(g) + \frac{1}{2}O_2(g)$
b. $NO_2(g) \rightarrow \frac{1}{2}N_2(g) + O_2(g)$
c. $NH_3(g) \rightarrow \frac{1}{2}N_2(g) + \frac{3}{2}H_2(g)$

Solution

Recall that entropy and free energy are both **state functions**.

$$\Delta S°_{reaction} = \Sigma S°_{products} - \Sigma S°_{reactants}$$
$$\Delta G°_{reaction} = \Sigma \Delta G°_{f(products)} - \Sigma \Delta G°_{f(reactants)}$$

For entropy,

a. $\Delta S° = [\Delta S°_{N_2(g)} + \frac{1}{2}\Delta S°_{O_2(g)}] - \Delta S°_{N_2O(g)} = [192\ \text{J/K mol} + \frac{1}{2}(205\ \text{J/K mol})] - [220\ \text{J/K mol}]$
   $\Delta S°_{reaction} = \textbf{74.5 J/K mol}$

b. $\Delta S° = [\frac{1}{2}\Delta S°_{N_2(g)} + \Delta S°_{O_2(g)}] - \Delta S°_{NO_2(g)} = [\frac{1}{2}(192\ \text{J/K mol}) + 205\ \text{J/K mol}] - [240\ \text{J/K mol}]$
   $\Delta S°_{reaction} = \textbf{61 J/K mol}$

c. $\Delta S° = [\frac{1}{2}\Delta S°_{N_2(g)} + \frac{3}{2}\Delta S°_{H_2(g)}] - \Delta S°_{NH_3(g)} = [\frac{1}{2}(192\ \text{J/K mol}) + \frac{3}{2}(131\ \text{J/K mol})] - [193\ \text{J/K mol}]$
   $\Delta S°_{reaction} = \textbf{100 J/K mol}$

For free energy, recall that $\Delta G°$ of elements in their standard states = 0.

a. $\Delta G° = 0 - \Delta G°_{N_2O(g)} = \textbf{−104 kJ/mol}$
b. $\Delta G° = 0 - \Delta G°_{NO_2(g)} = \textbf{−52 kJ/mol}$
c. $\Delta G° = 0 - \Delta G°_{NH_3(g)} = \textbf{+17 kJ/mol}$

Notice that $\Delta G = \Delta H - T\Delta S$ will serve as a double check in each case!

3.   Why are nitrogen-based explosives so effective (with regard to thermodynamics and gas volume)?
4.   Why does your textbook say that the thermodynamics and kinetics of the Haber process are in opposition?
5.   What are the conditions under which the Haber process is performed?
6.   Define **nitrogen fixation**.
7.   How does nitrogen fixation occur in an automobile?
8.   List two natural mechanisms of nitrogen fixation.
9.   Why is there such an interest in **nitrogen-fixing bacteria**?
10.   What is **denitrification**?
11.   What is the problem with accumulating excess nitrogen in soil and bodies of water?

Nitrogen Hydrides

12.   Why does ammonia have a much lower boiling point than water?
13.   What is the structure of hydrazine?
14.   Why is hydrazine a useful chemical in the space program?

Example 20.8 B   *Hydrazine as a Propellant*

The reaction between the rocket fuel monomethylhydrazine and the oxidizer dinitrogen tetroxide is:

$$5N_2O_4(g) + 4N_2H_3(CH_3)(g) \rightarrow 12H_2O(g) + 9N_2(g) + 4CO_2(g)$$

How many liters of carbon dioxide are formed at 1.00 atm and 25°C from the reaction of 3.00 grams of N_2O_4 with 5.00 g of $N_2H_3(CH_3)$?

Strategy

This problem combines a limiting reactant calculation with a gas law problem. The general strategy is

- Determine how much product is formed by each reactant.
- Determine how much product is formed by the limiting reactant.
- Convert moles of product to liters of product using the ideal gas law.

Solution

$$\text{mol } CO_2 \text{ from } N_2O_4 = 3.00 \text{ g } N_2O_4 \times \frac{1 \text{ mol } N_2O_4}{92.0 \text{ g } N_2O_4} \times \frac{4 \text{ mol } CO_2}{5 \text{ mol } N_2O_4} = \mathbf{0.0261 \text{ mol } CO_2}$$

$$\text{mol } CO_2 \text{ from } N_2H_3(CH_3) = 5.00 \text{ g } N_2H_3(CH_3) \times \frac{1 \text{ mol } N_2H_3(CH_3)}{46.0 \text{ g } N_2H_3(CH_3)} \times \frac{4 \text{ mol } CO_2}{4 \text{ mol } N_2H_3(CH_3)}$$

$$= 0.109 \text{ mol } CO_2$$

N_2O_4 is the limiting reactant, and **0.0261 mol CO_2 is formed**.

$$V_{CO_2} = \frac{nRT}{P} = \frac{(0.0261 \text{ mol})(0.08206 \text{ L atm/K mol})(298.2 \text{ K})}{1.00 \text{ atm}}$$

$$= \mathbf{0.639 \text{ L } CO_2 \text{ formed}}$$

15. List two uses of hydrazine in addition to that of a rocket propellant.

Nitrogen Oxides

16. List the formula, and determine the oxidation state of the nitrogen in each nitrogen oxide listed in <u>Table 20.14 in your textbook</u>.
17. List several uses of nitrous oxide.
18. What is the role of nitrous oxide in the Earth's climate control?
19. How is nitric oxide prepared?
20. Why does NO turn brown in air?
21. Why does NO^+ have a higher bond energy than NO?
22. What happens to NO_2 at low temperatures?

Oxyacids of Nitrogen

23. List some uses of nitric acid.
24. Give the reactions involved in the **Ostwald process**.
25. How can the concentration of nitric acid be increased from 68% to 95%?
26. Why does a nitric acid solution turn yellow upon constant exposure to sunlight?
27. How is nitrous acid prepared?

Multiple Choice Questions:

28. 6 g of N_2O_5 is dissolved in enough water to prepare 700. mL of solution. Calculate the molarity of the nitric acid solution produced.

    A. 0.159   B. 1.26   C. 0.630   D. 0.252

29. 5.0 g of NO_2 are produced in the atmosphere and then dissolved in 500.0 L of rainwater. What is the pH of the rain, assuming no volume change due to addition?

    A. 3.96   B. 7.00   C. 3.48   D. 6.50

30. Dynamite is produced by the absorption of which of the following substances in porous silica?

    A. $C_7H_5N_3O_7$ B. $C_3H_5N_3O_9$ C. $C_9H_4N_2O_8$ D. $C_7H_5N_3$

31. Which of the following is an example of nitrogen fixation?

    A. Production of N_2
    B. Absorption of N_2 and its transformation into elemental nitrogen
    C. Absorption of nitric acid and its transformation into N_2
    D. Absorption of N_2 and its transformation into NH_3

32. In the Haber process, a high temperature is required to:

    A. Increase the equilibrium constant of the reaction.
    B. Decrease the equilibrium constant of the reverse reaction.
    C. Help break hydrogen gas molecules into elemental hydrogen.
    D. Help speed the rate of the reaction by breaking nitrogen molecules.

33. Hydrazine can be used as a rocket propellant because:

    A. It is a very powerful oxidizing agent and therefore can generate large amounts of energy.
    B. It requires a large amount of heat to start the propellant reaction, thus it is safe.
    C. It is a very powerful reducing agent and therefore can generate large amounts of energy.
    D. It does not readily react with oxygen, thus it is safe in handling.

Answers to review questions are at the end of this study guide chapter.

20.9   The Chemistry of Phosphorus

The following questions and the example will help you review the material in this section.

1. List four reasons for the differences in chemical properties between nitrogen and phosphorus.
2. What are the three allotropic forms of phosphorus? How are they different from one another?
3. How is red phosphorus made?
4. How is black phosphorus obtained from either red or white phosphorus?
5. What is the oxidation state of phosphorus in phosphides? Give some examples of phosphides.
6. Why are phosphonium salts uncommon?
7. Why is phosphine viewed as an exception to the VSEPR model?
8. Discuss the use of tripolyphosphoric acid in detergents.
9. Your textbook says that H_3PO_3 and H_3PO_2 are diprotic and monoprotic, respectively. Why aren't they triprotic?
10. Why isn't naturally occurring phosphorus in soil easily usable by plants?
11. What is the oxidation state of phosphorus in PF_3? PF_5?

12. How many tons of P_4O_{10} are required to produce 900 gallons of 42.5 % (w/w) phosphoric acid? The density of phosphoric acid = 1.689 g/mL.

    A. 71.9   B. 3.1   C. 0.78   D. 0.39

13. Phosphine is a weaker base than ammonia because:

    A. It is less soluble in water.
    B. Phosphorus has a lower affinity for hydrogen atoms.
    C. Phosphorus is more electronegative than nitrogen.
    D. It has a lower affinity for hydrogen nuclei.

Example 20.9   Reactions of Phosphorus

Give the products of each of the following reactions (discussed in this section of your textbook):

    a. $2Na_3P(s) + 6H_2O(l) \rightarrow$?
    b. $P_4O_{10}(s) + 6H_2O(l) \rightarrow$?
    c. $P_4(s) + 3O_2(g; \text{limited}) \rightarrow$?
    d. $P_4(s) + 5O_2(g; \text{excess}) \rightarrow$?
    e. $P_4O_6(s) + 6H_2O(l) \rightarrow$?

Solution

    a. $2Na_3P(s) + 6H_2O(l) \rightarrow 2PH_3(g) + 6Na^+(aq) + 6OH^-(aq)$
    b. $P_4O_{10}(s) + 6H_2O(l) \rightarrow 4H_3PO_4(aq)$
    c. $P_4(s) + 3O_2(g; \text{limited}) \rightarrow P_4O_6(s)$
    d. $P_4(s) + 5O_2(g; \text{excess}) \rightarrow P_4O_{10}(s)$
    e. $P_4O_6(s) + 6H_2O(l) \rightarrow 4H_3PO_3(aq)$

Answers to review questions are at the end of this study guide chapter.

20.10, 20.11 & 20.12 The Chemistry of Group 6A Elements Focusing on Oxygen and Sulfur

The following questions and examples will help you review the material in these sections.

Example 20.10 A Electron Configurations

Write the shorthand configurations for each of the following elements:

    a. oxygen   b. selenium   c. polonium

Solution

    a. oxygen: **[He] $2s^2 2p^4$**
    b. selenium: **[Ar] $4s^2 3d^{10} 4p^4$**
    c. polonium: **[Xe] $6s^2 4f^{14} 5d^{10} 6p^4$**

Note that all Group 6A elements have the configuration $ns^2 np^4$.

1. What is the most common oxidation state of Group 6A elements in ionic compounds?
2. Give some examples of Group 6A covalent hydrides.
3. Why do Group 6A elements (other than oxygen) form compounds with more than eight electrons around the central atom?

Example 20.10 B VSEPR Review

Predict the geometry of TeI_4.

Solution

We learned about the VSEPR model in Chapter 8. Our first step is to draw the Lewis structure.

    1. # valence electrons in the system = 6 for Te + 4 × 7 for I = 34 electrons.
    2. # electrons if happy = 8 for Te + 4 × 8 for I = 40 electrons.
    3. # bonds = (40 − 34)/2 = 3 bonds.

This is an exception! We must therefore draw the Lewis structure and put extra electron pairs around the central atom. (We have omitted electrons around the iodine atoms for clarity.)

There are five effective electron pairs around the central atom. The structure is **based on a trigonal bipyramid**. The lone pair will be in the equatorial plane. This is a **seesaw** structure.

4. Why don't Group 6A elements form +6 cations?
5. Why has there been growing interest in the chemistry of selenium?
6. Why might polonium be related to cancer in smokers?
7. What are some of the important uses of oxygen in our world?
8. What percent of the Earth's atmosphere is oxygen?
9. How is oxygen isolated from air?
10. How can we demonstrate the paramagnetism of oxygen?
11. What is the VSEPR geometry of ozone?
12. Why is the bond angle in ozone less than 120°?
13. How is ozone prepared in the laboratory?
14. Why might ozone be useful in municipal water purification?
15. What is the importance of the ozone layer of the Earth's atmosphere?
16. Name some minerals that contain sulfur.
17. Describe the **Frasch Process**.
18. Why does elemental oxygen exist as O_2 whereas sulfur exists as S_8?
19. What is the structural difference between rhombic and monoclinic sulfur?
20. Why is O_2 more stable than SO?
21. What role might dust play in the conversion of SO_2 to SO_3?

22. Which one of the following electron configurations does Se^{2-} have?

    A. [Ne]   B. [Kr]   C. [Xe]   D. [Rn]

23. Which one of the following species cannot exist?

    A. O^{2+} B. Te^{2+} C. TeH_2 D. Po^{4+}

24. How many isotopes of polonium are known?

    A. 12   B. 28   C. 37   D. 27

25.   Which of the following statements about selenium is not true?

    A. It is important in vitamin E synthesis.
    B. It provides protection against some types of cancers.
    C. It is not toxic.
    D. Its deficiency in the body can lead to congestive heart failure.

26.   Which one of the following is an action of ozone?

    A. It is an eye irritant.
    B. It can be used to disinfect water.
    C. It blocks gamma radiation from the sun.
    D. It causes plastic materials to become brittle.

27.   Which one of the following is not a known oxoanion of sulfur?

    A. SO^{2-} B. SO_3^{2-} C. $S_2O_3^{2-}$ D. $S_4O_6^{2-}$

28.   Which one of the following is not an oxidation state of sulfur?

    A. 0   B. +6   C. +4   D. +8

29.   Which one of the following is not true about sulfur compounds?

    A. Sulfuric acid has a very high affinity for water.
    B. SO is the most stable sulfur oxide.
    C. Sulfuric acid is a strong oxidizing agent.
    D. Thiosulfate ion is used in photography to form a complex with silver ions.

Example 20.12   Reactions of Sulfur

Based on the material in this section, predict the products of each of the following reactions:

    a. $2S_2O_3^{2-}(aq) + I_2(aq) \rightarrow$?
    b. $C_{12}H_{22}O_{11}(s) + 11H_2SO_4(conc) \rightarrow$?
    c. $SO_2(aq) + H_2O(l) \rightarrow$?
    d. $2SO_2(g) + O_2(g) \rightarrow$?

Solution

    a. $2S_2O_3^{2-}(aq) + I_2(aq) \rightarrow S_4O_6^{2-}(aq) + 2I^-(aq)$
    b. $C_{12}H_{22}O_{11}(s) + 11H_2SO_4(conc) \rightarrow 12C(s) + 11H_2SO_4{\cdot}H_2O(l)$
    c. $SO_2(aq) + H_2O(l) \rightarrow H_2SO_3(aq)$
    d. $2SO_2(g) + O_2(g) \rightarrow 2SO_3(g)$

Answers to review questions are at the end of this study guide chapter.

20.13 and 20.14   The Group 7A and 8A Elements

The following questions and examples will help you review the material in these sections.

Example 20.13 A Electron Configurations

Write the shorthand configurations for each of the following elements:

    a. bromine   b. iodine

Solution

a.   bromine: [Ar] $4s^2 3d^{10} 4p^5$
b.   iodine: [Kr] $5s^2 4d^{10} 5p^5$

Note that the halogens have the configuration $ns^2 np^5$.

Example 20.13 B The Decomposition of Astatine

Astatine has been used for cancer therapy because of its short half-life. The longest-lived isotope, ^{210}At, has $t_{1/2} = 8.3$ hours. If you have a 1.000-g sample of ^{210}At, how much will remain after 24.0 hours? (The radioactive decomposition of astatine obeys a **first-order** kinetic rate law.)

Solution

For a first-order decay, the rate constant

$$k = 0.693/t_{1/2} = 0.693/8.3 \text{ hr} = \mathbf{8.35 \times 10^{-2} \ hr^{-1}}$$

$$\ln\left(\frac{N}{N_0}\right) = -kt$$

where $N_0 = 1.000$ g
        $N = ?$
        $k = 8.35 \text{ H } 10^{-2} \text{ hr}^{-1}$
        $t = 24.0$ hr

$$\ln\left(\frac{N}{1.000}\right) = -8.35 \times 10^{-2} \text{ hr}^{-1} (24.0 \text{ hr})$$

$$\ln\left(\frac{N}{1.000}\right) = -2.00$$

Taking the antilog of both sides,

$$\frac{N}{1.000} = 0.135$$

$N = 0.135$ g of astatine remaining at time t

You can see that after just one day, the majority of a sample of astatine will decompose.

1.   What kind of bonds do halogens tend to form with nonmetals? Metals in lower oxidation states?
2.   What about bonding with metals in higher oxidation states?
3.   Why does HF have such a high boiling point relative to other hydrogen halides?
4.   Why can't the relative strength of hydrogen halides as acids be assessed in water?
5.   What is the order of acid strength of hydrogen halides in acetic acid?
6.   What is the rationale for HF being the least acidic hydrogen halide?
7.   Why is hydrochloric acid such an important industrial chemical?
8.   Why must $HOClO_3$ be handled so carefully?
9.   Define **"disproportionation reaction."**
10. List some uses of chlorate salts.
11. Why is OF_2 called oxygen difluoride rather than difluorine oxide?

Example 20.13 C Reactions of Halogens

Predict the products of the following reactions:

a. $H_2(g) + Cl_2(g) \xrightarrow{\text{U. V. light}} ?$
b. $SiO_2(s) + 4HF(aq) \rightarrow ?$
c. $4F_2(g) + 3H_2O(l) \xrightarrow{\text{NaOH}} ?$
d. $Cl_2(aq) + H_2O(l) \rightleftharpoons ?$

Solution

a. $H_2(g) + Cl_2(g) \xrightarrow{\text{U. V. light}} 2HCl(g)$
b. $SiO_2(s) + 4HF(aq) \rightarrow SiF_4(g) + 2H_2O(l)$
c. $4F_2(g) + 3H_2O(l) \xrightarrow{\text{NaOH}} 6HF(aq) + OF_2(g) + O_2(g)$
d. $Cl_2(aq) + H_2O(l) \rightleftharpoons HOCl(aq) + H^+(aq) + Cl^-(aq)$

12. What is the major source of helium on earth?
13. What are the major uses of helium, neon, and argon?
14. What are some xenon and krypton compounds that have been prepared?

15. Which one of the following is the strongest acid?

A. HF   B. HCl   D. HBr   D. HI

16. HF is a weak acid, because:

A. It has a low enthalpy of hydration.
B. F^- has a much higher entropy factor than other halide anions.
C. HF has a weak bond.
D. The enthalpy of hydration opposes dissociation of HF.

17. Which one of the halogens does not form any known oxoacids?

A. Chlorine   B. Fluorine   C. Iodine   D. Astatine

18. Which of the following oxidation states is present in any of the chlorines in the following reaction?

$$Cl_2(aq) + H_2O(l) \rightarrow HOCl(aq) + H^+(aq) + Cl^-(aq)$$

A. 0   B. !2   C. !5   D. +2

19. Which one of the following interhalogen compounds would probably not exist?

A. ClF_5 B. FCl_5 C. IF_7 D. ClF

20. The nucleus of which one of the following elements forms the alpha particle?

A. He   B. Ne   C. Kr   D. Xe

21. Which one of the following is not a possible compound of xenon?

A. XeO_6 B. XeO_3F_2 C. XeO_3 D. XeO_2F_4

Answers to review questions are at the end of this study guide chapter.

Answers to Review Questions

Section 20.2

1. The "ns^1" configuration is common to each of the elements in this group.
2. This illustrates that reaction rate and thermodynamics are not directly related.
3. They react so well together because they lose electrons to form M^+ ions.
4. D   5. C

Section 20.3

1. Hydrogen is a nonmetal that is very small and can either gain or lose an electron, but does so covalently rather than ionically.
2. Hydrogen has low boiling and melting points because of its low molecular weight and nonpolarity.
3. $2H_2(g) + O_2(g) \rightarrow 2H_2O(l)$
4. $CH_4(g) + H_2O(g) \rightarrow CO(g) + 3H_2(g)$
5. It isn't very practical because of the high cost of electricity.
6. Hydrogen is used in the production of ammonia and in hydrogenating unsaturated vegetable oil.
7. **hydride:** Binary compounds containing hydrogen.
   ionic hydride: Hydrogen combines with Group 1A or 2A metals.
   covalent hydride: Hydrogen combines with nonmetals.
   interstitial hydride: Hydrogen atoms occupy holes ("interstices") in a metal's crystal structure.
8. Hydrogen diffuses through the palladium metal wall (see "interstitial hydride") leaving impurities behind.
9. Interstitial hydrides lose absorbed ("stored") hydrogen when heated.
10. B   11. A   12. D   13. D

Section 20.4

1. They are called **alkaline earth** metals because their oxides form bases in water.
2. Beryllium displays different properties because of its relatively small size and high electronegativity.
3. For example, magnesium is used in flash bulbs and in structural materials.
4. Hard water is that which contains large concentrations of Ca^{2+} and Mg^{2+} ions.
5. In a nutshell, hard water cations become bound to sites on a polymer resin which releases sodium ions as a result, thus softening the water.
6. A   7. B   8. A

Section 20.5

1. Boron is very small and has a relatively high electronegativity, thus forming mostly covalent compounds.
2. B_2H_6 is the simplest stable borane.
3. Boranes have one or more three-center two-electron bonds similar to that in $BeCl_2$. The remaining bonds are normal covalent bonds.
4. As with beryllium, boranes are highly electron-deficient, which accounts for their reactivity.
5. Covalent bonding is responsible for the amphoteric nature of aluminum.
6. Gallium is useful in thermometers because, of all the elements, it has the widest temperature range in which it is a liquid.
7. C   8. B

Section 20.6

1. The central atom is sp^3 hybridized.
2. There are no d electrons available in carbon to allow it to exceed an octet.
3. Carbon p orbitals have significant overlap with other **similar-size** Period 2 elements. Other Group 4A elements cannot have this degree of overlap.
4. The bond energy of the Si−O bond is substantially higher than that of the Si−Si bond.
5. An allotrope has the same element but different structures i.e., S, S_2 and S_8, or O_2 and O_3.

6. Graphite, buckminsterfullerene, and diamond are allotropic forms of carbon.
7. The major use of silicon is in semiconductors. The major use of germanium is also for semiconductors.
8. White tin is stable at ambient temperatures. Gray tin is stable below 13.2°C. Brittle tin is stable above 161°C.
9. Tin disease is the conversion of white tin to powdery gray tin in cold temperatures.
10. Galena (PbS) is the common source of lead.
11. High concentrations of lead are toxic. Significant levels of lead have been found in bones from the Roman era.
12. Lead in gasoline is a significant contributor to lead poisoning.
13. Lead(IV) reduces to lead(II) in the presence of bromide or iodide ions, as well as at temperatures above 100°C.
14. C

Section 20.7

1. Too much energy is required to remove all five electrons from the neutral atom.
2. NH_3 and PH_3 have lone pairs that they can donate.
3. The hybridization is sp^3.
4. Nitrogen has no available d orbitals, so it cannot exceed the octet.
5. Bonding in PCl_5 is an exception to the octet rule. It is trigonal bipyramidal hybridization.
6. C

Section 20.8

1. N_2 is very stable because it has a very high bond strength.
2. Nitrogen is used as an "inert atmosphere" because it is so unreactive.
3. Nitrogen-based explosives are effective because they produce large volumes of gas, and the decomposition to N_2 (gas) is a thermodynamically favorable process.
4. In the Haber process, the equilibrium lies in the direction of NH_3 at room temperature. Because the reaction is exothermic, increasing the temperature to increase the reaction rate decreases the equilibrium constant.
5. High pressure and moderately high temperature.
6. The process of transforming N_2 to other nitrogen-containing compounds.
7. Nitrogen from the air reacts with oxygen to form NO and then NO_2.
8. They are lightning and nitrogen-fixing bacteria.
9. Such bacteria produce ammonia in the soil at standard conditions.
10. Denitrification is the return of nitrogen-containing compounds to the atmosphere as N_2 gas.
11. Algae and other undesirable organisms may accumulate because of the presence of too much fixed nitrogen.
12. Water contains two polar bonds and two lone pairs. It can exhibit much more substantial hydrogen bonding than ammonia.
13. See Figure 20.14 in your textbook.
14. Hydrazine is an excellent reducing agent that produces a great deal of energy when reacted with oxygen.
15. It is used as a blowing agent in plastics and is also used in the production of agricultural pesticides.
16. See Table 20.14 in your textbook.
17. It is used by dentists as an anesthetic. It is also used as a propellant in aerosol cans of whipped cream.
18. It strongly absorbs infrared radiation, thus helping with climate control.
19. NO is prepared by the reaction of copper with dilute (6 M) nitric acid.
20. NO is oxidized in air to brown NO_2.
21. The M.O. bond order for NO^+ is 3 and for NO is 2.5. Therefore, NO^+ has a higher bond energy than NO.
22. NO_2 can dimerize to form N_2O_4.
23. It is used in the manufacture of fertilizer and explosives, among other things.
24. a. $4NH_3(g) + 5O_2(g) \rightarrow 4NO(g) + 6H_2O(g)$
    b. $2NO(g) + O_2(g) \rightarrow 2NO_2(g)$
    c. $3NO_2(g) + H_2O(l) \rightarrow 2HNO_3(aq) + NO(g)$

25. The concentration is increased by treatment with sulfuric acid, a dehydrating agent.
26. HNO_3 decomposes in sunlight to give $NO_2(g)$, a brown gas that colors the solution yellow.
27. It is prepared by acidifying the nitrite ion, which is made by mixing NO, NO_2, and OH^- ion.
28. A   29. A   30. B   31. D   32. D   33. C

Section 20.9

1. a. Nitrogen can form stronger bonds.   c. Phosphorus is larger.
    b. Nitrogen is more electronegative.   d. Phosphorus has available d orbitals.
2. White, red and black phosphorus differ in structure as outlined in <u>Figure 20.18 in your textbook</u>.
3. Red phosphorus is formed by heating white phosphorus in the absence of air.
4. Black phosphorus is made by heating white or red phosphorus at high pressures.
5. The oxidation state is -3. Examples are Na_3P and Ca_3P_2.
6. Phosphonium salts are uncommon because phosphine is a very weak base.
7. The bond angle is far smaller ($94°$) than expected for a tetrahedral structure with one lone pair.
8. The sodium salt of tripolyphosphoric acid is used in detergents because its anion forms complexes with Ca^{2+} and Mg^{2+}, which otherwise interfere with detergent action.
9. They are not triprotic because hydrogens attached to phosphorus are not acidic.
10. Phosphorus in soil is often present in insoluble (and thus unusable) minerals.
11. The oxidation state is $+3$ in PF_3 and $+5$ in PF_5.
12. A   13. D

Section 20.10, 20.11 and 20.12

1. The most common oxidation state is -2.
2. Examples are H_2S and H_2O.
3. Other than oxygen, all Group 6A elements have available d orbitals.
4. It takes far too much energy to remove six electrons from these elements.
5. It is possible that selenium has a role in cancer protection.
6. Polonium is found in tobacco and is an α emitter.
7. In general, we need oxygen for respiration and for combustion.
8. Oxygen makes up 21% of the Earth's atmosphere.
9. Nitrogen distillation isolates oxygen.
10. We can pour liquid oxygen between poles of a magnet.
11. The VSEPR structure is trigonal planar.
12. The lone pair requires more room than bonded pairs.
13. Ozone is prepared by passing an electric charge through pure oxygen gas.
14. Ozone is a strong oxidant.
15. Ozone absorbs ultraviolet radiation.
16. Galena, cinnabar, pyrite, and gypsum, among others.
17. The Frasch process is used to obtain sulfur from underground deposits. Superheated water is pumped in to melt sulfur, which is recovered by air pressure.
18. Oxygen can form π-bonds, thus stabilizing O_2. Sulfur can only form σ-bonds, thus stabilizing the large form S_8.
19. The difference has to do with the way the rings are stacked.
20. O_2 has stronger bonding than SO.
21. Dust acts as a catalyst for the conversion.
22. B   23. A   24. D   25. C   26. B   27. A   28. D   29. B

Section 20.13 and 20.14

1. Halogens form polar covalent bonds with nonmetals and ionic compounds with metals in lower oxidation states.
2. Bonds with metals in higher oxidation states tend to be polar covalent.
3. HF can hydrogen-bond (due to the small size and high electronegativity of fluorine).
4. Hydrogen halides (except HF) completely dissociate in water, so they appear "equally strong."
5. HI > HBr > HCl >> HF
6. The H−F bond strength is higher than the other H−X acids.
7. HCl is used for, among other things, cleaning steel.
8. $HOClO_3$ is an incredibly strong oxidizing agent that reacts explosively with many organic compounds.
9. A disproportionation reaction occurs when an element is both oxidized and reduced in the same reaction.
10. Among the uses are as weed killers and as oxidizers in fireworks.
11. Fluorine is the anion. IUPAC nomenclature rules dictate that it gets the "ide" ending.
12. The major sources are natural gas deposits.
13. Helium is used as a coolant, among other things; neon and argon are used in lighting.
14. Examples are $XePtF_6$, XeF_4, KrF_4, and KrF_2.
15. D   16. D   17. B   18. A   19. B   20. A   21. A

Chapter 21

Transition Metals and Coordination Chemistry

Transition metals show a wonderful and remarkable variety of bonding properties that make them useful to our bodies and our world. This chapter explores how and why they act as they do.

21.1   The Transition Metals: A Survey

When you finish this section you will be able to:

- Write electron configurations for first-row transition metal ions.
- Analyze trends in periodic properties of transition metals.

This section begins by noting that transition metal chemistry shows **much greater consistency** than that of the representative elements as we go across the periodic table. This is because the filling of transition metals involves *d*- or *f*-block electrons, which are core electrons. These do not participate as easily in bonding as *s* and *p* electrons. Read the information on General Properties in your textbook and then try the following example.

Example 21.1 A   Properties of Transition Metals

Answer the following questions regarding properties of first-row transition metals.

1. List the properties that they have in common.
2. What is the trend with regard to

   a. density?
   b. electrical conductivity?
   c. atomic radius?
   d. common oxidation states?

Solution

1. Properties in common: metallic luster, high electrical conductivity, high thermal conductivity.
2. a. The density shows a steady increase with the exception of zinc.
    b. In general, electrical conductivity increases.
    c. Atomic radius decreases, then increases after iron. The key argument here is the balance between increased nuclear charge and electron-electron repulsion. In any case the radii are not drastically different across the period.
    d. Vanadium, chromium, and manganese have many common oxidation states due to the availability of s and d electrons. Atoms such as nickel, copper, and zinc have available only s electrons (it would require too much energy to strip d electrons away), so these elements have fewer oxidation states available.

You learned about electron configurations in Chapter 7. In this section, your textbook points out that where exceptions to filling rules occur (such as chromium, which is $4s^1 3d^5$) it is because the **4s and 3d levels are close together**, and electron-electron repulsions can be minimized by this unusual configuration.

However, your textbook also points out that, **IN IONS**, the energy of the $3d$ level is significantly lower than that of the $4s$ level. This means that the $3d$ energy level may be occupied while the $4s$ will be unoccupied for the first-row transition metal ions.

Example 21.1 B   Electron Configurations of Ions

Write the shorthand electron configurations for each of the following:

     a. V   b. V^{2+} c. V^{3+} d. V^{5+}

Solution

     a. V: [Ar] $4s^2 3d^3$ c. V^{3+}: [Ar] $3d^2$
     b. V^{2+}: [Ar] $3d^3$ d. V^{5+}: [Ar]

The discussion that closes out this section deals with the $4d$ and $5d$ transition series. The **lanthanide contraction** is the slight shrinkage in atomic size that occurs across these periods due to extra nuclear-electron charge attraction as $4f$ or $5f$ electrons are added. The lanthanide contraction also accounts for the similarity in properties between $4d$ and $5d$ elements in a particular group.

21.2   The First-Row Transition Metals

The following questions will help you review your knowledge of the materials in this section.

1. What elements are being discussed? What do they have in common?
2. What is the most common oxidation state for scandium in its compounds?
3. Why are scandium compounds generally colorless and diamagnetic?
4. How is scandium prepared?
5. What is the major industrial use of scandium?
6. What properties make titanium a useful structural material?
7. Why is titanium used for reaction vessels in the chemical industry?
8. List some uses of titanium(IV) oxide.
9. What are the main natural sources of titanium?
10. Write the reactions for the purification of TiO_2 from its ores.
11. What are the major industrial uses of vanadium?
12. How is pure vanadium prepared?

13.   What is vanadium steel?
14.   Why do V^{5+} and V^{4+} exist in solution as VO_2^+ and VO^{2+}?

Example 21.2   A Bit of Detective Work

An aqueous solution of ammonium vanadate (NH_4VO_3) was yellow. It was added to a "Jones Reductor," which is a Zn-Hg amalgam. The solution was swirled over the reductor.

1.   An aliquot of the solution was removed. It was **blue**.
2.   The remaining solution was swirled over the reductor.
3.   Another aliquot was removed. It was **green**.
4.   The remaining solution was swirled over the reductor.
5.   A **final** aliquot was removed. It was **violet**.
6.   Upon standing in air, this **final** aliquot turned **green**.

Describe, in terms of the oxidation states of vanadium, the chemistry that occurred.

Solution

a.   The original solution (yellow) was vanadium(V) (as VO_2^+)
b.   When swirled, the vanadium was reduced to V^{4+} (as VO^{2+}), which is blue.
c.   When further swirled, VO^{2+} was reduced to V^{3+}, which is green.
d.   A final mixing reduced V^{3+} to V^{2+} (violet).
e.   Upon sitting, V^{2+} was air oxidized to V^{3+} (green).

15.   How is chromium prepared?
16.   What are the common oxidation states of chromium?
17.   Write the half-reaction for the reduction of the dichromate ion in acidic solution.
18.   What are the structural differences between chromium(VI) in acidic and basic solutions?
19.   What is "cleaning solution" composed of? What is it used for?
20.   What are the most common uses of manganese?
21.   What are "manganese nodules"?
22.   What are the common oxidation states of manganese?
23.   Why is MnO_4^- such a powerful oxidizing agent?
24.   What are the important oxidation states of iron?
25.   Why are iron(II) solutions light green?
26.   Why are iron(III) solutions yellow?
27.   Why does $Fe(H_2O)_6^{3+}$ behave as an acid in aqueous solution?
28.   What are some commercial uses of cobalt?
29.   What species causes the rose color of cobalt in aqueous solution?
30.   Why is nickel used for plating active metals?
31.   Why is copper so useful?
32.   What is the major use of copper?
33.   What species causes the characteristic blue color of aqueous copper solutions?
34.   What is the main industrial use of zinc?

21.3   Coordination Compounds

When you finish this section you will be able to interchange between the formula and name of coordination compounds.

This section begins with several important definitions:

1. **coordination compound**: It consists of a **complex ion** (a transition metal with attached ligands) and **counter ions** (anions and cations as needed to balance charge).
2. **ligand**: a neutral molecule or ion **having a lone pair** that can be used to form a bond to the central metal ion.
3. **monodentate ligand**: Can form one bond to a metal ion.
4. **bidentate ligand**: Can form two bonds to a metal ion.
5. **polydentate ligand**: Can form more than two bonds to a metal ion.

Table 21.13 in your textbook gives examples of each kind of ligand.

Nomenclature of coordination compounds can get sticky, so let's do a variety of problems together. **Memorize** the Rules for Naming Coordination Compounds in your textbook. Let's try to name $[Cr(H_2O)_4Cl_2]Cl$.

The coordination compound is composed of a **complex ion**. $[Cr(H_2O)_4Cl_2]^+$, which is a cation, and a **counter anion, Cl^-**. According to the rules of nomenclature, the cation (complex ion in this case) will be named first.

Naming the cation: The ligands are named before the metal ion. The two different ligands are **water** ("aqua") and **chlorine** ("chloro"). The ligands are placed **alphabetically,** so **aqua goes before chloro**. This is done before adding any prefixes to the ligands.

<u>aqua</u>   <u>chloro</u>   _____

There are **four** waters ("tetra") and **two** chlorines ("di") so we have

<u>tetraaqua</u>   <u>dichloro</u>   _____

We must determine the oxidation state of the chromium ion. Water is a neutral base and each chlorine has a -1 charge. The chromium is therefore in the $+3$ oxidation state.

tetraaqua   **dichloro**   <u>**chromium(III)**</u>

The complex ion is treated as one word.

tetraaquadichlorochromium(III)

Naming the anion: The anion is chloride. This finishes the naming of the entire compound. Remember the cation and anion are named separately.

tetraaquadichlorochromium(III) chloride

Example 21.3 A   Naming Coordination Compounds

Name each of the following compounds:

a. $K_2[Ni(CN)_4]$
b. $K_4[Fe(CN)_6]$
c. $(NH_4)_2[Fe(H_2O)Cl_5]$
d. $[Co(NH_3)_2(en)_2]Cl_2$
e. $[Ag(NH_3)_2]Cl$

Solution

a. The complex ion is an **anion**. You know this because potassium, the counter ion, must be a cation. Therefore, potassium is listed first. We **don't** say **di**potassium because we can explicitly determine the number of potassiums necessary based on the complex ion charge.

>   **potassium** _____

The complex ion has four cyanide ions ("tetracyano").

>   **potassium tetracyano** _____

The nickel is in the **+2 oxidation state** because 4 CN = −4 and 2 K = +2. We put "ate" on the end of a complex **an**ion.

>   **potassium tetracyanonickelate(II)**

b. The complex ion is again an anion. The cation is named first.

>   **potassium** _____

There are six cyanides in the complex cation ("hexacyano").

>   **potassium hexacyano** _____

The iron ("ferrate") has an oxidation state of +2. (Can you see why?) The compound name is

>   **potassium hexacyanoferrate(II)**

c. The counter ion, ammonium, is a cation. It is named first.

>   **ammonium** _____

The complex ion has one water ("aqua") and five chlorines ("pentachloro"). The iron ("ferrate") is in the +3 oxidation state ($NH_4 = +1 \times 2 = +2$ and $Cl_5 = -5$).

>   **ammonium aquapentachloroferrate(III)**

d. The counter ion, chloride, is an anion. It is named last.

>   _____ **chloride**

The complex ion has two ammonia ligands ("diammine") and two ethylenediamines ("bis(ethylenediamine)"). The cobalt is in the +2 oxidation state (en and NH_3 are neutral).

>   **diamminebis(ethylenediamine)cobalt(II) chloride**

e. Again, the counter ion is an anion, chloride.

>   _____ **chloride**

The complex ion has two NH_3 ("diammine") and silver in the +1 oxidation state.

>   **diamminesilver(I) chloride**

Let's try the reverse procedure.

Example 21.3 B   Writing Coordination Compound Formulas

Write the formulas for each of the following compounds.

>   a. potassium pentacyanocobaltate(II)
>   b. tris(ethylenediamine)nickel(II) sulfate
>   c. potassium dicarbonyltricyanocobaltate(I)

Solution

a. pentacyano = $(CN)_5$ (-1 per CN $\times$ 5 = -5)
    cobaltate(II) = Co (oxidation state = $+2$)

The complex anion is $Co(CN)_5$. The total charge = -3. We must therefore have three potassium ions to balance electrically.

$$K_3[Co(CN)_5]$$

b. tris(ethylenediamine) = $(en)_3$ (neutral Lewis base)
    nickel(II)   = Ni (oxidation state = $+2$)
    sulfate   = SO_4 (-2 charge)

$$[Ni(en)_3]SO_4$$

c. dicarbonyl = $(CO)_2$ (neutral Lewis base)
    tricyano   = $(CN)_3$ (-1 per CN $\times$ 3 = -3)
    cobaltate(I) = Co (oxidation state = $+1$)

The complex anion is $Co(CO)_2(CN)_3$. The total charge = -2. We must therefore have two potassium ions to balance electrically.

$$K_2[Co(CN)_3(CO)_2]$$

21.4   Isomerism

When you finish this section you will be able to define the basic terms and identify examples of isomerism.

There are several new definitions introduced in this section:

1. **isomerism:** Same chemical formula, different compounds.
2. **structural isomerism:** Same atoms, different bond arrangement.
    a. **coordination isomerism:** complex ion and counter ion interchange members.
    b. **linkage isomerism:** point of attachment of ligand to metal is different.
3. **stereoisomerism:** Same bond arrangement, different spatial arrangement.
    a. **geometrical isomers:** atoms or groups of atoms assume different positions around the central atom (*cis*- or *trans*- isomerism).
    b. **optical isomers:** rotate plane polarized light in different directions. Such molecules have chiral centers.

Your textbook gives some examples of each kind of isomerism. After reading the section, try the following example.

Example 21.4   Recognizing Isomers

Identify the type of isomerism exhibited by each pair of substances.

a. *trans*-$[RuCl_2(H_2O)_4]^+$ and *cis*-$[RuCl_2(H_2O)_4]^+$
b. $[Co(en)_2(NH_3)Cl]^{2+}$ exists in one form that can rotate plane-polarized light to the left and one that can rotate it to the right.
c. $[Co(NH_3)_5(NO_2)]^{2+}$ and $[Co(NH_3)_5(ONO)]^{2+}$

Solution

a.   geometrical isomers
b.   optical isomers
c.   linkage isomers

21.5   Bonding in Complex Ions: The Localized Electron Model

Your textbook points out that the VSEPR model fails when considering a coordination number of four, which can mean a tetrahedral <u>or</u> square planar arrangement. The localized electron model can be used to rationalize general bonding schemes, but does not predict key properties of complex ions.

21.6   The Crystal Field Model

When you finish this section you will be able to:

- State the underlying assumptions behind the crystal field model.
- Use the model to assess magnetism and color of complex ions.

The purpose of the crystal field model is to attempt to **explain the magnetism and colors of complex ions**. The underlying assumptions of the model are:

1.   That ligands are **"negative point charges."**
2.   That the bonding between ligands and the central atom is **totally ionic**. That is, electrostatic interactions keep ligands bonded to the central atom. (There are no electrons shared or donated by the ligands.)

Regarding **octahedral complex ions**, your textbook discusses two types of field splitting between t_{2g} and e_g states for the d orbitals on the metal. "Δ" is the splitting energy.

- **low-spin** = strong field = large energy difference ("Δ") = **minimum** number of unpaired electrons. (See <u>Figure 21.22 in your textbook</u>.)
- **high-spin** = weak field = small Δ = **maximum** number of unpaired electrons.

In octahedral complex ions, two things determine the type of "spin state" (low or high).

- The spectrochemical series (given just below <u>Example 21.4 in your textbook</u>).
- The **charge on the metal ion**. (A higher charge causes stronger field-splitting.)

Example 21.6 A Crystal Field Splitting

The complex ion $Mn(H_2O)_6^{3+}$ is observed to have a high-spin state. How many unpaired electrons are in the complex?

Solution

Mn has an [Ar] $4s^2 3d^5$ configuration. The Mn^{3+} is $4s^0 3d^4$. Having a **high-spin state** means that water induces **weak field** splitting in the complex ion. That implies a small Δ.

There are four unpaired electrons in the complex.

Example 21.6 B   Determining Splitting from Unpaired Electrons

The complex ion $Co(NH_3)_6^{3+}$ is observed to have no unpaired electrons. Is it a low- or high-spin complex?

Solution

Cobalt (III) is a **$3d^6$** ion. The two options are

OR

large E, low-spin   **small E, high-spin**

The high-spin complex would result in four unpaired electrons. The low spin complex would result in none. This complex is therefore a **low-spin complex.**

Color of the Complex

Your textbook points out that it takes energy to promote electrons between the t_{2g} and e_g states. The **higher the energy**, the **shorter the wavelength** of absorbed color. That is, higher energy means blue and violet will be absorbed (you will see a reddish compound). Lower energy means red will be absorbed (you will see a bluish compound).

If you look at Table 21.17 in your textbook, you can see that as the number of NH_3 groups diminishes, the color goes toward violet, indicating more red is being absorbed (a lower Δ value).

Example 21.6 C   Colors and Field Splitting

A solution of $[Cu(en)_2]^{2+}$ is green. The color of a $[CuBr_4]^{2-}$ solution is violet. What does this tell you about the relative crystal field splitting energies? Which ligand causes the greater splitting, **en** or **Br^-**?

Solution

Because the $[CuBr_4]^{2-}$ solution is violet, it means that lower energy radiation is being absorbed (so that violet is emitted). This indicates a relatively small splitting energy. The $[Cu(en)_2]^{2+}$ solution transmits green, indicating that it absorbs blue, a relatively high-energy radiation (along with red and orange). This means that the **splitting energy is relatively large**.

The ethylenediamine ligand **causes greater splitting than the bromide ligand** (all other things being equal).

21.7   *The Biological Importance of Coordination Complexes*

The following questions will help you review the material in this section.

1.   What are some biological uses of coordination complexes?
2.   Why are coordination complexes ideal for biological applications?
3.   What are the principal sources of energy in mammals?

4.   What is the **respiratory chain**?
5.   What are **cytochromes**? What are their two main parts?
6.   What is the structure of chlorophyll?
7.   How can Fe^{2+} have room to attach to myoglobin?
8.   What is the mechanism for the oxidation of Fe^{2+} in heme?
9.   Describe the transport of O_2 in blood.
10. What biochemical mechanism leads to sickle cell anemia?
11. What does Le Châtelier's principle have to do with altitude sickness?
12. Describe how CO and CN^- interact with iron to cause death.
13. Define **respiratory inhibitor**.

21.8   Metallurgy and Iron and Steel Production

The following questions and the example will help you to review material in this section.

1.   Why are almost all metals found in ores?
2.   Define **metallurgy**.
3.   List each of the steps involved in preparing a metal for use.
4.   What is "gangue"?
5.   Why are silicate minerals not often used as a metal source?

Example 21.9   Minerals and Oxidation States

Determine the oxidation state of each of the elements in the following materials (see <u>Table 21.19 in your textbook</u>).

   a.   galena   b.   bauxite   c.   siderite

Solution

a.   Galena is **PbS**. Recalling your rules for assigning oxidation states (see Section 4.9),

   $$Pb = +2 \qquad\qquad S = -2$$

b.   Bauxite is **Al_2O_3**. The oxygen, being very electronegative and in group VI, gets the assignment -2. This forces the aluminum to be $+3$ to balance electrically.

   $$Al = +3 \qquad\qquad O = -2$$

c.   Siderite is **$FeCO_3$**. The carbonate anion should be familiar to you as CO_3^{2-}. This means that $O = -2$ and $C = +4$. Iron must be $+2$ to balance electrically.

   $$Fe = +2 \qquad\qquad C = +4 \qquad\qquad O = -2$$

6.   Describe the flotation process.
7.   What is the function of roasting?
8.   Why is it both useful and necessary to collect SO_2 gas from the roasting process?
9.   Describe the smelting process.
10. Describe the process of zone refining.
11. Why does zone refining purify metals (discuss the crystal lattice)?
12. What are some of the problems with pyrometallurgy?
13. Define leaching.
14. Why is water not often useful as a leaching agent?
15. Why is iron recovered from pyrite not suitable for use in steel?

16. How are iron ore particles separated from gangue?
17. How does a blast furnace work?
18. Write the reactions for the reduction of iron oxide in a blast furnace.
19. What is slag? How is it formed?
20. What is pig iron composed of?
21. What is the advantage of a direct reduction furnace?
22. What is steel?
23. What is the basic chemical difference between the production of iron and steel?
24. Describe the "open hearth" process of steel making.
25. What is the advantage of the "basic oxygen" process of steel making?
26. Describe the "electric arc" method of steel making.
27. What are the different crystal forms of iron? How are they different?
28. Why is "tempering" useful?

Exercises

Section 21.1

1. Write electron configurations for each of the following metals:

    a. Mn   b. Pd   c. Zr   d. Zn   e. Rh

2. Write electron configurations for each of the following ions:

    a. Cr^{6+} b. Cr^{2+} c. Fe^{6+} d. Fe^{3+} e. Mn^{2+}

3. How do you explain the fact that transition metals commonly exhibit several oxidation states?

4. Suggest an explanation for the fact that, for the compounds of any one element which contain H and O, the acidity increases with increasing oxidation state.

5. Classify each of the following as metal, nonmetal, or borderline: tellurium, hafnium, radon, sulfur, zirconium, lutetium, germanium, helium.

6. Consider the relative positions of the nine elements in the small square area of the periodic table defined by the atomic numbers 19, 20, 21, 37, 38, 39, 55, 56, and 57. Assuming that the trends in Family IIIB are similar to those in IA and IIA, predict:

    a. Which of the nine elements has the smallest ion, which the largest.
    b. Which has the smallest electronegativity, which the greatest.
    c. Which has the greatest ionization energy, which the least.

Section 21.3

7. What is the coordination number on the metal in each of the following complex ions?

    a. $[CuF_4]^{2-}$ c. $[Fe(en)(H_2O)_4]^{2+}$
    b. $[CuF_6]^{3-}$ d. $[Fe(en)_3]^{2+}$

8. What is the coordination number of the central atom in the following ions?

    a. $Cu(NH_3)_4^{2+}$ b. $Co(NO_2)_6^{3-}$ c. SF_6 d. $Ag(NH_3)_2^{+}$

9. Gadolinium has the electron configuration $[Xe]\,6s^2 5d^1 4f^7$. Why is this configuration more energetically favorable than $[Xe]\,6s^2 5d^0 4f^8$?

10. The final step in the preparation of cobalt is the reduction of Co_3O_4 with aluminum:

    $$3Co_3O_4(s) + 8Al(s) \rightarrow 9Co(s) + 4Al_2O_3(s)$$

    How much cobalt can be obtained from 125.0 g of Co_3O_4?

11. Iron in aqueous solution can undergo a series of reactions with ethylenediamine ("en").

    $$[Fe(H_2O)_6]^{2+} + en \rightarrow [Fe(en)(H_2O)_4]^{2+} + 2H_2O$$
    $$[Fe(en)(H_2O)_4]^{2+} + en \rightarrow [Fe(en)_2(H_2O)_2]^{2+} + 2H_2O$$
    $$[Fe(en)_2(H_2O)_2]^{2+} + en \rightarrow [Fe(en)_3]^{2+} + 2H_2O$$

    a. Name each complex ion.
    b. What is the overall geometry of the complexes?
    c. What is the coordination number of the iron?
    d. Will the crystal field splitting increase or decrease as these reactions proceed?

12. What is the oxidation state of the metal ion in each of the following complex ions?

    a. $[Mn(CO)_5]^+$
    b. $Pt(CO)Cl_2$
    c. $[Co(CN)_5OH]^{3-}$

13. Name each of the complex ions in the previous problem.

14. Name the following compounds and ions:

    a. $CrO_4^{2!}$
    b. $Ni(CO)_4$
    c. $[Co(NH_3)_6]Cl_3$
    d. $LiAlH_4$

15. Is there a most common oxidation state of the transition metals? Is the periodic table of much value in predicting the oxidation states of these elements? Explain your answer.

16. Give the formula for each of the following coordination compounds.

    a. potassium tetrahydroxynickelate(II)
    b. tetraaquamanganese(II) sulfate
    c. tris(ethylenediamine)cobalt(III) chloride

Section 21.6

17. How many unpaired electrons will be present in each of the following complex ions?

    a. $[Fe(NH_3)_6]^{2+}$ (high-spin)
    b. $[Co(NH_3)_6]^{3+}$ (low-spin)

18. Determine the spin of $[Ru(H_2O)_6]^{3+}$ if it has one unpaired electron.

19. Based on the spectrochemical series, predict whether each of the following will be high- or low-spin.

    a. $[Fe(CN)_6]^{4-}$
    b. $[CoF_6]^{3-}$
    c. $[MnCl_6]^{4-}$

20. How many unpaired electrons will be present in each of the complex ions in Problem 19?

21. Arrange the following colors in order of increasing wavelength.

    red, violet, green, blue, yellow

22. Arrange the colors in Problem 21 in terms of increasing energy.

23. Two compounds are synthesized. One is red. The other is green. Which compound has the larger value for Δ? Why?

24. An electron is excited from the t_{2g} to the e_g state by radiation of wavelength 620 nm.

    a. What color light did the electron absorb?
    b. What is the energy of the absorbed light?

25. The complex ion $[Cu(H_2O)_6]^{2+}$ has an absorption maximum at around 800 nm. When four ammonias replace water, $[Cu(NH_3)_4(H_2O)_2]^{2+}$, the absorption maximum shifts to around 600 nm. What do these results mean in terms of the relative field splittings of NH_3 and H_2O?

Multiple Choice Questions

26.   Which one of the following statements is not true about the lanthanide series?

    A.   Going from left to right, the atomic size decreases.
    B.   Going from right to left, the atomic size decreases.
    C.   Going from left to right, the lanthanide elements are filling their $4f$ orbitals.
    D.   Due to the lanthanide elements, the $4d$ and $5d$ elements in a vertical group have similar properties.

27.   Which one of the following metals is the best reducing agent?

    A.   Ni   B.   Co   C.   Ti   D.   Cr

28.   Which one of the following statements about scandium is not true?

    A.   The most common oxidation state of scandium is +3.
    B.   Its chemistry strongly resembles that of lanthanides.
    C.   Its electron configuration is $[Ar]4s^2 3d^1$.
    D.   Scandium is prepared by electrolysis of molten $ScCl_5$.

29.   Which one of the following statements about vanadium is not true?

    A.   Its electron configuration is $[Ar]4s^2 3d^2$.
    B.   It can be used to produce a steel that is hard and corrosion resistant.
    C.   Its principal oxidation state is +5.
    D.   Its higher oxidation states do not exist as hydrated ions of the type V^{n+}. They cause the attached waters to become very acidic.

30.   Which one of the following statements about chromium is not true?

    A.   Chromite, produced by reacting carbon with ferrochrome, is added to iron in the steelmaking process.
    B.   Its most common oxidation state is +2.
    C.   Chromium(VI) species are excellent oxidizing agents.
    D.   The lower the pH, the higher the oxidizing abilities of chromium(VI).

31.   Which one of the following statements about copper is not true?

    A.   Copper is second best (after silver) in conducting heat and electricity.
    B.   It is very prone to corrosion, especially when oxidized.
    C.   It is used in many alloys such as sterling silver and brass.
    D.   Aqueous solutions of copper(II) are typically blue in color due to the presence of the $Cu(H_2O)_6^{2+}$ ion.

32.   Which one of the following ligands is unable of acting in a bidentate manner?

    A. S^{2-} B. CN^- C. Br^- D. NH_3

33.   The ligands in a complex ion act as:

    A.   Lewis acids   B.   Lewis bases   C.   Arrhenius bases   D.   oxidizing agents

34.   If a ligand acts in an octadentate manner, how many electrons does it donate?

    A.   4   B.   6   C.   8   D.   16

35.   A complex ion is found to contain 31.56 % zinc and 68.44 % chlorine. Which one of the following geometries is consistent with this data?

    A.   linear   B.   square planar   C.   tetrahedral   D.   octahedral

36.   What is the formula for the following compound, tetrachlorodiaquaferrate(III) ion?

    A. $[Fe(H_2O)_2Cl_2]^-$ B. $[Fe(H_2O)_2Cl_4]^-$ C. $[Fe(H_2O)_2Cl_4]^{2-}$ D. $[FeH_2OCl_4]^{2+}$

37. What is the name of the nonpolar complex ion, $[Cr(H_2O)_4I_2]^+$?

   A. trans-tetraaquadiiodochromium(III) ion
   B. cis-tetraaquadiiodochromium(III) ion
   C. trans-tetraaquadiiodochromium ion
   D. cis-tetraaquadiiodochromium(I) ion

38. What is the name of the nonpolar complex ion, $[CuCl_2F_2]^{2-}$?

   A. trans-dichlorodifluorocopper(II) ion
   B. trans-dichlorodifluorocuprate(II) ion
   C. cis-dichlorodifluorocuprate(II) ion
   D. cis-dichlorodifluorocopper(II) ion

39. Which one of the following pairs are coordination isomers?

   A. trans-dichlorodifluorocuprate(II) ion and cis-dichlorodifluorocuprate(II) ion
   B. $[Cr(H_2O)_4I_2]Cl$ and $[Cr(H_2O)_4Cl_2]I$
   C. trans-dichlorodifluorocuprate(II) ion and trans-dichlorodifluorocopper(II) ion
   D. $[Cu(H_2O)_6]^{2+}$ and $[Cu(H_2O)_6]I_2$

40. Which one of the following metal ions should exhibit the largest crystal field splitting energy?

   A. Fe^{3+} B. Ir^{3+} C. Mn^{3+} D. Co^{3+}

41. How many unpaired electrons are there in the elevated d-orbitals of $[FeCl_4]^-$?

   A. 0   B. 1   C. 2   D. 3

42. If a complex ion is square planar, which d-orbital is highest in energy?

   A. $d_{x^2-y^2}$ B. d_{z^2} C. d_{xy} D. d_{zz}

43. In myoglobin:

   A. Fe^{2+} is oxidized to Fe^{3+} when it transports oxygen.
   B. Fe^{3+} is bound to a porphyrin ring containing bromine as a ligand.
   C. Fe^{2+} is not oxidized to Fe^{3+} when it transports oxygen.
   D. The iron-containing heme is bound to a protein.

44. In the process of roasting, sulfur can be separated from zinc by:

   A. Oxidizing sulfur and removing elemental zinc.
   B. Oxidizing both sulfur and zinc.
   C. Oxidizing zinc and removing elemental sulfur.
   D. Oxidizing zinc and removing hydrated sulfur.

45. The process of cyanidation:

   A. Is a process of pyrometallurgy.
   B. Dissolves elemental gold by forming $Au(CN)_2^-$.
   C. Purifies $Au(CN)_2^-$ by reacting it with Zn to yield elemental gold.
   D. Purifies elemental gold from $Au(CN)_2^-$ by reacting it with oxygen at high temperatures.

Answers to Exercises

1. a. Mn: [Ar] $4s^2 3d^5$
    b. Pd: [Kr] $5s^0 4d^{10}$
    c. Zr: [Kr] $5s^2 4d^2$
    d. Zn: [Ar] $4s^2 3d^{10}$
    e. Rh: [Kr] $5s^1 4d^8$

2. a. Cr^{6+}: [Ar]
    b. Cr^{2+}: [Ar] $3d^4$
    c. Fe^{6+}: [Ar] $3d^2$
    d. Fe^{3+}: [Ar] $3d^5$
    e. Mn^{2+}: [Ar] $3d^5$

3. Transition elements often have *s* and *d* electrons available that allow more oxidation states.

4. Increasing oxidation states means loss of electrons. Increasing acidity means a higher hydronium ion concentration. As electrons are lost from a molecule, the molecule acquires an increasingly positive character, increasing the leaving ability of the proton in order to return the molecule to a more neutral state.

5. Borderline, metal, nonmetal, nonmetal, metal, metal, borderline, nonmetal.

6. a. Smallest - 21, largest - 55   c. Greatest - 21, least - 55
    b. Smallest - 55, greatest - 21

7. a. 4   b. 6   c. 6   d. 6

8. a. 4   b. 6   c. 0   d. 2

9. Although electron configurations are hard to nail down unequivocally this deep in the periodic table, it would seem that the 4*f* and 5*d* energy levels are close enough together to allow electron crossover. The half-filled 4*f* helps minimize the electron-electron repulsion that would occur if the configuration were $5d^0 4f^8$.

10. 91.78 g cobalt

11. a. $[Fe(H_2O)_6]^{2+}$ = hexaaquairon(II) ion
        $[Fe(en)(H_2O)_4]^{2+}$ = tetraaquaethylenediamineiron(II) ion
        $[Fe(en)_2(H_2O)_2]^{2+}$ = diaquabis(ethylenediamine)iron(II) ion
        $[Fe(en)_3]^{2+}$ = tris(ethylenediamine)iron(II) ion
    b. All complexes are octahedral.
    c. The iron has a coordination number of 6.
    d. According to the spectrochemical series, the splitting will increase (larger D).

12. a. +1   b. +2   c. +3

13. a. pentacarbonylmanganese(I) ion
    b. carbonyldichloroplatinum(II)
    c. pentacyanohydroxycobaltate(III) ion

14. a. Chromate ion   c. Hexaamine cobalt(III) chloride
    b. Nickel tetracarbonyl   d. Lithium tetrahydridealuminate(III) (lithium aluminum hydride)

15.   Most of the transition metals tend to show several oxidation states. None of these elements has a common oxidation state of less than +2, and most of the elements in group VIII B have a maximum oxidation state of +4. The periodic table is useful in predicting oxidation states of these elements because of the apparent trends.

16.   a. $K_2[Ni(OH)_4]$ b. $[Mn(H_2O)_4]SO_4$ c. $[Co(en)_3]Cl_3$

17.   a.   4   b.   0

18. Ru^{3+} is $4d^5$. One unpaired electron corresponds to **low-spin** (strong field case).

19.   a.   low-spin   b.   high-spin   c.   high-spin

20.   a.   0   b.   4   c.   5

21.   (lowest) violet, blue, green, yellow, red (highest)

22.   (lowest) red, yellow, green, blue, violet (highest)

23.   The red compound has the higher Δ value because it absorbs blue light. The green compound absorbs red and yellow light, which are of lower energy.

24.   a.   red light   b. $E = hc/\lambda = 3.2 \text{ H } 10^{-19}$ J/photon

25.   An absorption maximum shift to 600 nm means Δ is larger (600 nm represents greater radiation than 800 nm). Therefore, a stronger field splitting is occurring. This is consistent with the spectrochemical series.

26.	B	27.	C	28.	D	29.	A	30.	A	31.	B
32.	D	33.	B	34.	C	35.	C	36.	B	37.	A
38.	B	39.	B	40.	B	41.	D	42.	A	43.	D
44.	B	45.	B									

Chapter 22

Organic and Biological Molecules

Organic chemistry is the study of compounds that contain carbon. Such compounds are ubiquitous because carbon **forms strong bonds** to hydrogen, oxygen, and nitrogen, among others. It also has the **unique** ability to form **chains and rings** with other carbon atoms. This chapter serves as a simple introduction to the tens of thousands of known organic compounds.

22.1   Alkanes: Saturated Hydrocarbons

When you finish this section you will be able to:

- Draw isomers of simple alkanes.
- Name isomers of alkanes.

Alkanes are a group of **saturated hydrocarbons. Saturated** means that the carbon is bound to **four atoms** (each by a single bond). Each carbon is sp^3 hybridized. Alkanes have the general formula C_nH_{2n+2}. Table 22.1 in your textbook gives names, formulas, and some properties of straight-chain ("n") alkanes.

Example 22.1 A   Alkanes

Give the formula and name for the straight-chain alkanes with $n = 6$ and $n = 8$. Compare their boiling and melting points. Justify the difference.

Solution

$$\text{For } n = 6, \ C_nH_{2n+2} \ = \ \textbf{C}_6\textbf{H}_{14} \ = \ \textbf{hexane}$$
$$\text{For } n = 8, \ C_nH_{2n+2} \ = \ \textbf{C}_8\textbf{H}_{18} \ = \ \textbf{octane}$$

The boiling point of octane is 58° higher than hexane. The melting point is 38° higher. Octane is considerably heavier and longer. Its London forces are more extensive, thus more energy is required to break intermolecular bonds.

Isomers are compounds with the **same formula** but **different structures**. Let's draw some of the isomers for heptane, C_7H_{16}. The key is to make sure that if the main chain lengths are the same, **the type** or **location** of groups on the chain are **unique.**

Given the straight chain,

$$H-\underset{\underset{H}{|}}{\overset{\overset{H}{|}}{C}}-\underset{\underset{H}{|}}{\overset{\overset{H}{|}}{C}}-\underset{\underset{H}{|}}{\overset{\overset{H}{|}}{C}}-\underset{\underset{H}{|}}{\overset{\overset{H}{|}}{C}}-\underset{\underset{H}{|}}{\overset{\overset{H}{|}}{C}}-\underset{\underset{H}{|}}{\overset{\overset{H}{|}}{C}}-\underset{\underset{H}{|}}{\overset{\overset{H}{|}}{C}}-H \qquad\qquad H_3C!CH_2!CH_2!CH_2!CH_2!CH_2!CH_3 \qquad\qquad C_7H_{16}$$

An example of an isomer is

$$H_3C-\underset{\underset{CH_3}{|}}{\overset{\overset{H}{|}}{C}}-\overset{H_2}{C}-\overset{H_2}{C}-\overset{H_2}{C}-CH_3$$

- The formula is still **C_7H_{16}.**
- The longest chain has **6 carbons, a hexane.**
- The numbering of carbons puts "the functional group" closest to the "#1" carbon. Therefore the **methyl** group is attached to the **"#2" carbon.**
- The name of this isomer of heptane is **2-methylhexane.**
- Carefully examine the nomenclature rules listed next to Table 22.2 in your textbook.

Another isomer is

$$H_3C-\overset{H_2}{C}-\underset{\underset{\underset{H}{|}}{\underset{H_3C-\overset{|}{C}-CH_3}{}}}{\overset{H}{C}}-CH_3$$

The longest chain looks like 4 carbons. It is actually 5 (eliminating hydrogens for clarity):

$$C_5-C_4-C_3-C$$
$$\underset{C-C_2-C_1}{\overset{|}{}}$$

This is 2,3-dimethylpentane

Remember that what you **draw on paper** is only a **shorthand representation** of a three dimensional structure. Look very hard for the longest chain.

Example 22.1 B   Isomers

There are **9 isomers** of heptane, C_7H_{16}. We have drawn and named 3. **Draw and name the other 6.** (You may eliminate hydrogens for clarity if you wish.)

Solution

Structure	Name
4. $C-C-C-C-C$ with C branches below 2nd and 4th carbons	2-4-dimethylpentane
5. $C-C-C-C-C-C$ with C branch below 3rd carbon	3-methylhexane

6.
$$
\begin{array}{c}
C \\
| \\
C\!-\!C\!-\!C\!-\!C\!-\!C \\
| \\
C
\end{array}
$$
2,2-dimethylpentane

7.
$$
\begin{array}{c}
CC \\
|| \\
C\!-\!C\!-\!C\!-\!C \\
| \\
C
\end{array}
$$
2,2,3-trimethylbutane

8.
$$
\begin{array}{c}
C \\
| \\
C\!-\!C\!-\!C\!-\!C\!-\!C \\
| \\
C
\end{array}
$$
3,3-dimethylpentane

9.
$$
\begin{array}{c}
C\!-\!C \\
| \\
C\!-\!C\!-\!C\!-\!C\!-\!C
\end{array}
$$
3-ethylpentane

Let's try naming some compounds and drawing some structures.

Example 22.1 C   Nomenclature

Give IUPAC names for the following structures:

a.
$$
\begin{array}{c}
H_3C\!-\!CH_2\!-\!CH\!-\!CH_2\!-\!CH\!-\!CH_2\!-\!CH_3 \\
|| \\
CH_3H_2C\!-\!CH_2\!-\!CH_3
\end{array}
$$

b.
$$
\begin{array}{c}
HHH_2 \\
||| \\
H_3C\!-\!C\!-\!C\!-\!C\!-\!CH_3 \\
|| \\
ClCl
\end{array}
$$

c.
$$
\begin{array}{c}
CH_3 \\
| \\
H_2HH_2H \\
|||| \\
H_3C\!-\!C\!-\!C\!-\!C\!-\!C\!-\!C\!-\!CH_3 \\
|||H \\
CH_3H_2C\!-\!C\!-\!CH_3 \\
H_2
\end{array}
$$

Solution

a. The longest chain has 8 members, an octane.

$$
\begin{array}{c}
C_1\!-\!C\!-\!C\!-\!C\!-\!C\!-\!C\!-\!C \\
|| \\
CC\!-\!C\!-\!C_8
\end{array}
$$

The number would start on the **left-hand carbon** because the methyl group is closest to that side (in the #3 position). The name of the structure is **3-methyl-5-ethyloctane.**

b. **2,3-dichloropentane**

c.   The longest chain has 8 members, an octane.

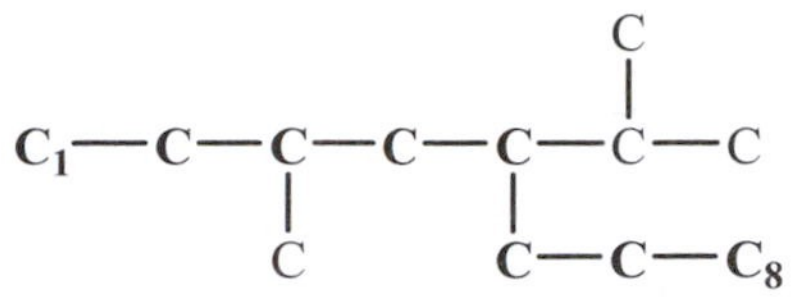

$$C_1-C-C-C-C-\overset{\overset{\displaystyle C}{|}}{C}-C$$
$$\quad\ \ \underset{C}{|}\qquad\ \ \underset{C-C-C_8}{|}$$

The closest group to a #1 carbon is the methyl group in the 3 position. The name of the structure is **3-methyl-5-isopropyloctane.**

Example 22.1 D   Drawing Structures

Draw structures for the following compounds:

   a.   3,4-dimethyl-4-ethylnonane
   b.   1-chloro-2-bromobutane

Solution

a. 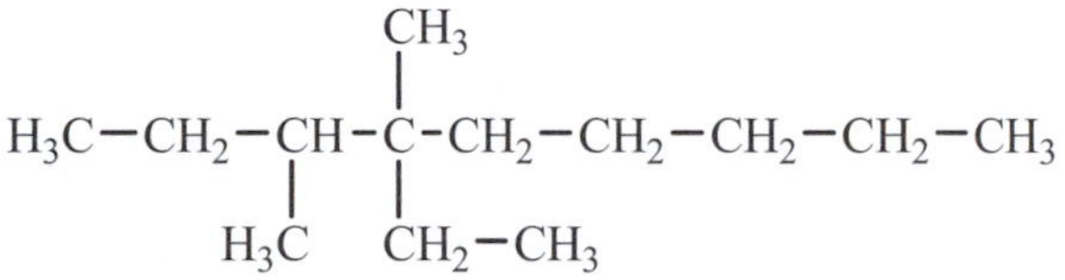 $H_3C-CH_2-CH-\overset{\overset{\displaystyle CH_3}{|}}{\underset{\underset{\displaystyle H_3C\quad CH_2-CH_3}{|\qquad\ |}}{C}}-CH_2-CH_2-CH_2-CH_2-CH_3$

b. 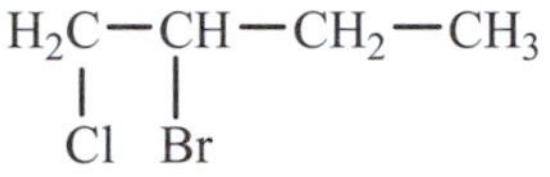 $H_2C-\underset{\underset{\displaystyle Cl}{|}}{CH}-\underset{\underset{\displaystyle Br}{|}}{CH_2}-CH_3$

Cyclic alkanes have the general formula C_nH_{2n} (for example, C_6H_{12} is cyclohexane). Note the use of the **shorthand notation** for each of the cyclic compounds in your textbook. Notice also that, as with straight chain alkanes, numbering is done so that **substituents are attached to carbons with the lowest possible numbers.**

Example 22.1 E   Cyclic Alkanes

Name the following compounds:

   a.    Cl   c. 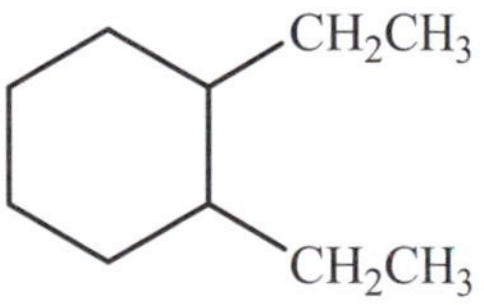 CH_2CH_3

                                                                CH_2CH_3

   b. 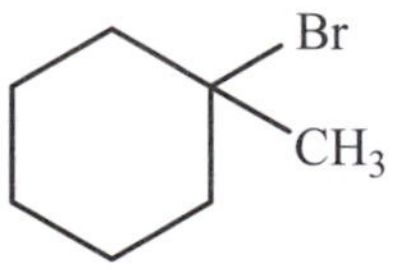   Br

                        CH_3

Solution

   a.   chlorocyclopropane
   b.   1-bromo-1-methylcyclohexane
   c.   1,2-diethylcyclohexane

22.2 Alkenes and Alkynes

When you finish this section you will be able to name and draw structures for simple alkenes and alkynes.

Straight-chain **alkenes** have the general formula C_nH_{2n}. They are characterized by the presence of **at least one carbon-carbon double bond.** The bond is formed by sharing p-orbitals. The rules for naming alkenes are the same as for alkanes with the following exceptions:

- **-ane** is changed to **-ene** (i.e., hex**ane** becomes hex**ene**).
- The position of the **double bond** has **highest priority** in terms of nomenclature. For example

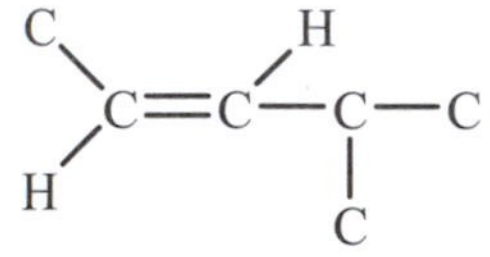

is **4-methyl-trans-2-pentene** NOT 2-methyl-trans-4-pentene.

- **cis** and **trans** isomers exist because rotation around a carbon-carbon double bond is **restricted** due to p-p orbital interaction between carbons.

Example 22.2 A   Naming Alkenes

Name the following alkenes:

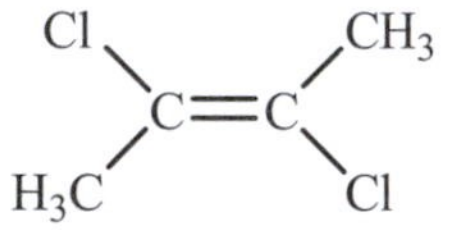

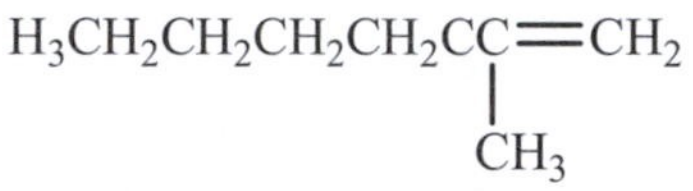

Solution

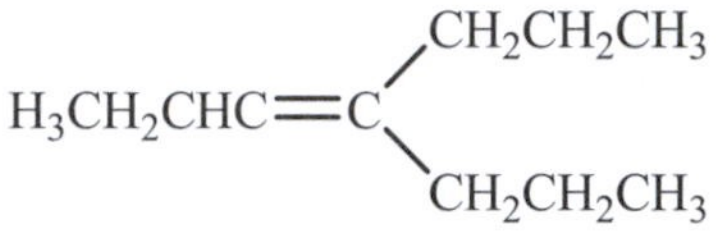

a.   The longest straight chain is a butene. In this case it is **2-butene.** The chlorines are **cis** to each other and in the **2 and 3 positions.**

2,3-dichloro-cis-2-butene

b.   Everything is the same except the chlorines are **trans** to one another.

2,3-dichloro-trans-2-butene

c.   The longest straight chain is a 7-membered, or **heptene** chain. The double bond is in the 3 position. The attached group is a **propyl** group.

4-propyl-3-heptene

d.   The longest chain here is a **heptene.** The double bond is in the 2 position. The methyl group is in the 2 position.

2-methyl-1-heptene

Example 22.2 B   Drawing Alkenes

Draw the following alkenes:

- a.   3-methyl-1-hexene
- b.   1-chloro-4-ethyl-3-hexene
- c.   4-methyl-cis-2-pentene

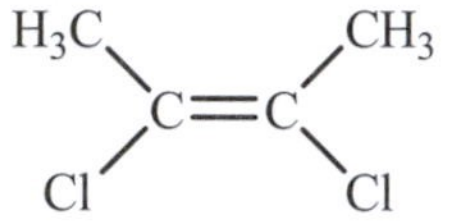

Solution

a. 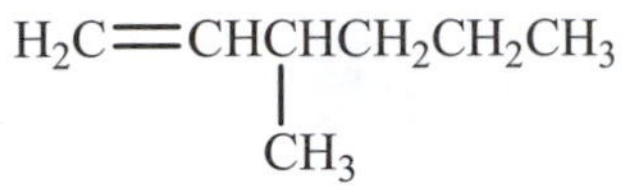 $H_2C{=}CHCHCH_2CH_2CH_3$
$\qquad\qquad\quad |$
$\qquad\qquad\quad CH_3$

b. 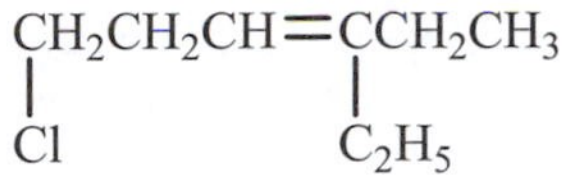 $CH_2CH_2CH{=}CCH_2CH_3$
$\quad\;\; |\qquad\qquad\; |$
$\quad\;\; Cl\qquad\qquad C_2H_5$

c.

$\qquad H\qquad\qquad H$
$\qquad\;\backslash\qquad\quad\;/$
$\qquad\quad C{=}C$
$\qquad/\qquad\quad\;\backslash$
$\quad H_3C\qquad\qquad CHCH_3$
$\qquad\qquad\qquad\quad |$
$\qquad\qquad\qquad\quad CH_3$

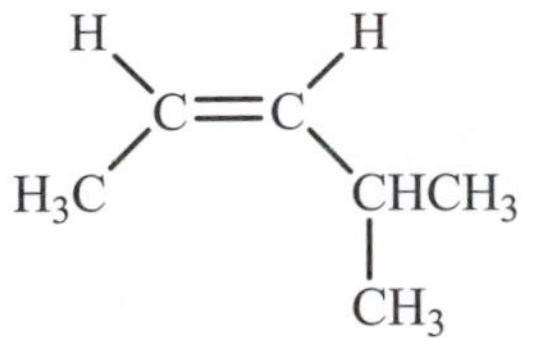

Alkynes are molecules that contain triple bonds ($1s$, 2 bonds involving 2 carbons). The nomenclature follows the same strategy as always except the compound ends in **"yne."**

Example 22.2 C   Naming Alkynes

Name the following:

a. $HC{\equiv}C(CH_2)_6CH_3$

b. 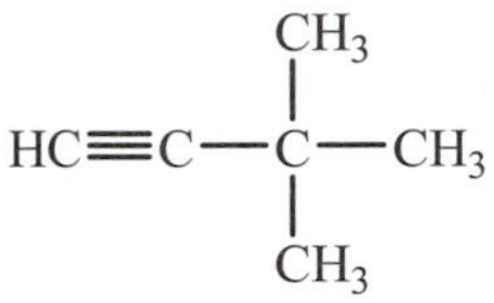
$\qquad\qquad\qquad CH_3$
$\qquad\qquad\qquad |$
$HC{\equiv}C{-}C{-}CH_3$
$\qquad\qquad\qquad |$
$\qquad\qquad\qquad CH_3$

Solution

a.   1-nonyne

b.   3,3-dimethyl-1-butyne

Example 22.2 D   Just for Fun

Name this compound:

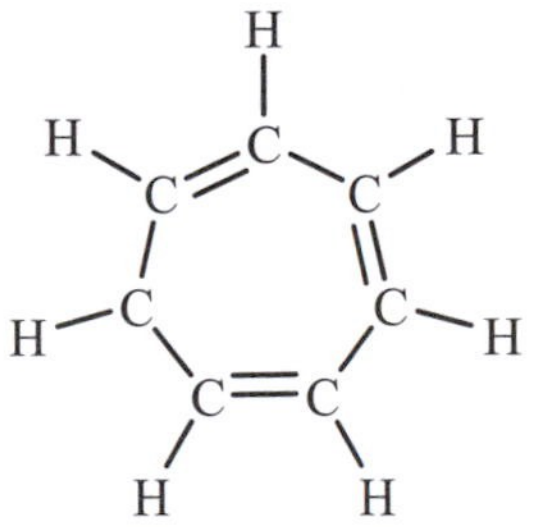

Solution

The three double bonds are in the **1, 3, and 5** positions. This is a **7-membered ring**.

1,3,5-cycloheptatriene (also called **Tropilidene**)

22.3   Aromatic Hydrocarbons

When you finish this section you will be able to name and draw structures for simple benzene derivatives.

The main idea in this section is that **electrons** in benzene and benzene-related compounds are **delocalized** (can move freely around the molecule). This makes benzene family compounds unreactive to addition, but reactive instead to **substitution** (where hydrogen atoms are replaced by other atoms).

The system of naming benzene-related compounds is shown before <u>Figure 22.12 in your textbook.</u> Notice that the numbering system is similar to that with alkanes, alkenes, and alkynes. Note as well that when there is a substituent in the "1" position, the substituent in the "2" position is called "ortho." The "3" position is "meta" and the "4" position is "para."

Example 22.3 A   Naming Aromatic Compounds

Name the following compounds (NO_2 = "nitro"):

a.

c.

b.

Solution

a.   1-nitro-2,4,5-trichlorobenzene
b.   hexabromobenzene
c.   1,3,5-trimethyl-2,4-dinitrobenzene

Example 22.3 B   Drawing Aromatic Compounds

Draw the following compounds:

a.   1-nitro-2,3,6-triiodobenzene
b.   1-ethyl-2-methylbenzene ("2-ethyltoluene")
c.   1,4-bis(dibromomethyl)benzene

Solution

a.

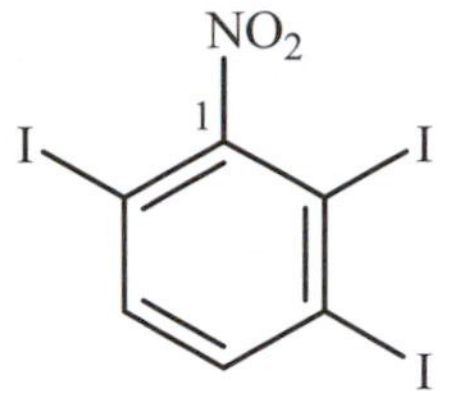

b.

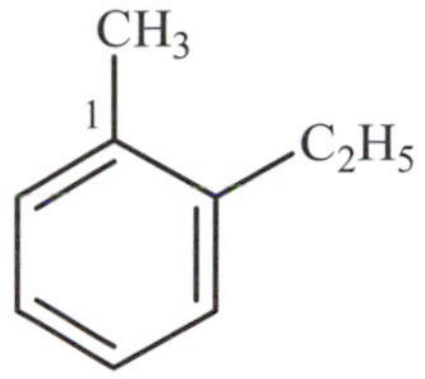

c. **"bis"** implies that the dibromomethyl **appears twice**, once in the 1 and once in the 4 position.

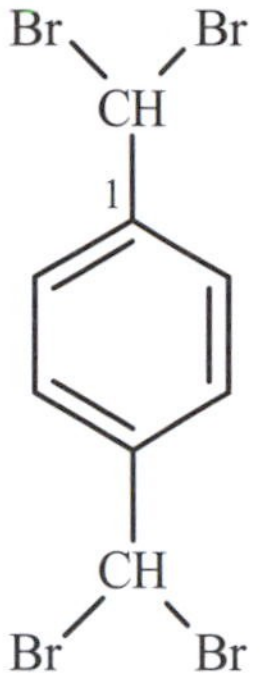

22.4   *Hydrocarbon Derivatives*

When you finish this study section you will be able to name compounds containing various functional groups.

Hydrocarbon derivatives are molecules that have substituents (**functional groups**) that contain some atoms that are **not carbon or hydrogen**. Your textbook discusses the properties of several functional groups in this section. You should know the properties of

- alcohols
- aldehydes and ketones
- carboxylic acids and esters
- amines

The functional groups are summarized in Table 22.4 in your textbook. Let's try some naming and drawing exercises (remembering that when **benzene** is an attached group rather than the main focus of the molecule, it is called a "**phenyl**" group).

Example 22.4 A   Functional Groups

Name the following compounds:

a.

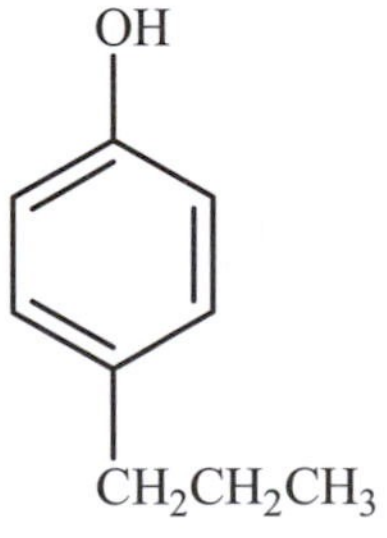

b.

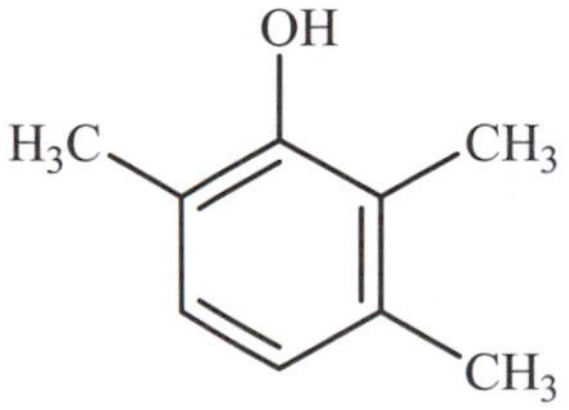

c.

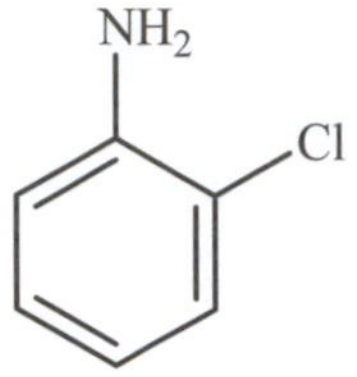

NH_2

Cl

d.

O

$CH_2CH_2CCH_3$

Cl

(See Table 22.6
in your textbook.)

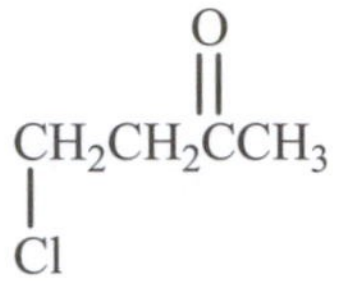

Solution

a.   4-propylphenol
b.   2,3,6-trimethylphenol
c.   2-chloroaniline (or "*o*-chloroaniline")
d.   4-chloro-2-butanone

Example 22.4 B   More Functional Groups

Draw the following compounds:

a.   1,2-pentanediol
b.   3-fluorobenzoic acid
c.   1,2-cyclopentanedicarboxylic acid

Solution

a. $H_3CH_2CH_2CH_2C$—CH_2
           | |
          OH   OH

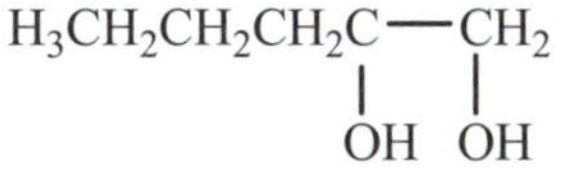

b.

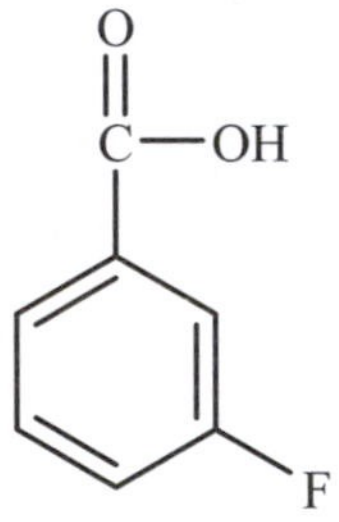

O

$\overset{\|}{C}$—OH

F

c.

O

$\overset{\|}{C}$—OH

O

$\overset{\|}{C}$—OH

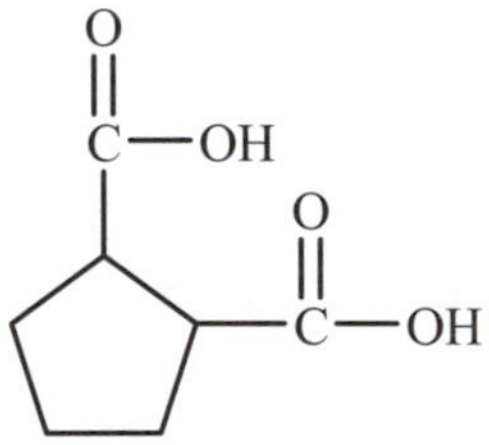

Example 22.4 C   Revenge of the Functional Groups

List all of the functional groups in each of the following molecules:

a.

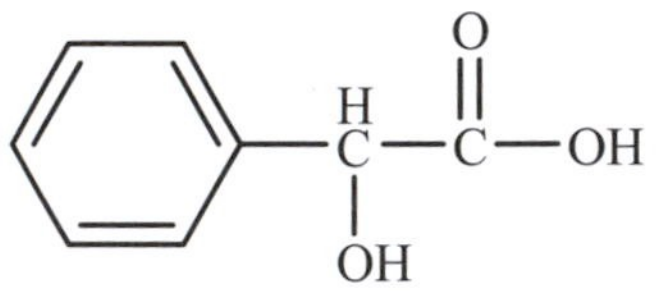

$\overset{\text{H}}{\underset{\text{OH}}{C}}$—$\overset{O}{\overset{\|}{C}}$—OH

c.

$N\overset{CH_3}{\diagdown}_{CH_3}$

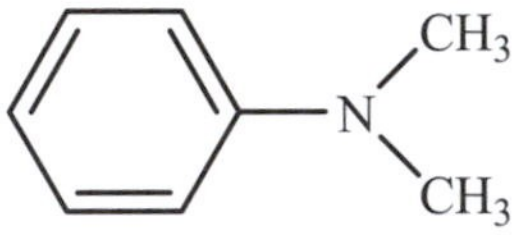

b. $H_3CH_2CH_2CH_2C$—N—CH_2CH_3
                            |
                           H

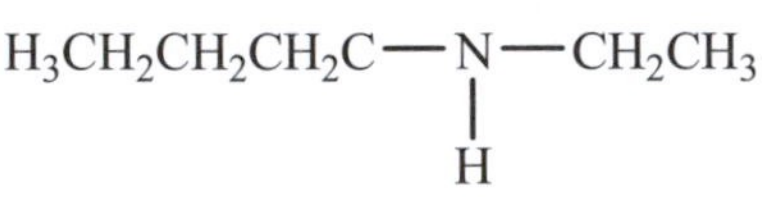

d.

H—$\overset{O}{\overset{\|}{C}}$

CH_2

CH_2

N

H

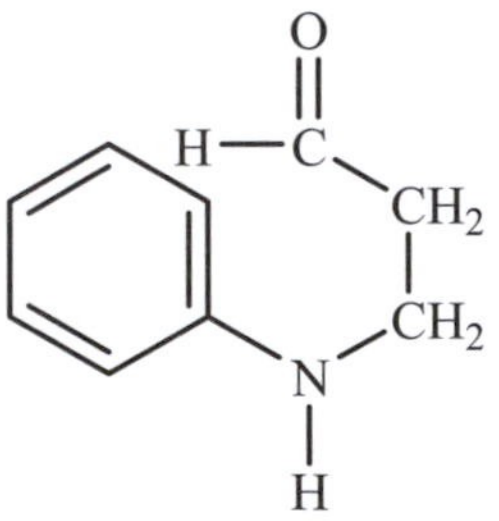

Solution

a. alcohol, carboxylic acid
b. secondary amine (2 hydrogens have been substituted)
c. tertiary amine (3 hydrogens have been substituted)
d. secondary amine, aldehyde

22.5   Polymers

The following questions will help you test your knowledge of the material in this section.

1. Define **polymer**.
2. What are polymers made from?
3. How can you vary properties when making polymers?
4. What is Teflon made of?
5. Why is Teflon so widely used?
6. Define **addition polymerization**.
7. What is a free radical?
8. Define **condensation polymerization**.
9. Why is nylon called a copolymer?
10. Use equations to show how water is a product when nylon is formed.
11. Why is Dacron a "polyester"?

22.6   Natural Polymers

The following questions and exercises will help to review your understanding of the nature of proteins.

1. What are **proteins**?
2. What range of molar masses can proteins have?
3. Why are the acids that comprise proteins called **α-amino** acids? That is, what does "α" mean? What does "amino" mean?

Example 22.6 A   The Common Protein Amino Acids

Match the name of the amino acid in the left-hand column with its "R" group in the right-hand column.

Amino Acid		R group
1. Glycine	a.	$-\overset{H_2}{C}-\overset{H_2}{C}-\overset{\overset{O}{\|\|}}{C}-NH_2$
2. Methionine	b.	$-\overset{H_2}{C}-\overset{H_2}{C}-\overset{\overset{O}{\|\|}}{C}-OH$
3. Glutamic Acid	c.	$-CH_2-(C_6H_5)$
4. Glutamine	d.	$-(CH_2)_4-NH_2$
5. Lysine	e.	$-CH_2-CH_2-S-CH_3$
6. Phenylalanine	f.	$-H$

Solution

Amino Acid		R group
1.	Glycine	f. –H
2.	Methionine	e. $-CH_2-CH_2-S-CH_3$
3.	Glutamic Acid	b.
4.	Glutamine	a.
5.	Lysine	d. $-(CH_2)_4-NH_2$
6.	Phenylalanine	e. $-CH_2-(C_6H_5)$

For item b:
$$-\underset{H_2}{C}-\underset{H_2}{C}-\overset{O}{\underset{\|}{C}}-OH$$

For item a:
$$-\underset{H_2}{C}-\underset{H_2}{C}-\overset{O}{\underset{\|}{C}}-NH_2$$

4. What is a peptide linkage?
5. Define **dipeptide**.

Example 22.6 B   Peptide Linkage

Draw the structure of the dipeptide formed from the condensation reaction between **leucine** and **phenylalanine**.

Solution

As pointed out in your textbook, the peptide linkage occurs between the **carboxylic acid** end of one molecule and the **amine** end of the other. Also, as "standard procedure," the terminal amine is always put on the left, and the terminal carboxylic acid is put on the right.

Leucine (Leu) + Phenylalanine (Phe) → (Leu-Phe)

Example 22.6 C   Practice with Peptide Linkages

Draw the structure of the polypeptide with the sequence **Trp-Ser-Asp**.

Solution

This is formed as a result of two peptide linkages, one between **Trp and Ser** and one between **Ser and Asp**.

Tryptophan (Trp) + Serine (Ser) + Aspartic Acid (Asp) ⟶

(Trp-Ser-Asp)

6.   How many sequences can possibly exist for a polypeptide chosen from ten unique amino acids?
7.   What are the four levels of structure in proteins?
8.   What kinds of bonding interactions are responsible for each level of structure?
9.   What type of structure is an α-helix?
10. Give some practical examples of primary, secondary, and tertiary structures.
11. Discuss the role of the **disulfide** linkage in getting "permanent waves" of hair.
12. Define **denaturation**.
13. List some causes of denaturation.

The following questions will help you review the material on carbohydrates.

14. Why are carbohydrates so named?
15. What are monosaccharides?
16. What is a **hexose**?
17. What is necessary for **optical isomerism** to exist in a molecule?

Example 22.6 D   Chiral Carbons

How many chiral carbons are there in D-Glucose?

```
        1
        CHO
        |
     2  |
H —— C —— OH
        |
     3  |
HO —— C —— H
        |
     4  |
H —— C —— OH
        |
     5  |
H —— C —— OH
        |
     6  |
        CH₂OH
```

Solution

A chiral carbon has **4 different substituents** attached to it.

Carbon **#1** has only 3 substituents. It is not chiral.

Carbon **#2 is chiral**.

```
        CHO
        |
H ——  C  —— OH
        |
HO —— C —— H
        |
H —— C —— OH
        |
H —— C —— OH
        |
        CH₂OH
```

Carbon **#3 is chiral**.

```
        CHO
        |
H —— C —— OH
        |
HO —— C —— H
        |
H —— C —— OH
        |
H —— C —— OH
        |
        CH₂OH
```

Carbon **#4 is chiral**.

```
        CHO
        |
H —— C —— OH
        |
HO —— C —— H
        |
H ——  C  —— OH
        |
H —— C —— OH
        |
        CH₂OH
```

Carbon **#5 is chiral**.

```
        CHO
        |
H —— C —— OH
        |
HO —— C —— H
        |
H —— C —— OH
        |
H ——  C  —— OH
        |
        CH₂OH
```

Carbon #6 has two identical groups. It **is not chiral**.

In summary, **carbons 2, 3, 4, and 5 are chiral**. Hexoses have $2^4 = 16$ optical isomers.

18.   What are the bonds involved in cyclizing pentoses? hexoses?
19.   Define **disaccharide**.
20.   What is a **glycoside linkage**?
21.   What is the function of **α-amylase**?
22.   Describe the structure of **starch**.
23.   Why is it advantageous for "fuel" storage to have starch as one long molecule instead of many small ones?
24.   Why is cellulose **indigestible** by humans, but can be digested by certain animals?

The following questions will help you review the material on nucleic acids.

25.   What are the functions of DNA?
26.   List the basic parts that make up nucleotides.
27.   Why is a double-helix structure important to the function of DNA?
28.   Cytosine and guanine form hydrogen-bonding pairs. What in their structure makes this possible?
29.   Outline the process for replication of DNA.

Example 22.6 E   Complimentary Sequences

A single strand of DNA contains the nucleotide sequence

A - A - G - T - T - G - C - C - A - T

List its complimentary strand.

Solution

Adenine (A) and thymine (T) form complimentary pairs as do cytosine (C) and guanine (G). The complimentary strand would be

old A - A - G - T - T - G - C - C - A - T
new T - T - C - A - A - C - G - G - T - A

30.   Define **gene**.
31.   What is a **codon**? **anticodon**?
32.   Describe the functions of mRNA and tRNA to protein construction.

Answers to review questions are at the end of this study guide chapter.

Exercises

Section 22.1

1. Name the following compounds using IUPAC nomenclature.

    a. 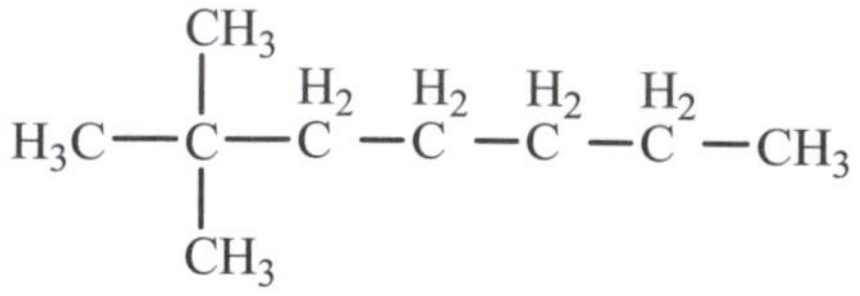
    $$H_3C-\underset{\underset{CH_3}{|}}{\overset{\overset{CH_3}{|}}{C}}-\overset{H_2}{C}-\overset{H_2}{C}-\overset{H_2}{C}-\overset{H_2}{C}-CH_3$$

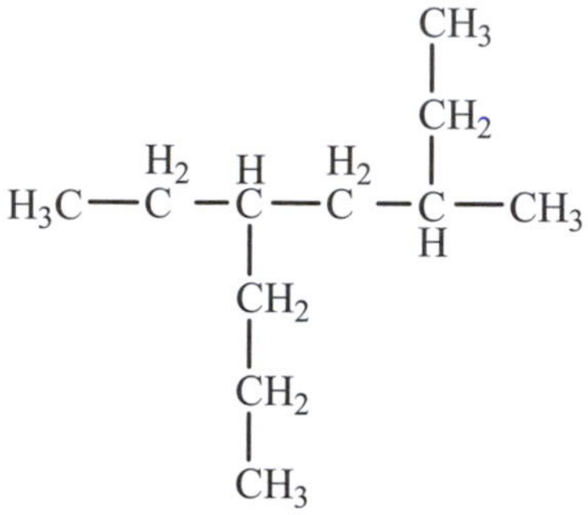

    b.
    $$H_3C-\overset{H_2}{C}-\underset{\underset{\underset{\underset{CH_3}{|}}{\overset{CH_2}{|}}}{\overset{H}{C}}}{\overset{}{C}}-\overset{H_2}{C}-\underset{\overset{}{H}}{\overset{\overset{\overset{CH_3}{|}}{CH_2}}{C}}-CH_3$$

    c.
    $$H_3C-\overset{\overset{H}{|}}{C}-CH_3$$
    $$H_3C-\underset{\underset{CH_3}{|}}{C}-CH_3$$

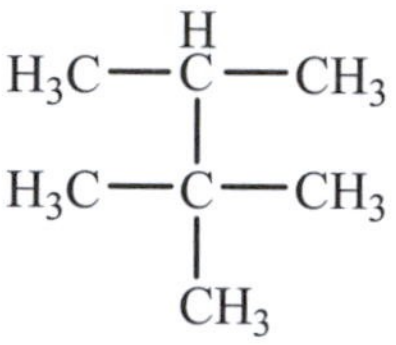

    d. 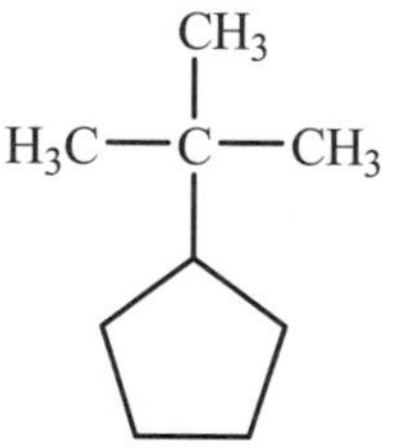
    $$H_3C-\underset{\underset{}{|}}{\overset{\overset{CH_3}{|}}{C}}-CH_3$$

2. Write structures for the following systematic names.

    a. 1-ethyl-3-propylcyclohexane
    b. 1,1,2-trichloroethane

3. Are saturated hydrocarbons (alkanes) soluble in water?

Section 22.2

4. Match the following structures with their correct systematic names.

    a. 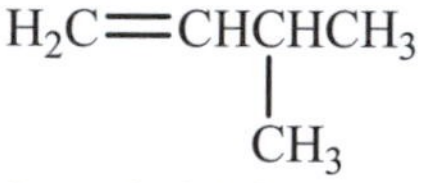 $H_2C=CHCHCH_3$
        $\quad\quad\quad\quad\overset{|}{CH_3}$

    b. 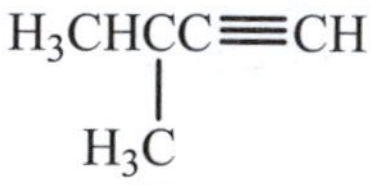 $H_3CHCC\equiv CH$
        $\quad\quad\overset{|}{H_3C}$

    c. 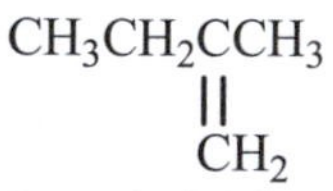 $CH_3CH_2CCH_3$
        $\quad\quad\quad\quad\overset{||}{CH_2}$

    1. 3-methyl-1-butyne
    2. 3-methyl-1-butene
    3. 2-methyl-1-butene

5. What products would form after hydrogenation of 2-methyl-1-butene? After halogenation (Cl_2)?

6. Why, in general, are alkenes more reactive in addition reactions than alkanes?

Section 22.3

7. What are the products of the following reactions?

a.  + Br_2 / $FeBr_2$ →

b.  + Br_2 / CCl_4 →

c.  + $H_3C\overset{O}{\underset{||}{-C}}-Cl$ / $AlCl_3$ →

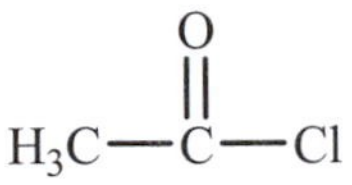

d.  + H_2SO_4 (fuming) →

Section 22.4

8. Name the functional group(s) in each of the following compounds:

a. $CH_3–O–CH_2–CH_2–OH$
   (1)   (2)

b. 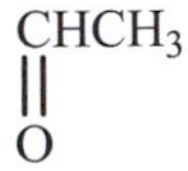 $\underset{(1)}{CHCH_3}$ with $\overset{||}{O}$

c. 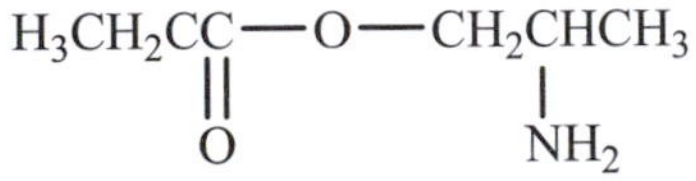 $H_3CH_2CC\overset{||}{\underset{O}{(1)}}-O-CH_2\underset{NH_2 \atop (2)}{CHCH_3}$

d. 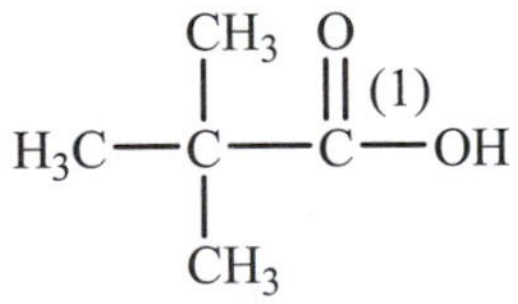 $H_3C-\underset{CH_3}{\overset{CH_3}{C}}-\overset{O}{\overset{||}{C}}-OH$ (1)

9. Name the following compounds or give their structure:

a. 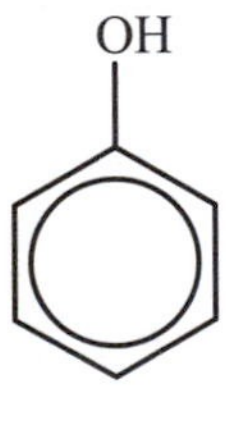 OH

b. 3-chlorobenzaldehyde

c. 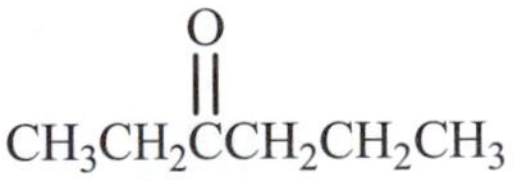 $CH_3CH_2\overset{O}{\overset{||}{C}}CH_2CH_2CH_3$

d. 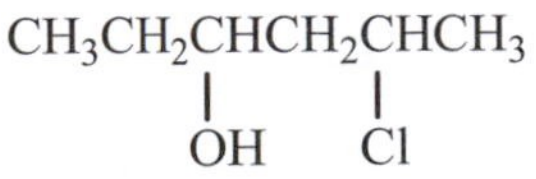 $CH_3CH_2\underset{OH}{\overset{}{C}}HCH_2\underset{Cl}{\overset{}{C}}HCH_3$

10. Arrange the molecules in order from lowest to highest boiling point:

$$CH_3CH_2CH_2OH \qquad CH_3CH_2\overset{\overset{\displaystyle O}{\|}}{C}H \qquad CH_3\overset{\overset{\displaystyle O}{\|}}{C}CH_3$$

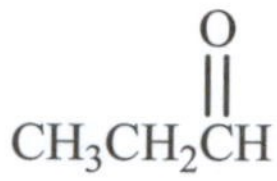

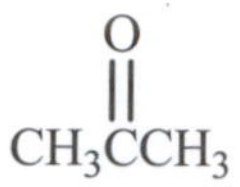

11. Name the reactants in each equation below. Give the structure of the products that would form.

a. 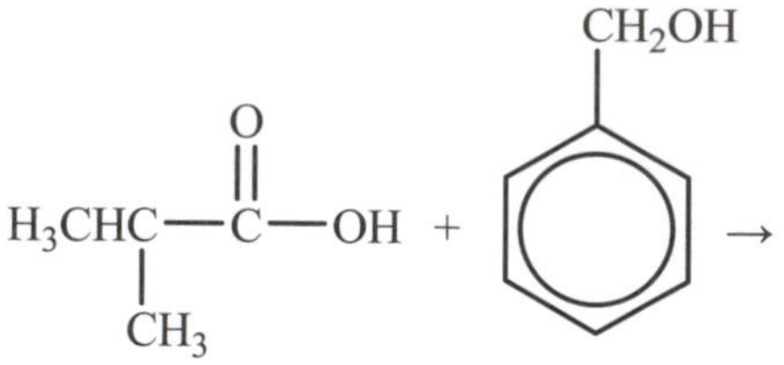 $H_3CHC\overset{\displaystyle CH_3}{\underset{\displaystyle |}{|}}\overset{\overset{\displaystyle O}{\|}}{C}-OH$ + (benzene ring with CH$_2$OH substituent) →

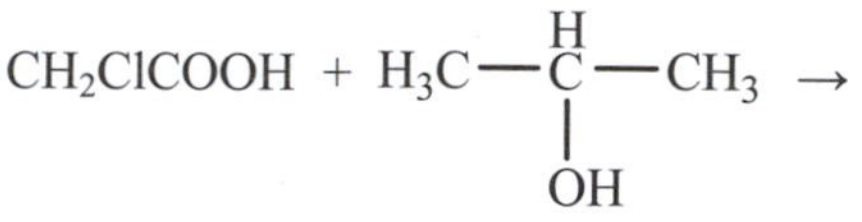

b. $CH_2ClCOOH$ + $H_3C\overset{\overset{\displaystyle H}{|}}{\underset{\underset{\displaystyle OH}{|}}{C}}-CH_3$ →

c. $H_3C\overset{\overset{\displaystyle CH_3}{|}}{\underset{\underset{\displaystyle CH_3}{|}}{C}}-CH_2CH_2OH \xrightarrow{KMnO_4\,(aq)}$

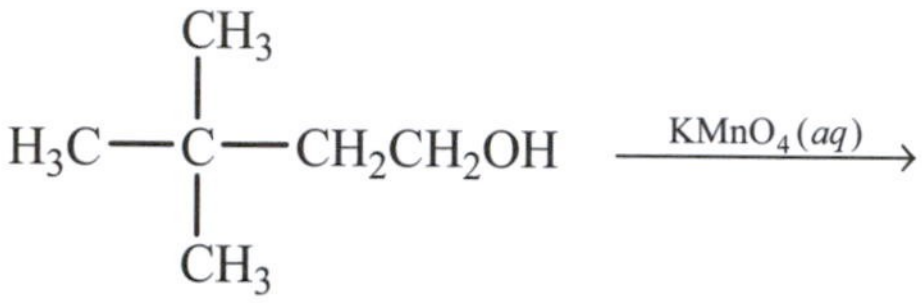

12. Label the following amines as 1°, 2°, or 3°.

a. $N(CH_2CH_3)_3$

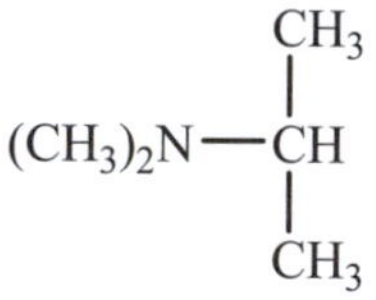

c. $CH_3CH_2NHCH_3$

b. $(CH_3)_2N\overset{\overset{\displaystyle CH_3}{|}}{\underset{\underset{\displaystyle CH_3}{|}}{C}}H$

d. 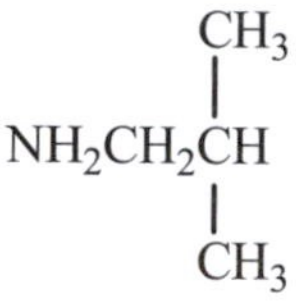 $NH_2CH_2\overset{\overset{\displaystyle CH_3}{|}}{\underset{\underset{\displaystyle CH_3}{|}}{C}}H$

Section 22.5

13. Define or explain the following terms:

a. dimer
b. free radical
c. copolymer
d. homopolymer
e. polymer

14. Distinguish between addition polymerization and condensation polymerization.

15. Write the *cis-* and *trans-* chair conformations of 1,2-dichlorocyclohexane.

16. Arrange the following alkenes from most stable to least stable. (Hint: Stability is directly related to the substitution of the double bond.)

a. 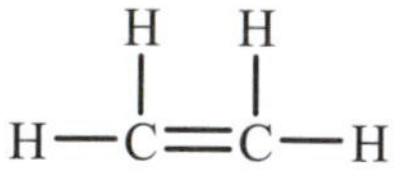 $H-\overset{\overset{\displaystyle H}{|}}{C}=\overset{\overset{\displaystyle H}{|}}{C}-H$

c. $H-\overset{\overset{\displaystyle CH_3}{|}}{C}=\overset{\overset{\displaystyle H}{|}}{C}-H$

b. $H_3CH_2C-\overset{\overset{\displaystyle H}{|}}{C}=\overset{\overset{\displaystyle CH_2CH_2CH_3}{|}}{C}-H$

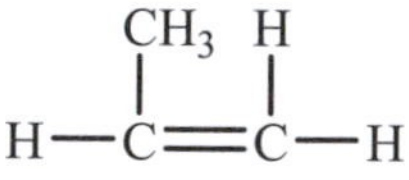

d. 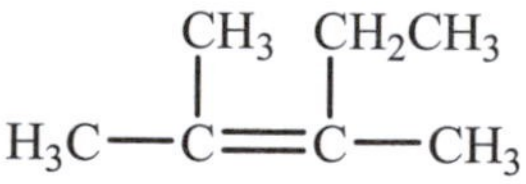 $H_3C-\overset{\overset{\displaystyle CH_3}{|}}{C}=\overset{\overset{\displaystyle CH_2CH_3}{|}}{C}-CH_3$

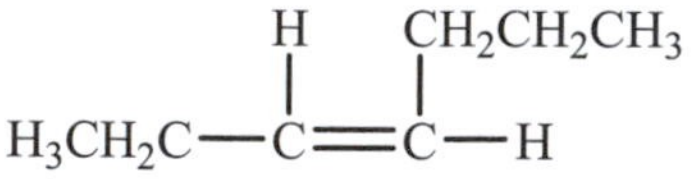

17.   Alcohols are capable of forming strong hydrogen bonds to each other that make them polar. Why is ethyl alcohol greatly soluble in water while heptyl alcohol is almost insoluble in water?

18.   Arrange the following amines from the highest to the lowest boiling point. Give an explanation of your answer.

   a. $(CH_3-CH_2)_2N-H$ b. $CH_3-CH_2-N(CH_3)_2$ c. $CH_3(CH_2)_3-NH_2$

Multiple Choice Questions

19.   What is the number of possible isomers for C_4H_8?

   A. 6   B. 3   C. 5   D. 2

20.   1,1,2-trimethylcyclopentane is an isomer of which one of the following compounds?

   A. nonane   B. isoheptane   C. 2-isopropyl-pentane   D. isohexane

21.   Which one of the following compounds can react with chlorine gas to produce 1,2-dichlorocyclohexane?

   A. hexane   B. cyclohexene   C. 3-methylcyclohexane D. 2-methylhexane

22.   When ethane is converted to ethylene (CH_2CH_2), the carbon atoms:

   A. are oxidized   B. are reduced   C. act as oxidizers   D. are unchanged

23.   What is the bond angle between H−C−C in acetylene?

   A. 180°   B. 90°   C. 109°   D. 120°

24.   What is the proper name of the following compound?

   A. 4-ethyl-2-methylcyclohexene   C. 4-ethyl-2-methylcyclohex-1-ene
   B. 5-ethyl-1-methyl-cyclohexene   D. 2-methyl-5-ethylcyclohex-1-ene

25.   What is the proper name of the following compound?

   A. cis-1,2-dichlorobutene   C. trans-1,2-dichloroethene
   B. trans-1,2-dichlorobutane   D. cis-1,2-dichloroethane

26.   With what would you react 2,2,3-trichlorononadiene in order to convert it to 2,2,3-trichlorononane?

   A. oxygen   B. hydrogen gas   C. chlorine gas   D. water

27.   A benzene compound with bromine in the 1 and 3 positions has the common name of:

   A. o-dibromobenzene   B. p-dibromobenzene   C. m-dibromobenzene   D. dibromobenzene

28.   The process by which hexane is converted into methylcyclopentane is known as:

   A. esterification   B. pyrolysis   C. catalytic reforming   D. isomerization

29. Which one of the following processes is not used to increase octane rating?

   A. polymerization   B. alkylation   C. isomerization   D. esterification

30. Which one of the following alcohols would you expect to have the highest boiling point?

   A. methanol   B. propanol   C. decanol   D. hexanol

31. Oxidation of which one of the following compounds would lead to an aldehyde?

   A. cyclohexanol   B. 2-butanol   C. methanol   D. phenol

32. What functional group(s) are present in this compound: $CH_3CHOHCOOH$?

   A. carboxylic acid   C. ketone, carboxylic acid
   B. alcohol, carboxylic acid   D. ether, carboxylic acid

33. The following compound can be prepared by reacting which one of the following pairs of reagents?

$$CH_3CH_2CH_2COOCH_2CH_3$$

   A. butyric acid with ethanol   C. ethanoic acid with butyrate
   B. butyraldehyde with ethanoic acid   D. 2-butanone with acetaldehyde

34. Which one of the following amines is a primary amine?

   A. diethylamine   B. 1-aminohexane   C. trimethylamine   D. diphenylamine

Answers to Exercises

1. a. 2,2-dimethylheptane
    b. 5-ethyl-3-methyloctane
    c. 2,2,3-trimethylbutane
    d. t-butylcyclopentane

2. a.

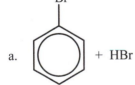

    b. $Cl-\overset{\overset{\displaystyle H}{|}}{C}-\underset{\underset{\displaystyle Cl}{|}}{C}H_2$
        $\qquad \underset{Cl}{|}$

3. Alkanes are almost totally insoluble in water. This is due to their nonpolar nature (water is polar) and their inability to form hydrogen bonds.

4. a. (2)   b. (1)   c. (3)

5. Hydrogenation:

    $$H_2C\!=\!\overset{\overset{\displaystyle H_2}{}}{C}\!-\!\underset{\underset{\displaystyle CH_3}{|}}{C}\!-\!CH_3 + H_2 \rightarrow H_3C\!-\!\overset{\overset{\displaystyle H}{|}}{C}\!-\!\underset{\underset{\displaystyle CH_3}{|}}{\overset{\overset{\displaystyle H_2}{}}{C}}\!-\!CH_3 \quad (2\text{-methylbutane})$$

    Halogenation:

    $$H_2C\!=\!\overset{\overset{\displaystyle H_2}{}}{C}\!-\!\underset{\underset{\displaystyle CH_3}{|}}{C}\!-\!CH_3 + Cl_2 \rightarrow H_2C\!-\!\overset{\overset{\displaystyle Cl}{|}}{\underset{\underset{\displaystyle CH_3}{|}}{C}}\!-\!\overset{\overset{\displaystyle Cl}{|}}{\overset{\overset{\displaystyle H_2}{}}{C}}\!-\!CH_3 \quad (1,2\text{-dichloro-2-methylbutane})$$

6. Alkenes have a carbon-carbon double bond consisting of a C–C σ bond and C–C π. Alkanes consist of C–C σ bonds. Thus the presence of the π bond and its exposed electrons make alkenes more susceptible to addition reaction than alkanes.

7. a. (bromobenzene) + HBr

    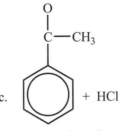

    c. (methylphenylketone) + HCl

    b. No Reaction. (No Lewis acid catalyst)

    d. (benzenesulfonic acid) + H_2O

8. a. (1) ether, (2) 1° alcohol
    b. (1) aldehyde
    c. (1) ester, (2) 1° amine
    d. (1) carboxylic acid

9. a. phenol

    c. 3-hexanone

    b. H—C (with =O above, attached to benzene ring with Cl)

    d. 2-chloro-4-hexanol

10. CH_3CH_2CH (with =O) $< CH_3CCH_3$ (with =O) $< CH_3CH_2CH_2OH$

11. a. 2-methylpropanoic acid + benzyl alcohol → H_3C—$\overset{H}{\underset{CH_3}{C}}$—$\overset{O}{C}$—$O$—$\overset{H_2}{C}$—⟨benzene ring⟩

     b. chloroethanoic acid + 2-propanol (isopropyl alcohol) → H_2CClC (with =O) —O—$\overset{H}{\underset{CH_3}{C}}$—$CH_3$ + H_2O

     c. 3,3-dimethyl-1-butanol $\xrightarrow{KMnO_4(aq)}$ H_3C—$\overset{CH_3}{\underset{CH_3}{C}}$—$CH_2COOH$

12. a. 3°   b. 3°   c. 2°   d. 1°

13. a. Two identical monomers joining together to form a molecule.
     b. A species with an unpaired electron.
     c. More than one type of monomer combining to form the chain in a polymer.
     d. Identical monomers combining to form the chain in a polymer.
     e. Large 1, 2, or 3-dimensional molecules consisting of large numbers of repeating monomer units.

14. **Addition polymerization** involves formation of the polymer by the free radical mechanism. The only product formed is the polymer, and the polymerization stops when two radicals react to form a bond without producing any other radicals.

     Condensation polymerization produces a product other than the polymer itself. The side products are most commonly small molecules such as water or alcohols.

15.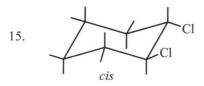

     cis   trans

16. d > b > c > a

17.   Heptyl alcohol has a seven-carbon chain while ethyl alcohol has a two-carbon chain. Due to this fact, the nonpolar characteristics in the seven-carbon chain dominate over the polar characteristics and make it insoluble in water.

18.   c > a > b due to hydrogen bonding. A 1° amine has a greater amount of hydrogen bonding when compared to a 2° amine. A 3° amine has little hydrogen bonding when compared to a 1° and 2° amine.

Answers to Multiple Choice Self-Test

19.	A	20.	C	21.	B	22.	A	23.	A	24.	B
25.	C	26.	B	27.	C	28.	C	29.	D	30.	C
31.	C	32.	B	33.	A	34.	B						

Answers to Section 22.6

1.   Proteins are large, amino acid-based "natural polymers" in our bodies that perform a variety of biological functions.
2.   The molecular weights can range from 6,000 to over 1,000,000.
3.   They are called "α-amino acids" because the amino group is attached to the α-carbon. Amino means an NH_2 group.
4.   A peptide linkage occurs when the carboxyl group of a carboxylic acid interacts with a hydrogen from an amine group.
5.   A dipeptide involves 2 amino acids in a peptide linkage.
6. $10! = 10 \times 9 \times 8 \times 7 \times 6 \times 5 \times 4 \times 3 \times 2 \times 1 = 3.6 \times 10^6$ sequences.
7.   The levels are primary, secondary, tertiary, and quaternary.
8.   Primary = peptide linkages; secondary = hydrogen bonding; tertiary = a variety of interactions including hydrogen bonding, ionic bonds, and covalent bonds, among others; quaternary = bonding between individual subunits.
9.   An α-helix represents a secondary structure.
10.   See the discussion in the text for some examples.
11.   The disulfide linkage in hair is broken and reformed to shape the hair. (See Figure 22.25 in your textbook.)
12.   Denaturation involves breaking down the three-dimensional structure of a protein, thus rendering it inactive.
13.   Heat and intense radiation are two causes.
14.   Historically, they were thought to be hydrates of carbon. For example, $C_{12}H_{22}O_{11}$ was thought to be $C_{12} \cdot 11H_2O$.
15.   Monosaccharides are simple sugars.
16.   A hexose is a sugar with 6 carbon atoms.
17.   Optical isomerism requires a chiral carbon.
18.   The oxygen of the terminal OH group combines with the carbon of the ketone group.
19.   A disaccharide is a combination of 2 simple sugars.
20.   A C–O–C linkage between rings of glucose and fructose is a glycoside linkage.
21.   The enzyme α-amylase is found in saliva. It catalyzes the decomposition of starch.
22.   Starch is a polymer of α-glucose. (See Figure 22.32 in your textbook.)
23.   There is less stress on the plant's internal structure. (See the discussion on osmotic pressure in Chapter 11.)
24.   We do not have the necessary enzymes, β-glycosidases, to break down cellulose.
25.   DNA stores and transmits genetic information.
26.   a.   A five-carbon sugar.
       b.   A nitrogen-containing organic base.
       c.   A phosphoric acid molecule.

27.   The double helix structure allows DNA to produce complementary strands.
28.   Polar C=O and N−H bonds lead to hydrogen bonding, which leads to the formation of the double helix.
29.   See <u>Figures 22.36 - 22.38 in your textbook</u>.
30.   A gene is a segment of DNA that contains the code for a specific protein.
31.   A codon consists of three bases and codes for a specific amino acid. An anticodon is a part of tRNA that decodes the "genetic message" from mRNA.
32.   mRNA migrates from DNA to the cell cytoplasm where protein synthesis occurs. tRNA decodes the genetic message from mRNA.